Advances in Microlocal Analysis

NATO ASI Series

Advanced Science Institutes Series

A series presenting the results of activities sponsored by the NATO Science Committee, which aims at the dissemination of advanced scientific and technological knowledge, with a view to strengthening links between scientific communities.

The series is published by an international board of publishers in conjunction with the NATO Scientific Affairs Division

A Life Sciences B Physics	Plenum Publishing Corporation London and New York
C Mathematical and Physical Sciences	D. Reidel Publishing Company Dordrecht, Boston, Lancaster and Tokyo
D Behavioural and Social Sciences E Engineering and Materials Sciences	Martinus Nijhoff Publishers The Hague, Boston and Lancaster
F Computer and Systems Sciences G Ecological Sciences	Springer-Verlag Berlin, Heidelberg, New York and Tokyo

Series C: Mathematical and Physical Sciences Vol. 168

Advances in Microlocal Analysis

edited by

H. G. Garnir †

University of Liège, Liège, Belgium

D. Reidel Publishing Company

Dordrecht / Boston / Lancaster / Tokyo

Published in cooperation with NATO Scientific Affairs Division

Proceedings of the NATO Advanced Study Institute on
Advances in Microlocal Analysis
Castelvecchio-Pascoli (Lucca), Italy
September 2-12, 1985

Library of Congress Cataloging in Publication Data

NATO Advanced Study Institute on Advances in Microlocal Analysis (1985: Castelvecchio Pascoli, Italy)
Advances in microlocal analysis.

(NATO ASI series. Series C, Mathematical and physical sciences; vol. 168)
Papers chiefly in English; 3 papers in French.
"Proceedings of the NATO Advanced Study Institute on Advances in Microlocal Analysis, Castelvecchio-Pascoli (Lucca), Italy, September 2-12, 1985"-T.p. verso.
"Published in cooperation with NATO Scientific Affairs Division."
Includes index.
1. Mathematical analysis-Congresses. I. Garnir, H. G. (Henri G.), 1921- . II. North Atlantic Treaty Organization. Scientific Affairs Division. III. Title. IV. Series: NATO ASI series. Series C, Mathematical and physical sciences; vol. 168.
QA299.6.N37 1985 515 85-31188
ISBN 90-277-2195-5

Published by D. Reidel Publishing Company
P.O. Box 17, 3300 AA Dordrecht, Holland

Sold and distributed in the U.S.A. and Canada
by Kluwer Academic Publishers,
190 Old Derby Street, Hingham, MA 02043, U.S.A.

In all other countries, sold and distributed
by Kluwer Academic Publishers Group,
P.O. Box 322, 3300 AH Dordrecht, Holland

D. Reidel Publishing Company is a member of the Kluwer Academic Publishers Group

Printed in The Netherlands

TABLE OF CONTENTS

PREFACE

The 1985 Castelvecchio-Pascoli NATO Advanced Study Institute is aimed to complete the trilogy with the two former institutes I organized : "Boundary Value Problem for Evolution Partial Differential Operators", Liège, 1976 and "Singularities in Boundary Value Problems", Maratea, 1980.

It was indeed necessary to record the considerable progress realized in the field of the propagation of singularities of Schwartz Distributions which led recently to the birth of a new branch of Mathematical Analysis called Microlocal Analysis.

Most of this theory was mainly built to be applied to distribution solutions of linear partial differential problems. A large part of this institute still went in this direction. But, on the other hand, it was also time to explore the new trend to use microlocal analysis in non linear differential problems.

I hope that the Castelvecchio NATO ASI reached its purposes with the help of the more famous authorities in the field.

The meeting was held in Tuscany (Italy) at Castelvecchio-Pascoli, little village in the mountains north of Lucca on September 2-12, 1985. It was hosted by "Il Ciocco" an international vacation Center, in a comfortable hotel located in magnificent mountain surroundings and provided with all conference and sport facilities.

There were 91 participants : 74 from NATO countries (Belgium : 13, Canada : 1, Denmark : 1, France : 19, West Germany : 6, Italy : 19, Portugal : 2, Turkey : 5, U.K. : 2, USA : 5) and 18 from non NATO countries (Australia : 1, Brasil : 2, Switzerland : 1, India : 1, Iran : 2, Japan : 9, Sweden : 2).

The lectures were held by 15 lecturers : 1 with a four-hour course, 9 with a three-hour course, 2 with a two-hour course and 3 with a one-hour course. Moreover 25 advanced seminars were organized by the participants and devoted to the discussion of their contribution in the field.

The creator of microlocal analysis, Prof. L. Hörmander of the University of Lund, Sweden, made us the honour to attend the meeting and delivered a brillant extra one-hour talk. His presence among us was a great encouragement.

I wish to express my gratitude to NATO which was the main sponsor of this Castelvecchio-Pascoli meeting. My thanks go to the Scientific Affairs Division for its efficient help and specially to Dr. C. SINCLAIR, Scientific Officer in charge of the ASI Program.

It is my pleasure to mention also the institutions which supported financially this meeting : the European Research Office of the U.S. Army, the Fondation Francqui of Belgium and the Belgian Fonds National de la Recherche Scientifique.

I thank heartly these institutions, which contribute greatly to the success of the meeting by allowing us to support the non-NATO participants.

For the location of the meeting and the publication of the Proceedings, I benefitted by the advices of the Association "International Transfer of Science and Technology" and specially from its director Mrs. B. KESTER to whom I am very grateful.

I again was efficiently helped by Prof. Léonard of the University of Liège who was, as before, a perfect codirector. To him my friendly thanks !

H.G. GARNIR
Director of the Institute

On the 18th of November 1985, the mathematical community suddenly lost one of its eminent members.
Professor Garnir's memory shall remain in our mind.
He was a great scientist, an appreciated professor and a clever organizer but, besides the mathematician, we will never forget the warm-hearted man, the friend we met.
To him, director of this institute, we all owe much gratitude.

Liège, December 16, 1985
P. LEONARD
Codirector of the Institute

It is with a profound sense of sorrow that this book is dedicated to the memory of Professor Garnir who died shortly after the meeting that is reported here.
It is hoped that the volume will be a sincere, if inadequate, tribute to his professional work in this field and to his outstanding human qualities.

NATO SCIENTIFIC AFFAIRS DIVISION

LIST OF PARTICIPANTS

GARNIR H.G. : Inst. de Math. Univ. de Liège
15, avenue des Tilleuls / B-4000 Liège
BELGIUM

LEONARD P. : Inst. de Math. Univ. de Liège
15, avenue des Tilleuls / B-4000 Liège

BENGEL G : Mathematisches Institut der Universität
Roxeler Strasse, 64
44 Münster / GERMANY

BONY J-M. : Université de Paris-Sud
Département de Mathématiques
91405 Orsay Cedex
FRANCE

CATTABRIGA L. : Universita di Bologna
Istituto Matematico "Salvatore Pincherle"
Piazza di Porta San Donato, 5
40127 Bologna / ITALY

DUFF G. : University of Toronto
Department of Mathematics
Toronto, 181 / CANADA

FRIEDLANDER F.G. : University of Cambridge
Department of Mathematics
Silver Street
Cambridge CB3 9EW / ENGLAND

KOMATSU H. : University of Tokyo
Faculty of Science
Department of Mathematics
Hongo, Tokyo / 113 JAPAN

LAUBIN P. : Université de Liège
Institut de Mathématique
15, avenue des Tilleuls
B-4000 Liège / BELGIUM

LEBEAU G. : C.M.A. Ecole Normale Supérieure
45, rue d'Ulm
750230 Paris Cedex 05
FRANCE

MELROSE R. : Massachussetts Institute of Technology
Department of Mathematics
Math. 2 - 171
Cambridge Mass. 02139 / USA

OAKU T. : University of Tokyo
Department of Mathematics
Hongo, Tokyo / 113 JAPAN

SCHAPIRA P. : Université de Paris-Nord - CSP
Avenue J.B. Clément
93430 Villetaneuse / FRANCE

SJOSTRAND J. : Université de Lund
Département de Mathématiques
Box 118, S-22100 Lund / SUEDE

TSUJI M. : Kyoto Sangyo University
Department of Mathematics
Kamigamo, Kita-ku
Kyoto 603 / JAPAN

VAILLANT J. : Unité C.N.R.S. 761
Université Pierre et Marie Curie (PARIS VI)
Mathématiques, tour 45-46, 5ème étage
4, place Jussieu
75230 Paris Cedex 05 / FRANCE

WAKABAYASHI S. : University of Tsukuba
Institute of Mathematics
Ibaraki 305 / JAPAN

ADAMOU A. : Université de Liège
Institut de Mathématique
15, avenue des Tilleuls
B-4000 Liège / BELGIQUE

AOKI T. : Kinki University
Department of Mathematics
Higashi-Osaka,
Osaka 577 / JAPAN

AYFER K. : Hacettepe University
Department of Mathematics
Beytepe, Ankara / TURQUIE

BASSANNELLI G. : University of Trento
Department of Mathematics
38050 Povo (Trento) / ITALY

BERNARDI E. : Universita di Bologna
Dipartimento di Matematica
Piazza di Porta San Donato, 5
40126 Bologna / ITALY

BOVE A. : Universita di Bologna
Dipartimento di Matematica
Piazza di Porta San Donato, 5
40126 Bologna, ITALY

BROS J. : Centre d'Etudes Nucléaires de Saclay
Institut de Recherche Fondamentale
Service de Physique Théorique
91191 Gif-sur-Yvette / FRANCE

CARDOSO F. : Universidade Federal de Pernambuco
Departamento de Matematica
50000 Recife, Pernambuco / BRASIL

CARVALHO e SILVA J. : Universidade de Coimbra
Departamento de Matematica
3000 Coimbra / PORTUGAL

CICOGNANI M. : Universita di Bologna
Dipartimento di Matematica
Piazza di Porta San Donato, 5
40126 Bologna / ITALY

CODEGONE M. : Politechnico di Torino
Dipartimento di Matematica
Corso Duca degli Abruzzi, 24
10129 Torino / ITALY

CORLI A. : Universita di Ferrara
Dipartimento di Matematica
Via Machiavelli, 35
44100 Ferrara / ITALY

DE JONGE J. : Université du Travail
Département de Mathématiques
Rue Paul Pastur
6000 Charleroi / BELGIUM

DELANGHE R. : University of Ghent
Department of Mathematics
Krijgslaan, 271
9000 Ghent / BELGIUM

DE MORAES L. : University of Rio de Janeiro
Department of Mathematics
Caixa Postal 1835
ZC-00, 20.000 Rio de Janeiro / RJ BRASIL

DENCKER N. : University of Lund
Matematiska Institutionen
Box 118
22100 Lund / SWEDEN

DIEROLF P. : FB IV, Mathematik der Universität Trier
Postfach 3825
5500 Trier / WEST GERMANY

ERDEM D. : Hacettepe University
Department of Mathematics
Beytepe
Ankara / TURKEY

ERDEM M. : Hacettepe University
Department of Mathematics
Beytepe
Ankara / TURKEY

ESSER P. : Université de Liège
Institut de Mathématique
15, avenue des Tilleuls
B-4000 Liège / BELGIQUE

ETIENNE J. : Université de Liège
Institut de Mathématique
15, avenue des Tilleuls
B-4000 Liège / BELGIQUE

FRANCHI B. : University of Bologna
Dipartimento di Matematica
Piazza di Porta San Donato, 5
40126 Bologna / ITALY

GERARD C. : Université d'Orsay
Département de Mathématiques, Bât. 425
Orsay 91 / FRANCE

GOBBO L. : University of Torino
Dipartimento di Matematica
Via Carlo Alberto, 10
10123 Torino / ITALY

GOBERT J. : Université de Liège
Institut de Mathématique
15, avenue des Tilleuls
B-4000 Liège / BELGIQUE

GODIN P. : Université Libre de Bruxelles
Département de Mathématique
Campus de la Plaine, CP 214
Boulevard du Triomphe
1050 BRUXELLES / BELGIQUE

GOURDIN D. : Université de Lille
Département de Mathématiques
59655 Villeneuve d'Ascq / FRANCE

GRIGIS A. : Ecole Polytechnique
Centre de Mathématiques
91128 Palaiseau Cedex / FRANCE

GRUBB G. : University of Copenhague
Matematisk Institut
Universitetsparken, 5
2100 Copenhagen / DENMARK

von GRUDZINSKI O. : Mathematisches Seminar der Universität
Olshausenstr. 40,
2300 Kiel 1, WEST GERMANY

GURARIE D. : Case Western Reserve University
Department of Mathematics
Cleveland, O.H. 44106 / USA

HANSEN S. : FB 17 Mathematik
Gesamthochschule Paderborn
Warburger Str. 10
4790 Paderborn / GERMANY

HELFFER B. : Université de Nantes
Département de Mathématiques
38, Boulevard Michelet
B.P. 1044
44037 Nantes Cedex / FRANCE

HORMANDER L. : Institut Mittag-Leffler
Auravägen, 17
18262 Djursholm / SWEDEN

IAGOLNITZER D. : Centre d'Etudes Nucléaires de Saclay
Institut de Recherche Fondamentale
Service de Physique Théorique
91191 Gif-sur-Yvette Cedex / FRANCE

KARACAY T. : Hacettepe Universitesi
Matematik Bölümü
Beytepe
Ankara / TURQUIE

LAURENT Y. : Université de Paris Sud
Centre d'Orsay
Département de Mathématique, Bât. 425
91405 Orsay Cedex / FRANCE

LEICHTNAM E. : Ecole Normale Supérieure
45, rue d'Ulm
75230 Paris Cedex 05 / FRANCE

LEWIS J.E. : University of Bologna
Dipartimento di Matematica
Porta San Donato, 5
40126 Bologna / ITALY

LIESS O. : Universität Bonn
FB Mathematik
61 Darmstadt / WEST GERMANY

MAIRE H.M. : Université de Genève
Section de Mathématiques
Case Postale 240
1211 Genève 24 / SUISSE

MAMOURIAN A. : University Faculty Educational Center
Department of Mathematics
P.O. Box 14155 -
4838 Tehran / IRAN

MARCHIONI G. : Università Cattolica del Sacro Cuore
Istituto di Matematica
Via Trieste, 17
25121 Brescia / ITALY

MARTINEZ A. : Université de Paris Sud
Mathématique, Bât. 425
91405 Orsay / FRANCE

MAZZEO R. : M.I.T.
Massachussetts Avenue, 77
Cambridge, Ma. 02139 / USA

MERIC R.A. : Research Institut for Basic Sciences
Department of Applied Mathematics
The Scientific and Technical Research Council of Turkey
Gebze, Kocaeli / TURKEY

MISRA O.P. : University of New Delhi
Indian Institute of Technology
New Delhi / INDIA

MONTEIRO-FERNANDES M.T. : University of Lisboa
Department of Mathematics
Lisboa Codex / PORTUGAL

MUNSTER M. : Université de Liège
Institut de Mathématique
15, avenue des Tilleuls
B-4000 Liège / BELGIQUE

MURTHY M.K. : Universita di Pisa
Dipartimento di Matematica
Via Buonarroti, 2
56100 Pisa / ITALY

NACINOVICH M. : Universita di Pisa
Dipartimento di Matematica
Via Buonarroti, 2
56100 Pisa / ITALY

NAKANE S. : Tokyo Institute of Polytechnics
1583 Iiyama, Atsugi-shi
Kanagawa 243-02 / JAPAN

NARDINI F. : Universita di Bologna
Dipartimento di Matematica
Piazza di Porta San Donato, 5
40126 Bologna / ITALY

NOURRIGAT J.F. : Université de Rennes
UER Mathématiques et Informatique
Avenue du Général Leclerc
Rennes, Beaulieu
35042 Rennes Cedex / FRANCE

OUCHI S. : Sophia University
Faculty of Science and Technology
7, Kioicho, Chiyoda-ku
Tokyo 102 / JAPAN

PARENTI C. : Universita di Bologna
Dipartimento di Matematica
Piazza di Porta San Donato, 5
40126 Bologna / ITALY

RADULESCO N. : Université de Paris Nord
Département de Mathématiques
Avenue J.B. Clément
Villetaneuse / FRANCE

RODINO L. : Universita di Torino
Dipartimento di Matematica
Via Carlo Alberto, 10
10123 Torino / ITALY

ROULEUX M. : Université de Paris XI
Département de Mathématiques
Centre d'Orsay, Bât. 425
91405 Orsay Cedex / FRANCE

SANKARAN S. : Queen Elizabeth College
Department of Mathematics
Campden Hill Road
London W8 7AH / ENGLAND

SCHNEIDERS J.P. : Université de Liège
Institut de Mathématique
15, avenue des Tilleuls
B-4000 Liège / BELGIQUE

SEGALA F. : Universita di Ferrara
Dipartimento di Matematica
Via Machiavelli, 35
44100 Ferrara / ITALY

SERRA E. : Universita di Bologna
Dipartimento di Matematica
Piazza di Porta San Donato, 5
40126 Bologna / ITALY

SILBERSTEIN J.P.O. : University of Western Australia
Department of Mathematics
Nedlands / WESTERN AUSTRALIA 6009

SUNG C.H. : University of San Diego
Department of Mathematics
San Diego
California / USA

TAHARA H. : Sophia University
Department of Mathematics
Kioicho, Chiyoda-ku
Tokyo 102 / JAPAN

TOSE N. : Tokyo University
Department of Mathematics
7 Hongo
Tokyo 113 / JAPAN

TOUGERON M. : Université de Rennes I
U.E.R. Informatique
Campus de Beaulieu
35042 Rennes Cedex / FRANCE

ULRICH K. : Universität Hannover
Lehrgebiet für Angewandte Analysis
Welfengarten 1
3000 Hannover 1 / GERMANY

WAGSCHAL C. : Ecole des Ponts et Chaussées
22, rue Brissard
92140 Clamart / FRANCE

WUIDAR J. : Université de Liège
Institut de Mathématique
15, avenue des Tilleuls
B-4000 Liège / BELGIQUE

ZAFARANI J. : University of Isfahan
Department of Mathematics
Isfahan / IRAN

ZAMPIERI G. : Universita di Padova
Istituto di Analisi dell' Universita
Via Belzoni, 7
35131 Padova / ITALY

ZANGHIRATI L. : Universita di Ferrara
Istituto di Matematico
Via Machiavelli, 35
44100 Ferrara / ITALY

CONVERGENCE OF FORMAL SOLUTIONS OF SINGULAR PARTIAL DIFFERENTIAL EQUATIONS

Gunter Bengel
Mathematisches Institut der WWU
Einsteinstrasse 62
D-4400 Münster
West Germany

ABSTRACT. We consider partial differential operators of the form

$$P(x,x\frac{\partial}{\partial x}) = \sum_{|\ell|\leqslant d} a_\ell(x)(x\frac{\partial}{\partial x})^\ell$$

and give conditions when the equation $Pu = f$, f analytic near the origin, has a power series solution which converges in a neighbourhood of the origin. More generally we consider non linear equations of the form $Pu = F(x,u)$ and give some applications. This is a report on the results of [2].

1. INTRODUCTION

We employ the usual notations concerning multiindices e.g. we set

$$(x\frac{\partial}{\partial x})^\ell = (x_1\frac{\partial}{\partial x_1})^{\ell_1}\ldots(x_n\frac{\partial}{\partial x_n})^{\ell_n}$$

if $\ell = (\ell_1,\ldots,\ell_n)$. If $u = \sum_\ell u_\ell x^\ell$ is a formal or convergent power series with complex coefficients and P is an operator of the form

$$(1.1)\qquad P(x,x\frac{\partial}{\partial x}) = \sum_{|\ell|\leqslant d} a_\ell(x)(x\frac{\partial}{\partial x})^\ell$$

where $a_\ell(x) \in \mathcal{O}$, the ring of convergent power series, the action of P on u is given by

$$(1.2)\qquad Pu = \sum_{|m|=0}^{\infty}\Big(\sum_{k\leqslant m} P_k(m-k)u_{m-k}\Big)x^m \quad .$$

In fact, using the expansion $a_\ell(x) = \sum_k a_{\ell,k}x^k$ we get

H. G. Garnir (ed.), Advances in Microlocal Analysis, 1–14.

$$P(x,x\frac{\partial}{\partial x})u = \sum_{|\ell|\leq d}\sum_{k,j} a_{\ell,k}\, x^k (x\frac{\partial}{\partial x})^{\ell}\, u_j x^j$$

$$= \sum_{|\ell|\leq d}\sum_{k,j} a_{\ell,k}\, u_j j^{\ell} x^{k+j}$$

$$= \sum_{k,j} P_k(j) u_j x^{k+j}$$

which gives (1.2) with $P_k(j) = \sum_{|\ell|\leq d} a_{\ell,k} j^{\ell}$.

Special cases are operators with "constant coefficients"

$$(1.3) \quad P(x\frac{\partial}{\partial x}) = \sum_k a_k (x\frac{\partial}{\partial x})^k$$

so here $P_k = 0$ for $k \neq 0$, and vector fields τ of the form

$$(1.4) \quad \tau = \sum_{j=1}^{n} (\lambda_j x_j + f_j(x))\frac{\partial}{\partial x_j}$$

where f_j vanishes of the second order at 0. Here

$$P_0(m) = \sum_{j=1}^{n} \lambda_j m_j = \langle\Lambda,m\rangle , \ \Lambda = (\lambda_1,\ldots,\lambda_n) .$$

In the second section we give conditions that each formal power series u which solves $Pu = f$, $f \in \mathcal{O}$, converges ; and more generally we prove that also the solutions of the nonlinear equation $Pu = F(x,u)$, F analytic near the origin, converge under appropriate conditions. In the case of a vector field (1.4) this gives as an application a theorem of Poincaré on normal forms of vector fields. In the third section we consider equations with parameters. The proofs are almost the same as in section 2. However there are cases, where we cannot prove that every formal solution converges but only that if there is a formal solution then there exists also a convergent solution which coincides with the given series up to arbitrary high order. As an application we give a theorem of S. Kaplan (Cor.3.4). In the last section we consider a problem where small denominators occur and prove a theorem of Siegel about normal forms of certain vector fields.

For the convergence proofs we introduce suitable norms on $\mathcal{O}$ or $\mathcal{O}^q$. If $u \in \mathcal{O}^q$, $u = \sum_k u_k x^k$ there exists a polydisc given by the polyradius $t = (t_1,\ldots,t_n)$, $t_j > 0$ and ρ , $0 < \rho < 1$, such that

$$(1.5) \quad \|u\|_{\rho} = \sum_{|k|=0}^{\infty} |u_k| \rho^{|k|} t^k < \infty$$

where $|.|$ is any norm on $\mathbb{C}^q$. Given t and ρ we denote by B^q_{ρ} the space of power series $u \in \mathcal{O}^q$ for which $\|u\|_{\rho}$ is finite. $\|.\|_{\rho}$ is a norm on B^q_ρ and B^q_ρ is a Banach space. For $q = 1$, $B_\rho = B^1_\rho$ is a Banach algebra. The dependence on t is not indicated since in general t will be kept fixed and it is no restriction to suppose in most cases $t = (1,\dots,1)$. M is the maximal ideal in $\mathcal{O}$ and $M^q_\rho = M^q \cap B^q_\rho$.

2. CONVERGENCE PROOFS

First we treat the case of constant coefficient operators of the form (1.3). If P is such an operator, we have

$$Pu = \sum_m P_0(m) u_m x^m$$

and if $Pu = f = \sum f_m x^m$, we get $P_0(m)u_m = f_m$. If $P_0(m) \neq 0$ for all m

$$u_m = \frac{f_m}{P_0(m)} .$$

The same is done with obvious changes if $u, f \in \mathcal{O}^q$ and $P_0(m)$ are invertible $q\times q$-matrices. To prove convergence we impose lower bounds on $P_0(m)$. To simplify notations we take $t = (1,\dots,1)$.

<u>Proposition 2.1</u>.

Let P be an operator with constant coefficients of the form (1.3) and let C, γ be positive constants.

a) If P satisfies the Poincaré condition

$$(2.1) \quad |P_0(m)| \geqslant C|m|^\gamma \quad \text{for all } m \neq 0$$

then $P^{-1} : M_\rho \to M_\rho$ is continuous.

b) If P satisfies the Siegel conditions

$$(2.2) \quad |P_0(m)| \geqslant C|m|^{-\gamma} \quad \text{for all } m \neq 0$$

then $P^{-1} : M_\rho \to M_\sigma$ is continuous for all $\sigma < \rho$ and we have the estimate

$$(2.3) \quad \|P^{-1}u\|_\sigma \leqslant \frac{C}{(\rho-\sigma)^\gamma} \|u\|_\rho \quad , \quad C > 0 .$$

c) If P satisfies the exponential condition

$$(2.4) \quad |P_0(m)| \geqslant C|\gamma|^{-|m|} \quad \text{for all } m \neq 0$$

then $P^{-1} : M_\rho \to M_\sigma$ is continuous for $\sigma = \frac{\rho}{\gamma}$.

<u>Proof</u>. a) $\|P^{-1}f\|_\rho = \Sigma|P_0(m)^{-1}f_m|\rho^{|m|}$

$$\leqslant \frac{1}{C} \sum_m |m|^{-\gamma} |f_m| \rho^{|m|} \leqslant \frac{1}{C} \|f\|_\rho \quad \text{since} \quad |m|^{-\gamma} \leqslant 1 .$$

c) $$\|P^{-1} f\|_\rho = \Sigma |P_0(m)^{-1} f_m| \rho^{|m|}$$
$$\leqslant \frac{1}{C} \Sigma |f_m| \gamma^{-|m|} \rho^{|m|} = \frac{1}{C} \Sigma |f_m| \sigma^{|m|} = \|f\|_\sigma \text{ for } \sigma = \gamma^{-1}\rho .$$

b) $$\|P^{-1} f\|_\rho = \Sigma |P_0(m)^{-1} f_m| \rho^{|m|}$$
$$\leqslant \frac{1}{C} \Sigma |f_m| |m|^\gamma \rho^{|m|} \leqslant \frac{1}{C} \left(\frac{\gamma}{e}\right)^\gamma \frac{1}{(\rho-\sigma)^\gamma} \|f\|_\sigma$$

since by elementary calculus the maximum of $f(x) = x^\gamma \eta^x$ is

$$\left(\frac{\gamma}{e}\right)^\gamma \frac{1}{(-\log \eta)^\gamma} , \quad \text{with } \eta = \frac{\sigma}{\rho} < 1 , \text{ so}$$

$$|m|^\gamma \eta^{|m|} \leqslant \left(\frac{\gamma}{e}\right)^\gamma \frac{1}{(\log \rho - \log \sigma)^\gamma} \leqslant \left(\frac{\gamma}{e}\right)^\gamma \frac{1}{(\rho-\sigma)^\gamma}$$

by the mean value theorem.

<u>Remarks</u>. 1) This applies especially to vector fields of the form (1.4) with $f_j = 0$.

2) If $P_0(m) \neq 0$ for all m , but decreases faster than exponentially along some sequence m_k there exist counterexamples showing that the solution of $Pu = f \in \mathcal{O}$ need not converge. Such a counterexample is e.g. the vector field

$$\tau = x \frac{\partial}{\partial x} - \lambda y \frac{\partial}{\partial y} \quad \text{where} \quad \lambda > 0$$

is a transcendental number which is extremely well approximable by rational numbers, cf.[3].

General operators (1.1) can now be treated as perturbations of constant coefficient operators.

<u>Proposition 2.2</u>. Let $P = P(x, x \frac{\partial}{\partial x})$ be a singular differential operator of the form (1.1) and suppose P_0 satisfies the Poincaré condition

$$(2.4) \qquad |P_0(m)| \geqslant C|m|^d , \; m \neq 0 ,$$

where d is the degree of the operator, then P is invertible and $P^{-1} : M_\rho \to M_\rho$ is continuous for sufficiently small ρ .

<u>Proof</u>. Set $P = P_0 + \widetilde{P}$, where

$$P_0(x \frac{\partial}{\partial x}) = \sum_{|\ell| \leqslant d} a_\ell(0) (x \frac{\partial}{\partial x})^\ell .$$

In order to solve the equation $P_0 u + \widetilde{P} u = f$ in M_ρ we solve $u + P_0^{-1} \widetilde{P} u = P_0^{-1} f \in M_\rho$. For this it is sufficient to show that

$$\| P_0^{-1} \tilde{P}u \|_\rho \leqslant C \| u \|_\rho \quad \text{with} \quad C < 1 .$$

With the notation of the proof of (1.2) we have

$$|P_k(j)| = |\sum_{|\ell| \leqslant d} a_{\ell,k} j^\ell| \leqslant C_k |j|^d \quad \text{with} \quad C_k = \sum_\ell |a_{\ell,k}| .$$

Since a_ℓ is analytic there is a ρ_0 such that

$$\phi(\rho) = \sum_{k>0} C_k \rho^{|k|} \quad \text{is convergent for} \quad \rho \leqslant \rho_0$$

and tends to zero for $\rho \to 0$. By (1.2) we get

$$\begin{aligned} \| P_0^{-1} \tilde{P}u \|_\rho &= \sum_m |P_0^{-1}(m) \sum_{0 \leqslant k \leqslant m} P_k(m-k) u_{m-k}| \rho^{|m|} \\ &\leqslant \frac{1}{C} \sum_m \sum_{0<k} \frac{C_k |m-k|^d}{|m|^d} |u_{m-k}| \rho^{|m|} \\ &\leqslant \frac{1}{C} \sum_m \sum_{0<k} C_k \rho^{|k|} |u_{m-k}| \rho^{|m-k|} \leqslant \frac{1}{C} \phi(\rho) \| u \|_\rho \end{aligned}$$

and since $\phi(\rho) \to 0$ for $\rho \to 0$, $\frac{1}{C} \phi(\rho) < 1$ for $\rho \leqslant \rho_0$.

Remarks. 1) By a similar perturbation argument we can also treat vector fields of the form

$$(2.5) \qquad \tau u = \sum_{j=1}^{n} \lambda_j x_j \frac{\partial}{\partial x_j} + \sum_{j=1}^{n-1} a_j x_{j+1} \frac{\partial}{\partial x_j} = \tau_s + \tau_n$$

where the eigenvalues λ_j satisfy the Poincaré condition

$$|\langle \Lambda, m \rangle| = |\Sigma \lambda_j m_j| \geqslant C|m| \quad \text{for} \quad |m| \geqslant 2 .$$

The nilpotent part

$$\tau_n = \sum_{j=1}^{n-1} a_j x_{j+1} \frac{\partial}{\partial x_j}$$

is treated as a perturbation. This is not an operator of the form (1.2), but we have

$$\tau_n u = \sum_{j,m} a_j x_{j+1} u_m \frac{\partial}{\partial x_j} x^m = \sum_{j,m} a_j u_m m_j x^{m+e_j} ,$$

with $e_j = (0,\dots,-1,1,0,\dots,0)$ and the -1 is in the j-th place. So we get

$$\| \tau_s^{-1} \tau_n u \|_\rho \leqslant \sum_{j,m} |\langle \Lambda, m+e_j \rangle|^{-1} |a_j u_m| m_j \rho^m t^m t_j^{-1} t_{j+1}$$

$$\leqslant \frac{1}{C} \sum_{j,m} |a_j u_m| \rho^{|m|} t^{|m|} t_j^{-1} t_{j+1}$$

$$\leqslant C \|u\|_\rho \sum_{j=1}^{n-1} t_j^{-1} t_{j+1} \quad .$$

This can be made as small as we want if the ratios $t_j^{-1} t_j$ we chosen successively small enough. This is the only case where the choice of t in the norms $\|.\|_\rho$ is essential.

2) An exponent $\gamma < d$ in (2.4) will in general not be sufficient as the following counterexample shows.

$$P = (x \frac{\partial}{\partial x} - 1)^2 - (y \frac{\partial}{\partial y} - 1)^2 + 1 - xy(x \frac{\partial}{\partial x} + y \frac{\partial}{\partial y}) \ .$$

We have $|P_0(n,m)| = |(n+1)^2 - (m+1)^2 + 1| \geqslant 1$. If we calculate the formal solution u of $Pu = v$

$$v = (1 - xy)^{-1} = \sum_m (xy)^m$$

we get $u_{n,m} = 0$ for $n \neq m$ and $u_{n,n} = 1 + 2(n-1) + 2^2(n-1)(n-2) + \ldots + 2^{n-1}(n-1)!$ so $\Sigma\, u_{n,m} x^n y^m$ is divergent.

Now we want to solve the nonlinear equation $Pu = f(x,u)$, where P is of the form (1.1) and f is a (possibly vectorvalued) function $f : \mathbb{C}^n \times \mathbb{C}^q \to \mathbb{C}^q$.

<u>Proposition 2.3</u>. Suppose $f(0,0) = 0$, $\frac{\partial f}{\partial u}(0,0) = 0$. If P_0 satisfies the Poincaré condition (2.4) the equation $Pu = f(x,u)$ has an analytic solution u in the neighbourhood of the origin. This solution is unique.

<u>Proof</u>. We use Newton's method to solve $Pu - f(x,u) = F(x,u) = 0$. If u_0 is an approximate solution we try to get a better one $u_1 = u_0 + \Delta$. Taylor formula gives

$$F(x,u_1) = F(x,u_0+\Delta) = F(x,u_0) + (P - \frac{\partial f}{\partial u}(x,u_0))\Delta + \mathcal{O}(\Delta^2) \ .$$

So, if Δ is a solution of

$$\text{(2.6)} \qquad (P - \frac{\partial f}{\partial u}(x,u_0))\Delta = - F(x,u_0) \ ,$$

we get $F(x,u_0) = \mathcal{O}(\Delta^2)$. The solvability of (2.6) is given by

<u>Lemma 2.4</u>. There exist $\rho, r > 0$ such that for any function $v : \mathbb{C}^n \to \mathbb{C}^q$ analytic near 0 with $v(0) = 0$ and $\|v\|_\rho \leqslant r$, the operator

$$P - \frac{\partial f}{\partial u}(x,v) : B_\rho^q \to B_\rho^q$$

is invertible. The norm of the inverse depends on ρ , r and the norm of P^{-1} but not on v .

Proof. We use again a perturbation argument and we have to estimate

$$\|P^{-1}(\frac{\partial f}{\partial x}(x,v)u\|_\rho .$$

By the proof of proposition (2.2) we have

$$\|P^{-1}\|_\rho \leqslant \frac{C}{1-\phi(\rho)}$$

which tends to C as $\rho \to 0$. Since

$$\frac{\partial f}{\partial u}(x,v) = \sum_j g_j(x)v^j = \sum_{j,k} g_{j,k}x^k v^j$$

converges in some polydisc we get

$$\left\|\frac{\partial f}{\partial u}(x,v(x))\right\|_\rho \leqslant \sum_j \|g_j(x)\|_\rho \|v(x)\|_\rho^{|j|} \leqslant \sum_{j,k} |g_{j,k}|\rho^{|k|}r^{|j|}$$

$$= \sum_{k\neq 0}(\sum_j |g_{j,k}|r^{|j|})\rho^{|k|} .$$

This series converges for some r , ρ and tends to zero as $\rho \to 0$ since there is no constant term.

To prove the proposition we define now inductively

$$u_{k+1} = u_k + \Delta_k , \quad u_0 = 0 ,$$

$$(2.7) \qquad (P - \frac{\partial f}{\partial u}(x,u_k))\Delta_k = - F(x,u_k) .$$

Since $f(0,0) = 0$ and $F(x,u_0) = f(x,0)$, the norm $\|F(x,u_0)\|_\rho = \varepsilon$ can be made arbitrarily small by choosing ρ small enough. As in the lemma we see that $\|F''(x,v(x))\|_\rho \leqslant C$ for $\|v\|_\rho \leqslant r$ where the constant C depends only on r but not on v . By (2.7) we get

$$\|\Delta_k\|_\rho \leqslant K\|F(x,u_k)\|_\rho$$

and

$$\|F(x,u_{k+1})\|_\rho = \left|\left\langle\int_0^1 (1-s)F''(u_k+s\Delta_k)ds\ \Delta_k,\Delta_k\right\rangle\right|$$

$$\leqslant \frac{C}{2}\|\Delta_k\|_\rho^2 \leqslant \frac{CK^2}{2}\|F(x,u_k)\|_\rho^2$$

if $\|u_k + s\Delta_k\|_\rho \leqslant r$. Choose ε so small that

$$\frac{2}{CK}\sum_{k=0}^{\infty} (\frac{CK^2}{2}\varepsilon)^{2^k} < r ,$$

then we can show by induction that

$$\|u_k\|_\rho \leqslant r \ , \ \|\Delta_k\|_\rho \leqslant \frac{2}{CK}\left(\frac{CK^2}{2}\,\varepsilon\right)^{2^k} \ , \ \|F(x,u_k)\|_\rho \leqslant \frac{2}{CK^2}\left(\frac{CK^2}{2}\,\varepsilon\right)^{2^k} .$$

This proves that $u = \Sigma\,\Delta_k$ converges in B^q_ρ and that $F(x,u_k)$ tends to zero, so $F(x,u) = 0$. The solution is unique since the formal solution is unique.

<u>Corollary 2.5.</u> If

$$\tau = \sum_{j=1}^{n} a_j(x)\frac{\partial}{\partial x_j}$$

is a vector field with $a_j \in M$ such that the Jacobi matrix $J(a_1,\dots,a_n)$ is semi simple with eigenvalues $\lambda_1,\dots,\lambda_n$ satisfying the Poincaré condition

$$|\Sigma_j\,\lambda_i m_j - \lambda_k| > C|m| \ , \text{ for } \ |m| \geqslant 2 \ ,$$

then there exists an analytic transformation $x_j = y_j + \phi_j(y)$, $\phi_j \in M^2$ in the neighbourhood of 0 which transforms τ to its linear part

$$\tilde{\tau} = \Sigma_j\,\lambda_j y_j \frac{\partial}{\partial y_j} \quad .$$

<u>Proof</u>. By a linear change a variable τ can be diagonalized i.e.

$$\tau = \Sigma(\lambda_j x_j + f_j(x))\frac{\partial}{\partial \lambda_j} \ , \ f_j \in M^2 \ .$$

An easy calculation shows that ϕ_j has to satisfy $\tilde{\tau}\phi_k - \lambda_k\phi_k = f_k(y+\phi)$. So proposition 2.3 gives the existence of the ϕ_j .

3. EQUATIONS WITH PARAMETERS

We consider now equations $Pu = F(x,y,u)$ containing an extra parameter $y \in \mathbb{C}^p$. If $\rho, r > 0$, $B^q_{\rho,r}$ is the space of power series

$$\Sigma_k\, u_k(y)x^k$$

such that $u_k \in B^q_r$ and

$$\Sigma_k\, \|u_k\|_r\ \rho^{|k|}t^k < \infty \ , \ M^\mu_{\rho,r}$$

is the subspace of those $u \in B_{\rho,r}$ with $u_k = 0$ for $|k| \leqslant \mu$. We consider operators P of the form

$$P = \sum_{|\ell| \leq d} a_\ell(x,y)\left(x\frac{\partial}{\partial x}\right)^\ell$$

which can be written as

$$(3.1) \qquad Pu = \sum_{|m|=0}^{\infty} \left(\sum_{k \leq m} P_k(m-k,y)u_{m-k}(y)\right)x^m$$

where $P_k(m,y)$ is a $q\times q$ matrix with entries in B_r polynomial with respect to m. The same proof as for proposition 2.3 gives

Proposition 3.1. Let P be an operator of the form (3.1) such that $P_0(m,0)$ is invertible for all m and satisfies the Poincaré condition

$$|P_0(m,0)^{-1}| \leq C|m|^{-d} \quad \text{for} \quad |m| \geq 1$$

and let $f = (f_1,\dots,f_q)$ be analytic in the neighbourhood of the origin in $\mathbb{C}^{n+p+q}$ with

$$f(0,0,0) = 0 \quad , \quad \frac{\partial f}{\partial u}(0,0,0) = 0 \; .$$

Then if r and ρ are chosen sufficiently small the equation $Pu = f(x,y,u)$ has a solution in $B^q_{\rho,r}$. More generally we have

Proposition 3.2. Let f, P be as above but suppose that $P_0(m,0)$ is invertible and satisfies the Poincaré condition only for $|m| \geq \mu$. If the equation $Pu = f(x,y,u)$ has a formal solution

$$u(x,y) = \sum_k u_k(y)x^k \quad \text{with} \quad u_k \in B^q_r$$

for some r and $u_0(0) = 0$ this series converges.

Proof. Since $P_0(m,0)$ and therefore $P_0(m,y)$ is invertible for $|m| \geq \mu$ the coefficients $u_k(y)$ are uniquely determined for $|k| \geq \mu$ if they are given for $|k| < \mu$. Set

$$w = \sum_{|k|<\mu} u_k x^k \quad \text{and} \quad v = u - w \; ,$$

then we get a new equation $Pv = f_1(x,y,v)$ which can be solved in $(M^\mu_{\rho,r})^q$ by (3.1).

Next we shows that if the equation has a formal solution there is also a formal solution $v = \sum v_k(y)x^k$ with $v_k \in B^q_r$ for some r.

Proposition 3.3. Let P, f be as in proposition 3.2. If the equation $Pu = f(x,y,u)$ has a formal solution $u = \sum u_k(y)x^k$ where the u_k are formal power series in y, $u_0(0) = 0$, then the equation has a formal solution $v = \sum v_k(y)x^k$ with $v_k \in B^q_r$ for some r and by proposition 3.2 this solution converges. Moreover given any ν the v_k can be chosen such that their power series expansion coincides with that of u_k up to order ν .

Proof. Expand f in a series with respect to x,u

$$f(x,y,u) = \sum_{k,m} f_{k,m}(y)x^k u^m$$

and rearrange

$$\sum_{k,m} f_{k,m}(y)x^k(\Sigma\, u_\ell(y)x^\ell)^m$$

with respect to the powers of x . The coefficient of x^m contains only those $u_\ell(y)$ with $\ell \leqslant m$ and this coefficient is an analytic function of y and the u_ℓ , $\ell \leqslant m$. Comparing coefficients in $Pu = f(x,y,u)$ we get

$$\sum_{k \leqslant m} P_k(m-k,y)u_{m-k}(y) = \text{coefficient of } x^m .$$

This is a finite system of analytic equations for u_m , $|m| \leqslant \mu$, which has by hypothesis a formal solution. M. Artin, theorem [1] shows that there exists convergent solution v_m coinciding with u_m up to order ν . The rest of the coefficients is now uniquely determined since $P_0(m,y)$ is invertible for $|m| > \mu$ as in the preceding proposition.

As a corollary we get a theorem of S. Kaplan [3] . We consider vector fields

$$(3.2) \qquad \tau = \sum_{j=1}^{n} X_j(z)\frac{\partial}{\partial z_j}$$

X_j analytic near 0 such that the germ

$$S = \{z_j X_j(z)=0, j=1\ldots n\}$$

is regular of dimension s_j set $d = n-s$. By an analytic change of coordinates $\zeta = \zeta(z)$ we may suppose

$$S = \{z;\zeta_j(z)=0, j=1,\ldots,d\} \quad .$$

τ is then transformed to

$$\Sigma\, Y_j(\zeta)\frac{\partial}{\partial \zeta_j} \quad \text{with} \quad Y_j(0,\ldots,0,\zeta_{d+1},\ldots,\zeta_n) = 0 .$$

This gives

$$\frac{\partial Y_j}{\partial \zeta_k}(0) = 0 \quad \text{for} \quad k = d+1,\ldots,n$$

and since S is regular of dimension s the Jacobian $\frac{\partial Y}{\partial \zeta}(0)$ has rank d . Let $\lambda_1,\ldots,\lambda_d$ be the nonzero eigenvalues. By a linear change of variables τ can be transformed to a vector field such that the Jacobian has Jordan normal form :

$$\left(\begin{array}{ccccc|c} \lambda_1 & a_1 & 0 & & & \\ 0 & \lambda_2 & \cdot & \cdot & & \\ & & \cdot & \cdot & \cdot & \\ & & & \cdot & \cdot \;\; a_{d-1} & 0 \\ & & & & \cdot \;\; \lambda_d & \\ \hline & 0 & & & & \\ & & & 0 & & 0 \end{array}\right)$$

So the linear part of τ has the form

$$\sum_{j=1}^{n} \lambda_j x_j \frac{\partial}{\partial x_j} + \sum_{j=1}^{n-1} a_j x_{j+1} \frac{\partial}{\partial x_j}$$

already considered in the remark following proposition 2.2.

Corollary 3.4. Let τ be a vector field (3.2) such that S defined above is regular and such that the nonzero eigenvalues λ_j satisfy the Poincaré condition

$$(3.3) \qquad |\langle\Lambda,m\rangle| \geqslant C|m| \quad \text{for} \quad |m| \geqslant 2 .$$

If the equation $\tau u = f(x,u)$, $f(0,0) = 0$, has a formal solution u then there exists a convergent solution v which coincides with u up to any given order. For the proof we refer to [2].

4. SMALL DENOMINATORS

In this last part we treat a case, where small denominators are present. Consider a vector field

$$\tau = \sum_{j=1}^{n} X_j(x)\frac{\partial}{\partial x_j} , \quad X_j(0) = 0$$

such that the Jacobian matrix $\frac{\partial X_j}{\partial x_k}(0)$ can be diagonalized, i.e. after a linear change of coordinates we may assume

$$\frac{\partial X_j(0)}{\partial x_k} = \Lambda \quad , \Lambda \text{ diagonal with eigenvalues } \lambda_1,\ldots,\lambda_n .$$

So τ can be written in the form

$$(4.1) \qquad \tau = \sum_{j=1}^{n} (\lambda_j x_j + f_j(x)) \frac{\partial}{\partial x_j}$$

where $f_j \in M^2$. Suppose further that for some $\gamma > 0$ the λ_j satisfy the Siegel condition

$$(4.2) \qquad |\langle\Lambda,m\rangle - \lambda_k| \geqslant C|m|^{-\gamma} \quad \text{for} \quad |m| \geqslant 2 .$$

Without restriction we assume $\gamma > 1$. Since $\langle\Lambda,m\rangle - \lambda_k \neq 0$ there is a formal change of variables $x_j = y_j + \phi_j(y)$ which transforms τ to its linear part

$$\tilde{\tau} = \sum_{j=1}^{n} \lambda_j y_j \frac{\partial}{\partial y_j} .$$

C.L. Siegel [5] has proved that this transformation converges. We will give a proof using singular partial differential equations :

Proposition 4.1. If the eigenvalues λ_j of the vector field τ in (4.1) satisfy the Siegel condition (4.2), there exists an analytic change of variables $x_j = y_j + \phi_j(y)$, $\phi_j \in M^2$ transforming τ to its linear part

$$\tilde{\tau} = \sum_{j=1}^{n} \lambda_j y_j \frac{\partial}{\partial y_j} .$$

Proof : (Sketch) As in Corollary 2.5 we see that $u = (\phi_1,\ldots,\phi_n)$ has to satisfy

$$(4.3) \qquad \tilde{\tau}u - \Lambda u = f(y+u)$$

where $f = (f_1,\ldots,f_n)$. Set $F(x,u) = (\tilde{\tau}-\Lambda)u - f(x,u)$. We have to solve $F(x,u) = 0$ and we do this by Newton method. So set $u_0 = 0$, $u_{k+1} = u_k + \Delta_k$ and

$$(4.4) \qquad (\tilde{\tau} - \Lambda)\Delta_k - \frac{\partial f}{\partial u}(x+u_k) = - F(x,u_k) .$$

However this equation cannot be solved by a perturbation argument as in lemma 2.4 since the inverse of $\tilde{\tau}-\Lambda$ is continuous only as an operator from B_ρ to B_σ with $\sigma < \rho$. Following H. Rüssmann [4] we set

$$(4.5) \qquad \Delta_k = J(x+u_k)\cdot E_k$$

where $J(v)$ is the Jacobian of v with respect to λ and E_k a column vector to be determined. We get the commutation relation

$$(4.6) \qquad J(\tau v) = \tau J v - J(v)\cdot\Lambda$$

and omitting for the moment the index k equation (4.4) becomes

$$\begin{aligned}
-F(x,u) &= (\tilde{\tau}-\Lambda)J(x+u)E - \frac{\partial f}{\partial u}(x+u)J(x+u)E \\
&= (\tilde{\tau}J(x+u))\cdot E + J(x+u)\tilde{\tau}E - \Lambda J(x+u)E - (Jf(x+u))\cdot E \\
&= (J\tilde{\tau}(x+u))\cdot E - J(x+u)\Lambda E - J(\Lambda(x+u))\cdot E \\
&\quad + J(x+u)\tilde{\tau}E - J(f(x+u))\cdot E \\
&= J((\tilde{\tau}-\Lambda)(x+u) - f(x+u))\cdot E + J(x+u)(\tilde{\tau}-\Lambda)E \\
&= J(F(x,u))\cdot E + J(x+u)(\tilde{\tau}-\Lambda)\cdot E
\end{aligned}$$

where $J(F(x,u))$ is the total Jacobian of $F(x,u(x))$ with respect to x. Now if u is an approximate solution $F(x,u)$ is small and $J(F(x,u))$ should also be small. We expect also Δ to be small and since $J(x+u)$ is invertible if u is small enough E should be small by (4.5). So, the term $J(F(x,u))\cdot E$ is small of second order, and since we omitted already a second order term in deriving (4.4) we might as well forget this one. The remaining equation

$$J(x+u)(\tilde{\tau}-\Lambda)E = -F(x,u)$$

can be solved since $J(x+u) = I + J(u)$ is invertible if u is small enough. This leads to the iteration scheme

$$\text{(4.8)}\qquad \begin{aligned} &u_0 = 0\ ,\ u_{k+1} = u_k + \Delta_k\ ,\ \Delta_k = J(x+u_k)E_k\ , \\ &E_k = -(\tau-\Lambda)^{-1}J(x+u_k)^{-1}F(x,u_k)\ . \end{aligned}$$

This can be solved in B_{ρ_k} for a decreasing sequence ρ_k and if everything is carefully estimated we can take

$$\rho_k \geqslant \frac{\rho_0}{2}\ .$$

So, we finally find a solution of our problem in $B_{\frac{\rho_0}{2}}$. For the details we refer to [2].

REFERENCES

[1] Artin, M. : 'On the solution of analytic equations'. *Invent.Math.* **5**, 277-291,(1968).

[2] Bengel, G., Gérard, R. : 'Formal and convergent solutions of singular partial differential equations'. *Manuscripta math.* **38**,343-373,(1982).

[3] Kaplan, S. : 'Formal and convergent power series solutions of singular partial differential equations'. *Trans.Amer.Math.Soc.* **256**, 163-183,(1979).

[4] Rüssmann, H. : 'Kleine Nenner II, Bemerkungen zur Newtonschen Methode'. *Nachr.Akad.Wiss.Göttingen,Math.Phys.Kl.*,1-10,(1972).

[5] Siegel, C.L. : 'Über die Normalform analytischer Differentialgleichungen in der Nähe üner Gleichgewichtslösung'. *Nachr.Akad.Wiss.Göttinger,Math.Phys.Kl.*,21-30,(1952).

SINGULARITES DES SOLUTIONS DE PROBLEMES DE CAUCHY HYPERBOLIQUES NON LINEAIRES

J-M. Bony
Université de Paris-Sud
Département de Mathématiques
91405 Orsay Cedex
France

RESUME. Pour une équation strictement hyperbolique semi-linéaire, on détermine la localisation des singularités d'une solution connaissant celle des données de Cauchy. Pour une équation d'ordre quelconque, on étudie le cas où les données sont singulières en un point ou sur une hypersurface; pour une équation d'ordre deux, le cas où elles sont singulières sur une sous-variété ou - en dimension deux d'espace - sur plusieurs courbes concourantes. Les démonstrations reposent d'une part sur l'étude des espaces de distributions involutives et conormales, d'autre part sur la théorie de la seconde microlocalisation.

0. INTRODUCTION

Nous considérons le problème de Cauchy, dans $\mathbb{R}^{n+1}$

$$\begin{aligned} &P(x,t,D_x,D_t)\, u = F(x,t,u,\ldots,\nabla^{m-1} u) \\ &\gamma_j\, u = (\partial/\partial t)^j\, u(x,0) = \varphi_j(x) \ , \end{aligned} \tag{0.1}$$

pour une équation semi-linéaire strictement hyperbolique d'ordre m. Nous cherchons à déterminer les singularités d'une solution $u \in H^s$ $(s > n/2+m)$ pour t petit, connaissant les singularités des φ_j. Notons que, pour ces valeurs de s, l'existence de solutions pour t petit est bien connue (voir [11]), et que, si la solution se prolonge pour t grand, l'étude de ses singularités relève des théorèmes de propagation et d'interaction ([1], [2], [4], [5], [7], [10]).

En dimension 1 d'espace, le problème est essentiellement résolu par J. Rauch et M. Reed [12]. En dimension quelconque, connaissant la régularité microlocale des φ_j, on peut en déduire des résultats d'appartenance à H^σ pour $\sigma < 2s - s_0$ en général, et $\sigma < 3s - s_1$ lorsque l'équa-

H. G. Garnir (ed.), Advances in Microlocal Analysis, 15–39.

tion est d'ordre 2 [1]. Ces résultats sont exposés au §.1.

Pour obtenir des régularités supérieures, et notamment l'appartenance de u à C^{∞} en dehors de certaines hypersurfaces, des hypothèses supplémentaires doivent être faites. Nous supposerons ici que les φ_j appartiennent à des espaces de distributions $H^s(\mathcal{N},\nu)$ définis comme suit : ces distributions appartiennent à H^s, et y restent lorsqu'on leur applique, au plus ν fois, des opérateurs pseudo-différentiels d'ordre 1 appartenant à un certain ensemble $\mathcal{N}$. Le cas le plus simple (distributions conormales) est celui où $N \in \mathcal{N}$ si son symbole principal s'annule sur le conormal d'une sous-variété V.

Des théorèmes généraux (théorèmes 4.1. et 4.10.) assurent que u appartient à des espaces $H^s(\mathcal{M},\nu)$ du même type, pourvu que soient satisfaites 3 conditions : une condition de commutation (C.C.) reliant $\mathcal{M}$ et P, une condition de trace reliant $\mathcal{M}$ et $\mathcal{N}$, et une "propriété de Leibniz" portant sur $\mathcal{M}$ et garantissant que les $H^s(\mathcal{M},\nu)$ sont des algèbres. On en déduit le cas où m=2 et où les données de Cauchy sont singulières sur une sous-variété (théorème 4.7., voir aussi [13]) et le cas, pour m quelconque, de données singulières sur une hypersurface (théorème 4.9.).

Le cas où les données sont singulières en un point (théorème 7.1.) nécessite l'usage du calcul 2-microdifférentiel introduit dans [6], [7], qui est l'analogue différentiel du calcul introduit par Y. Laurent [9] dans le cadre analytique. Les résultats en sont rappelés aux paragraphes 5 et 6.

Enfin, nous esquissons brièvement la démonstration du théorème 8.1. relative au cas où, pour l'équation des ondes en dimension 2 d'espace, les données de Cauchy sont singulières sur plusieurs courbes concourantes. Il s'agit de l'analogue, pour le problème de Cauchy, du théorème démontré dans [7] (voir aussi [10]) sur l'interaction de trois ondes.

Les solutions sont alors régulières hors des surfaces caractéristiques issues des courbes et du cône d'onde issu de leur point de concours. Toute la démonstration, et la définition même des espaces auxquel u va appartenir, fait appel au calcul 2-microdifférentiel.

1. THEOREMES GENERAUX DE PROPAGATION

Nous considérons, au voisinage de 0 dans $\mathbb{R}^{n+1} = \mathbb{R}^n_x \times \mathbb{R}_t$ une équation semi-linéaire

$$P(x,t,D_x,D_t)\, u(x,t) = F(x,t,u,\dots,\nabla^{m-1} u) \tag{1.1}$$

où P est un opérateur strictement hyperbolique d'ordre m, et F une fonction réelle C^∞ des coordonnées et des dérivées de u jusqu'à l'ordre $m-1$.

Compte tenu de la propagation à vitesse finie, et du caractère local des résultats cherchés, on ne restreint pas la généralité en supposant, ce que nous ferons désormais, que P est défini dans $\mathbb{R}^n \times]-T,+T[$, et que $\operatorname{Supp} u \subset K \times]-T,T[$, avec K compact.

Pour des raisons techniques, nous aurons à introduire la généralisation suivante de (1.1), où U est éventuellement à valeurs vectorielles $U = (u_1(x,t),\dots,u_N)$

$$P\,U = \mathcal{L}(U,\dots,\nabla^{m-1} U) \tag{1.2}$$

c'est-à-dire

$$P\,u_j = \mathcal{L}_j(u_1,\dots,u_N,\dots,\nabla^{m-1} u_N) \tag{1.2'}$$

où les $\mathcal{L}_j$ sont des expressions pseudo-différentielles non linéaires d'ordre 0 (epdnℓ0) au sens de la définition suivante.

1.1. <u>Définition</u> : On appelle epdnℓ0 une expression $\mathcal{L}(u,v,\dots,w)$ obtenue en composant un nombre fini de fois des fonctions non linéaires, et des opérateurs pseudo-différentiels d'ordre 0.

Par exemple :

$$\mathcal{L}(u,v) = F[x,t,AG(x,t,u,v)\ ,\ B\,H(x,t,C\,u,v)]$$

est une epdnℓ0 si $A,B,C \in \psi^o(\mathbb{R}^{n+1})$ et $F,G,H \in C^\infty(\mathbb{R}^{n+3})$. Ici et dans la suite, $\psi^k(\mathbb{R}^n)$ désignera l'ensemble des opérateurs pseudo-différentiels (classiques, i.e. à symboles polyhomogènes, sauf mention explicite du contraire) d'ordre k dans $\mathbb{R}^n$.

Une conséquence fondamentale du calcul paradifférentiel (voir [2]) est que, si u (ou U) est une solution de (1.1) (ou 1.2) appartenant à H^s_{loc}, avec

$$s = (n+1)/2 + m-1+\rho\ ,\quad \rho > 0\ , \tag{1.3}$$

on a

$$PU + RU = G \tag{1.4}$$

où G est une fonction (vectorielle) appartenant à $H^{s+\rho-m+1}_{loc}$, et où R est un opérateur paradifférentiel d'ordre m-1 : $R \in Op(\Sigma^{m-1}_{\rho})$.

1.2. *Théorème : Soit U une solution de (1.4), avec $U \in H^s$ (s vérifiant (1.3)), et $G \in H^{s+\rho-m+1}$ (c'est le cas notamment pour une solution de (1.1) ou (1.2)). Soit $\sigma \leqslant s+\rho$, et supposons que l'on ait $\gamma_j U \in H^{\sigma-j}$, $j=0,\dots,m-1$, microlocalement en $(x_o,\xi_o) \in \mathbb{R}^n \times (\mathbb{R}^n \setminus \{0\})$. Soient τ_ℓ $(\ell = 1,\dots,m)$ les racines de $p_m(x_o,0;\xi_o,\tau)=0$, et G_ℓ les bicaractéristiques issues de $(x_o,0,\xi_o,\tau_\ell)$. Alors U appartient à H^σ microlocalement sur $\bigcup_1^m G_j$.*

Nous n'utiliserons, et ne démontrerons ici, que le résultat global suivant, qui est bien sûr un corollaire immédiat du théorème 1.2.

1.3. *Théorème : Soit U une solution de 1.4. appartenant à H^{s-1} $(s=(n+1)/2+m+\rho,\ \rho>0)$. On suppose que $R \in Op(\Sigma^{m-1}_{\rho})$, $G \in H^{s-m+1}$ et que $\gamma_j U \in H^{s-j}$, $j=0,\dots,m-1$. Alors on a $U \in H^s$.*

Le point-clef réside en la démonstration d'une inégalité a priori, valable pour $U \in H^{\sigma+1}_{loc}(\mathbb{R}^n \times]-T,T[)$ à support compact en x et pour δ assez petit :

$$\begin{aligned} \|\varphi_\delta U\|_\sigma \leqslant\ & c(\delta)\,\|\varphi_\delta(P+R)U\|_{\sigma-m+1} + \\ & + M(\delta)\Big\{\sum_0^{m-1} \|\gamma_j U\|_{\sigma-j} + \|\phi(P+R)U\|_{\sigma-m} + \|\phi U\|_{\sigma-\gamma} + \\ & + \|\chi(x,t,D_x,D_t)U\|_{\sigma+1}\Big\} \end{aligned} \tag{1.5}$$

où $\gamma = \mathrm{Min}(\rho,1)$; où φ_δ est une famille de fonctions $\varphi_\delta(t)$ de classe C^∞ vérifiant $\varphi_\delta = 1$ près de 0, à support dans $|t| \leqslant \delta$; où $\phi(t) \in C^\infty_o$ et est égal à 1 pour t assez petit ; où $\chi \in \psi^o$ et a un symbole égal à 1 pour t assez petit et pour $|\xi| \leqslant \varepsilon_1 |\tau|$; et enfin $c(\delta) \to 0$ pour $\delta \to 0$.

Il faut remarquer que compte tenu du terme $\|\chi u\|_{\sigma+1}$, on peut remplacer toutes les normes dans $H^{\sigma'}(\mathbb{R}^{n-1})$ figurant dans (1.5) par les normes dans $L^2((-T,T)\,;H^{\sigma'}(\mathbb{R}^n))$.

En utilisant un argument classique de régularisation tangentielle, tout à fait analogue à la démonstration du théorème 6.2. de [2], on déduit de (1.5) le résultat suivant : si U appartient à $H^{\sigma-\gamma}$ et appartient microlocalement à $H^{\sigma+1}$ près du support de χ, si $\gamma_j U \in H^{\sigma-j}$, et $(P+R)U \in H^{\sigma-m+1}$, alors $\varphi_\delta U \in L^2(]-T,T[\,, H^\sigma(\mathbb{R}^n))$ et donc à $H^\sigma(\mathbb{R}^{n+1})$.

Le théorème 1.3. en résulte facilement. L'opérateur $P+R$ appartenant à $Op(\Sigma^m_{1+\rho})$, et étant elliptique pour $|\xi| \leqslant \varepsilon_1 |\tau|$, on en déduit que $u \in H^{s+\gamma}$ microlocalement en ces points. Les hypothèses précédentes sont donc vérifiées pour $\sigma = \mathrm{Min}(s, s+\rho-1)$ et on obtient $U \in H^s$ ou $H^{s+\rho-1}$ pour t assez petit, et donc partout par propagation des singularités. Si $\rho > 1$, la démonstration est terminée, sinon on recommence avec $\sigma = s+2\rho-1$ et on obtient $U \in H^s$ au bout d'un nombre fini d'étapes.

1.4. <u>Démonstration de l'inégalité (1.5)</u>.

Si (1.5) est valable pour P, elle est valable pour $(P+R)$. En effet, le terme $c(\delta) \| \varphi_\delta R U \|_{\sigma-m+1}$ peut s'estimer par $1/2 \, \| \varphi_\delta U \|_\sigma$ pour δ assez petit et être absorbé au premier membre, modulo des termes rentrant dans l'accolade. Nous supposerons donc $R = 0$, et $U = u$ scalaire.

Nous partirons de l'estimation (23.2.3) de ([B] tome 3) :

$$\| D_t^j u(\cdot,t) \|_{\sigma-j} \leqslant C \int_0^t \| P u(\cdot,t) \|_{\sigma-m+1} + C \sum_0^{m-1} \| \gamma^k u \|_{\sigma-k} .$$

Si $\varphi_\delta(t)$ est une fonction décroissante de $|t|$, à support dans $[-\delta, +\delta]$, il en résulte que

$$\sup_t \| \varphi_\delta(t) u(\cdot,t) \|_\sigma \leqslant C \int_0^t \| \varphi_\delta(t) P u(\cdot,t) \|_{\sigma-m+1} + C \sum_0^{m-1} \| \gamma^k u \|_{\sigma-k} .$$

En majorant la norme de $\varphi_\delta u$ dans $L^2((-T,T), H^\sigma)$ par sa norme dans $L^\infty((-T,T), H^\sigma)$, et celle de $\varphi_\delta Pu$ dans $L^1((-T,T), H^{\sigma-m+1})$ par sa norme dans $L^2((-T,T), H^{\sigma-m+1})$, il apparaît des constantes en $\sqrt{\delta}$, et on obtient

$$\| \varphi_\delta u \|_{L^2(H^\sigma)} \leqslant c(\delta) \, \| \varphi_\delta P u \|_{L^2(H^{\sigma-m+1})} + C' \sum_0^{m-1} \| \gamma^k u \|_{\sigma-k}$$

on en déduit (1.5) pour P en remplaçant la norme $L^2(H^\sigma)$ par la norme $H^\sigma(\mathbb{R}^{n+1})$, moyennant l'introduction d'un terme $C \| \chi u \|_{\sigma+1}$ au membre de droite.

1.5. Remarque.

La démonstration du théorème 1.2. nécessite une combinaison des arguments précédents et des arguments de la démonstration du théorème 6.2. de [2]. La démonstration du théorème 1.2. pour les solutions de (1.1) est plus simple. Nous en avons esquissée une dans [5] (théorème 3), et c'est également un cas (très) particulier des résultats de M. Sablé-Tougeron [14] sur la réflexion des singularités. La présence dans (1.4) d'opérateurs pseudo- et para-différentiels en D_x et D_t ne permet pas d'adapter ici ces méthodes.

Le résultat bien connu suivant nous sera utile.

1.6. Proposition : Supposons $u \in H^s_{loc}$ et $Pu \in H^{s-m+1}_{loc}$, avec $s > m-1$. Alors $\gamma^j u \in H^{s-j}_{loc}$, $j = 0, \ldots, m-1$.

On a en effet localement $D^j_t u \in L^2(\mathbb{R}, H^{s-j}) \cap C(\mathbb{R}, H^{s-j-1/2})$ et $Pu \in L^2(\mathbb{R}, H^{s-m+1})$. Pour une (en fait presque toute) valeur t_o, on a donc $D^j_t u \big|_{t=t_o} \in H^{s-j}$. En utilisant le théorème 23.2.2 de [8], d'existence dans $\bigcap_0^{m-1} C^j(\mathbb{R}, H^{s-j})$ et d'unicité dans $\bigcap_0^{m-1} C^j(\mathbb{R}, H^{s-j-1/2})$ pour le problème de Cauchy avec données en t_o, on obtient $u \in \bigcap_0^{m-1} C^j(\mathbb{R}, H^{s-j})$ et le résultat.

Pour l'équation des ondes (ou une équation hyperbolique d'ordre 2) semi-linéaire :

$$\Box u = \frac{\partial^2 u}{\partial t^2} - \sum_1^n \frac{\partial^2 u}{\partial x_i^2} = F(t,x,u) \ , \qquad (1.6)$$

on a le résultat plus précis suivant, dû à M. Beals [1].

1.7. Théorème : Soit u une solution de 1.4. appartenant à H^s, $s > \frac{n+1}{2}$. Soit $\sigma < 3s-n+1$. On suppose que l'on a $\gamma_j u \in H^{\sigma-j}$ $(j = 0,1)$ microlocalement en un point (x_o, ξ_o) de $\omega \times (\mathbb{R}^n \setminus \{0\})$. Alors u appartient à H^σ microlocalement sur les deux bicaractéristiques issues de $(x_o, 0; \xi_o, \pm|\xi_o|)$.

Par contre, un résultat dû également à M. Beals [1] montre que, même pour l'équation $\Box u = \beta(x,t) \cdot u^3$, la connaissance des singularités microlocales de $\gamma_o u$ et $\gamma_1 u$ ne suffit plus pour déterminer l'appartenance de u à H^{σ}, pour $\sigma > 3s - n+2$.

Pour pouvoir dépasser les limitations imposées à σ par les théorèmes 1.2. et 1.7., nous devrons faire des hypothèses sur les données de Cauchy plus précises que l'appartenance microlocale à un espace de Sobolev, et introduire des espaces de "distributions conormales".

2. DISTRIBUTIONS INVOLUTIVES ET CONORMALES

Nous désignerons dans ce qui suit par $\mathcal{M}$ un sous-ψ^{o}-module de ψ^{1}, contenant ψ^{o}, vérifiant :

> $\mathcal{M}$ est localement de type fini : il existe au voisinage de chaque point une famille finie $(M_{\alpha}) \subset \mathcal{M}$ telle que :
> $\forall M \in \mathcal{M} \quad M = \Sigma A_{\alpha} M_{\alpha} + A_{o} \quad ; \quad A_{o}, A_{\alpha} \in \psi^{o}$. (2.1)
> On dira que les M_{α} forment un système de générateurs de $\mathcal{M}$.

Pour $M_1, M_2 \in \mathcal{M}$, on a $[M_1, M_2] \in \mathcal{M}$. (2.2)

L'exemple typique de tels sous-modules est l'ensemble des $M \in \psi^{1}$ dont le symbole principal s'annule sur une sous-variété involutive donnée.

2.1. <u>Définition</u> (espaces de distributions involutives).

Pour $\mathcal{M}$ vérifiant (2.1) et (2.2), $s \in \mathbb{R}$, $k \in \mathbb{N}$, on pose

$$H^{s}(\mathcal{M}, \nu) = \{u \in H^{s} \mid M^{A} u \in H^{s}, |A| \leqslant \nu\} .$$

On note ici, et dans la suite, $A = (\alpha_1, \dots, \alpha_{\ell})$;

$$M^{A} = M_{\alpha_1} \circ \dots \circ M_{\alpha_{\ell}} \quad ; \quad |A| = \ell .$$

2.2. <u>Exemple</u> (distributions lagrangiennes et conormales)

Si Λ est une variété lagrangienne conique lisse, on note $\mathcal{M}_{\Lambda}$ l'ensemble des $M \in \psi^{1}$, dont le symbole principal s'annule sur Λ. On pose alors

$$H^{s,\nu}(\Lambda) = H^{s}(\mathcal{M}_{\Lambda}, \nu) .$$

Si Λ est le conormal d'une sous-variété lisse V de $\mathbb{R}$, $\Lambda = T^{*}_{V}\mathbb{R}$, on notera $H^{s,k}_{V} = H^{s,k}(T^{*}_{V}\mathbb{R})$. Un élément de $H^{s,k}_{V}$ appartient à H^{s+k} localement hors de V, et microlocalement hors de $T^{*}_{V}\mathbb{R}^{n}$.

2.3. Exemple.

Une situation courante est le cas suivant. On se donne une réunion $\Lambda = \Lambda_1 \cup \ldots \cup \Lambda_p$ de sous-variétés lagrangiennes coniques lisse de $T^*\mathbb{R}^n$ qui s'intersectent seulement 2 à 2. On suppose que $\Lambda_j \cap \Lambda_k$ est de dimension $n-1$ (si elle est non vide) et que l'intersection est nette.

On note encore dans ce cas $\mathcal{M}_\Lambda = \{M \in \psi^1 \mid \sigma_1(M) = 0 \text{ sur } \Lambda\}$, et on pose $H^{s,\nu}(\Lambda) = H^s(\mathcal{M}_\Lambda, \nu)$.

On peut bien entendu, par dualité et interpolation, définir les espaces $H^{s,s'}(\Lambda)$ pour $s' \in \mathbb{R}$.

2.4. Remarque.

On définit de façon évidente l'appartenance microlocale aux espaces $H^s(\mathcal{M},\nu)$. Il est clair que ces espaces sont transformés en espaces $H^s(\mathcal{M}',\nu)$ du même type sous l'effet d'un opérateur intégral de Fourier elliptique d'ordre 0, les symboles principaux des $\mathcal{M}'$ se déduisant de ceux de $\mathcal{M}$ par la transformation canonique associée.

En particulier, dans le cas de l'exemple 2.2., on peut se ramener microlocalement au cas où $\Lambda = T^*V$ par une telle transformation. Dans le cas de l'exemple 2.3., pour $(x_o,\xi_o) \in \Lambda_j \cap \Lambda_k$, on peut se ramener au cas où Λ_j est définie par $x = 0$, et Λ_k par $x_1 = x_{n-1} = \xi_n = 0$, pour $x_o = 0$, $\xi_o = (1,0,\ldots,0)$.

2.5. *Définition : On dit que $\mathcal{M}$ possède la propriété de Leibniz si pour un (et donc pour tout) système de générateurs (M_α) de M, et pour toute fonction $F(x,u,v)$ de classe C^∞ de plusieurs variables, on a*

$$M_\alpha F(x,u,v) = \mathcal{L}_\alpha(u,v,M^\beta u, M^\beta v) \tag{2.3}$$

où $\mathcal{L}_\alpha$ est une epdnℓ0 (Définition 1.1.).

2.6. *Théorème : Dans un ouvert de $\mathbb{R}^n$, si $\mathcal{M}$ possède la propriété de Leibniz, et si $s > n/2$, les espaces $H^s(\mathcal{M},\nu)$ sont des algèbres. En outre, pour $u,v \in H^s(\mathcal{M},\nu)$ et $|A| \leqslant \nu$, on a alors*

$$M^A F(u,v) = \mathcal{L}_A(M^B u, \ldots, M^C v) \in H^s \tag{2.4}$$

où B et C parcourent les multi-indices de longueur $\leqslant \nu$.

On déduit en fait (2.4) de (2.3) par une récurrence évidente, en utilisant le fait que H^s est stable par les opérateurs pseudo-différentiels d'ordre 0, et par les opérations non linéaires $u,v \to F(u,v)$ pourvu que $s > n/2$.

Il est évident que, si $\mathcal{M}$ possède localement un système de générateurs constitué de champs de vecteurs, $\mathcal{M}$ possède la propriété de Leibniz. Par contre, il n'est ni nécessaire, ni suffisant, pour que les $H^s(\mathcal{M},\nu)$ soient des algèbres, que $\mathcal{M}$ soit engendré microlocalement par des champs de vecteurs.

3. COMMUTATION ET TRACES

Soit $\mathcal{M}$, un sous-ψ^o-module de $\psi^1(\mathbb{R}^{n+1})$ vérifiant les conditions (2.1), (2.2), de générateurs (M_α), et P comme au §.1.

3.1. Définition : On dit que $\mathcal{M}$ vérifie la condition de commutation (C.C.) par rapport à P, si on a

$$(C.C.) \qquad [P,M_\alpha] \equiv \sum_\beta A_{\alpha\beta} M_\beta + B_\alpha P \quad (mod\ \psi^{m-1}),$$

avec $A_{\alpha\beta} \in \psi^{m-1}$, $B_\alpha \in \psi^o$.

Soit maintenant $\mathcal{N}$ un sous-ψ^o-module de $\psi^1(\mathbb{R}^n)$, de générateurs $N_\beta(x,D_x)$, vérifiant (2.1), (2.2).

3.2. Définition : On notera trace $(\mathcal{M}) \subset \mathcal{N}$ la relation suivante

$$\forall M \in \mathcal{M}, \ M \equiv t\,M' + \Sigma A_\beta N_\beta \ (mod\ \psi^o(\mathbb{R}^{n+1})) \text{ microlocalement pour } |\xi| \neq 0, \text{ avec } M' \in \psi^1(\mathbb{R}^{n+1}) \text{ et } A_\alpha \in \psi^o(\mathbb{R}^{n+1}). \tag{3.1}$$

3.3. Théorème : On suppose trace $(\mathcal{M}) \subset \mathcal{N}$. Soit $u \in H^s(\mathbb{R}^{n+1})$; $s > m-1/2$, tel que $Pu \in H^{s-m+1}$ et $P^2 u \in H^{s-2m+2}$. Alors, pour $M \in \mathcal{M}$, on a :

$$\gamma_j Mu \equiv \sum E_{j,k,\beta} N_\beta \gamma_k u \quad (mod\ H^s(\mathbb{R}^n)) \tag{3.2}$$

pour $j,k = 0,\ldots,m-1$, et $E_{jk\beta} \in \psi^{j-k}(\mathbb{R}^n)$.

Il est clair que, pour $E \in \psi^o$, Eu vérifie les mêmes hypothèses que u. Si le symbole de E est concentré près de $|\xi| = 0$, et égal à 1 pour $|\xi| \leqslant \varepsilon |\tau|$, l'opérateur P étant inversible en ces points, on a

$E\,u \in H^{s+2}$, et donc $\gamma_j\, M E\, u \in H^{s+1/2-j} \subset H^{s-j}$. On peut donc, en remplaçant u par $(I-E)u$, supposer que $WF(u)$ ne rencontre pas $\{(x,t,\xi,\tau) \mid |\xi| = 0\}$.

On a alors, d'après (3.1)

$$M\,u \equiv t\,M'\,u + \Sigma\, A_\beta\, N_\beta\, u \quad (\text{mod } C^\infty) . \tag{3.3}$$

On a $\gamma_j\, t\,M'\,u = \sum_{k<j} c_{jk}\, \gamma_k\, M'\,u$.

D'après la proposition 1.6., (on a $M'\,u \in H^{s-1}$ et $P\,M'\,u \in H^{s-1-m+1}$) on a $\gamma_k\, M'\,u \in H^{s-1-k}(\mathbb{R}^n) \subset H^{s-j}(\mathbb{R}^n)$.

Enfin, en désignant les symboles principaux par des minuscules, on a

$$a_\beta(x,t,\xi,\tau) = e_\beta(x,t,\xi,\tau)\; p(x,t,\xi,\tau) + \sum_0^{m-1} b_{k\beta}(x,t,\xi)\tau^k$$

la partie polynomiale provenant du théorème d'interpolation de Lagrange, pour $\xi \neq 0$. En désignant par E_β un élément de ψ^{-m} dont le symbole est e_β pour $|\xi| > \varepsilon|\tau|$, et par $B_{\beta k}$ des éléments de $\psi^{-k}(\mathbb{R}^n)$ dépendant de t de manière C^∞, et de symbole principal $b_{\beta k}$, on déduit de (3.3) :

$$\gamma_j\, M\,u \equiv \Sigma\, \gamma_j\, E_\beta\, P\, N_\beta\, u + \gamma_j\, B_{\beta k}\, D_t^k N^\beta u \quad (\text{mod } H^{s-j}) . \tag{3.4}$$

On a $E_\beta\, P\, N_\beta\, u = E_\beta\, N_\beta\, P\,u + E_\beta[P,N_\beta]u \in H^s$

$$P(E_\beta\, P\, N_\beta\, u) = E_\beta\, N_\beta\, P^2 u + K_1\, P\,u + K_2\, u \in H^{s-m+1}$$

avec des commutateurs K_1 d'ordre 0 et des bicommutateurs K_2 d'ordre $m-1$. On a donc, d'après la proposition 1.6., $\gamma_j\, E_\beta\, P\, N_\beta\, u \in H^{s-j}$.

Il ne reste donc à considérer, dans (3.4) que les valeurs, pour $t = 0$, de $D_t^j\, B_{\beta k}\, D_t^k\, N_\beta\, u$ que l'on peut réécrire :

$$\gamma_j\, u \equiv \sum_{\ell=0}^{2m-2} C_{j\ell\beta}(x,D_x) N_\beta\, D_t^\ell u \Big|_{t=0} + \Sigma\, D_{j\ell}(x,D_x) D_t^\ell u \Big|_{t=0} \quad (\text{mod } H^{s-j}) \tag{3.5}$$

avec $C_{j\ell\beta} \in \psi^{j-\ell}$ et $D_{j\ell} \in \psi^{j-\ell-1}$.

On montre comme précédemment que $D_{j\ell}\, D_t^\ell u \in H^{s-j}$ et que $P\, D_{j\ell}\, D_t^\ell u \in H^{s-j-m+1}$.

Dans la première sommation, si $\ell = m+q$ est supérieur ou égal à m, on peut écrire en utilisant la forme de P :

$$P = D_t^m + \Sigma\, p_{m-k}(x,D_x) D_t^k$$

$$C_{j\ell\beta} N_\beta D_t^{m+q} u = C_{j\ell\beta} N_\beta D_t^q P u - \Sigma C_{j\ell\beta} N_\beta P_k D_t^{k+q} \ . \qquad (3.6)$$

Le premier terme de (3.6) se traite comme les premiers termes de (3.4) et a une trace d'ordre 0 dans H^{s-j}. Les autres termes de (3.6) se remettent sous la forme $\Sigma C'_{j\ell\beta} N_\beta D_t^\ell$, mais avec $\ell \leqslant 2m-3$, modulo des commutateurs qui se traitent comme les derniers termes de (3.5).

En itérant ce procédé, on obtient :

$$\gamma_j u \equiv \sum_{\ell=0}^{m-1} C''_{j\ell\beta} N_\beta D_t^\ell u \Big|_{t=0} \quad (\mathrm{mod}\ H^s)$$

et (3.2) est démontré.

3.4. <u>Corollaire</u> : On suppose que trace $(\mathcal{M}) \subset \mathcal{N}$ *et que* $\mathcal{M}$ *vérifie (C.C). On suppose que, pour* $\nu \in \mathbb{N}$, *on a*

$$u \in H^s(\mathcal{M},\nu) \ ; \ Pu \in H^{s-m+1}(\mathcal{M},\nu) \ ; \ P^2 u \in H^{s-2m+2}(\mathcal{M},\nu) \ , \qquad (3.7)$$

et $\gamma^j u \in H^{s-j}(\mathcal{N},\nu+1)$; $j = 0,\dots,m-1$.

Alors, pour tout multi-indice A avec $|A| \leqslant \nu+1$, *on a*

$$\gamma^j M^A u = \Sigma E_{ABjk}(x,D_x) N^B \gamma_k u \in H^{s-j} \qquad (3.8)$$

où $|B| \leqslant \nu+1$, *et* $E_{ABjk} \in \psi^{j-k}$.

On démontre d'abord aisément, par récurrence sur ν, et en utilisant la condition de commutation (C.C.) que les conditions (3.7) entraînent $M^A u \in H^s$, $P M^A u \in H^{s-m+1}$, $P^2 M^A u \in H^{s-2m+2}$ pour $|A| \leqslant \nu$.

Supposons maintenant (3.8) établi au rang $(\nu-1)$, et appliquons le théorème 3.3. aux fonctions $M^A u$, pour $|A| \leqslant \nu$

$$\gamma_j M_\alpha M^A u \equiv \Sigma E_{jk\alpha A\beta} \cdot N_\beta \gamma_k M^A u \quad (\mathrm{mod}\ H^s) \ .$$

En remplaçant dans le membre de droite les $\gamma_k M^A u$ par leurs expressions en vertu de (3.8) au rang $\nu-1$, on obtient la décomposition voulue des $\gamma_j M_\alpha M^A u$, pour tout (α,A) de longueur $\nu+1$.

Signalons d'autre part la conséquence suivante de la condition de commutation (C.C.), qui se démontre par une récurrence évidente sur $|A|$.

3.5. <u>Proposition</u> : On suppose que $\mathcal{M}$ *vérifie (C.C.). Alors, pour tout multi-indice A, on a :*

$$[P,M^A] = \sum_{|B|\leq|A|} K_{AB} M^B + \sum_{|C|\leq|A|-1} L_{AC} M^C P \qquad (3.9)$$

avec $K_{AB} \in \psi^{m-1}$ *et* $L_{AC} \in \psi^o$.

4. REGULARITE DES SOLUTIONS DU PROBLEME DE CAUCHY

Nous allons donner, pour l'équation (1.1), un théorème général de régularité des solutions. Nous considérerons ensuite deux applications : le cas d'une équation du second ordre avec données singulières sur une sous-variété lisse de $\mathbb{R}^n$, le cas d'une équation d'ordre quelconque avec données singulières sur une hypersurface.

4.1. <u>*Théorème*</u> *: Soient* $\mathcal{M}$ *et* $\mathcal{N}$*, vérifiant (2.1) et (2.2), des sous-modules de* $\psi^1(\mathbb{R}^{n+1})$ *et* $\psi^1(\mathbb{R}^n)$*. On suppose que* $\mathcal{M}$ *vérifie la condition de commutation (C.C.), la condition de Leibniz, et que trace* $\mathcal{M} \subset \mathcal{N}$ *(voir définitions 2.5., 3.1., 3.2.). Soit u une solution de (1.1) appartenant à* H^s $(s > (n+1)/2 + m)$ *telle que*

$$\gamma_j u \in H^{s-j}(\mathcal{N},\nu) \quad , \quad j = 0,\dots,m-1 . \qquad (4.1)$$

Alors on a $u \in H^s(\mathcal{M},\nu)$.

Nous allons introduire les fonctions à valeurs vectorielles U_μ, de composantes $(M^A u)$ pour $|A| \leq \mu$, et démontrer par récurrence sur μ que, pour $\mu \leq \nu$, on a $U_\mu \in H^s$. Nous supposons donc dans la suite cette propriété vérifiée au rang $\mu-1$, ce qui entraîne $U_\mu \in H^{s-1}$.

4.2. <u>Equation vérifiée par U_μ</u>.

Pour $|A| \leq \mu$, on peut écrire

$$P M^A u = [P,M^A]u + M^A P u$$

et, d'après (3.9)

$$P M^A u = \Sigma K_{AB} M^B u + \Sigma L_{AC} M^C P u + M^A P u$$

on peut, d'après (1.1) remplacer $M^A P u$ (ou $M^C P u$) par $-M^A F(x,t,u,\dots,\nabla^{m-1} u)$, et développer ce terme d'après (2.4) ce qui est légitime puisque $\nabla^{m-1} u \in H^{s-m}$, et $s-m > (n+1)/2$. On obtient, compte tenu de $L_{AC} \in \psi^o$:

$$P M^A u = \Sigma K_{AB} M^B u + \mathcal{L}_A(\dots, M^B \partial^\beta u, \dots) \quad , \qquad (4.2)$$

où dans l'expression de $\mathcal{L}_A$, B parcourt les multi-indices tels que $|B| \leqslant \mu$, et $|\beta| \leqslant m-1$. On peut réexprimer chaque $M^B \partial^\beta u$ comme somme de $\partial^\gamma M^C u$, et on obtient en mettant (4.2) sous forme vectorielle

$$P\, U_\mu - K_\mu U_\mu = \mathcal{L}_\mu(U_\mu, \ldots, \nabla^{m-1} U_\mu) \tag{4.3}$$

où $K_\mu \in \psi^{m-1}$ (matriciel), et où $\mathcal{L}$ est une epdnℓC.

4.3. Régularité des traces de U_μ.

Nous allons appliquer le corollaire 3.4., où $(\mu-1)$ jouera le rôle de ν. Nous savons que $U_{\mu-1}$ appartient à H^s et vérifie une équation (4.3) :

$$P\, U_{\mu-1} - K_{\mu-1} U_{\mu-1} = \mathcal{L}_{\mu-1}(\ldots, \nabla^{m-1} U_{\mu-1}) . \tag{4.4}$$

En posant $s = (n+1)/2 + m + \rho$, on peut écrire le membre de droite de (4.4) sous la forme $L\, U_{\mu-1} + G$, où L est un opérateur paradifférentiel d'ordre $m-1$, $L \in \mathrm{Op}\, \Sigma^{m-1}_{1+\rho}$ et $G \in H^{s-m+2+\rho}$. On a donc

$$P\, U_{\mu-1} = (K+L)\, U_{\mu-1} + G \in H^{s-m+1}$$

$$P^2 U_{\mu-1} = [P, (K+L)]\, U_{\mu-1} + (K+L) P\, U_{\mu-1} + P\, G \in H^{s-2m+2} .$$

Les conditions d'application du corollaire 3.4. sont donc satisfaites, et on obtient $\gamma_j M^A u \in H^{s-j}$ pour $|A| = \mu$, c'est-à-dire

$$\gamma_j U_\mu \in H^{s-j} \quad \text{pour} \quad j = 0, \ldots, m-1 . \tag{4.5}$$

4.4. Fin de la démonstration du théorème 4.1.

Compte tenu de $U_\mu \in H^{s-1}$, l'équation (4.3) se met sous la forme

$$P\, U_\mu - (K+L)\, U_\mu = G$$

avec $L \in \mathrm{Op}(\Sigma^{m-1}_\rho)$, et $G \in H^{s-m+1+\rho}$. Compte tenu de (4.5), les hypothèses du théorème 1.3. sont satisfaites, et on obtient $U_\mu \in H^s$ ce qui achève la démonstration.

Le résultat suivant sera utile pour vérifier les hypothèses du théorème 4.1.

4.5. Théorème : *Soit V une sous-variété lisse de $\mathbb{R}^n$. Soit $\mathcal{N}$ le sous-module formé des éléments de $\psi^1(\mathbb{R}^n)$ dont le symbole principal s'annule sur $T^*_V \mathbb{R}^n$. Soit $\Lambda_o = T^*_V \mathbb{R}^{n+1}$, et soit Λ_1 la variété lagrangienne* *constituée de la réunion des bicaractéristiques de P issues de* $\Lambda_o \cap \{p_m = 0\}$. *Soit* $\mathcal{M}_\Lambda \subset \psi^1(\mathbb{R}^{n+1})$ *constitué des* $M \in \psi^1$ *dont le symbole principal s'annule sur* $\Lambda_o \cup \Lambda_1$. *Alors* $\mathcal{M}_\Lambda$ *vérifie (C.C.) et trace* $(\mathcal{M}_\Lambda) \subset \mathcal{N}$.

Cette dernière propriété est évidente. On peut supposer V défini par $x_1 = \ldots = x_p = 0$. Si $M \in \mathcal{M}_\Lambda$ (il suffit que son symbole principal m s'annule sur Λ_o), on a

$$m = m' t + a_1 |\xi| x_1 + \ldots + a_p |\xi| x_p + a_{p+1} \xi_{p+1} + \ldots + a_n \xi_n$$

pour $|\xi| \neq 0$ où m' et les a_j sont des symboles d'ordres 1 et 0, ce qui est la forme requise pour (3.1).

La condition (C.C.) est microlocale et se lit sur les symboles principaux. Au voisinage d'un point de $\Lambda_o \cap \{p_m = 0\}$, on peut faire une transformation canonique envoyant Λ_o sur $\{x = t = 0\}$, et $\{p_m = 0\}$ sur $\{\tau = 0\}$ au voisinage du point $x = t = 0$; $\xi_1 = 1$, $\xi_2 = \ldots = \xi_n = \tau = 0$. La réunion des bicaractéristique de D_t passant par $\Lambda_o \cap \{\tau = 0\}$ est bien une variété lagrangienne Λ_1 définie par $x = 0$, $\tau = 0$. Les symboles principaux des éléments de $\mathcal{M}_\Lambda$ sont engendrés par les $m_i = x_i \xi_1$ et $m_o = t\tau$, et il est évident que $\{\tau, m_j\}$ est combinaison de τ et des m_j $(j = 0, \ldots, n)$.

4.6. Problème de Cauchy pour une équation d'ordre 2, avec données singulières sur une sous-variété.

On se donne donc une sous-variété V de $\mathbb{R}^n$ de dimension p et on suppose P d'ordre m=2 dans (1.1). Le théorème 4.1 nous fournit un résultat de propagation des singularités dans la ligne de ceux que nous avons démontrés dans [3] et [5] pour $p = n-1$ et 0. Cela dit nous obtiendrons un résultat moins précis que celui de N. Ritter [13] (pour p quelconque, les termes non linéaires étant d'ordre 0) qui suppose seulement $\gamma^j u \in H_V^{s-j,\nu}$ par $s > n+1-p/2$, $s + \nu > n+1/2$.

Les lagrangiennes Λ_o et Λ_1 étant définies comme dans le théorème 4.5., il résulte de [13] que Λ_1 est le conormal d'une hypersurface Γ cylindro-conique difféomorphe à $\{t^2 = x_{p+1}^2 + \ldots + x_n^2\}$. Cela permet, comme dans [13], de voir que $\mathcal{M}_\Lambda$ est engendré par des champs de vecteurs et possède donc la propriété de Leibniz. Les propriétés (C.C.) et trace $\mathcal{M}_\Lambda \subset \mathcal{N}$ résultent du théorème 4.5., et le théorème 4.1. nous fournit le corollaire suivant.

4.7. *Théorème : Soit $u \in H^s$, $s > (n+1)/2 + 2$, solution de (1.1) avec $m = 2$. On suppose que*

$$\gamma_j u \in H_V^{s-j,\nu} \qquad j = 0, 1 .$$

Alors $u \in H^s(\mathcal{M}_\Lambda, \nu)$ et en particulier $u \in H^{s+\nu}$ localement hors de Γ et microlocalement hors des conormaux à Γ et V.

4.8. Problème de Cauchy pour une équation d'ordre m, avec données singulières sur une hypersurface.

On se donne donc V hypersurface de $\mathbb{R}^n$. La lagrangienne Λ_1 est alors la réunion (disjointe) des variétés conormales $T^*_{\Sigma_j}\mathbb{R}^{n+1}$ aux m hypersurfaces caractéristiques $\Sigma_1, \dots, \Sigma_m$ issues de V. On note encore $\Lambda = \Lambda_0 \cup \Lambda_1 = T^*_V\mathbb{R}^{n+1} \cup \bigcup_j T^*_{\Sigma_j}\mathbb{R}^{n-1}$, la situation étant celle décrite dans l'exemple 2.3.

Nous renvoyons à [4] pour la démonstration du fait que $\mathcal{M}_\Lambda$ possède la propriété de Leibniz. $\mathcal{M}_\Lambda$ n'est pas engendré localement par des champs de vecteurs. La propriété qui joue un rôle-clef est le fait que, pour tout couple de points (x_0,ξ_1) et (x_0,ξ_2) on peut trouver une algèbre $\mathcal{M}' \supset \mathcal{M}_\Lambda$, engendrée par des champs de vecteurs, et qui coïncide avec $\mathcal{M}_\Lambda$ microlocalement près de (x_0,ξ_1) et (x_0,ξ_2). Compte tenu du théorème 4.5., on peut appliquer le théorème 4.1. et on obtient le corollaire suivant.

4.9. *Théorème : Soit u une solution de (1.1) appartenant à H^s $(s > (n+1)/2 + m)$ telle que*

$$\gamma^j u \in H_V^{s-j,\nu} \quad , \quad j = 0, \dots, m-1 .$$

Alors $u \in H^{s,\nu}(T^*_V\mathbb{R}^{n+1} \cup \bigcup_k T^*_{\Sigma_k}\mathbb{R}^{n+1})$. En particulier u appartient à $H^{s+\nu}$ hors des Σ_k, et est une distribution conormale $(u \in H^{s,\nu}_{\Sigma_k})$ près de $\Sigma_k \setminus V$.

Nous allons enfin donner une variante du théorème 4.1. pour les équations dont la non-linéarité est plus faible.

$$P(x,t,D_x,D_t)u = F(x,t,u,\dots,\nabla^{m-2}u) \ . \qquad (4.6)$$

La propriété de Leibniz n'est alors plus nécessaire, il suffit de savoir que les $H^s(\mathcal{M},\nu)$ sont des algèbres (stables par opérations non linéaires $u,v \to F(u,v)$). Nous laissons au lecteur le soin d'énoncer les analogues des théorèmes 4.7. et 4.9. dans ce cas.

4.10. <u>Théorème</u> : *Soient $\mathcal{M}$ et $\mathcal{N}$ comme dans le théorème 4.1. On suppose que $\mathcal{M}$ vérifie (C.C.), que trace $\mathcal{M} \subset \mathcal{N}$, et que les $H^s(\mathcal{M},\nu)$ sont des algèbres pour $s > (n+1)/2$. Soit u une solution de (4.6) appartenant à H^s $(s > (n+1)/2 + m - 2)$ telle que*

$$\gamma^j u \in H^{s-j}(\mathcal{M},\nu) \quad , \quad j = 0, \dots, m-1 \ .$$

Alors on a $u \in H^s(\mathcal{M},\nu)$.

La démonstration suit celle du théorème 4.1. En supposant $u \in H^s(\mathcal{M},\mu-1)$, on obtient immédiatement $Pu \in H^{s-m+2}(\mathcal{M},\mu-1)$ et $(P-K)U_\mu \in H^{s-m+1}$ à la place de (4.3). On vérifie très facilement que $PU_{\mu-1} \in H^{s-m+1}$ et $P^2 U_{\mu-1} \in H^{s-2m+2}$ et on obtient donc (4.5) et le résultat en vertu du théorème 1.3. (qu'il suffirait d'établir pour des opérateurs pseudo-différentiels).

5. SECONDE MICROLOCALISATION

Nous ne ferons ici que rappeler, sans démonstration, les résultats de [7] (voir aussi [6]). A l'exception de la remarque 5.7., nous ne considérerons que la seconde microlocalisation relative à la variété conormale de l'origine 0 dans $\mathbb{R}^n$. Nous noterons $H^{s,k}$ (et $H^{s,s'}$ pour $s,s' \in \mathbb{R}$, les espaces qui s'en déduisent par dualité et interpolation) les espaces notés $H_0^{s,k}$ au n° 2.2. Tous les résultats seront locaux pour x voisin de 0.

5.1. <u>Symboles</u>.

On note $\Sigma^{m,m'}$ l'espace des fonctions $a(x,\xi)$, définies et de classe C^∞ pour $|x||\xi| \geq 1$ et vérifiant

$$|D_\xi^\alpha D_x^\beta a(x,\xi)| \leq C_{\alpha\beta} |\xi|^{m-|\alpha|+|\beta|} (|x||\xi|)^{m'-|\beta|} \ .$$

5.2. <u>Opérateurs</u>.

On note $Op(\Sigma^{m,m'})$ l'espace des opérateurs A appliquant $H^{s,s'}$ dans

$H^{s-m,s'-m'}$ tels que, pour chaque famille finie de $P_i \in \psi^{p_i}$, pour chaque famille finie de $Q_j \in \psi^{q_j}$ dont le symbole principal s'annule pour $x=0$, on ait

$$(\mathrm{ad}\,P)^I (\mathrm{ad}\,Q)^J A : H^{s,s'} \to H^{s-m-\Sigma p_i - \Sigma q_j + |J|, s'-m'-|J|} .$$

Par exemple, un élément de ψ^m appartient à $Op(\Sigma^{m,o})$, et à $Op(\Sigma^{m-1,1})$ si son symbole principal s'annule pour $x = 0$.

5.3. <u>Théorème</u> *(Calcul symbolique)*

Il existe une application σ, *envoyant chaque* $Op(\Sigma^{m,m'})$ *dans* $\Sigma^{m,m'}$ *telle que*

a) σ *induit un isomorphisme de* $Op(\Sigma^{m,m'})/Op(\Sigma^{m,-\infty}) \tilde{\to} \Sigma^{m,m'}/\Sigma^{m,-\infty}$.

b) Pour $A_j \in Op(\Sigma^{m_j,m'_j})$, $j = 1,2$, *on a :*

$$\sigma(A_1 A_2) \equiv \sigma(A_1) \# \sigma(A_2) \pmod{\Sigma^{m_1+m_2,-\infty}}$$

où $a \# b = \Sigma 1/\alpha!\ \partial_\xi^\alpha a\, D_x^\alpha b$.

c) $\sigma(A_1^*) \equiv \Sigma 1/\alpha!\ (i D_x \cdot D_\xi)^\alpha \overline{\sigma(A_1)} \pmod{\Sigma^{m_1,-\infty}}$.

d) Si $A \in \psi^m$, *alors* $A \in Op(\Sigma^{m,o})$, *et son symbole est le symbole usuel (modulo* $\Sigma^{m,-\infty}$*)*.

5.4. <u>Front d'onde 2-microlocal.</u>

Si $u \in H^{s,-\infty}$, son front d'onde d'ordre (s,s') peut être défini comme un sous-ensemble de l'éclaté de $T^*\mathbb{R}^n \setminus \{0\}$ le long de la lagrangienne $\{x = 0\}$, c'est-à-dire

$$\left\{(x,\xi) \mid x \neq 0 , \xi \neq 0\right\} \cup \left\{(0,\xi,\delta x) \mid \delta x \in \mathbb{R}^n \setminus \{0\},\ \xi \neq 0\right\} . \qquad (5.1)$$

On dit que $u \in H^{s,s'}$ 2-microlocalement en un point (x_o,ξ_o) du premier type si $u \in H^{s+s'}$ microlocalement en (x_o,ξ_o). On dit que $u \in H^{s,s'}$ 2-microlocalement en $(0,\xi_o,\delta x_o)$ si on a

$$A \circ \chi(D)(\varphi u) \in H^{s,s'}$$

pour $\varphi \in C_o^\infty$ à support près de 0, $\chi \in \psi^o$ à symbole concentré près de ξ_o, et $A \in Op(\Sigma^{o,o})$ de symbole $a(x)$ concentré dans un petit voisinage conique de δx_o.

On peut définir également le sous-ensemble de (5.1) où un opérateur $A \in \Sigma^{m,m'}$ a son symbole inversible. En un tel point, si $Au \in H^{s-m,s'-m'}$ 2-microlocalement, on a $u \in H^{s,s'}$ 2-microlocalement à condition que u appartienne déjà à $H^{s,-\infty}$.

5.5. Invariance par transformation canonique.

Si χ est une transformation canonique envoyant $T_0^*\mathbb{R}^n$ sur lui-même, et si $\mathcal{F}$ est un opérateur intégral de Fourier d'ordre 0 inversible associé à χ, alors $\mathcal{F}$ applique $H^{s,s'}$ dans $H^{s,s'}$ et $A \to \mathcal{F}^{-1} A \mathcal{F}$ applique $Op(\Sigma^{m,m'})$ dans lui-même. On a $\sigma(\mathcal{F}^{-1} A \mathcal{F}) = \sigma(A) \circ \chi$ modulo $\Sigma^{m,m'-1}$.

5.6. Invariance par difféomorphismes singuliers.

Soit ϕ un homéomorphisme de $\mathbb{R}^n$ sur $\mathbb{R}^n$ (près de 0) tel que $\phi(0) = 0$ et que ϕ soit un difféomorphisme de $\mathbb{R}^n \setminus \{0\}$ sur $\mathbb{R}^n \setminus \{0\}$ vérifiant ainsi que son inverse $|D_x^\beta \phi(x)| \leqslant C_\beta |x|^{1-|\beta|}$ (c'est le cas par exemple pour ϕ homogène de degré 1). On peut alors lui associer un opérateur ϕ^*, de $H^{s,s'}$ dans $H^{s,s'}$ tel que $(\phi\psi)^* - \phi^* \circ \psi^*$ applique $H^{s,s'}$ dans $H^{s,\infty}$, et que, pour ϕ de classe C^∞ à l'origine, $u \to \phi^* u - u \circ \phi$ applique $H^{s,s'}$ dans $H^{s,\infty}$.

Si $P \in Op(\Sigma^{m,m'})$, alors $(\phi^{-1})^* P \phi^* \in Op(\Sigma^{m,m'})$ et son symbole se calcule (modulo $\Sigma^{m,-\infty}$) par la formule habituelle de changement de variable.

5.7. Remarque.

Ce qui se précède se généralise au cas où $T_0^*\mathbb{R}^n$ est remplacé par une lagrangienne Λ, et les espaces $H_0^{s,s'}$ par $H^{s,s'}(\Lambda)$. Compte tenu des résultats d'invariance précédents, il suffit de transporter toute la théorie par un opérateur intégral de Fourier inversible associé à une transformation canonique échangeant $T_0^*\mathbb{R}^n$ et Λ (voir [7] pour les définitions explicites des symboles, opérateurs, fronts d'onde, ... dans ce cas).

6. CHAMPS DE VECTEURS SINGULIERS ET ALGEBRES ASSOCIEES

Nous notons toujours $H^{s,s'}$ les espaces de distributions conormales associés à l'origine dans $\mathbb{R}^n$. Ces espaces sont des algèbres pourvu que $s > n/2$ et $s + s' > n/2$.

L'intérêt essentiel de la seconde microlocalisation sera ici qu'il existe de nouveaux opérateurs, appartenant à $Op(\Sigma^{o,1})$, dont le symbole est linéaire, et que ces opérateurs jouiront des bonnes propriétés que possèdent les champs de vecteurs vis-à-vis des opérations de produit et de composition.

6.1. Définition : On appelle champ de vecteurs singulier un élément Z de $Op(\Sigma^{o,1})$ dont le symbole (modulo $\Sigma^{o,-\infty}$) est linéaire en ξ.

On a donc dans ce cas :

$$\sigma(Z) \equiv \Sigma\, a_j(x)\, \xi_j \pmod{\Sigma^{o,-\infty}}$$

$$|D^\beta a_j(x)| \leqslant C_\beta\, |x|^{1-|\beta|} .$$

Le cas typique est celui où les a_j sont homogènes de degré 1, et C^∞ en dehors de 0. Nous renvoyons à [7] pour le théorème suivant.

6.2. Théorème (Propriété de Leibniz)

Soit Z un champ de vecteurs singulier, et $F(u,v)$ une fonction C^∞ de plusieurs variables. Soient $u,v \in H^{s,s'}$ avec $s > n/2$, $s + s' > n/2$. Il existe alors des opérateurs M, N appartenant à $Op(\Sigma^{o,o})$ (dépendant de Z, u, v) tels que

$$Z\,F(u,v) \equiv \frac{\partial F}{\partial u}\cdot Zu + \frac{\partial F}{\partial v}\cdot Zv + Mu + Nv \pmod{H^{s,\infty}} . \qquad (6.1)$$

Nous désignerons dans ce qui suit par $\mathfrak{Z}$ une sous-algèbre de Lie de type fini de l'ensemble des champs de vecteurs singuliers, c'est-à-dire :

Il existe une famille finie (Z_α) de champs de vecteurs singuliers, tels que, pour tout $Z \in \mathfrak{Z}$, on ait : (6.2)

$$\sigma(Z) \equiv \Sigma\, a_\alpha(x)\, \sigma(Z_\alpha) \pmod{\Sigma^{o,-\infty}}$$

avec $a_\alpha(x) \in \Sigma^{o,o}$.

Pour $Z_1, Z_2 \in \mathfrak{Z}$, on a $[Z_1, Z_2] \in \mathfrak{Z}$. (6.3)

6.3. Définition : Soit $\mathcal{Z}$ *vérifiant (6.2) et (6.3). Nous poserons, pour* $\nu \in \mathbb{N}$:

$$H^{s,s'}(\mathcal{Z},\nu) = \left\{ u \in H^{s,s'} \mid Z^A u \in H^{s,s'},\ |A| \leqslant \nu \right\} .$$

6.4. Théorème : Pour $s > n/2$ *et* $s + s' > n/2$, *les espaces* $H^{s,s'}(\mathcal{Z},\nu)$ *sont des algèbres. En outre, pour* $F(x,u,v)$ *de classe* C^∞, *pour* $u,v \in H^{s,s'}(\mathcal{Z},\nu)$, *et pour* $|A| \leqslant \nu$, *on a, modulo* $H^{s,\infty}$,

$$Z^A F(x,u,v) \equiv \mathcal{L}(\ldots Z^B u\, ,\, Z^C v \ldots) \in H^{s,s'}$$

avec $|B| \leqslant \nu$, $|C| \leqslant \nu$, *et où* $\mathcal{L}$ *est une expression obtenue en composant un nombre fini de fois des fonctions non-linéaires, et des éléments de* $Op(\Sigma^{0,0})$.

La démonstration, par récurrence sur ν, est évidente à partir de (6.1).

7. PROBLEME DE CAUCHY POUR UN OPERATEUR D'ORDRE m, AVEC DONNEES SINGULIERES EN UN POINT

On considère, dans $\mathbb{R}^{n+1}$ une équation du type (1.1) dont les termes non linéaires sont d'ordre $\leqslant m-2$

$$P(x,t,D_x,D_t) = F(x,t,u,\ldots,\nabla^{m-2} u) \ . \qquad (7.1)$$

On notera $\tilde{P}$ l'opérateur à coefficients constants $P(0,0,D_x,D_t)$. On pose $\Lambda_o = T_0^*\mathbb{R}^{n+1}$, on appelle Λ_1 la réunion des bicaractéristiques issues de $\Lambda_o \cap \{p_m = 0\}$, et Γ la projection de Λ_1 sur $\mathbb{R}^{n+1}_{x,t}$ (le " cône " d'onde). On note $\tilde{\Lambda}_1$ et $\tilde{\Gamma}$ les ensembles correspondants relatifs à $\tilde{P}$. Nous ferons l'hypthèse suivante.

Le cône d'onde $\tilde{\Gamma}$ est lisse en dehors de 0. (7.2)

On voit facilement qu'il existe au voisinage de 0 un homéomorphisme de $\mathbb{R}^{n+1}$ sur lui-même, de classe C^∞ hors de 0, et vérifiant les conditions du n° 5.6 qui échange Γ et $\tilde{\Gamma}$. Cela résulte du fait que les projections sur $\mathbb{R}^{n+1}$ des bicaractéristiques de P et $\tilde{P}$ issues d'un point de $\Lambda_o \cap \Lambda_1$ sont tangentes en 0.

En posant $\Lambda = \Lambda_o \cup \Lambda_1$, on a le résultat suivant.

7.1. Théorème : Sous l'hypothèse (7.2), soit u une solution de (7.1) appartenant à H^s, $s > (n+1)/2 + m - 2$. *On suppose que*

$$\gamma_j u \in H_0^{s-j,\nu} \quad , \quad j = 0,\dots,m-1 .$$

Alors $u \in H^{s,\nu}(\Lambda)$. *En particulier u appartient à* $H^{s+\nu}$ *hors du "cône" d'onde* Γ, *et est une distribution conormale* ($u \in H_\Gamma^{s,\nu}$) *près de* $\Gamma \setminus \{0\}$.

Nous sommes ici dans les conditions d'application du théorème 4.10., avec $\mathcal{M}$ et $\mathcal{N}$ associés à Λ et $T_0^*\mathbb{R}^n$ respectivement. D'après le théorème 4.5. les propriétés (C.C.) et trace $(\mathcal{M}_\Lambda) \subset \mathcal{N}$ sont satisfaites. Il ne reste plus qu'à démontrer que les $H^{s,\nu}(\Lambda)$ sont des algèbres, ce qui sera fait au corollaire 7.4., et le théorème 7.1. sera démontré.

7.2. Définition : Nous noterons $\mathcal{Z}$ *l'ensemble des champs de vecteurs singuliers tangents à* Γ *pour* $(x,t) \neq 0$. *Les espaces* $H^{s,s'}(\mathcal{Z},\nu)$ *sont définis au n° 6.3.*

Nous noterons $\mathcal{T}$ *le sous-espace de* $Op(\Sigma^{0,1})$ *constitué des éléments T dont un symbole principal (i.e. un élément de* $\sigma(T) + \Sigma^{0,0}$) *s'annule sur* Λ_1 *pour* $(x,t) \neq 0$. *On posera*

$$H^{s,s'}(\mathcal{T},\nu) = \left\{ u \in H^{s,s'} \mid T^A u \in H^{s,s'} , \ T_\alpha \in \mathcal{T} , \ |A| \leqslant \nu \right\}.$$

7.3. Théorème :

a) $H^{s,s'}(\mathcal{T},\nu) = H^{s,s'}(\mathcal{Z},\nu)$

b) $H^{s,0}(\mathcal{T},\nu) = H^s(\mathcal{M}_\Lambda,\nu)$.

a) Il est clair qu'il suffit de prouver que, pour $T \in \mathcal{T}$, on a

$$\sigma(T) \equiv \Sigma \lambda_i \sigma(Z_i) \mod (\Sigma^{0,0}) \tag{7.3}$$

avec $\lambda_i \in \Sigma^{0,0}$ et $Z_i \in \mathcal{Z}$. En utilisant une partition de l'unité par des fonctions $\varphi(x,t)$ homogènes de degré 0, C^∞ hors de 0, à support dans de petits cônes, on est ramené à démontrer (7.3) dans un petit voisinage conique G de chaque $(x_0,t_0) \neq 0$.

Si G est loin de Γ, les fonctions $x_i\xi_j, x_i\tau, t\xi_j, t\tau$ coïncident dans G avec le symbole d'un élément de $\mathcal{Z}$, et engendrent $\Sigma^{0,1}$.

Si G coupe Γ, on peut trouver un difféomorphisme singulier envoyant $\Gamma \cap G$ sur l'hyperplan $t=0$. D'après le n° 5.6., les éléments de

$Op(\Sigma^{o,1})$ sont transformés en eux-mêmes par conjugaison par ce difféomorphisme singulier, et il en est de même des champs de vecteurs singuliers d'après la formule de changement de variables. Dans l'image de G, les fonctions $t\tau$, $t\xi_j$, $x_i\xi_j$ sont des symboles d'éléments de l'image de $\mathcal{Z}$, et ils engendrent ceux de l'image de $\mathcal{C}$. Cela établit a).

b) Il suffit maintenant de prouver que, pour $T \in \mathcal{C}$, on a

$$\sigma(T) \equiv \Sigma\, \lambda_i\, \sigma(M_i) \mod (\Sigma^{o,o}) \tag{7.4}$$

avec $\lambda_i \in \Sigma^{o,o}$ et $M_i \in \mathcal{M}_\Lambda$, et il suffit de le démontrer microlocalement dans un voisinage conique W de $(0,0;\xi_o,\tau_o)$. Si $W \cap \Lambda_1 = \phi$, les $x_i\xi_j, x_i\tau, t\xi_j, t\tau$ coïncident dans W avec des éléments de $\mathcal{M}_\Lambda$, et elles engendrent $\Sigma^{o,1}$ entier dans W.

Si W coupe Λ_1, on peut faire une transformation canonique conservant Λ_o, et envoyant $\Lambda_1 \cap W$ sur $x = \tau = 0$ (voir la démonstration du théorème 4.5.). En transformant la situation par un opérateur intégral de Fourier inversible associé (n° 5.5.) il est clair que microlocalement les fonctions $t\tau$, $x_i\tau$, $x_i\xi_j$ sont des symboles d'éléments de l'image de $\mathcal{M}_\Lambda$, et engendrent ceux de l'image de $\mathcal{C}$. Cela établit b).

7.4. *Corollaire : Les $H^{s,\nu}(\Lambda)$ sont des algèbres pour $s > (n+1)/2$.*

On a en effet $H^{s,\nu}(\Lambda) = H^{s,o}(\mathcal{C},\nu) = H^{s,o}(\mathcal{Z},\nu)$ et ce dernier espace est une algèbre d'après le théorème 6.4.

7.5. Remarque.

A l'exception du cas où $\tilde{P}$ est un produit d'opérateurs du second ordre, $\mathcal{M}_\Lambda$ n'est pas engendré - même microlocalement - par des champs de vecteurs. Les arguments du n° 4.6. ne peuvent pas s'étendre au cas d'équations d'ordre supérieur.

En utilisant la seconde microlocalisation relative à $T_V^*(\mathbb{R}^{n+1})$, on pourrait démontrer l'analogue du théorème 7.1. pour des données de Cauchy dans $H_V^{s-j,\nu}$, pour V de dimension quelconque dans $\mathbb{R}^n$ - toujours en supposant que la surface cylindro-conique projection de Λ_1 est lisse hors de V.

8. PROBLEME DE CAUCHY AVEC DONNEES SINGULIERES SUR PLUSIEURS COURBES

Nous considérons ici, en dimension 2 d'espace, l'équation des ondes

$$\square u = f(x,y,t,u) \tag{8.1}$$

avec $\square = D_t^2 - D_x^2 - D_y^2$. Soient $V_1,\dots,V_q$ des courbes de $\mathbb{R}^2$, passant par l'origine et ayant des tangentes distinctes en ce point. Nos données de Cauchy appartiendront aux espaces (intervenant au n° 4.3. en dimension n+1) $H^{s,\nu}(\Lambda) = H^s(\mathcal{N},\nu)$ où $\mathcal{N}$ est l'ensemble des éléments de $\psi^1(\mathbb{R}^2)$ dont le symbole principal s'annule sur $\Lambda = T_0^*\mathbb{R}^2 \cup \bigcup_1^q T_{V_j}^*\mathbb{R}^2$.

Par chaque V_j passent deux surfaces caractéristiques $\Sigma_j^\pm$, et nous noterons Γ le cône d'onde issu de 0. Nous désignerons par $\mathcal{Z}$ l'espace des champs de vecteurs singuliers dans $\mathbb{R}^3$ qui sont tangents aux $\Sigma_j^\pm$ et à Γ hors de l'origine.

8.1. <u>Théorème</u> : Soit u une solution de (8.1) appartenant à $H^s (s > 3/2)$. *On suppose que l'on a :*

$$\gamma_j u \in H^{s-j}(\mathcal{N},\nu) \quad , \quad j = 0,1 .$$

Alors on a, pour tout $s' < s$,

$$u \in H^{s'+1/2,-1/2}(\mathcal{Z},\nu) \quad .$$

En particulier, u appartient à $H^{s'+\nu}$ *en dehors de* Γ *et des* $\Sigma_j^\pm$, *et est une distribution conormale* $(u \in H_\bullet^{s',\nu})$ *près d'un point d'une de ces surfaces qui n'appartient pas aux autres.*

La démonstration de ce théorème utilise toute la souplesse du calcul 2-microdifférentiel et est très proche de la démonstration du théorème d'interaction des singularités de [7]. Nous n'indiquons ici que les principales idées de la preuve. La démonstration détaillée sera publiée ultérieurement.

On introduit les U_μ, fonctions vectorielles dont les composantes sont les $(Z^A u)$, pour $|A| \leqslant \mu$ et (Z_α) système de générateurs de $\mathcal{Z}$, et on cherche à prouver par récurrence sur μ, que $U_\mu \in H^{s'+1/2,-1/2}$.

Les arguments de [7] (n° 7.9) s'appliquent mot pour mot pour fournir l'équation

$$\square U_\mu + R_\mu U_\mu = F_\mu \in H^{s'-2,1/2} \qquad (8.1)$$

où $R_\mu \in Op(\Sigma^{2,-1})$ et appartient à $Op(\Sigma^{2,-2}) + Op(\Sigma^{1,o})$ 2-microlocalement près de $\{\tau^2 = \xi^2 + \eta^2 ; \delta x/\xi = \delta y/\eta = \delta t/\tau\}$.

En ramenant, grâce à l'opérateur d'aplatissement M (voir [7],§.6), les champs de vecteurs singuliers à de vrais champs de vecteurs dans $\mathbb{R}^3 \setminus \{0\}$, on peut montrer que

$$\gamma_j U_\mu \in H^{s'-j} \ , \ j = 0,1 \ . \qquad (8.2)$$

Il reste à démontrer l'équivalent des théorèmes de propagation des singularités de ([7],§.5), mais ici à partir des données de Cauchy, pour prouver que (8.1) et (8.2) entraînent $U_\mu \in H^{s',-1/2}$. Cela achève, par récurrence, la démonstration du théorème 8.1.

8.2. Remarque.

Les propriétés de commutation menant à l'équation (8.1) sont l'équivalent de la condition (C.C.) (qui serait fausse pour les opérateurs pseudo-différentiels dont le symbole principal s'annule sur la réunion des conormaux des strates de la variété singulière $\Gamma \cup \cup \Sigma_j^{\pm}$), et c'est la propriété de Leibniz pour les champs de vecteurs singuliers qui fournit la régularité du membre de droite. L'argument menant à (8.2) est l'équivalent du théorème 3.3. tandis que l'argument final est l'équivalent du théorème 1.3.

BIBLIOGRAPHIE

[1] M. Beals : 'Self spreading and strength of singularities for solutions to semilinear wave equations', *Annals of Math*, 118 (1983), 187-214.

[2] J.M. Bony : 'Calcul symbolique et propagation des singularités pour les équations aux dérivées partielles non linéaires', *Ann. Sci. Ec. Norm. Sup.*, 4ème série 14, (1981), 209-246.

[3] [4] J.M. Bony : 'Interaction des singularités pour les équations aux dérivées partielles non linéaires', *Séminaire Goulaouic-Meyer-Schwartz*, 1979-80, n° 22 et 1981-82 n° 2.

[5] J.M. Bony : 'Propagation et interaction des singularités par les solutions des équations aux dérivées partielles non linéaires', *Proc. Int. Cong. Math.*, Warszawa (1983), 1133-1147.

[6] J.M. Bony : 'Interaction des singularités pour les équations de Klein-Gordon non linéaires', *Sém. Goulaouic-Meyer-Schwartz*, 1983-84, n° 10.

[7] J.M. Bony : 'Second microlocalization and propagation of singularities for semi-linear hyperbolic equations', (à paraître).

[8] L. Hörmander : 'The Analysis of linear partial differential operators', *Springer Verlag*, 1985.

[9] Y. Laurent : 'Théorie de la deuxième microlocalisation dans le domaine complexe', *Progress in Math.*, vol. 53, Birkhäuser (1985).

[10] R. Melrose, N. Ritter : 'Interaction of non-linear progressing waves', (à paraître).

[11] S. Mizohata : 'Lectures on the Cauchy problem', *Tata Inst.*, Bombay, 1965.

[12] J. Rauch, M. Reed : 'Non linear microlocal analysis of semi-linear hyperbolic systems in one space dimension', *Duke Math. J.*, 49, (1982), 397-475.

[13] N. Ritter : 'Progressing wave solutions to non-linear hyperbolic Cauchy problems', *Ph. D. Thesis, M.I.T.*, (1984).

[14] M. Sablé-Tougeron : 'Réflexion des singularités pour des problèmes aux limites non linéaires', *Ann. Inst. Fourier*, Grenoble (à paraître).

ABSTRACT. For a strictly hyperbolic semi-linear equation, the localization of the singularities is known from knowing the one of the Cauchy data. For an equation of arbitrary order, we study the case where the data are singular at a point or on a hypersurface; for a second order equation, when they are singular on a subvariety or - in two dimension space - on several crossing curves. The proofs are based on the study of the spaces of involutive and conormal distributions and on the second microlocalization theory.

FOURIER INTEGRAL OPERATORS OF INFINITE ORDER ON GEVREY SPACES. APPLICATIONS TO THE CAUCHY PROBLEM FOR HYPERBOLIC OPERATORS.

L. Cattabriga
Department of Mathematics
University of Bologna
Piazza di Porta S. Donato, 5
40127 Bologna / Italy

L. Zanghirati
Department of Mathematics
University of Ferrara
Via Machiavelli, 35
44100 Ferrara / Italy

ABSTRACT. The aim of these lectures is to prove results on the Gevrey wave front set of the solution of the Cauchy problem with data in spaces of Gevrey ultradistributions for hyperbolic operators with characteristics of constant multiplicity. These results are obtained by constructing a parametrix, with ultradistribution kernel, represented by means of Fourier integral operators of infinite order defined on Gevrey spaces.

0. INTRODUCTION

In these lectures we are concerned with the problem of propagation of Gevrey singularities for solutions of pseudo-differential equations with characteristics of constant multiplicity and with the Cauchy problem for weakly hyperbolic operators with data in spaces of functions or ultradistributions of Gevrey type. As suggested by the Paley-Wiener theorem in these spaces proved by H. Komatsu [17], we use here as a main tool in the study of these problems Fourier integral operators with amplitudes of infinite order, i.e. of suitable exponential growth in the dual space variables. By means of these operators we can prove the existence of a parametrix with ultradistribution kernel for the above mentioned Cauchy problem and prove herewith results on the existence and propagation of singularities in Gevrey spaces for this solution.

Results of this type have been proved by the energy method by S. Mizohata [23] and, by constructing a fundamental solution, by K. Taniguchi [29] when the data are ordinary distributions. A fundamental solution for the Cauchy problem considered here, but with weaker regularity properties, has also been obtained by K. Kajitani [14].

Pseudo-differenzial operators of infinite order on Gevrey spaces have been defined and studied in [32].

Analogous operators of finite order were considered by F. Trèves [30], G. Metivier [22], P. Bolley - J. Camus - G. Metivier [2] in the analytic case and in the general case by S. Hashimoto - T. Matsuzawa - Y. Morimoto [10]. Pseudo-differential operators of finite order on

H. G. Garnir (ed.), Advances in Microlocal Analysis, 41–71.

Gevrey spaces were also studied by different methods by L. Boutet de Monvel - P. Krée [6], L.R. Volevic [31] and V. Iftimie [12]. A calculus for analytic pseudo-differential operators of infinite order has been developed by L. Boutet de Monvel [5] and by T. Aoki [1].

In section 1 we describe some properties and state some calculus rules for the Fourier integral operators of infinite order to be used in the applications, starting from the calculus for pseudo-differential operators of infinite order developed in [32].

In section 2 we show some applications to the Cauchy problem for a hyperbolic operator with the most simple principal part as given in [7] and to the propagation of singularities of solutions of a certain type of pseudo-differential equations with characterics of constant multiplicity treated in [26].

In section 3 we use the results of the previous sections for studying the Cauchy problem for hyperbolic operators with characteristics of constant multiplicity.

The authors thank A. Bove, C. Parenti, L. Rodino for useful discussions with them on the subject of these lectures.

0.1. Main Notations

In the following $\mathbb{Z}_+$ will denote the set of all non negative integers; for $x = (x_1,\dots,x_n) \in R^n$ we set $D_x = (D_1,\dots,D_n)$, $D_j = -i\partial/\partial x_j$, $j = 1,\dots,n$, and for $\alpha = (\alpha_1,\dots,\alpha_n) \in \mathbb{Z}_+^n$

$$|\alpha| = \sum_{i=1}^{n} \alpha_i, \quad D_x^\alpha = D_1^{\alpha_1}\dots D_n^{\alpha_n} .$$

Let Ω be an open set in R^n, K a compact subset of Ω and let $\sigma > 1$, $A > 0$. We denote by $G^{(\sigma),A}(K)$ the set of all complex valued functions $\varphi \in C^\infty(\Omega)$ such that

$$\|\varphi\|_{K,h} = \sup_{\alpha \in \mathbb{Z}_+^n} A^{-|\alpha|}\alpha!^{-\sigma} \sup_{x \in K} |D_x^\alpha \varphi(x)| < +\infty.$$

We set

$$G^{(\sigma)}(K) = \varinjlim_{A\to+\infty} G^{(\sigma),A}(K) , \quad G^{(\sigma)}(\Omega) = \varprojlim_{K\to\Omega} G^{(\sigma)}(K) ,$$

$$G_0^{(\sigma)}(K) = G^{(\sigma)}(K) \cap C_0^\infty(K) , \quad \text{and}$$

$$G_0^{(\sigma)}(\Omega) = \varinjlim_{K\to\Omega} \varinjlim_{A\to+\infty} G^{(\sigma),A}(K) \cap C_0^\infty(K) .$$

$G_0^{(\sigma)'}(\Omega)$ and $G^{(\sigma)'}(\Omega)$ will denote the dual spaces of $G_0^{(\sigma)}(\Omega)$ and $G^{(\sigma)}(\Omega)$ respectively. They are also called spaces of ultradistributions of Gevrey type. $G^{(\sigma)'}(\Omega)$ can be identified with the subspace of ultradistributions of $G_0^{(\sigma)'}(\Omega)$ with compact support. We usually indicate as "Gevrey spaces of order σ" the spaces of functions or ultradistributions defined above.

The Fourier-Laplace transform $\tilde{u}(\zeta)$, $\zeta \in \mathbb{C}^n$, is defined by

$$\tilde{u}(\zeta) = \int_{R^n} e^{i\langle x,\zeta\rangle} u(x)dx, \qquad \zeta \in \mathbb{C}^n, \qquad \langle x,\zeta\rangle = \sum_{j=1}^{n} x_j\zeta_j ,$$

if $u \in G_0^{(\sigma)}(\Omega)$ and by $\tilde{u}(\zeta) = \langle u, \exp(-i\langle x,\zeta\rangle)\rangle$ if $u \in G^{(\sigma)'}(\Omega)$, where $\langle u,\varphi\rangle$ denotes the value of u at $\varphi \in G^{(\sigma)}(\Omega)$ (for more details on these spaces see for example [16] and [17]).

Finally if V is a topological vector space and I an interval, we write $f \in \mathscr{B}_I^m(V)$ to indicate that f is a function defined on I and with values in V, which is bounded in V together with its derivatives up to the order m.

1. FOURIER INTEGRAL OPERATORS OF INFINITE ORDER ON GEVREY SPACES

1.1. Symbols of infinite order and oscillatory integrals

Definition 1.1.1. Let X be an open set of R^ν and let $\sigma > 1$ and $\mu \in [1,\sigma]$. If $\tilde{K}$ is a compact subset of X and $A > 0$, $B \geq 0$, we denote by $S^{\infty,\sigma,\mu}(X \times R^N;\tilde{K},A,B)$ the space of all complex valued functions $a \in C^\infty(X \times R^N)$ such that for every $\varepsilon > 0$.

$$\|a\|_{\tilde{K},\varepsilon}^{A,B} = \sup_{\substack{\alpha \in \mathbb{Z}_+^N \\ \beta \in \mathbb{Z}_+^\nu}} \sup_{\substack{x \in \tilde{K} \\ |\xi| \geq B|\alpha|^\sigma}} A^{-|\alpha|-|\beta|}\alpha!^{-\mu}\beta!^{-\sigma}(1+|\xi|)^{|\alpha|}\exp(-\varepsilon|\xi|^{1/\sigma}) \times$$
$$\times |D_\xi^\alpha D_x^\beta a(x,\xi)| < +\infty, \tag{1.1.1}$$

endowed with the topology defined by the set of seminorms $\|\ \|_{\tilde{K},\varepsilon}^{A,B}$, $\varepsilon > 0$. The space $S^{\infty,\sigma,\mu}(X \times R^N)$ of symbols of infinite order considered here is defined as

$$S^{\infty,\sigma,\mu}(X \times R^N) = \varinjlim_{\tilde{K}\to X} S^{\infty,\sigma,\mu}(X \times R^N;\tilde{K}) = \varprojlim_{\tilde{K}\to X} \varinjlim_{A,B\to+\infty} S^{\infty,\sigma,\mu}(X \times R^N;\tilde{K},A,B).$$

Concerning the convergence in the space of symbols we shall make use of the following result

<u>Proposition</u> 1.1.2. Let $\{a_k\}$ be a bounded sequence in $S^{\infty,\sigma,\mu}(X\times R^N)$ which is also bounded as a sequence in $C^{\infty}(X\times R^N)$. If a_k is pointwise convergent on $X\times R^N$ to a function a, then $a\in S^{\infty,\sigma,\mu}(X\times R^N)$ and $a_k\to a$ also in the topology of $S^{\infty,\sigma,\mu}(X\times R^N)$.

<u>Examples</u> 1.1.3. i) Let $\chi\in C^{\infty}(R^N)$ be such that $\chi(0)=1$ and there exist positive constants C_0, C_1 and h such that for every $\alpha\in\mathbb{Z}_+^N$

$$|D_\xi^\alpha\chi(\xi)|\leq C_0C_1^{|\alpha|}\alpha!(1+|\xi|)^{-|\alpha|}\exp(-h|\xi|^{1/\sigma})\ ,\qquad \xi\in R^N\ .$$

For $a\in S^{\infty,\sigma,\mu}(X\times R^N)$ the set of the functions $\{a_\rho(x,\xi)=\chi(\rho\xi)a(x,\xi),\ \rho>0\}$ is bounded in $C^{\infty}(X\times R^N)$ and in $S^{\infty,\sigma,\mu}(X\times R^N)$ and $a_\rho(x,\xi)\to a(x,\xi)$ pointwise in $X\times R^N$ and hence in $S^{\infty,\sigma,\mu}(X\times R^N)$ as $\rho\to 0_+$.

ii) Starting from a sequence $g_j\in C_0^{\infty}(R^N)$ such that $g_j(\xi)=1$ when $|\xi|\leq 2$, $g_j(\xi)=0$ when $|\xi|\geq 3$ and

$$|D^\alpha g_j(\xi)|\leq (cj)^{|\alpha|}\qquad \text{for } |\alpha|\leq j\ .$$

c a positive constant, consider the partition of unity defined by the functions

$$\psi_0(\xi)=g_1(\xi/R)\ ,\quad \psi_j(\xi)=g_{j+1}(\xi/R(j+1)^\sigma)-g_j(\xi/Rj^\sigma)\qquad (1.1.2)$$

where R is a positive constant (see [22], [2] , [10], [32]). If $a\in S^{\infty,\sigma,\mu}(X\times R^N)$ it is easy to prove that the sequence

$$a_k(x,\xi)=\sum_{j=0}^{k}\psi_j(\xi)a(x,\xi)$$

has the properties required by Proposition 1.1.2, whence $a_k\to a$ in $S^{\infty,\sigma,\mu}(X\times R^N)$.

The following definitions of formal series of symbols and of equivalence of formal series of symbols given in [32] when $\mu=1$, are needed.

<u>Definition</u> 1.1.4. A series $\sum_{j\geq 0}a_j(x,\xi)$, $a_j\in S^{\infty,\sigma,\mu}(X\times R^N)$ is called a formal series of symbols in $S^{\infty,\sigma,\mu}(X\times R^N)$ if for every compact set $\tilde{K}\subset X$ there exist constants $A>0$, $B\geq 0$ such that for every $\varepsilon>0$

$$\sup_{j\in\mathbb{Z}_+}\ \sup_{\substack{\alpha\in\mathbb{Z}_+^N\\ \beta\in\mathbb{Z}_+^\nu}}\ \sup_{\substack{x\in\tilde K\\ |\xi|\geq B(j+|\alpha|)^\sigma}} A^{-|\alpha|-|\beta|-j}\alpha!^{-\mu}(\beta!j!)^{-\sigma}(1+|\xi|)^{|\alpha|+j}\exp(-\varepsilon|\xi|^{1/\sigma})\times$$
$$\times\,|D_\xi^\alpha D_x^\beta a_j(x,\xi)| < +\infty\ .$$

The set of all formal series of simbols in $S^{\infty,\sigma,\mu}(X\times R^N)$ will be denoted by $SF^{\infty,\sigma,\mu}(X\times R^N)$.

Definition 1.1.5. We shall say that two series $\sum_{j\geq 0} a_j$, $\sum_{j\geq 0} b_j$ in $SF^{\infty,\sigma,\mu}(X\times R^N)$ are equivalent and write $\sum_{j\geq 0} a_j \sim \sum_{j\geq 0} b_j$, if for every compact set $\tilde K\subset X$ there exist constants $A>0$ and $B\geq 0$ such that for every $\varepsilon>0$

$$\sup_{s\in\mathbb{Z}_+}\ \sup_{\substack{\alpha\in\mathbb{Z}_+^N\\ \beta\in\mathbb{Z}_+^\nu}}\ \sup_{\substack{x\in\tilde K\\ |\xi|\geq B(s+|\alpha|)^\sigma}} A^{-|\alpha|-|\beta|-s}\alpha!^{-\mu}(\beta!s!)^{-\sigma}(1+|\xi|)^{|\alpha|+s}\exp(-\varepsilon|\xi|^{1/\sigma})\times$$
$$\times\,|D_\xi^\alpha D_x^\beta(\sum_{j<s}[a_j(x,\xi)-b_j(x,\xi)])| < +\infty\ .$$

Since $\sum_{j\geq 0} a_j \in SF^{\infty,\sigma,\mu}(X\times R^N)$ when $a_0\in S^{\infty,\sigma,\mu}(X\times R^N)$ and $a_j=0$ for every $j>0$, we shall consider $S^{\infty,\sigma,\mu}(X\times R^N)$ as a subset of $SF^{\infty,\sigma,\mu}(X\times R^N)$. The following remark will be useful.

Remark 1.1.6. Let $a\in S^{\infty,\sigma,\mu}(X\times R^N)$ and let $a\sim 0$, then for every compact $\tilde K\subset X$ there exist two positive constants A and h such that

$$\sup_{\beta\in\mathbb{Z}_+^\nu}\ \sup_{\substack{x\in\tilde K\\ \xi\in R^N}} A^{-|\beta|}\beta!^{-|\sigma|}\exp(h|\xi|^{1/\sigma})|D_x^\beta a(x,\xi)| < +\infty\ .$$

Every element of the factor space $SF^{\infty,\sigma,\mu}(X\times R^N)/_\sim$ contains an element of $S^{\infty,\sigma,\mu}(X\times R^N)$, in fact we have

Theorem 1.1.7. If $\sum_{i\geq 0} a_j \in SF^{\infty,\sigma,\mu}(X\times R^N)$, then for every open set $X'\subset\subset X$ there exists $a_{X'}\in S^{\infty,\sigma,\mu}(X'\times R^N)$ such that $a_{X'}\sim\sum_{j\geq 0} a_j|_{X'}$ in $SF^{\infty,\sigma,\mu}(X'\times R^N)$.

From this theorem with the aid of partition of unity in $G_0^{(\sigma)}(X)$ related to a locally finite covering of X by relatively compact open subsets we obtain

Corollary 1.1.8. For every $\sum_{j\geq 0} a_j \in SF^{\infty,\sigma,\mu}(X \times R^N)$ there exists $a \in S^{\infty,\sigma,\mu}(X \times R^N)$ such that $a \sim \sum_{j\geq 0} a_j$ in $SF^{\infty,\sigma,\mu}(X \times R^N)$.

Definition 1.1.9. $S^{m,\sigma,\mu}(X \times R^N)$, $m \in R$, $\sigma > 1$, $\mu \in [1,\sigma]$ will denote the set of all $a \in C^\infty(X \times R^N)$ such that for every compact set $\tilde{K} \subset X$ there exist two constants $A > 0$, $B \geq 0$ such that

$$|a|_{\tilde{K}}^{A,B} = \sup_{\substack{\alpha \in \mathbb{Z}_+^N \\ \beta \in \mathbb{Z}_+^\nu}} \ \sup_{\substack{x \in \tilde{K} \\ |\xi| \geq B|\alpha|^\sigma}} A^{-|\alpha|-|\beta|}\alpha!^{-\mu}\beta!^{-\sigma}(1+|\xi|)^{-m+|\alpha|}|D_\xi^\alpha D_x^\beta a(x,\xi)| < +\infty.$$

m is called the order of a.

Definition 1.1.10. A real valued function $\phi \in S^{1,\sigma,\mu}(X \times R^N)$, homogeneous of degree one with respect to ξ when $|\xi| \geq B_\phi$, B_ϕ a positive constant, and such that

$$\nabla_{x,\xi}\phi(x,\xi) \neq 0 \qquad \text{for } (x,\xi) \in X \times \{|\xi| \geq B_\phi\}$$

will be called a phase function.

If ϕ is a phase function, the operator

$$L = \sum_{j=1}^N a_j \partial_{\xi_j} + \sum_{h=1}^\nu b_h \partial_{x_h} + c ,$$

where $a_j = i|\xi|^2\theta\partial_{\xi_j}\phi$, $b_h = i\theta\partial_{x_h}\phi$, $c = \sum_{j=1}^N \partial_{\xi_j} a_j + \sum_{h=1}^\nu \partial_{x_h} b_h$,

$$\theta = \left[|\xi|^2 \sum_{j=1}^N (\partial_{\xi_j}\phi)^2 + \sum_{h=1}^\nu (\partial_{x_h}\phi)^2 \right]^{-1} , \tag{1.1.3}$$

with the property that ${}^tLe^{i\phi} = e^{i\phi}$ is well defined together with its iterates

$$L^k = \sum_{|\alpha|+|\beta|\leq k} l_{\alpha,\beta}^{(k)}(x,\xi)\partial_\xi^\alpha\partial_x^\beta , \qquad \text{for } x \in X , \ |\xi| \geq B_\phi .$$

It is easily seen that for every compact $\tilde{K} \subset X$ there exist positive constants Λ_ϕ, A_ϕ such that for every $\alpha,\gamma \in \mathbb{Z}_+^N$, β, $\delta \in \mathbb{Z}_+^\nu$

$$|D_\xi^\gamma D_x^\delta l_{\alpha,\beta}^{(k)}(x,\xi)| \leq \Lambda_\phi^k A_\phi^{k+|\gamma|+|\delta|}(k-|\alpha|-|\beta|+|\gamma|+|\delta|)!^\sigma|\xi|^{-(k-|\alpha|+|\gamma|)}$$

when $(x,\xi) \in \tilde{K} \times \{|\xi| \geq B_\phi\}$. As a consequence of these properties, if $a \in S^{\infty,\sigma,\mu}(X \times R^N)$ and $u \in G_0^{(\sigma)}(X)$ the number R in (1.1.2) can be chosen so as the series

$$\sum_{j \geq 0} |\iint e^{i\phi(x,\xi)} L^j(a(x,\xi)\psi_j(\xi)u(x))dx đ\xi| \,, \qquad đ\xi = (2\pi)^{-N}d\xi$$

is convergent. Thus if we define the oscillatory integral

$$I_\phi(au) = Os\text{-}\iint e^{i\phi(x,\xi)} a(x,\xi)u(x)dx đ\xi \tag{1.1.4}$$

$$= \sum_{j \geq 0} \iint e^{i\phi(x,\xi)} a(x,\xi)\psi_j(\xi)u(x)dx đ\xi$$

$a \in S^{\infty,\sigma,\mu}(X \times R^N)$, $u \in G_0^{(\sigma)}(X)$, we have

<u>Proposition</u> 1.1.11. Let $u \in G_0^{(\sigma),A_u}(\tilde{K})$, $a \in S^{\infty,\sigma,\mu}(X \times R^N;\, \tilde{K},\, A_a,\, B_a)$ and let ϕ be a phase function. Then there exist ε, $c > 0$ dependent only on $\tilde{K}$, A_a, B_a, A_u, ϕ such that

$$|I_\phi(au)| \leq c\|a\|^{A_a,B_a}_{\tilde{K},\varepsilon} \|u\|_{\tilde{K},A_u} \,.$$

Moreover if ϕ_ν is a sequence of phase functions bounded in $S^{1,\sigma,\mu}(X \times R^N)$, pointwise convergent to a phase function ϕ on $X \times R^N$ and such that the corresponding functions (1.1.3) are uniformly bounded on $\tilde{K} \times \{|\xi| = B\}$, B independent of ν, then $I_{\phi_\nu}(au) \to I_\phi(au)$.

Note that in wiew of the example 1.1.3 ii) the map

$$a \to I_\phi(au)$$

defined on $S^{\infty,\sigma,\mu}(X \times R^N)$ by (1.1.4) for fixed u and ϕ is the unique continuous extension to $S^{\infty,\sigma,\mu}(X \times R^N)$ of the map

$$a \to \iint e^{i\phi(x,\xi)} a(x,\xi)u(x)dx đ\xi$$

defined for $a \in S^{\infty,\sigma,\mu}(X \times R^N) \cap L^1(X \times R^N)$.

Note also that from Proposition 1.1.11 and the example 1.1.3 i) it follows that

$$I_\phi(au) = \lim_{\rho \to 0+} I_\phi(a_\rho u)$$

for every $a \in S^{\infty,\sigma,\mu}(X \times R^N)$, $u \in G_0^{(\sigma)}(X)$.

Another representation for $I_\phi(au)$ follows from

Proposition 1.1.12. Let $u \in G_0^{(\sigma),A_u}(\tilde{K})$, $a \in S^{\infty,\sigma,\mu}(X \times R^N; \tilde{K}, A_a, B_a)$ and let ϕ be a phase function such that $\nabla_x\phi(x,\xi) \neq 0$ for $(x,\xi) \in \tilde{K} \times \{|\xi| \geq B_\phi\}$. Then there exists $c > 0$ only dependent on $\tilde{K}$, A_a, A_u, ϕ such that for every $\varepsilon > 0$ and $|\xi| \geq B_\phi$

$$|\int e^{i\phi(x,\xi)}a(x,\xi)u(x)dx| \leq \|a\|_{\tilde{K},\varepsilon}^{A_a,B_a}\|u\|_{\tilde{K},A_u} \exp((\varepsilon-c)|\xi|^{1/\sigma}) .$$

From this result, choosing $\varepsilon \in]0,c[$, it follows that if $\nabla_x\phi \neq 0$ for $(x,\xi) \in \tilde{K} \times \{|\xi| \geq B_\phi\}$ we have

$$I_\phi(au) = \int đ\xi \int e^{i\phi(x,\xi)}a(x,\xi)u(x)dx , \tag{1.1.5}$$

$a \in S^{\infty,\sigma,\mu}(X \times R^N)$, $u \in G_0^{(\sigma)}(\tilde{K})$.

A definition analogous to (1.1.4) can be given for the oscillatory integral

$$I_\phi(a)(x) = \text{Os-}\int e^{i\phi(x,\xi)}a(x,\xi)đ\xi \tag{1.1.6}$$

$$= \sum_{j\geq 0} \int e^{i\phi(x,\xi)}a(x,\xi)\psi_j(\xi)đ\xi \quad , \quad a \in S^{\infty,\sigma,\mu}(X \times R^N) ,$$

when $\nabla_\xi\phi(x,\xi) \neq 0$ for $|\xi| \geq B_\phi$. The following result analogous to Proposition 1.1.11 holds

Proposition 1.1.13. Let $X_\phi = \{x \in X; \nabla_\xi\phi(x,\xi) \neq 0, |\xi| \geq B_\phi\}$ and let $\tilde{K}$ be a compact subset of X_ϕ. Then

$$|I_\phi(a)(x)| \leq c\|a\|_{\tilde{K},\varepsilon}^{A_a,B_a}$$

for every $x \in \tilde{K}$ and $a \in S^{\infty,\sigma,\mu}(X \times R^N; \tilde{K}, A_a, B_a)$, where ε, c are positive constants only dependent on $\tilde{K}, A_a, B_a$ and ϕ. Moreover $I_{\phi_\nu}(a)(x) \to I_\phi(a)(x)$ uniformly on $\tilde{K}$ if ϕ_ν is a sequence of phase functions bounded in $S^{1,\sigma,\mu}(X \times R^N)$ pointwise convergent to a phase function ϕ and such that $X_{\phi_\nu} \supset X_\phi$ and $|\nabla_\xi\phi_\nu|^{-1}$ are uniformly bounded on $\tilde{K} \times \{|\xi| \geq B_\phi\}$.

For $u \in G_0^{(\sigma)}(X_\phi)$, $a \in S^{\infty,\sigma,\mu}(X \times R^N)$ we have also

$$I_\phi(au) = \int I_\phi(a)(x)u(x)dx \quad . \tag{1.1.7}$$

1.2. Fourier integral operators of infinite order

Let now $a \in S^{\infty,\sigma,\mu}(\Omega\times\Omega\times R^N)$, Ω an open subset of R^n, and let $\varphi \in S^{1,\sigma,\mu}(\Omega\times\Omega\times R^N)$ be a phase function such that

$$\nabla_{y,\xi}\varphi(x,y,\xi) \neq 0 \quad \forall\ (x,y,\xi) \in \Omega\times\Omega\times\{|\xi| \geq B_\phi\} \tag{1.2.1}$$

where B_ϕ is a positive constant.

In view of (1.1.4) we define the operator A on $G_0^{(\sigma)}(\Omega)$ by setting

$$(Au)(x) = \text{Os-}\iint e^{i\varphi(x,y,\xi)}a(x,y,\xi)u(y)dy\,đ\xi, \quad x \in \Omega . \tag{1.2.2}$$

Using Proposition 1.1.12 and (1.1.5) it is immediately seen that

$$(Au)(x) = \int đ\xi \int e^{i\varphi(x,y,\xi)}a(x,y,\xi)u(y)dy , \qquad x \in \Omega ,$$

if $\nabla_y\varphi(x,y,\xi) \neq 0$ for $(x,y,\xi) \in \Omega\times\Omega\times\{|\xi| \geq B_\phi\}$ (in particular this is the case when $\varphi(x,y,\xi) = \langle x-y,\xi\rangle$, i.e. when A is a pseudo-differential operator; see [32]).

Next we note that for every $\beta \in \mathbb{Z}_+^n$

$$D_x^\beta(e^{i\varphi(x,y,\xi)}a(x,y,\xi)) = e^{i\varphi(x,y,\xi)}a_\beta(x,y,\xi)$$

where, for every compact subset H of Ω, $\{a_\beta(x,\cdot,\cdot),\ x \in H\}$ is a bounded set in $S^{\infty,\sigma,\mu}(\Omega\times R^N)$ and in $C^\infty(\Omega\times R^N)$. Thus an application of Proposition 1.1.2 and Proposition 1.1.11 shows

<u>Lemma</u> 1.2.1. Let $a \in S^{\infty,\sigma,\mu}(\Omega\times\Omega\times R^N)$ and let $\varphi(x,y,\xi)$ be a phase function satisfying (1.2.1). Then (1.2.2) defines a continuous linear map from $G_0^{(\sigma)}(\Omega)$ to $G^{(\sigma)}(\Omega)$.

If φ is a phase function such that

$$\nabla_{x,\xi}\varphi(x,y,\xi) \neq 0 \qquad \forall\ (x,y,\xi) \in \Omega\times\Omega\times\{|\xi| \geq B_\phi\} \tag{1.2.1'}$$

the same conclusion of Lemma 1.2.1 holds for the transposed ${}^tA$ of A defined on $G_0^{(\sigma)}(\Omega)$ by

$$({}^tAv)(y) = \text{Os-}\iint e^{i\varphi(x,y,\xi)}a(x,y,\xi)v(x)dx\,đ\xi , \quad y \in \Omega .$$

Thus we have

<u>Theorem</u> 1.2.2. Let $a \in S^{\infty,\sigma,\mu}(\Omega\times\Omega\times R^N)$ and let $\varphi(x,y,\xi)$ be a phase function satisfying (1.2.1) and (1.2.1'). Then the operator A from $G_0^{(\sigma)}(\Omega)$ to $G^{(\sigma)}(\Omega)$ defined by (1.2.2) extends to a continuous linear map from

$G^{(\sigma)'}(\Omega)$ to $G_0^{(\sigma)'}(\Omega)$ with kernel $K_A \in G_0^{(\sigma)'}(\Omega \times \Omega)$ defined by

$$K_A(w) = \mathrm{Os}\text{-}\iiint e^{i\varphi(x,y,\xi)} a(x,y,\xi) w(x,y)\, dx\, dy\, đ\xi, \qquad w \in G_0^{(\sigma)}(\Omega\times\Omega). \tag{1.2.3}$$

Set now

$$R_\varphi = \{(x,y) \in \Omega\times\Omega\ ;\ \nabla_\xi \varphi(x,y,\xi) \neq 0, \quad |\xi| \geq B_\varphi\} \tag{1.2.4}$$

and according to (1.1.6) consider the oscillatory integral

$$I_\varphi(a)(x,y) = \mathrm{Os}\text{-}\int e^{i\varphi(x,y,\xi)} a(x,y,\xi)\, đ\xi\ , \qquad (x,y) \in R_\varphi\ .$$

Arguing as for proving Lemma 1.2.1 it can be seen that $I_\varphi(a) \in G^{(\sigma)}(R_\varphi)$. Moreover in view of (1.1.7) $I_\varphi(a) = K_A|_{R_\varphi}$, whence σ-singsupp $K_A \subset \complement R_\varphi$. This leads to

<u>Theorem</u> 1.2.3. Let a and φ be as in Theorem 1.2.2. Then for every $u \in G^{(\sigma)'}(\Omega)$

$$\sigma\text{-singsupp } Au \subset \complement R_\varphi \circ \sigma\text{-singsupp } u$$

$$= \{x \in \Omega;\ \exists y \in \sigma\text{-singsupp } u,\ (x,y) \in \complement R_\varphi\}$$

where $\complement R_\varphi$ is the complement of the set R_φ defined by (1.2.4), and σ-singsupp Au denotes the smallest closed subset of Ω such that Au is in $G^{(\sigma)}$ in the complement.

In particular note the case when $\varphi(x,y,\xi) = \langle x-y,\xi\rangle$, i.e. when A is a pseudo-differential operator, where $\complement R_\varphi = \{(x,y) \in \Omega\times\Omega\ ;\ x = y\}$ and σ-singsupp $Au \subset \sigma$-singsupp u (see [32]), and the case when $R_\varphi = \Omega\times\Omega$, where A maps $G^{(\sigma)'}(\Omega)$ continuously into $G^{(\sigma)}(\Omega)$.

<u>Definition</u> 1.2.4. A continuous linear operator from $G_0^{(\sigma)}(\Omega)$ to $G^{(\sigma)}(\Omega)$ is called σ-regularizing if it extends to a continuous linear map from $G^{(\sigma)'}(\Omega)$ to $G^{(\sigma)}(\Omega)$.

From Remark 1.1.6 and (1.2.3) if follows

<u>Proposition</u> 1.2.5. If $a \sim 0$ in $SF^{\infty,\sigma,\mu}(\Omega\times\Omega\times R^N)$, then the kernel K_A of the operator A given by (1.2.2.) is in $G^{(\sigma)}(\Omega\times\Omega)$ and A is σ-regularizing.

Note that the singularities of the kernel of A given by (1.2.3) only depend on the behaviour of the function a in an open neighborhood C' of the conic set $C = \{(x,y,\xi) \in \Omega\times\Omega\times\{|\xi| \geq B_\varphi\};\ \nabla_\xi\varphi(x,y,\xi) = 0\}$. Thus $K_A \in G^{(\sigma)}(\Omega\times\Omega)$ if $ha \sim 0$ in $SF^{\infty,\sigma,\mu}(\Omega\times\Omega\times R^N)$ for a suitable cut-off function h equal to one in C and with support in C' (for more details see

2.2).

Definition 1.2.6. (See H. Kumano-go [19]). $\mathscr{P}$ will denote the set of all real valued functions $\phi \in S^{1,\sigma,\mu}(\Omega\times R^n)$, homogeneous of degree one with respect to $\xi \in R^n$ when $|\xi| \geq B_\phi$, B_ϕ a positive constant, and such that for every compact set $K \subset \Omega$ there exists $\tau_k \in [0,1[$ such that

$$\sum_{|\alpha+\beta|\leq 2} \sup_{\substack{x \in K \\ |\xi|\geq B_\phi}} \{ |D_\xi^\alpha D_x^\beta[\phi(x,\xi) - \langle x,\xi\rangle]|/(1+|\xi|)^{1-|\alpha|}\} \leq \tau_k .$$

It is immediately seen that all $\varphi(x,y,\xi) = \phi(x,\xi) - \langle y,\xi\rangle$, $\phi \in \mathscr{P}$, satisfy the hypotheses of Theorems 1.2.2 and 1.2.3.

The following result on the composition of operators of type (1.2.2) will be used later.

Theorem 1.2.7. Let P_1 and P_2 be defined on $G_0^{(\sigma)}(\Omega)$ by

$$(P_1 u)(x) = \iint e^{i\langle x-y,\xi\rangle} p_1(x,\xi)u(y)dy đ\xi,$$

$$(P_2 u)(x) = \iint e^{i\phi(x,\xi)-i\langle y,\xi\rangle} p_2(x,\xi)u(y)dy đ\xi$$

$x \in \Omega$, $p_1 \in S^{\infty,\sigma,1}(\Omega\times R^n)$, $p_2 \in S^{\infty,\sigma,\mu}(\Omega\times R^n)$, $\mu \in [1,\sigma]$, $\phi \in \mathscr{P}$, and let P_1 be properly supported (i.e. the set $\{(x,y) \in \operatorname{supp} K_{P_1}, x \in H \text{ or } y \in H\}$ is compact for every compact set $H \subset \Omega$; K_{P_1} the kernel of P_1). Moreover suppose that one of the following conditions is satisfied:

a) the estimates for p_1 in (1.1.1) hold for every ξ with $|\xi|$ greater than a constant independent of $\alpha \in \mathbb{Z}_+^n$,

b) $\phi(x,\xi) = \langle x,\xi\rangle$.

Then $P_1 P_2 = Q+R$, where R is a σ-regularizing operator and Q is defined on $G_0^{(\sigma)}(\Omega)$ by

$$(Qu)(x) = \iint e^{i(\phi(x,\xi)-\langle y,\xi\rangle)} q(x,\xi)u(y)dy đ\xi$$

where

$$q(x,\xi) \sim \sum_{j\geq 0} q_j(x,\xi) \quad \text{in} \quad SF^{\infty,\sigma,\mu}(\Omega\times R^n) ,$$

$$q_j(x,\xi) = \sum_{|\alpha|=j} \alpha!^{-1} D_y^\alpha [\partial_\xi^\alpha p_1(x,\tilde{\nabla}_x\phi(x,y,\xi))p_2(y,\xi)]_{y=x}$$

and $\tilde{\nabla}_x\phi(x,y,\xi) = \int_0^1 \nabla_x\phi(y+\theta(x-y),\xi)d\theta$.

If in addition, p_1, p_2 and ϕ are continuous functions of a parameter z an a compact subset Z of R^ν with values in $S^{\infty,\sigma,1}(\Omega\times R^n)$, $S^{\infty,\sigma,\mu}(\Omega\times R^n)$ and $S^{1,\sigma,\mu}(\Omega\times R^n)$ respectively, then $q\in C(Z;\ S^{\infty,\sigma,\mu}(\Omega\times R^n))$ and R has a kernel continuous on Z with values in $G^\sigma(\Omega\times\Omega)$.

The following definition is well known

Definition 1.2.8. If $u\in G_0^{(\sigma)'}(\Omega)$ we denote by $WF^{(\sigma)}(u)$ the complement in $\Omega\times(R^n\setminus\{0\})$ of the set of (x_0,ξ_0) such that there exist a neighborhood U of $x_0\in\Omega$, a conic neighborhood Γ of ξ_0 and a function $\chi\in G_0^{(\sigma)}(\Omega)$ which is equal to one in U such that

$$|(\widetilde{\chi u})(\xi)| \leq c\exp(-h|\xi|^{1/\sigma}) , \qquad \xi\in\Gamma$$

for some positive constants c and h.

If $u\in G^{(\sigma)'}(\Omega)$ and Ω' is an open relatively compact convex subset of Ω such that supp $u\subset\Omega'$, then (see H. Komatsu [16]) there exists $u_0\in C_0(\overline{\Omega'})$ and an ultradifferentiable operator P(D) such that $P(D)u_0=u$. Using this property and the partition of unity (1.1.2) one can prove for operators of infinite order and for ultradistributions the same result that holds for distributions and finite order operators (see K.Taniguchi [29]).

Theorem 1.2.9. Let A be the operator defined by (1.2.2) with $\varphi=\phi(x,\xi)-\langle y,\xi\rangle$, $\phi\in\mathscr{P}$. Then for every $u\in G^{(\sigma)'}(\Omega)$

$$WF^{(\sigma)}(Au)\subset\{(x,\rho\nabla_x\phi(x,\xi));\ (\nabla_\xi\phi(x,\xi),\xi)\in WF^{(\sigma)}(u),\ |\xi| \text{ large},\ \rho>0\} .$$

2. APPLICATION OF PSEUDO-DIFFERENTIAL OPERATORS OF INFINITE ORDER

In this and in the following section we denote by $OPS^{\infty,\sigma,1}(\Omega)$ the set of all operators defined by (1.2.2) when $\varphi=\langle x-y,\xi\rangle$ and a, independent of y, is in $S^{\infty,\sigma,1}=S^{\infty,\sigma,1}(\Omega\times R^n)$.

2.1. Construction of a parametrix for a Cauchy problem

We consider here the Cauchy problem

$$\begin{cases} P(t,x,D_t,D_x)u = f(t,x) & (t,x)\in[0,T]\times\Omega \\ D_t^j u(s,x) = g_j(x) & x\in\Omega\ ;\ j=0,\dots,m-1 , \end{cases} \qquad (2.1.1)$$

$T > 0$, Ω an open subset of R^n, $s \in [0,T]$, for the hyperbolic operator of order m

$$P(t,x,D_t,D_x) = D_t^m + \sum_{j=1}^{m} a_j(t,x,D_x)D_t^{m-j} , \qquad (2.1.2)$$

where for every $t \in [0,T]$, $a_j(t,x,D_x)$, $j=1,\dots,m$, are pseudo-differential operators on Ω of order ρj, $\rho \in]0,1[$. We suppose that:

i) the simbols $a_j(t,x,\xi)$ of the operators $a_j(t,x,D_x)$, $j = 1,\dots,m$, are continuous on $[0,T]$ with values in $C^\infty(\Omega \times R^n)$ and for a given $\sigma \in]1,1/\rho[$, $a_j \in \mathscr{B}^0_{[0,T]}(S^{\rho j,\sigma,1})$ where $S^{\rho j,\sigma,1} = S^{\rho j,\sigma,1}(\Omega \times R^n)$ (see 0.1 and Definition 1.1.9).

What follows clearly shows that we can always suppose that for every $t \in [0,T]$, $a_j(t,x,D_x)$ are properly supported operators on $G_0^{(\sigma)}(\Omega)$, uniformly with respect to t since this can be obtained up to σ-regularizing operators.

For every $(s,t) \in [0,T] \times [0,T]$ we want to find $E(t,s) \in OPS^{\infty,\sigma,1}(\Omega)$ such that its symbol $e(t,s) \in \mathscr{B}^m_{[0,T]^2}(S^{\infty,\sigma,1}) \cap C^m([0,T]^2 ; C^\infty(\Omega \times R^n))$ and

$$\begin{cases} P(t,x,D_t,D_x)E(t,s) = R(t,s) & t \in [0,T] \\ D_t^j E(s,s) = 0 , & j = 0,\dots,m-2 , \\ D_t^{m-1}E(s,s) = iI & \end{cases} \qquad (2.1.3)$$

where $R(t,s)$ is a σ-regularizing operator for every s, t and I is the identity operator.

In view of Corollary 1.1.8, we can obtain the symbol $e(t,s;x,\xi)$ of $E(t,s)$ through a formal series of symbols

$$\sum_{h \geq 0} e_h(t,s;x,\xi), \ e_h \in \mathscr{B}^m_{[0,T]^2}(S^{\infty,\sigma,1}) \cap C^m([0,T]^2; C^\infty(\Omega \times R^n)).$$

By Theorem 1.2.7 $P(t,x,D_t,D_x)E(t,s)$ is equal, up to an operator $R(t,s)$ mapping $C([0,T] ; G^{(\sigma)'}(\Omega))$ into $C([0,T]^2; G^{(\sigma)}(\Omega))$, to an operator $Q(t,s) \in OPS^{\infty,\sigma,1}(\Omega)$ with symbol $q(t,s;x,\xi)$ such that $q(t,s;x,\xi) \sim \sum_{h \geq 0} q_h(t,s;x,\xi)$ in $SF^{\infty,\sigma,1}(\Omega \times R^n)$ uniformly for $(t,s) \in [0,T]^2$, where

$$q_h(t,s;x,\xi) = D_t^m e_h(t,s;x,\xi) + \sum_{j=1}^{m} \sum_{|\alpha|+k=h} \alpha!^{-1}\partial_\xi^\alpha a_j(t,x,\xi) \times$$
$$\times D_x^\alpha D_t^{m-j} e_k(t,s;x,\xi) .$$

Thus, by Proposition 1.2.5, the first equation in (2.1.3) will be satisfied if the functions $e_h(t,s;x,\xi)$ are determined so that $q_h(t,s;x,\xi) \equiv 0$ for every h. Hence the functions e_h must be solutions of the following Cauchy problems

$$\begin{cases} (D_t^m + \sum_{j=1}^{m} a_j(t,x,\xi)D_t^{m-j})e_0 = 0 & (t,x) \in [0,T] \times \Omega, \\ D_t^j e_0(s,s;x,\xi) = 0 , \quad x \in \Omega & j = 0,\dots,m-2 \\ D_t^{m-1} e_0(s,s;x,\xi) = i , \quad x \in \Omega , & \end{cases} \qquad (2.1.4_\circ)$$

and

$$\begin{cases} (D_t^m + \sum_{j=1}^{m} a_j(t,x,\xi)D_t^{m-j})e_h = - \sum_{j=1}^{m} \sum_{\ell=1}^{h} \sum_{|\alpha|=\ell} \alpha!^{-1}\partial_\xi^\alpha a_j(t,x,\xi) \times \\ \qquad \times D_x^\alpha D_t^{m-j} e_{h-\ell}(t,s;x,\xi), \\ D_t^j e_h(s,s;x,\xi) = 0 \qquad j = 0,\dots,m-1 \end{cases} \qquad (2.1.4_h)$$

$h = 1,\dots$

Starting from the classical estimates for the solution of the problem $(2.1.4_\circ)$ and arguing by induction, one can prove that the solutions e_h of the problems $(2.1.4_h)$, $h = 0,1,\dots$ give rise to a series $\sum_{h\geq 0} e_h \in SF^{\infty,\sigma,1}(\Omega\times R^n)$, uniformly for $(t,s) \in [0,T]^2$ together with all the series $\sum_{h\geq 0} D_t^j e_h$, $j = 0,\dots,m$.

Thus we prove

<u>Theorem</u> 2.1.1. For $(t,s) \in [0,T]^2$ there exist

$$e \in \mathscr{B}^m_{[0,T]^2}(S^{\infty,\sigma,1}) \cap C^m([0,T]^2;\ C^\infty(\Omega \times R^n))$$

and an operator $R(t,s)$ mapping $C([0,T];\ G^{(\sigma)'}(\Omega))$ into $C([0,T]^2;G^{(\sigma)}(\Omega))$ such that for every $\varphi \in C([0,T];\ G_0^{(\sigma)}(\Omega))$ the function

$$\begin{aligned} u(t,x) &= \int_s^t E(t,s')\varphi(s',\cdot)ds' \\ &= \int_s^t ds' \int_{R^n} e^{i\langle x,\xi\rangle} e(t,s';x,\xi)\tilde{\varphi}(s',\xi)d\xi \end{aligned} \qquad (2.1.5)$$

is in $C^m([0,T]; G^{(\sigma)}(\Omega))$ and satisfies the equations

$$\begin{cases} P(t,x,D_t,D_x)u = \varphi(t,x) + \int_s^t R(t,s')\varphi(s',\cdot)ds', & (t,x) \in [0,T]\times\Omega, \\ D_t^j u(s,x) = 0 \quad x \in \Omega, & j = 0,\dots,m-1 \end{cases}$$

where P is the operator (2.1.2) satisfying i).

Let now H be a compact subset of Ω and let $\chi \in G_0^{(\sigma)}(\Omega)$ be equal to one on H and with supp $\chi \subset \Omega'$, $\Omega' \subset\subset \Omega$. If we set

$$(R_\chi f)(t,x) = \chi(x)\int_s^t R(t,s')f(s',\cdot)ds', \qquad f \in C([0,T]; G_0^{(\sigma)}(\Omega)),$$

we can prove that the Neumann series $f(t,x) + \sum_{\nu=1}^{\infty}(-1)^\nu R_\chi^\nu f$ is convergent in $G_0^{(\sigma)}(\overline{\Omega'})$ uniformly in $[0,T]$ to a function $\varphi \in C([0,T]; G_0^{(\sigma)}(\overline{\Omega'}))$ which satisfies the equation

$$\varphi(t,x) + \int_s^t R(t,s')\varphi(s',\cdot)ds' = f(t,x) \qquad \text{for } (t,x) \in [0,T]\times H. \qquad (2.1.6)$$

Since $R(t,s)$ is a σ-regularizing operator, the same argument can be applied when $f \in C([0,T]; G^{(\sigma)'}(\Omega))$.

We conclude with

<u>Theorem</u> 2.1.2. Let $f \in C([0,T]; G_0^{(\sigma)}(\Omega))$ $(f \in C([0,T]; G^{(\sigma)'}(\Omega)))$ and let H be a compact subset of Ω. Then there exists

$$\varphi \in C([0,T]; G_0^{(\sigma)}(\overline{\Omega'}))(\varphi \in C([0,T]; G^{(\sigma)'}(\overline{\Omega'}))$$

with $WF^{(\sigma)}(\varphi(t,\cdot)) = WF^{(\sigma)}(f(t,\cdot))$ for every $t \in [0,T]$), $H \subset \Omega' \subset\subset \Omega$, such that (2.1.5) is a solution $C^m([0,T]; G^{(\sigma)}(\Omega))$ $(C^m([0,T]; G_0^{(\sigma)'}(\Omega)))$ in $[0,T]\times H$ of the problem (2.1.1) with $g_j = 0$, $j = 0,\dots,m-1$ for the operator (2.1.2) satisfying i).

If, we suppose that

i') condition i) holds for the derivatives with respect to t up to the order m-j of the operators a_j,

then Theorem 2.1.1 and 2.1.2 hold also for the transposed of the operator P. By a standard argument this implies

<u>Theorem</u> 2.1.3. If P satisfies conditions i'), then the Cauchy problem

(2.1.1) has at most one solution in $C^m([0,T]; G^{(\sigma)'}(\Omega))$.

We remark that if the functions $a_j(t,x,\xi)$ vanish when x is outside a compact set $K_0 \subset \Omega$, then supp $E(t,s)\varphi$ and supp $R(t,s)\varphi$ are contained in $K_0 \cup$ supp φ. From this it follows that there exists φ satisfying (2.1.6) on $[0,T] \times \Omega$ such that the function $u(t,\cdot)$ given by (2.1.5) has its values with compact support and is a solution in $[0,T] \times \Omega$ of the problem (2.1.1) with $g_j = 0$, $j=0,\dots,m-1$, $f \in C([0,T];G^{(\sigma)}(\Omega))$ $(f \in C([0,T];G^{(\sigma)'}(\Omega)))$.

Let now $u \in C^m([0,T]; G^{(\sigma)'}(\Omega))$ and suppose, without loss of generality, that there exists a compact set $K \subset \Omega$ such that supp $(Pu)(t,\cdot)$ (and supp $u(t,\cdot)) \subset K$ for every $t \in [0,T]$. Hence if

$$v=u-\sum_{j=0}^{m-1}(it)^j D_t^j u(0,\cdot)/j!,\quad Pv=P_\chi v,\quad P_\chi=D_t^m+\sum_{j=1}^{m}\chi(x)a_j(t,x,D_x)D_t^{m-j},\ \chi\in G_0^{(\sigma)}(\Omega),$$

$\chi \equiv 1$ on K. Then, by the previous remark and Theorem 2.1.3, v is given by (2.1.5) with $s = 0$, where the operators $E(t,s)$ are now as in Theorem 2.1.1 with P replaced by P_χ and φ is the solution in $[0,T] \times \Omega$ of (2.1.6) whit $f = Pv$. From this and Theorem 1.2.9 we obtain the following result on propagation of Gevrey singularities

<u>Theorem</u> 2.1.4. Let P satisfy i') and let $u \in C^m([0,T]; G^{(\sigma)'}(\Omega))$ be such that $Pu \in C([0,T]; G_0^{(\sigma)}(\Omega))$. Then for every $t \in [0,T]$

$$\bigcup_{j=0}^{m-1} WF^{(\sigma)}(D_t^j u(t,\cdot)) = \bigcup_{j=0}^{m-1} WF^{(\sigma)}(D_t^j u(0,\cdot)) .$$

2.2. Pseudodifferential operators with multiple characteristics and Gevrey singularities

Another application of the symbolic calculus for pseudo-differential operators of infinite order concernes the propagation of Gevrey-singularities for a certain class of pseudo-differential operators with multiple characteristics [26].

Let P be an analytic pseudo-differential operator whose principal symbol is assumed to vanish exactly of order k, $k \geq 2$, on a regular submanifold of codimension one in the cotangent space. As is known, this hypothesis is sufficient to conclude that the operator is non-analytic--hypoelliptic and that the analytic wave front set propagates along the multiple bicharacteristic strips (see Bony-Shapira [4]), whereas to obtain a similar result in the C^∞ category it is necessary to add the so-called Levi-condition on the lower order terms (see Chazarain [8] , Bony [3] , Sjöstrand [27]). In [26] is obtained a result of propagation for $G^{(\sigma)}$-singularities under a suitable Levi-condition which depends on

σ and is weaker than the usual C^∞ Levi-condition.

In the following we shall refer to the space of the σ-microfunctions defined on a conic open subset Γ of $T^*\Omega\setminus 0$, Ω an open subset of R^n. Precisely, writing $f\sim g$ for $f,g\in G_0^{(\sigma)'}(\Omega)$ to mean that $\Gamma\cap WF^{(\sigma)}(f-g)=\emptyset$, we shall denote by $M^{(\sigma)}(\Gamma)$ the factor space $G_0^{(\sigma)'}(\Omega)/\sim$. The σ-wave front set $WF^{(\sigma)}(u)$ of a microfunction $u\in M^{(\sigma)}(\Gamma)$ is a well defined conic closed subset of Γ.

We consider an analytic pseudo-differential operator P, with symbol $p(x,\xi)\sim\sum_{j=0}^{\infty} p_{m-j}(x,\xi)$, where $p_{m-j}(x,\xi)$ is homogeneous of degree m-j in ξ (i.e. P is classical).

Since our arguments will be microlocal in a conic neighborhood of a point $(x_0,\xi_0)\in T^*\Omega\setminus 0$, we shall suppose from the beginning that $p(x,\xi)$ is given only in such a neighborhood. The map $P:\mathscr{D}'(\Omega)\to\mathscr{D}'(\Omega)$ is then well defined modulo errors which are micro-analytic in a smaller conic neighborhood Γ of (x_0,ξ_0), and it extends for every σ, $1<\sigma<\infty$, to an operator acting on $M^{(\sigma)}(\Gamma)$ (Cfr. Boutet de Monvel-Krée [6]).

On the principal symbol $p_m(x,\xi)$ we make the following hypothesis:

In a conic neighborhood of Γ we may write $p_m(x,\xi)=q_{m-k}(x,\xi)a_1(x,\xi)^k$ for a fixed $k\in[1,m]$, where $q_{m-k}(x,\xi)$ is an elliptic symbol, homogeneous of order m-k, and the first order symbol $a_1(x,\xi)$ is real valued and of principal type, i.e. $d_{x,\xi}a_1(x,\xi)$ never vanishes and is not parallel to $\sum_{h=1}^{n}\xi_h dx_h$ on $\Sigma=\{(x,\xi)\in\Gamma, a_1(x,\xi)=0\}\neq\emptyset$. (2.2.1)

As for the lower order terms, we fix now p, $0<p<1$, and assume that the following p-Levi-condition is satisfied:

Let A be a classical analytic pseudo-differential operator whose principal symbol is given by the function $a_1(x,\xi)$ in (2.2.1); then P can be written in Γ in the form $P=\sum_{j=0}^{k} Q_j A^{k-j}$, where Q_j, $j=0,\dots,k$, are classical analytic pseudo-differential operators of order $\leq m-k+pj$. (2.2.2)

Let us assume $(x_0,\xi_0)\in\Sigma$ and write γ_0 for the restriction to Γ of the bicharacteristic strip through (x_0,ξ_0). Then we have

Theorem 2.2.1. Let P satisfy the conditions indicated above, in particular the assumption (2.2.1) and the p-Levi-condition (2.2.2) and let $\sigma \in\ 1,1/p$. Then, taking Γ sufficiently small:

(i) There exists $u \in M^{(\sigma)}(\Gamma)$ with $Pu = 0$ and $WF^{(\sigma)}(u) = \gamma_0$.

(ii) If u is in $M^{(\sigma)}(\Gamma)$ with $Pu = 0$, then $(x_0,\xi_0) \in WF^{(\sigma)}(u)$ implies $\gamma_0 \subset WF^{(\sigma)}(u)$.

(iii) For every $v \in M^{(\sigma)}(\Gamma)$ there exists $u \in M^{(\sigma)}(\Gamma)$ such that $Pu = v$.

We observe that the condition (2.2.1) may be equivalently expressed by writing:

$$C^{-1}d_\Sigma(x,\xi)^k \leq |p_m(x,\xi)| \leq Cd_\Sigma(x,\xi)^k$$

for a suitable constant C and for all $(x,\xi) \in \Gamma$, $|\xi| = 1$, where $d_\Sigma(x,\xi)$ is the distance from (x,ξ) to Σ.

Concerning the condition (2.2.2) we have the following

Proposition 2.2.2. The operator P satisfies (2.2.2) for a given p, $1/2 \leq p < 1$, if and only if

$$|p_{m-j}(x,\xi)| \leq Cd_\Sigma(x,\xi)^{k-j/(1-p)}, \quad 0 \leq j < k(1-p) \tag{2.2.3}$$

for some constant C and for all $(x,\xi) \in \Gamma$, $|\xi| = 1$.

For $0 < p < 1/2$ the equivalence of (2.2.3) with (2.2.2) fails.

For $k = 1$, i.e. when P is of principal type, (2.2.1) implies (2.2.2) for every p and the conclusions of Theorem 2.2.1 hold for all σ, $1 < \sigma < \infty$ (the result of propagation in this case is well known; cfr. Hörmander [11]). When $k \geq 2$, the conclusions of Theorem 2.2.1 fail in general for $1/p \leq \sigma < \infty$ and the study of the corresponding $G^{(\sigma)}$-regularity requires then a further analysis of the lower order terms. Some results of Gevrey hypoellipticity and solvability in this connection are contained in [26].

A result on propagation of Gevrey singularities intersecting (ii) in Theorem 2.2.1 has also been proved by Kessab [15].

The first step in the proof of Theorem 2.2.1 is a standard application of the classical theory of Fourier integral operators which allow us to argue on the operator:

$$P = D_{x_n}^k + \sum_{j=1}^{k} Q_j D_{x_n}^{k-j} \quad , \tag{2.2.4}$$

where the Q_j, $j = 1,\dots,k$ are classical analytic pseudo-differential oper ators of order $\leq pj$, defined on Γ that we may assume of the form $\Gamma = \Omega \times X$, with Ω a neighborhood of $x_0 = 0$ and X a conic neighborhood of $\xi_0 = (\xi_{1,0}, \ldots, \xi_{n-1,0}, 0)$.

As a second step we write the equation $Pu = v$ as a first order system:

$$\mathcal{P}U = D_{x_n}U + \mathcal{A}(x,D)U = V, \tag{2.2.5}$$

where $\mathcal{A}(x,D)$ is a $k \times k$ matrix of classical analytic pseudodifferential operators of order $\leq p$ such that the study of $P : M^{(\sigma)}(\Gamma) \to M^{(\sigma)}(\Gamma)$ in (2.2.4) is equivalent to the study of the operator $\mathcal{P} : M^{(\sigma)}(\Gamma,k) \to M^{(\sigma)}(\Gamma,k)$ in (2.2.5), (we denote by $M^{(\sigma)}(\Gamma,k)$ the space of the k-tuples of elements of $M^{(\sigma)}(\Gamma)$).

Finally, following standard arguments in the C^∞ category (cfr. Duistermat-Hörmander [9]) we may obtain easily the conclusions of Theorem (2.2.1) using

Proposition 2.2.3. Let $\mathcal{A}(x,D)$ be a $k \times k$ matrix of classical analytic pseudo-differential operators on Γ of order $\leq p$, with $0 < p < 1$, and define $\mathcal{P} = D_{x_n} + \mathcal{A}(x,D)$. Then, under the assumption $1 < \sigma < 1/p$, there exist two linear maps $Q, Q' : M^{(\sigma)}(\Gamma,k) \to M^{(\sigma)}(\Gamma,k)$, such that:

(j) Q, Q' are σ-microlocal, i.e. $WF^{(\sigma)}(QU) \subset WF^{(\sigma)}(U)$ and $WF^{(\sigma)}(Q'U) \subset WF^{(\sigma)}(U)$ for all $U \in M^{(\sigma)}(\Gamma,k)$ (the σ-wave front set of a vector is defined here as the union of the σ-wave front sets of the components).

(jj) $QQ' = Q'Q =$ identity in $M^{(\sigma)}(\Gamma,k)$.

(jjj) $Q'\mathcal{P}Q = D_{x_n}$ in Γ.

The proof of Proposition 2.2.3 is based on a microlocal version of the calculus for pseudo-differential operators of infinite order, that we now pass to describe briefly.

Let us begin by noting that for every $\xi_0 \in \mathbb{R}^n \setminus 0$, all given conic neighborhoods X, X' of ξ_0 with $X \subset\subset X'$, and every $R > 0$ we can find a function $h(\xi) \in C^\infty(\mathbb{R}^n \setminus 0)$ such that, for large $|\xi|$, $h = 1$ in X, $h = 0$ in $\mathbb{R}^n \setminus X'$, and $|D^\alpha h(\xi)| \leq C(C/R)^{|\alpha|/\sigma}(1 + |\xi|/R)^{1/\sigma}(1 + |\xi|)^{-(1-1/\sigma)|\alpha|}$, $|\alpha| \leq (|\xi|/(2R))^{1/\sigma}$. It easily follows that $(x_0,\xi_0) \in WF^{(\sigma)}u$ if and only if $h(D)u \in G^{(\sigma)}$ in a neighborhood of x_0, for a suitable choice of the conic neighborhoods X, X' of ξ_0 in the definition of $h(\xi)$.

Now let $p(x,\xi)$ be defined in $\Omega' \times X'$ with Ω' neighborhood of $\Omega \subset \mathbb{R}^n$, X' conic neighborhood of the conic set $X \subset \mathbb{R}^n$, $\Omega \subset\subset \Omega'$, $X \subset\subset X'$. Assume that there exist constants A and B such that for every $\varepsilon > 0$:

$$\sup_{\alpha,\beta\in\mathbb{Z}^n_+}\ \sup_{\substack{(x,\xi)\in\Omega'\times X'\\ |\xi|\geq B|\alpha|^\sigma}} A^{-|\alpha+\beta|}\alpha!^{-1}\beta!^{-\sigma}|D^\alpha_\xi D^\beta_x p(x,\xi)|(1+|\xi|)^{|\alpha|}\exp(-\varepsilon|\xi|^{1/\sigma})<\infty .$$

Let then h be defined as before with $h=1$ in a neighborhood of X and supp $h\subset X'$, and let $\varphi\in G^{(\sigma)}_0(\Omega')$ with $\varphi=1$ in a neighborhood of Ω. Define:

$$Ⓗ\, u(x) = \int \exp(i\langle x,\xi\rangle)p(x,\xi)h(\xi)(\varphi u)^{\sim}(\xi)đ\xi \ . \tag{2.2.6}$$

It is easy to prove that Ⓗ is well defined as a map from $G^{(\sigma)}(\Omega)$ to $G^{(\sigma)}(\Omega)$ and from $G^{(\sigma)'}_0(\Omega)$ to $G^{(\sigma)'}_0(\Omega)$ and that is independent of the choice of h and φ, modulo errors whose σ-wave front sets do not intersect $\Gamma=\Omega\times X$.

We shall write $S^{\infty,\sigma}(\Gamma)$ to denote the space of all symbols $p(x,\xi)$ just considered, and $OP\dot{S}^{\infty,\sigma}(\Gamma)$ to denote the corresponding class of operators $p(x,D)$ defined by (2.2.6), and acting on the space of the σ-microfunctions $M^{(\sigma)}(\Gamma)$. It turns out that:

$$WF^{(\sigma)}p(x,D)u\subset WF^{(\sigma)}u, \quad u\in M^{(\sigma)}(\Gamma) \ , \tag{2.2.7}$$

and that the usual rules of the symbolic calculus hold unchanged for operators in $OP\dot{S}^{\infty,\sigma}(\Gamma)$. In particular we have:

Proposition 2.2.4. Let $p_1(x,D)$, $p_2(x,D)\in OP\dot{S}^{\infty,\sigma}(\Gamma)$. Then $p_1(x,D)p_2(x,D)=q(x,D)\in OP\dot{S}^{\infty,\sigma}(\Gamma)$ and $q(x,\xi)\sim\sum_\alpha \alpha!^{-1}\partial^\alpha_\xi p_1(x,\xi)D^\alpha_x p_2(x,\xi)$.

For the proof of Proposition 2.2.3 we shall consider the class $S^{\infty,\sigma}(\Gamma,k)$ of all the $k\times k$ matrices of symbols $P(x,\xi)=(p_{jh}(x,\xi))_{j,h=1,\dots,k}$, $p_{jh}(x,\xi)\in S^{\infty,\sigma}(\Gamma)$.

The rules of symbolic calculus extend in a natural way to the class $OP\dot{S}^{\infty,\sigma}(\Gamma,k)$ of the corresponding operators $P(x,D)=(p_{jh}(x,D))_{j,h=1,\dots,k}$, $p_{jh}(x,D)\in OP\dot{S}^{\infty,\sigma}(\Gamma)$, defined on $M^{(\sigma)}(\Gamma,k)$. Also the space $SF^{\infty,\sigma}(\Gamma,k)$ of all $k\times k$ matrices of formal series is defined consequently.

The proof of Proposition 2.2.3 starts by proving that there exists $Q(x,D)\in OP\dot{S}^{\infty,\sigma}(\Gamma,k)$ such that:

$$(D_{x_n}+A(x,D))Q(x,D) = Q(x,D)D_{x_n} \quad \text{on } M^{(\sigma)}(\Gamma,k) \ , \tag{2.2.8}$$

where the symbol $Q(x,\xi)\in S^{\infty,\sigma}(\Gamma,k)$ of $Q(x,D)$ can be determined through its asymptotic expansion $\sum_{j\geq 0}Q_j(x,\xi)\in SF^{\infty,\sigma}(\Gamma,k)$. Regarding

$P = D_{x_n} + A(x,D)$ as an element of $O\dot{P}S^{\infty,\sigma}(\Gamma,k)$ and applying Proposition 2.2.4 we see that the matrices $Q_j(x,\xi)$ must satisfy the transport equations:

$$D_{x_n}Q_0 + A\,Q_0 = 0 \qquad (2.2.9)$$

$$D_{x_n}Q_j + AQ_j = -\sum_{1\leq \nu\leq j}\ \sum_{|\alpha|=\nu} \alpha!^{-1}\partial_\xi^\alpha A D_x^\alpha Q_{j-|\alpha|}\ ,\quad j = 1,2,\ldots \qquad (2.2.10)$$

If Q_0 is the solution of (2.2.9) satisfying $Q_0|_{x_n=0} = I$ = identity matrix and Q_j, $j = 1,2,\ldots$, the solution of (2.2.10) satisfying $Q_j|_{x_n=0} = 0$, then one can prove by induction, that $\sum_{j\geq 0} Q_j(x,\xi)$ belongs to $SF^{\infty,\sigma}(\Gamma,k)$. Hence $Q \sim \sum_{j\geq 0} Q_j$ satisfies (2.2.8) and, in view of (2.2.7), $WF^{(\sigma)}(QU) \subset WF^{(\sigma)}(U)$ for all $U \in M^{(\sigma)}(\Gamma,k)$. By analogous arguments we can prove the existence of $Q'(x,D) \in OPS^{\infty,\sigma}(\Gamma,k)$ such that:

$$Q'(x,D)(D_{x_n} + A(x,D)) = D_{x_n}Q'(x,D) \quad \text{on } M^{(\sigma)}(\Gamma,k)\ , \qquad (2.2.11)$$

and by using Proposition 2.2.4, that:

$$Q(x,D)Q'(x,D) = Q'(x,D)Q(x,D) = \text{identity on } M^{(\sigma)}(\Gamma,k). \qquad (2.2.12)$$

Now j), jj) and jjj) follow immediately from (2.2.8), (2.2.11), (2.2.12).

3. APPLICATION TO CAUCHY PROBLEM FOR HYPERBOLIC OPERATORS WITH CHARACTERISTICS OF CONSTANT MULTIPLICITY

3.1. Operators with multiple characteristic : a microlocal model

The results stated in 2.1 may be used to prove analogous results when

$$P(t,x,D_t,D_x) = (D_t-\lambda(t,x,D_x))^m + \sum_{j=1}^{m} a_j(t,x,D_x)(D_t-\lambda(t,x,D_x))^{m-j}. \qquad (3.1.1)$$

Here we suppose that for every $t \in [0,T]$, $\lambda(t,x,D_x)$ and $a_j(t,x,D_x)$ are pseudo-differential operators in an open set $\Omega \subset R^n$, of order one and pj respectively, $p \in [0,1[$, such that:

$\lambda(t,x,\xi)$ is a real valued function homogeneous of degree one in ξ for $|\xi|$ large, contained in $C^{m-1}([0,T];\ C^{\infty}(\Omega\times R^n))$ and for a given $\sigma\in]1,1/p[$ in $\mathscr{B}^{m-1}_{[0,T]}(S^{1,\sigma,1}(\Omega\times R^n))$; (3.1.2)

$a_j(t,x,\xi)$, $j=1,\dots,m$, are in $C([0,T];C^{\infty}(\Omega\times R^n))$ and in $\mathscr{B}^{0}_{[0,T]}(S^{pj,\sigma,1}(\Omega\times R^n))$ and satisfy condition a) of Theorem 1.2.7. (3.1.3)

Moreover, after the eventual addition to P of an operator $R(t,x,D_t,D_x)=\sum_{j=1}^{m} r_j(t,x,D_x)D_t^{m-j}$ with σ-regularizing operators r_j, we can suppose that the operators λ and a_j, $j=1,\dots,m$ are all properly supported.

If we consider the canonical equations

$$\begin{aligned} &dx/dt = -\nabla_{\xi}\lambda(t,x,\xi)\ , \quad d\xi/dt = \nabla_x\lambda(t,x,\xi) \\ &x(s,s;y,\eta) = y\ , \quad \xi(s,s,y,\eta) = \eta \end{aligned} \tag{3.1.4}$$

$y\in\Omega'\subset\subset\Omega$, $\eta\in R^n$, we see that there exists $T_0\in[0,T]$ such that for every $s\in[0,T_0]$ there exists a unique solution $x(t,s;y,\eta)\in\Omega'$, $\xi(t,s;y,\eta)\in R^n$ of the problem (3.1.4) in $[0,T_0]$. Moreover x and ξ are homogeneous with respect to η when $|\eta|$ is large, of degree zero and one respectively and $x\in\mathscr{B}^{0}_{[0,T_0]^2}(S^{0,\sigma,\sigma}(\Omega'\times R^n))$, $\xi\in\mathscr{B}^{0}_{[0,T_0]^2}(S^{1,\sigma,\sigma}(\Omega'\times R^n))$ (see [29]). Denoting with $y(t,s;x,\eta)$ the inverse function of $y\to x(t,s;y,\eta)$, we obtain as in [19] and [29] the solution $\phi(t,s;x,\xi)$ of the eikonal equation

$$\begin{cases} \phi_t(t,s;x,\xi) = \lambda(t,x,\nabla_x\phi(t,s;x,\xi)) \\ \phi(s,s;x,\xi) = \langle x,\xi\rangle \end{cases} \tag{3.1.5}$$

$x\in\Omega'$, $(s,t)\in[0,T_0]^2$, $\xi\in R^n$, and we see that ϕ is a real valued function homogeneous of degree one with respect to ξ when $|\xi|$ is large and that $\phi\in\mathscr{B}^{0}_{[0,T_0]^2}(S^{1,\sigma,\sigma}(\Omega'\times R^n))$ and $\phi\in\mathscr{P}$(see Definition 1.2.6) uniformly with respect to $(t,s)\in[0,T_0]^2$.

For every $(t,s)\in[0,T_0]^2$ we look for a Fourier integral operator $E(t,s)$ defined as (1.2.2) with phase function given by $\phi(t,s;x,\xi)-\langle y,\xi\rangle$ and amplitude $e(t,s;x,\xi)$ such that

$$e(t,s)\in C^m([0,T_0]^2;\ C^{\infty}(\Omega'\times R^n))\cap\mathscr{B}^{m}_{[0,T_0]^2}(S^{\infty,\sigma,\sigma}(\Omega'\times R^n))$$

and

$$\begin{cases} P(t,x,D_t,D_x)E(t,s) = R(t,s) \\ D_t^j E(s,s) = 0 \qquad j = 0,\dots,m-2 \\ D_t^{m-1}E(s,s) = iI \end{cases} \tag{3.1.6}$$

where $R(t,s)$ is a σ-regularizing operator for every $(t,s) \in [0,T_0]^2$. If we let

$$e'(t,s;y,\eta) = e(t,s,x(t,s;y,\eta),\eta)$$

then

$e'(t,s) \in \mathscr{B}^m_{[0,T_0]^2}(S^{\infty,\sigma,\sigma}(\Omega' \times R^n))$ and $e(t,s;x,\xi) = e'(t,s,y(t,s;x,\xi),\xi)$.

If we look e' as a formal series $\sum_{h\geq 0} e'_h$, we see by Theorem 1.2.7 and the eikonal equation (3.1.5) that, up to a regularizing operator, $(D_t-\lambda(t,x,D_x))E(t,s)$ is equal to a Fourier integral operator with the same phase function as E and with an amplitude $q^{(1)} \sim \sum_{h\geq 0} q_h^{(1)}$, where

$$q_0^{(1)} = (D_t+A_0(t,s;x(t,s;y,\xi),\xi))e'_0$$

$$q_h^{(1)}=(D_t+A_0(t,s;x(t,s;y,\xi),\xi))e'_h - \sum_{\ell=1}^{h} \sum_{|\beta|\leq \ell+1} A^{(1)}_{\beta,\ell}(t,s;x(t,s;y,\xi),\xi)D_y^\beta e'_{h-\ell}$$

$h \geq 1$, $A_0(t,s;x,\xi) = -i/2 \sum_{i,k} \partial^2_{\xi_i\xi_k}\lambda(t,x,\nabla_x\phi(t,s;x,\xi))\partial^2_{x_ix_k}\phi(t,s;x,\xi)$

and $A^{(1)}_{\beta,\ell}(t,s) \in \mathscr{B}^{m-1}_{[0,T_0]^2}(S^{-\ell,\sigma,\sigma}(\Omega' \times R^n))$.

In this way we obtain, up to a σ-regularizing operator, $P(t,x,D_t,D_x)E(t,s)$ as a Fourier integral operator with an amplitude given by a formal series. Letting the terms of this series to be identically zero, we find that e'_h must satisfy the equations

$$(D_t^m + \sum_{j=1}^m b_j(t,s,y,\xi)D_t^{m-j})e'_h = \sum_{j=1}^m \sum_{\ell=1}^h \sum_{|\beta|\leq \ell+1} A_{j,\ell,\beta}(t,s;y,\xi) \times$$
$$\times D_t^{m-j}D_y^\beta e'_{h-\ell}\ , \qquad h = 0,1,\dots, \tag{3.1.7}$$

(where the right hand side has to be thought identically zero when h=0) and the initial conditions

$$D_t^j e_0'(s,s) = 0 , \quad j = 0,\dots,m-2, \quad D_t^{m-1} e_0'(s,s) = i , \qquad (3.1.8_\circ)$$

$$D_t^j e_h'(s,s) = 0, \quad j = 0,\dots,m-1, \quad h \geq 1. \qquad (3.1.8_h)$$

Here $b_j(t) \in \mathscr{B}^0_{[0,T_0]}(S^{pj,\sigma,\sigma}(\Omega' \times R^n))$ and

$A_{j,\ell,\beta}(t,s) \in \mathscr{B}^0_{[0,T_0]^2}(S^{pj-\ell,\sigma,\sigma}(\Omega' \times R^n))$. Thus we conclude that $e_h'(t,s;y,\xi)$ must satisfy equations of the same type as $(2.1.4_\circ)$ and $(2.1.4_h)$. This leads to a solution $E(t,s)$ of (3.1.6) with the properties required above and hence to results analogous to Theorems 2.1.1 and 2.1.2 for the operator (3.1.1) satisfying conditions (3.1.2) and (3.1.3) (in the statements of the resuls for the operator (3.1.1) T must obviously be replaced by T_0).

If in addition to (3.1.2), we suppose that:

condition (3.1.3) holds for the derivative with respect to t up to the order m-j of the operators a_j, $j=1,\dots,m$, (3.1.3')

then the transposed ${}^tP$ of P may be written in the same form as (3.1.1) and satisfies conditions (3.1.2) and (3.1.3).

Thus the same results obtained for the operator P hold also for its transposed ${}^tP$ and the uniqueness Theorem 2.1.3 is also true for the operator (3.1.1) satisfying conditions (3.1.2) and (3.1.3').

Arguing as in 2.1 and using Theorem 1.2.9 we can prove the following result analogous to Theorem 2.1.4

Theorem 3.1.1. Let P satisfy (3.1.2) and (3.1.3') and let $u \in C^m([0,T]; G^{(\sigma)'}(\Omega))$ be such that $Pu \in C([0,T]; G_0^{(\sigma)}(\Omega))$. Then for every $t \in [0,T_0]$, T_0 sufficiently small

$$\bigcup_{j=0}^{m-1} WF^{(\sigma)}(D_t^j u(t,\cdot)) = \{(x(t,0;y,\eta),\rho\xi(t,0;y,\eta)),\ \rho>0,$$

$$|\eta| \text{ large};\ (y,\eta) \in \bigcup_{j=0}^{m-1} WF^{(\sigma)}(D_t^j u(0,\cdot))\} ,$$

where $(x(t,0;y,\eta),\ \xi(t,0;y,\eta))$ is the solution of the problem (3.1.4), whith $s = 0$.

The arguments used above and in 2.1 may be applied to the Cauchy problem

$$\begin{cases} LU = D_t U - \lambda(t,x,D_x)IU + A(t,x,D_x)U = F(t,x), & (t,x) \in [0,T]\times\Omega \\ U(0,x) = G(x), & x \in \Omega, \end{cases} \tag{3.1.9}$$

where $U = (u_1,\dots,u_m)$, $F = (f_1,\dots,f_m)$, $G = (g_1,\dots,g_m)$ and

> $\lambda(t,x,\xi)$ satisfies (3.1.2) with m=1, and, for every $t \in [0,T]$, A is an $m\times m$ matrix of pseudo-differential operators with symbols in $C([0,T];\ C^\infty(\Omega\times R^n))$ and in $\mathscr{B}^0_{[0,T]}(S^{p,\sigma,1}(\Omega\times R^n))$ satisfying condition a) of Theorem 1.2.7. (3.1.10)

In particular it can be proved (see Taniguchi [29] when U has its values in $\mathscr{E}'$)

Theorem 3.1.2. Let the operator L in (3.1.9) satisfy condition (3.1.10) and let $U \in C^1([0,T];\ G^{(\sigma)'}(\Omega))$ be solution of the Cauchy problem (3.1.9) with $F \in C([0,T];\ G_0^{(\sigma)}(\Omega))$. Then for every $t \in [0,T_0]$, T_0 sufficiently small

$$WF^{(\sigma)}(U(t,\cdot)) = \{(x,\xi);\ x = x(t,0;y,\eta),\ \xi = \rho\xi(t,0;y,\eta),\ \rho>0,\ |\eta| \text{ large};\ (y,\eta) \in WF^{(\sigma)}(G)\},$$

where $x(t,0;y,\eta)$, $\xi(t,0;y,\eta)$ are as in Theorem 3.1.1.

We can also obtain results for the singularities of the solutions of the Cauchy problem for the operator (3.1.1) or (3.1.9) as ultradistributions in $[0,T]\times\Omega$. Suppose in fact for example that

> in addition to (3.1.10) $\lambda \in S^{1,\sigma,1}([0,T]\times\Omega\times R^n)$ and the elements of A are in $S^{p,\sigma,1}([0,T]\times\Omega\times R^n)$. (3.1.11)

Then outside a small conic neighborhood of $\{\xi = 0\}$ in R^{n+1} the operator L is a pseudo-differential operator in (t,x) of the type considered in 2.2, except for the regularity with respect to x which is here of Gevrey type. Using the results contained in 2.2 with the variants necessary for the actual situation, we can prove as a consequence of Theorem 3.1.2

Theorem 3.1.3. Let the operator L in (3.1.9) satisfy condition (3.1.11) and let $U \in C^1([0,T_0];\ G^{(\sigma)'}(\Omega))$ be solution of (3.1.9) with $F = 0$. Assume that $\xi \neq 0$ when $(t,x,\tau,\xi) \in WF^{(\sigma)}(U)$. Then

$$WF^{(\sigma)}(U) = \{(t,x,\lambda(t,x,\xi),\xi);\ x = x(t,0;y,\eta),\ \xi = \rho\xi(t,0;y,\eta),$$
$$\rho>0,\ |\eta| \text{ large},\ t\in[0,T_0];\ (y,\eta)\in WF^{(\sigma)}(G)\}. \tag{3.1.12}$$

We can also prove

Theorem 3.1.4. Let the operator L in (3.1.9) satisfy condition (3.1.11) and let $U\in C^1([0,T_0];\ G^{(\sigma)'}(\Omega))$ and $U(0,\cdot)=0$. Assume that $\xi\neq 0$ when $(t,x,\tau,\xi)\in WF^{(\sigma)}(LU)$. Then

$$WF^{(\sigma)}(U)\subset WF^{(\sigma)}(LU)\cup\{(t,x,\lambda(t,x,\xi),\xi));\ x = x(t,s;y,\eta),$$
$$\xi = \rho\xi(t,s;y,\eta),\ \rho>0,\ |\eta| \text{ large};\ (s,y,\lambda(s,y,\eta),\eta)\in WF^{(\sigma)}(LU), s\geq 0\};$$

where $(x(t,s;y,\eta),\ \xi(t,s;y,\eta))$ is the solution of (3.1.4).

Analogous results with $WF^{(\sigma)}(LU)$ and $WF^{(\sigma)}(G)$ replaced by $WF^{(\sigma)}(Pu)$ and $\bigcup_{j=1}^{m-1} WF^{(\sigma)}(g_j)$ respectively and with $\subset$ in place of $=$ in (3.1.12), hold for $WF^{(\sigma)}(u)$ when u is solution of the Cauchy problem for the operator (3.1.1).

3.2. Hyperbolic operators with multiple characteristics

We now consider the Cauchy problem for the operator of order m

$$P(t,x,D_t,D_x) = P_\nu(t,x,D_t,D_x)\ldots P_1(t,x,D_t,D_x) \tag{3.2.1}$$

where, for $h=1,\ldots,\nu$,

$$P_h(t,x,D_t,D_x) = (D_t-\lambda_h(t,x,D_x))^{m_h}+\sum_{j=1}^{m_h} a_{h,j}(t,x,D_x)(D_t-\lambda_h(t,x,D_x))^{m_h-j},$$

and $m_1+\ldots+m_\nu = m$. Let $p\in[0,1[$ and assume that

$\lambda_h(t,x,\xi)$, $h=1,\ldots,\nu$, are real valued functions homogeneous of degree one with respect to ξ when $|\xi|$ is large and, for a given $\sigma\in]1,1/p[$, $\lambda_h\in S^{1,\sigma,1}([0,T]\times\Omega\times R^n)$. Moreover when $h\neq k$ $\lambda_h(t,x,\xi)\neq\lambda_k(t,x,\xi)$ for every (t,x,ξ), $|\xi|$ large; (3.2.2)

$a_{h,j}\in S^{pj,\sigma,1}([0,T]\times\Omega\times R^n)$, $h = 1,\ldots,\nu$, $j = 1,\ldots,m_h$, and satisfy condition a) of Theorem 1.2.7. (3.2.3)

As a consequence of the results in 3.1 (see for example [23]) we have

Theorem 3.2.1. Let P given by (3.2.1) satisfy conditions (3.2.2) and (3.2.3) and let $u \in C^m([0,T]; G^{(\sigma)'}(\Omega))$ be solution of the Cauchy problem (2.1.1) with $f = 0$ and $g_j \in G^{(\sigma)'}(\Omega)$. Then for every $t \in [0,T_0]$, $T_0 > 0$ sufficiently small

$$\bigcup_{j=0}^{m-1} WF^{(\sigma)}(D_t^j u(t,\cdot)) = \bigcup_{h=1}^{\nu} \{(x,\xi);\ x = x_h(t,0;y,\eta),\ \xi = \rho\xi_h(t,0;y,\eta),\ \rho>0,$$

$$|\eta| \text{ large};\ (y,\eta) \in \bigcup_{j=0}^{m-1} WF^{(\sigma)}(g_j)\},$$

where $(x_h(t,0;y,\eta),\ \xi_h(t,0;y,\eta))$, $h=1,\dots,\nu$, is the solution of the problem

$$dx/dt = -\nabla_\xi \lambda_h(t,x,\xi) \qquad d\xi/dt = \nabla_x \lambda_h(t,x,\xi)$$

$$x(t,0;y,\eta) = y, \qquad \xi(t,0;y,\eta) = \eta.$$

Moreover if $\xi \neq 0$ when $(t,x,\tau,\xi) \in WF^{(\sigma)}(u)$, then

$$WF^{(\sigma)}(u) \subset \bigcup_{h=1}^{\nu} \{(t,x,\lambda_h(t,x,\xi),\xi),\ x = x_h(t,0;y,\eta),\ \xi = \rho\xi_h(t,0;y,\eta),\ \rho>0,$$

$$|\eta| \text{ large},\ t \in [0,T_0];\ (y,\eta) \in \bigcup_{j=0}^{m-1} WF^{(\sigma)}(g_j)\}.$$

To end these lectures let

$$P(t,x,D_t,D_x) = D_t^m + \sum_{j=1}^{m} a_j(t,x,D_x) D_t^{m-j} \tag{3.2.4}$$

where $a_j(t,x,D_x) = \sum_{|\alpha| \leq j} a_{j\alpha}(t,x) D_x^\alpha$, $j = 1,\dots m$, are differential operators with coefficients $a_{j\alpha} \in G^{(\sigma),A}(K)$, A independent of the compact set $K \subset [0,T] \times R^n$. Let $a_j(t,x,\xi) = a_j^0(t,x,\xi) + a_j'(t,x,\xi)$ where a_j^0 is either a homogeneous polynomial in ξ of degree j or identically zero and a_j' a polynomial in ξ of degree $\leq$ j-1. Suppose that P has characteristics with constant multiplicity i.e. that

$$p_m(t,x,\tau,\xi) = \tau^m + \sum_{j=1}^{m} a_j^0(t,x,\xi)\tau^{m-j} = \prod_{h=1}^{\nu} (\tau - \lambda_h(t,x,\xi))^{m_h}, \tag{3.2.5}$$

where λ_h are distinct real valued functions homogeneous of degree one in ξ, and $m_1+\ldots+m_\nu = m$.

Results on the existence and uniqueness of solution of the Cauchy problem for the operator (3.2.4) satisfying condition (3.2.5) has been proved by several authors such as [13], [14], [18], [20], [21], [23], [25], [28]. A result by S. Mizohata (see [23] and [24]) states that if P satisfies all the conditions above, then P may be written up to a σ-regularizing operator, in the form (3.2.1) where $a_{h,j}$ are pseudo-differential operators of order $\leq j$, $j = 1,\ldots,m_h$. If $p_h = \max_{j=1,\ldots,m_h} \text{ord } a_{h,j}/j$, then

$$a_{h,j} \in S^{p_h j,\sigma,1}([0,T]\times R^n \times R^n),\ h = 1,\ldots,\nu,\ j = 1,\ldots,m_h,$$

$$p_h \leq (m_h-1)/m_h\ ,\quad h = 1,\ldots,\nu$$

and

$$\lambda_h \in S^{1,\sigma,1}([0,T]\times R^n \times R^n),\quad h = 1,\ldots,\nu\ .$$

Hence $\max_h p_h \leq (\max_h m_h - 1)/\max_h m_h$. Moreover if

$$p = \max_{j=1,\ldots,m} \text{ord } a_j'/j\ , \tag{3.2.6}$$

then $p_h \leq p$, $h = 1,\ldots,\nu$

Thus we conclude with the following result (see[23]) :

<u>Theorem</u> 3.2.2. Let P be a differential operator given by (3.2.4) and let p_m given by (3.2.5)be its principal symbol.Assume that the conditions on the coefficients of P and on λ_h indicated above are satisfied. Then Theorem 3.2.1 holds for such P and every $\sigma \in]1, \max_h m_h/(\max_h m_h - 1)[$. Furthermore Theorem 3.2.1 holds for every $\sigma \in]1,1/p[$, $p = \max_h p_h$ and in particular when p is defined by (3.2.6).

REFERENCES

[1] Aoki T., 'Calcul exponentiel des opérateurs microdifférentiels d'ordre infini I', <u>Ann. Inst. Fourier</u>, Grenoble, **33**, (1983), 227--250; 'Calcul exponentiel des opérateurs microdifferéntiels d'ordre infini II', to appear.

[2] Bolley P. - Camus J. - Metivier G., 'Regularité Gevrey et iterés pour une classe d'opérateurs hypoelliptiques', <u>Rend. Sem. Mat.Univ. Politecn. Torino</u>, numero speciale (1983).

[3] Bony J.M., 'Propagation des singularités différentiables pour une classe d'opérateurs différentiels à coefficients analytiques', Astérisque, **34-35** (1976), 43-91.

[4] Bony J.M. - Schapira P., 'Propagation des singularités analytiques pour les solutions des equations aux dérivées partielles', Ann. Inst. Fourier, Grenoble, **26** (1976), 81-140.

[5] Boutet de Monvel L., 'Opérateurs pseudo-différentiels analytiques et opérateurs d'ordre infini', Ann. Inst. Fourier, Grenoble, **22** (1972), 229-268.

[6] Boutet de Monvel L. - Krée P., 'Pseudo-differential operators and Gevrey classes', Ann. Inst. Fourier, Grenoble, **17** (1967), 295-323.

[7] Cattabriga L. - Mari D., 'Parametrix of infinite order for a Cauchy problem for certain hyperbolic operators', to appear.

[8] Chazarain I.J., 'Propagation des singularités pour une classe d'opé rateurs à caractéristiques multiples et résolubilité locale', Ann. Inst. Fourier, Grenoble, **24** (1974), 209-223.

[9] Duistermaat J.J. - Hörmander L., 'Fourier integral operators II', I, Acta Math., **128** (1972), 183-269.

[10] Hashimoto S. - Matsuzawa T. - Morimoto Y., 'Opérateurs pseudo différentiels et classes de Gevrey', Comm. Partial Differential Equations, **8** (1983), 1277-1289.

[11] Hörmander L., 'Uniqueness theorems and wave front sets for solutions of linear differential equations with analytic coefficients', Comm. Pure Appl. Math., **24** (1971), 671-704.

[12] Iftimie V., 'Opérateurs hypoelliptiques dans des espaces de Gevrey', Bull. Soc. Sci. Math. R.S. Roumanie, **27** (1983), 317-333.

[13] Ivrii V.Ja., 'Condizioni di correttezza in classi di Gevrey del problema di Cauchy per operatori non strettamente iperbolici', Sibirsk. Mat. Ž., **17** (1976), 547-563 = Siberian Math. J., **17** (1976), 422-435.

[14] Kajitani K., 'Leray-Volevich's system and Gevrey class', J. Math. Kyoto Univ., **21** (1981), 547-574.

[15] Kessab A., 'Propagation des singularités Gevrey pour des opérateurs à caractéristiques involutives', Thèse, Université de Paris-Sud, Centre d'Orsay, 1984.

[16] Komatsu H., 'Ultradistributions, I. Structure theorems and a charac

terization', J. Fac. Sci. Univ. Tokyo, Sect. I A Math. **20** (1973), 25-105.

[17] Komatsu H., 'Ultradistributions, II. The kernel theorem and ultradistributions with support in a submanifold', J. Fac. Sci. Univ. Tokyo, Sect. 1 A Math. **24** (1977), 607-628.

[18] Komatsu H., 'Linear hyperbolic equations with Gevrey coefficients', J. Math. pures et appl., **59** (1980), 145-185.

[19] Kumano-go H., Pseudodifferential operators, MIT Press, 1981.

[20] Leray J., 'Equations hyperboliques non strictes: contre-exemples du type De Giorgi aux théoremes d'existence et d'unicité', Math. Annalen, **162** (1966), 228-236.

[21] Leray J. - Ohya Y., 'Systémes linéaires, hyperboliques non-stricts', Centre Belg. de Rech. Math. Deuxième Colloq. sur l'Analyse fonctionelle, Liège, 1964, 105-144.

[22] Metivier G., 'Analytic hypoellipticity for operators with multiple characteristics', Comm. in Partial Differential Equations, **6** (1981), 1-90.

[23] Mizohata S., 'Propagation de la régularité au sens de Gevrey pour les opérateurs différentiels à multiplicité constante', J. Vaillant, Sèminaire Equations aux dérivées partielles hyperboliques et holomorphes, Hermann Paris, 1984.

[24] Mizohata S., 'On perfect factorizations in Gevrey classes', personal comunication.

[25] Ohya Y., 'Le problème de Cauchy pour les équations hyperboliques à caractéristique multiple', J. Math. Soc. Japan, **16** (1964), 268-286.

[26] Rodino L. - Zanghirati L., 'Pseudo differential operators with multiple characteristics and Gevrey singularities', to appear.

[27] Sjöstrand J., 'Propagation of singularities for operators with multiple involutive characteristics', Ann. Inst. Fourier, Grenoble, **26** (1976), 141-155.

[28] Talenti G., 'Un problema di Cauchy', Ann. Scuola Norm. Sup. Pisa, Cl. Sci. (3), **18** (1964), 165-186.

[29] Taniguchi K., 'Fourier integral operators in Gevrey class on R^n and the fundamental solution for a hyperbolic operator', Publ. RIMS, Kyoto Univ., **20** (1984), 491-542.

[30] Treves F., Introduction to pseudo differential and Fourier integral operators, vol. I, Plenum Press, 1981.

[31] Volevic L.R., 'Pseudo-differential operators with holomorphic symbols and Gevrey classes', Trudy Moskov Mat. Obšč., **24** (1971), 43-68 = Trans. Moscow Math. Soc., **24** (1974), 43-72.

[32] Zanghirati L., 'Operatori pseudodifferenziali di ordine infinito e classi di Gevrey', Seminario di Analisi Matematica, Dipartimento di Matematica dell'Università di Bologna, 1984.

SINGULARITIES, SUPPORTS AND LACUNAS

G. F. D. Duff
Department of Mathematics,
University of Toronto
Toronto, Canada
M5S 1A1

PREFACE. These notes are a survey of three aspects of the modern theory of linear partial differential equations, and its generalization to the microlocal analysis of pseudodifferential operators. The first chapter is a study of the propagation of singularities of partial and pseudo differential equations, beginning with a sketch of the extensive background of pseudo differential and Fourier integral operators and wave front sets - the machinery of microlocal analysis in phase space. Selected results on equations with multiple characteristics are then discussed, in the involutive and non-involutive cases. The second chapter is a description of the work of C. Fefferman and others on the approximate simultaneous diagonalization of differential operators with variable coefficients, regarded as algebraic operators in phase space. The uncertainty principle, a title borrowed from Heisenberg's quantum mechanics, limits the precision of this process, since a function and its Fourier transform cannot both have small supports. This area of investigation, which draws upon the full resources of microlocal analysis, appears to have interesting future prospects.

The third chapter is a description of the promising state of an older and quite famous problem, formulated by Hadamard for wave equations with variable coefficients: which wave equations satisfy Huygens' principle of clean cut wave propagation, with the support equal to the singular support? This study draws upon tensor analysis and differential geometry, and relies on series expansions to produce conditions that limit the Riemannian metric. In three space dimensions there is a nontrivial plane wave metric which disproves Hadamard's conjecture that only equations trivially reducible to constant coefficients have the Huygens' property. The recent work is described, including a result that settles the problem for metrics of Petrov class N in three space dimensions.

The author's thanks are due to Prof. H. G. Garnir for the opportunity to take part in this conference. Preparation of this paper was partially supported by NSERC grant A-3004.

H. G. Garnir (ed.), Advances in Microlocal Analysis, 73–133.

CHAPTER 1. THE PROPAGATION OF SINGULARITIES.

1.1 Introduction

For linear partial differential operators, the propagation of singularities has classically been studied in connection with hyperbolic equations and the related wave propagation theories of mathematical physics. The study of operators of elliptic, hypoelliptic, parabolic and other intermediate types, and of operators of principal type, as initiated by Hörmander in his thesis, and in the study of boundary value problems (42 , 43 , 47), gave rise to algebraic complexities that motivated the study of pseudo-differential operators (47 , 96). The study of characteristics, bi-characteristics, and wave propagation phenomena for pseudo-differential operators gave way in turn to the rise of the Fourier integral in a more general context, and to the theory of Fourier integral operators and wave front sets as introduced by Hörmander and Duistermaat (19 , 44).

For differential and pseudo-differential operators with constant coefficients, the study of characteristics, bicharacteristics, wave fronts and the singularities of solutions has an essential simplifying property due to the presence of an invariance under translation. Thus the multiplicities of characteristics do not change under translation and are involved only in terms of the problems posed and the nature of the data. That is, caustics may form through the focussing and convergence of wave fronts, but volumetric refraction, internal scattering, and spontaneous wave front formation will not occur in problems with constant coefficients.

On the existential side, problems with constant coefficients have had a satisfactory and universal positive resolution, typified by the construction of a fundamental solution by Malgrange and Ehrenpreis (47, vol. 1) in very general cases.

However it has been through the much more complicated problems of operators with variable coefficients that the present contemporary studies of pseudo differential operators, Fourier integral operators, wave front sets and propagation of singularities have emerged. These are also closely linked to problems of existence, uniqueness, and regularity of solutions, so that all aspects of the classical theories of partial differential equations are combined in the most recent generation of microlocal analysis.

The initiation of this period was signalled by three events: the work of Hörmanders' thesis, the initiation of the theory of pseudo-differential operators, and the discovery by H. Lewy of a non-existence result for a simple first order system with variable (linear) coefficients. It gathered momentum with the work of Hörmander and others (47, vols. 3,4) on nonexistence and through the further development and application of pseudo differential operators. Then, returning to basic classical themes of Hamiltonian mechanics and mathematical physics, the study of characteristics and bi-characteristics, the theory of Fourier integral operators was developed by Hörmander and Duistermaat (19 , 44 , 96), and the concept of wave front set introduced. Joined by a series of developments related to several complex variables, distributions and hyperfunctions, these theories gave rise to a more penetrating outlook on the analysis of singularities of which the basic theme is microlocal analysis: the simultaneous study of position and momentum or transform variables, as in the phase space of physics. In the past decade, many diverse results have appeared and the field of study has grown to a new level of sophistication and complexity. The analytic and the C^∞ or C^k cases of higher multiplicities, and the complications attendant upon boundary phenomena have all led to significant results, though not yet, for the most part, to simplifying general principles and conclusions. Perhaps for this reason Hörmander has, on the whole, not emphasized singularities and multiple characteristics in his recent four-volume work on the Analysis of Linear Partial Differential Operators (47). These waters are certainly deep, and one can perceive the effort and skill that will be needed to keep them clear so that the passing seafarer can see to the bottom! In this lecture is given only an introduction and a brief survey of some recent results, chosen according to one viewpoint only, with the hope that others may use it to travel onwards.

1.2 Background and techniques

Let X denote a neighbourhood or open set in R^n and f a real or complex valued continuous function on X. The support $S(f)$ of f is defined, following L. Schwartz, by the closure

$$\operatorname{supp} f = \{x \in X : f(x) \neq 0\}.$$

We set

$$D_j = -i\frac{\partial}{\partial x_j}, \quad j = 1,\dots,n,$$

with $i^2 = -1$, and $D^\alpha = D_1^{\alpha_1}\cdots D_n^{\alpha_n}$, with $\alpha = (\alpha_1,\dots,\alpha_n)$ and where

$$|\alpha| = \sum_{j=1} \alpha_j ,$$

with a similar notation for $x^\alpha = x_1^{\alpha_1}\cdots x_n^{\alpha_n}$.

A distribution u on X is defined as a continuous linear form on $C_0^\infty(X)$ relative to its standard topology. $D'(X)$ denotes the space of these distributions with the natural locally convex weak topology (47, vol. 1 ; 93, chap. 1). The support $S(u)$ of $u \in D'(X)$ is the smallest closed set (in X) in the complement of which $u = 0$ (that is, such that $\langle u, \phi \rangle = 0$ for $\phi \in C_0^\infty$ which vanish in the support of u). Likewise, u is equal to a function f in x if and only if

$$\langle u, \phi \rangle = \int_X f(x)\,\phi(x)\,dx$$

for all $\phi \in C_0^\infty(X)$.

The singular support $SS(u)$ is defined as the smallest relatively closed set (in X) in the complement of which u is a C^∞ function. These familiar concepts underlie the idea of a wave front set which will be introduced below.

The Fourier transform of a function f with domain R^n is defined by

$$\hat{f}(\xi) = \int_{R^n} e^{-ix \cdot \xi} f(x)\, dx$$

where

$$x \cdot \xi = \sum_{j=1}^{n} x_j \xi_j$$

is the Euclidean scalar product (47, chap. 7 ; 95). If f belongs to various classes, such as $L^p(R^n)$, the properties of $\hat{f}(\xi)$ can be deduced in many cases, and we have the inversion formula

$$f(x) = (2\pi)^{-n} \int_{R^n} e^{ix \cdot \xi} \hat{f}(\xi)\, d\xi$$

with various interpretations holding in various cases. Observe that the Fourier transform of $D_j f(x)$ is $\xi_j \hat{f}(\xi)$.

Now let $S(R^n)$ denote the set of all $\phi \in C^\infty(R^n)$ such that

$$\sup_x | x^\beta D^\alpha \phi(x) | < \infty$$

for all multi-indices α and β ; this defines a topology on $S(R^n)$ which is known as the space of rapidly decreasing functions on R^n. The dual $S'(R^n)$, i.e. the set of linear forms on S, with the natural locally convex weak topology, is known as the space of tempered distributions on R^n. Note that the Fourier transform is a topological isomorphism of S onto S and of S' onto S', where the Fourier transform on S' is defined by the natural duality

$$\langle \hat{u}, \phi \rangle = \langle u, \hat{\phi} \rangle \quad , u \in S' , \phi \in S.$$

If u is a distribution with compact support ($u \in E'$ in Schwartz notation) then $\hat{u}(\xi) = \langle u, e^{-ix \cdot \xi} \rangle$ where, on the right, $e^{-ix \cdot \xi}$ is considered as a C^∞ test function. This symbolic product may be extended as an entire analytic function of the complex variables ξ_j, or of $\zeta_j = \xi_j + i\eta_j$, and is then known as the Fourier-Laplace transform of u. Conversely, by the Paley-Wiener Schwartz theorem, an entire analytic function $U(\zeta)$ is the Fourier-Laplace transform of a C_0^∞ function with support in the ball $|x| \leq A$ if and only if for every positive integer N there is a constant C_N such that

$$|U(\zeta)| \leq C_N (1 + |\zeta|)^{-N} e^{A|\mathrm{Im}\zeta|}.$$

Given a function or distribution $u = u(x)$ and a C_0^∞ function $\phi(x)$, their convolution is defined as

$$u * \phi(x) = \langle u(y), \phi(x-y) \rangle = \int_{R^n} u(y)\, \phi(x-y)\, dy$$
$$= \phi * u(x)$$

We observe that the Dirac distribution $\delta = \delta(x)$ is the unit under convolution and that $\widehat{f*g}(\xi) = \hat{f}(\xi)\,\hat{g}(\xi)$.

Given a linear differential operator of order m,

$$P(x, D) = \sum_{|\alpha| \leq m} a_\alpha(x) D^\alpha$$

with C^∞ coefficients $a_\alpha(x)$ defined on $X \subseteq R^n$, then a distribution E in X such that $P(x, D)E = \delta$ is called a fundamental solution in X for P. Such a fundamental solution E always exists, by the Ehrenpreis-Malgrange theorem, for operators with constant coefficients (47 , vol.2). When the coefficients are all constant, $a_\alpha(x) \equiv a_\alpha$, then a fundamental solution E gives rise by convolution to a solution $u = E * f$ of the general nonhomogeneous equation $Pu = f$. Since differentiation does not in general commute with variable functions of x, this convolution result does not hold for equations with variable coefficients. However, by such devices as freezing coefficients $a_\alpha(x)$ at x_0, approximate or asymptotic fundamental solutions can be constructed in many cases. A partial differential operator or pseudodifferential operator which has a local fundamental solution is said to be solvable, since locally defined solutions of $Pu = f$ can be constructed for smooth right hand side functions f, as we shall discuss in the next section.

1.3 Pseudo differential operators

Given a linear partial differential operator

$$P(x, D) = \sum_{|\alpha| \leq m} a_\alpha(x) D^\alpha ,$$

we may represent $Pu(x)$ by means of the Fourier transform as

$$Pu(x) = (2\pi)^{-n} \int_{R^n} e^{ix\cdot\xi} p(x, \xi)\, \hat{u}(\xi) d\xi$$

where the symbol $p(x, \xi)$ is given by

$$p(x, \xi) = \sum_{|\alpha| \leq m} a_\alpha(x) \xi^\alpha .$$

Many important properties of this expression will remain valid if $p(x, \xi)$ is not only a polynomial in ξ but also a more general function of x and ξ that simply satisfies certain kinds of estimates. Thus, following Hörmander (44 , 47) we define the symbol class $S^m_{\rho,\delta}(X)$ as the set of all $p \in C^\infty(X \times R^n)$ such that for every compact set $K \subset X$ and all multi-indices α, β we have a constant $C_{\alpha,\beta,K}$ such that

$$|D_x^\alpha D_\xi^\beta p(x,\xi)| \leq C_{\alpha,\beta,K}(1+|\xi|)^{m-\rho|\beta|+\delta|\alpha|}$$

The union and intersection (for all m) of $S^m_{\rho,\delta}$ will be denoted by $S^\infty_{\rho,\delta}$ and $S^{-\infty}_{\rho,\delta}$ respectively, while $S^m_{1,0}(X)$ is denoted by $S^m(X)$, or S^m if $S = R^m$. When $\delta < 1$ the pseudo differential operator above gives a continuous linear mapping of $C_0^\infty(X)$ into $C^\infty(X)$ which can be extended to a continuous mapping of $E'(X)$ into $D'(X)$. The distribution kernel of $P(x,D)$ is a C^∞ function outside the diagonal in $X \times X$ and so $P(x,D)$ has the pseudolocal property that $SSP(x,D)u \subset SS(u)$.

For $u \in E'(X)$ we many define $P(x,D)u$ by

$$< P(x,D)u\, , v > = (2\pi)^{-n} \int_{R^n} Pv(\xi)\, \hat{u}\,(\xi)\, d\xi$$

where

$$Pv(\xi) = \int_{R^n} v(x)\, p(x\, , \xi\,)\, e^{ix\cdot\xi}\, dx\, , \qquad v \in C_0^\infty(X).$$

The Schwartz kernel $K_{x,y}$ of $P(x,D)$ is given for $x \neq y$ by

$$(x-y)^\alpha K_{x,y} = (2\pi)^{-n} \int_{R^n} e^{i(x-y)\cdot\xi}(-D_\xi)^\alpha p(x\, , \xi)\, d\xi\, ,$$

for α an arbitrary sufficiently large multi-index such that the integral converges absolutely. P is said to be "properly supported" in X if for $g \in C_0^\infty(X)$ both $g(x)K_{x,y}$ and $K_{x,y}g(y)$ have compact support in $X \times X$, in which case P maps each of $C_0^\infty(X)$, $E'(X)$ and $D'(X)$ into themselves.

If $K(x,y) = K_{x,y}$ is C^∞ then P can be defined by a symbol in $S^{-\infty}(X)$, namely

$$p(x\, , \xi) = (2\pi)^n e^{-i(x-y)\cdot\xi} K(x\, , y) X(\xi)$$

where $X \in C_0^\infty(R^n)$ and $\int_{R^n} X(\xi)\, d\xi = 1$. For then

$$Pu(x) \equiv (2\pi)^{-n} \int_{R^n} \int_{R^n} e^{i(x-y)\cdot\xi} p(x\, , \xi) u(y)\, dy d\xi$$

$$= \int_{R^n} K(x\, ,y)u(y)\, dy \int_{R^n} X(\xi)\, d\xi$$

$$= \int_{R^n} K(x\, ,y)u(y)\, dy$$

so that Pu has the C^∞ kernel $K(x\, , y)$. A given pseudo differential operator symbol $p(x\, , \xi) \in S^{n_0}(X)$ can often be expanded in a descending series corresponding to descending powers of ξ:

$$p \sim \sum_{j=0}^{\infty} p_{m_j}$$

where, commonly, $p_{m_j} = p_{m_0 - j}(x,\xi)$ is a C^∞ function on $X \times R^n\backslash 0$ which is homogeneous in ξ of degree $m - j$. This corresponds to the classical case of differential operators, at least for terms of positive degree. In studies of singularities one usually neglects any C^∞ part of a pseudo differential operator and writes $p_1 \sim p_2$ to signify $p_1 - p_2 \in S^{-\infty}(X)$. We shall adopt this convention henceforth.

A pseudo differential operator may be defined by an amplitude function $a(x, y, \xi)$ as follows: Let

$$Au(x) = (2\pi)^{-n} \int_{R^n} \int_{R^n} e^{i(x-y)\cdot\xi} a(x, y, \xi) u(y) dy d\xi \ , \ u \in C_0^\infty(X)$$

where the integration over y is performed first. If $a(x, y, \xi)$ is properly supported then A is a pseudo differential operator with symbol

$$p(x, \xi) = e^{-ix\cdot\xi} A(e^{ix\cdot\xi})$$

and there is an asymptotic expansion

$$p(x, \xi) \sim \sum_{|\alpha| \geq 0} \frac{1}{\alpha!} \partial_\xi^\alpha D_y^\alpha a(x, y, \xi) \Big|_{y=x}$$

where $\partial_\xi \equiv \frac{\partial}{\partial \xi}$

In particular, $S^m(X \times R^n)$ shall denote the space of C^∞ functions $a(x, \xi)$ in $X \times R^n$ such that for every differential operator P on X with C^∞ coefficients, there is a constant $C(P, \alpha, K) > 0$ such that

$$|P\partial_\xi^\alpha a(x, \xi)| \leq C(1 + |\xi|)^{m - |\alpha|}$$

where α is an arbitrary multiindex, $x \in K$ and $\xi \in R^n$.

A pseudo-differential equation $Pu = f$ is said to be locally solvable in X if every point $x_0 \in X$ has two open neighbourhoods $V \subset U$ such that for every $f \in C_0^\infty(V)$ there is a distribution u in X with $S(u) \subset U$ which satisfies $Pu = f$ in V . The equation is locally solvable at a point $x_0 \in X$ if it is solvable in some neighbourhood of x_0. Clearly partial differential operators with constant coefficients, which possess elementary solutions, are locally solvable.

As in the theory of partial differential equations, we must define characteristics as follows:

$$\text{Char}\, P = \{(x, \xi) \in T^*(X)\backslash 0 : p(x, \xi) = 0\}.$$

This is a closed conic subset of $T^*(X)\backslash 0$, i.e. $(x,\xi) \in \text{Char}P \Rightarrow (x, \lambda\xi) \in \text{Char}\, P$ for $\lambda > 0$.

If Char P is an empty set, then P is called elliptic.

A pseudo differential operator Q is called a right parametrix of P if $PQ = I + R_0$ where I is the identity and R_0 denotes an operator with C^∞ kernel. Thus $PQ \sim I$. Similarly, Q' is a left parametrix if $QP \sim I$. Note that if P has a right parametrix Q then P is locally solvable in C^∞ . This can be achieved in sufficiently small neighbourhoods in close analogy to the classical use of a parametrix for elliptic partial differential operators. Indeed it is not difficult to construct a right parametrix for an elliptic operator P ; for details see (96 , vol. 2, p. 332).

Elliptic operators, for which $\operatorname{char} P$ is empty, are locally solvable. Operators of principal type, for which $dp(x, \xi) \neq 0$ on $\operatorname{char} P$, are also locally solvable, at least if $p(x, \xi)$ is real; for the complex valued case see Hörmander (47, vol. 4 , Chap. 26). The characteristics of a principal type operator are necessarily <u>simple</u>.

If multiple characteristics are present, the principal symbol $p(x, \xi)$ may not alone determine the solvability of $p(x, \xi)$ and it is necessary to introduce the (invariant) subprincipal symbol

$$p - \frac{1}{2i} \sum_j \frac{\partial^2 p}{\partial x_j \partial \xi_j} .$$

If P is a classical pseudo differential operator with symbol

$$p \sim \sum_{j \geq 0} p_{m-j}$$

then the subprincipal symbol is given by

$$p_{m-1} - \frac{1}{2i} \sum_{j=1}^{n} \frac{\partial^2 p_m}{\partial x_j \partial \xi_j} ,$$

which is uniquely and invariantly determined modulo S^{m-2}.

1.4 The Wave Front Set

To study singularities from the phase space or microlocal viewpoint using conjugate variables $(x, \xi) \in X \times R^n/0$, Hörmander's concept of the wave front set will be presented (19 , 44 , 47 , 96). Note that iff $x_0 \notin SS(u)$, there is a $\phi \in C_0^\infty(X)$ with $\phi = 1$ in a neighbourhood of x_0, and there are constants C_N, $N = 1,2,\ldots$, such that $|\widehat{\phi u}(\xi)| \leq C_N(1 + |\xi|)^{-N}$. We leave the routine proof to the reader. Observe that multiplication of the Fourier transform $\widehat{\phi u}(\xi)$ by any monomial ξ^α does not change the form of these conditions, so that derivatives of ϕu have the same property.

The wave front set $WF(u)$ is a set of pairs (x_0, ξ_0) in $T^*X\backslash 0$ which is the complement of the set (x_0, ξ_0) for which there exists a $\phi \in C_0^\infty(X)$ with $\phi = 1$ in a neighbourhood of x_0, and a conic neighbourhood Γ_{ξ_0} of ξ_0, together with constants C_N such that $|\widehat{\phi u}(\xi)| \leq C_N(1 + |\xi|)^{-N}$, $N = 1,2,\ldots$ for $\xi \in \Gamma_{\xi_0}$.

Thus an element of the wave front set corresponds to a singular element of u having direction, or orientation, as well as location.

Hörmander also showed that (44)

$$WF(u) = \cap \{(x, \xi) \in T^*(X)\backslash 0 ; a(x, \xi) = 0\}$$

where the intersection is taken over all properly supported pseudo differential operators A of order zero, with homogeneous principal symbol $a(x, \xi)$, such that $Au \in C^\infty(X)$. Hence $WF(u)$ is a closed conic subset of $T^*(X)\backslash 0$. The projection $(x, \xi) \to x$ of $WF(u)$ is precisely the singular support $SS(u)$, and it can be shown that

$$\begin{aligned} WF(u_1 \otimes u_2) \subset\ & WF(u_1) \times WF(u_2) \\ & \cup \{WF(u_1) \times (\operatorname{supp} u_2 \times (0)\} \\ & \cup \{(\operatorname{supp} u_1 \times (0) \times WF(u_2)\}. \end{aligned}$$

If u_1, u_2 are distributions such that the set $WF(u_1) \oplus WF(u_2) = \{(x, \xi_1 + \xi_2) ; (x, \xi_1) \in WF(u_1) ; (x, \xi_2) \in WF(u_2)\} \subset T^*X\backslash 0$ then (44, p. 126) the product $u_1 u_2$ is defined and $WF(u_1 u_2) \subset WF(u_1) \cup WF(u_2) \cup [WF(u_1) \oplus WF(u_2)]$.

The Hamiltonian vector field of a pseudo-differential operator $p(x, \xi)$ is defined as

$$H_p = \sum_{j=1}^{n} \frac{\partial p}{\partial \xi_j} \frac{\partial}{\partial x_j} - \frac{\partial p}{\partial x_j} \frac{\partial}{\partial \xi_j}$$

while the bicharacteristic flow defined by H_p is the set of solution curves (or bicharacteristic strips) for the system

$$\frac{dx_j}{dt} = \frac{\partial p}{\partial \xi_j} ; \frac{d\xi_j}{dt} = -\frac{\partial p}{\partial x_j}, \quad j = 1, .., n.$$

Since $\frac{dp}{dt} = H_p(p) \equiv 0$ it follows that p is constant under this bicharacteristic flow. If $p = 0$ we call the flow <u>null</u>, corresponding to the classical null bicharacteristics.

For a pseudo differential operator p, multiplication by an elliptic operator q (such that $q(x, \xi) \neq 0$ in $T^*(X)\backslash 0$) does not alter the null bicharacteristics.

In the classical theory the singularities (but not necessarily the function values as we shall see in chapter 3 later) propagate along the bicharacteristics. In the present, microlocal setting, we have the property (45, p. 19)

$$WF(u) \subset WF(Pu) \cup \operatorname{char} P .$$

Note that there are some symbol classes $S^m_{\rho,\delta}(X)$ for which this general result does not hold (75).

Hörmander and Duistermaat (19) established the following general theorem on the propagation of singularities: Assume P is a properly supported pseudo differential operator with homogeneous real principal symbol p. If $u \in D'(X)$ and $Pu = f$ then $WF(u)\backslash WF(f) \subset p^{-1}(0)$ and is invariant under the Hamiltonian flow defined by H_p.

Existence theorems in Sobolev spaces were also derived by these authors for P with real homogeneous principal symbols, and more generally for operators of real principal type (47, vol. 4 , ch. 26 ;

93, p. 132)

One such result, is: Let $P \in S^m$ have real principal symbol, and suppose that no bicharacteristic strip lies over a compact set $K \subset X$. Then $u \in E'(X)$, $Pu \in C^\infty(X)$ imply $u \in C^\infty(X)$.

Recent work has concentrated on local solvability and propagation of singularities for operators with multiple characteristics. At such multiple curves or points the Hamiltonian flow H_p becomes singular, and we may expect the spreading of singularities such as occurs in conical refraction, at multiple roots of wave equations, and in higher order phenomena going beyond the hyperbolic case. Before describing results of this kind, we discuss reduction to canonical or standard forms of differential and pseudo differential operators by means of canonical transformations and Fourier integral operators.

1.5 Canonical forms and Fourier Integral Operators.

In classical mechanics and partial differential equations, canonical transformations play a double role. They can represent the flow of time and its effect on the state of an evolution system, or they can be used to transform to an apparently different, and hopefully simpler or standard system. These twin themes will recur throughout in the theory of pseudo differential equations and systems, and in the related theory of singularities and wave propagation. The freedom to multiply by an elliptic pseudo differential operator without changing the set of characteristics is of great advantage in singularity theory, for it permits the isolation and explicit display of the characteristic factors, often as differential operators. Hence much of the wave propagation theory for hyperbolic differential equations can be taken over and used, with minor adaptations, in the more general pseudo differential theory.

Motivated by this prospect, we define a Fourier Integral Operator as (19 , 96)

$$Ff(x) = \int e^{iS(x,\xi)} a(x,\xi)\,\hat{f}(\xi)\,d\xi$$

$$= \int\int e^{i\phi(x,y,\xi)} a(x,y,\xi)\,f(y)\,dy\,d\xi$$

where $\phi(x,y,\xi) = S(x,\xi) - y\cdot\xi$. It is believed to have been first noted by Egorov (21) that if P and Q are pseudo differential operators related by $PA = AQ$, then the principal symbols p and q are related by the canonical transformation with generating function $S(x,\xi)$.

The generalization present here, as compared to pseudo differential operators, lies in the phase function $\phi(x,y,\xi)$ in the exponent. It is customary when working locally to assume that

a) $\phi(x,y,\xi)$ is a real valued C^∞ function in $X \times X \times R^n \backslash 0$.

b) ϕ is positive homogeneous of degree one with respect to ξ

c) the two differentials, $d_{x,\theta}\phi$ and $d_{y,\theta}\phi$ do not vanish anywhere in $X \times X \times R^n \backslash 0$.

If $a \in S^m$, then this Fourier integral operator can be shown to define a continuous linear map from $C_0^\infty(X)$ to $C^\infty(X)$ which can (by duality) be extended to a continuous map from $E'(X)$ to $D'(X)$. (19).

With the aid of Fourier integral operators and canonical transformations, the study of pseudo differential equations can be systematically reduced to the study of certain standard types, based on the algebraic geometry of the characteristics. For example, if P is elliptic, it is equivalent under conjugation with a Fourier integral operator to a first order operator ξ_n plus a C^∞ operator of class $S^{-\infty}$ (45, p. 31). If this latter operator vanishes (as in the "analytic" case) or can be handled by solution of an integral equation with a smooth kernel, then an existence theorem, regularity properties, and even in explicit cases the construction of a solution can be made to follow. If P is a properly supported pseudo differential operator with real principal symbol $p(x, \xi)$, then analogous results with due allowance for the presence of characteristics will apply. For example, Hörmander has shown in this case that if $p \in S^m$, $u \in D'(X)$ and γ is an interval on a bicharacteristic strip where $Pu \in H_s$, then $u \in H_{s+m-1}$ on γ if this is true at any one point on γ. (44, 45).

For reasons of space we shall not enter here into the detailed and extensive study of canonical transformations and algebraic preparation theorems necessary for the completion of such a programme. Detailed accounts are given in (47, 93, 96). We shall rather proceed directly to the consideration of specific equations and systems which illustrate the extensive results obtained in these directions in recent years.

1.6 A Geometrical Optics Construction.

Following Taylor (93, p. 147) we study the prototype hyperbolic equation

$$\frac{\partial u}{\partial t} = i\,\lambda\,(t, x, D)\,u$$

where $u \in D'(X)$,

$$\lambda(t, x, \xi) = \lambda_1(t, x, \xi) + \lambda_0(t, x, \xi) + \dots$$

is a pseudo differential operator of order 1 with the above expansion in descending degrees of $|\xi|$, where $\lambda_j (j = 1, 0, -1, -2, \dots)$ is homogeneous of degree j in $|\xi|$. We postulate a solution of the form

$$u(t, x) = \int a(t, x, \xi)\, e^{i\phi(t, x, \xi)}\, \hat{u}_0(\xi)\, d\xi$$

where $\phi(t, x, \xi)$ is a real phase function, $a \in S^m_{\rho,\delta}$ with $\rho > 0$, $\delta < 1$ and a, ϕ are to be determined. We set $u(0, x) = u_0(x)$ with $u_0(x) \in E'(X)$. Where necessary, an integration by parts procedure

(96 , vol. 1 , p. 14 , vol. 2 , p. 326) can be used to give a definite, convergent form equivalent to the above Fourier integral operator. We find

$$\lambda(t , x , D)u = \int a(t , x , \xi)\, \lambda (t , x , \nabla_x\phi)\, e^{i\phi(t , x , \xi)}\hat{u}_0(\xi)\, d\xi ,$$

so that

$$\left[\frac{\partial}{\partial t} - i\lambda(t , x , D)\right] u = \int c(t , x , \xi)\, e^{i\phi(t , x , \xi)}\hat{u}_0(\xi)\, d\xi$$

where $c(t , x , \xi) = i\phi_t a + a_t - ia\lambda(t , x , \nabla\phi)$.

Expanding in a descending series, we write $a(t,x,\xi) \sim \Sigma_{j\leq 1} a_j(t,x,\xi)$ where a_j is homogeneous of degree j in $|\xi|$, and similarly for other terms. Equating to zero each successive term in the expression for $c(t , x , \xi)$ we first obtain the eikonal condition $c_1(t , x , \xi) = ia_0(\phi_t - \lambda_1(t , x , \nabla\phi) = 0$. This is a first order nonlinear partial differential equation which describes the characteristic surfaces; ϕ is obtained by solving this equation, at least for small t , with given values for $\phi(0 , x , \xi)$ which we suppose is homogeneous in ξ of first degree and has non-vanishing gradient with respect to x on the conic support of a . The simplest and most natural choice is $\phi(0 , x , \xi) = x\cdot\xi$.

The next term is

$$c_0(t , x , \xi) = Xa_0 - i\lambda a_0 - \sum_{|\alpha|=2} \frac{1}{\alpha!} \lambda_1^{(\alpha)} \phi_{(\alpha)} a_0$$

where

$$X \equiv \frac{\partial}{\partial t} - \sum_{i=1}^{n} \frac{\partial\lambda_1}{\partial\xi_i} \frac{\partial}{\partial x_i}$$

The vanishing of c_0 yields the <u>transport equation</u> for a_0 , in effect a kind of ordinary pseudo differential equation along bicharacteristic strips. We can solve these, at least locally, with $a_0(0 , x , \xi) = 1$.

Further terms in the series can be obtained in the same way:

$$c_j(t , x , \xi) = Xa_j - i \left(\lambda_0 + \sum_{|\alpha|=2} \frac{1}{\alpha!} \lambda_1^{(\alpha)} \phi_{(\alpha)}\right)a_j - d_j = 0 ,$$

where $d_j(t , x , \xi)$ can be expressed in terms of $\phi , a_0 , a_1, \ldots a_{j-1}$ and is now regarded as known at the jth stage. We solve the transport equation of order j with $a_j(0 , x , \xi) = 0$ for $j < 0$.

This procedure yields a formal local solution. If we are working in the analytic framework, then a convergence proof becomes necessary. Such proofs have been given by Sjöstrand (90) . If we are interested in the C^∞ scenario, then the above series is regarded as asymptotic and there is a permanent remainder term which belongs to $S^{-\infty}$. This can be obtained through the solution of a Volterra type integral equation with independent variable t. In fact this method could be used in a C^k framework where the expansion is broken off after a finite number of terms.

1.7 Operators with multiple characteristics.

Let us study second order operators of the form $P = X_0 + \sum_{j=1}^{k} X_j^2$ where $X_0, X_1, \ldots, X_k$, $k \le n$ are real vector fields on $X \subseteq R^n$. In 1967 Hörmander showed (43) that if the Lie algebra generated by $(X_0, X_1, \ldots, X_k)$ spans all vector fields over X, then P is hypoelliptic, and there is an $\epsilon > 0$ such that $Pu \in H^s_{loc}(X) \Rightarrow u \in H^{s+\epsilon}_{loc}(X)$. Thus u is as smooth as Pu, and there is said to be a "loss of $2-\epsilon$ derivatives" (as compared to the corresponding property for elliptic operators).

Suppose $P \in S^2$ with

$$p(x, \xi) = p_2(x, \xi) + p_1(x, \xi) + \ldots$$

Suppose $p_2(x, \xi) \ge 0$ and vanishes to second order on a smooth conic manifold $\Sigma \subset T^*(X)$ of codimension k. Then, by the Morse lemma, one has

$$p_2(x, \xi) = \sum_{j=1}^{k} a_j(x, \xi)^2$$

with each $a_j(x, \xi)$ homogeneous of degree 1 in ξ, and with the gradients of a_j linearly independent on Σ. Then

$$p(x, \xi) = \sum_{j=1}^{k} a_j(x, \xi)^2 + \sigma(x, \xi)$$

where, on Σ, we have

$$\sigma(x, \xi) = p_1(x, \xi) - \frac{1}{i} \sum_{|\alpha|=1} a_j^{(\alpha)}(x, \xi)\, a_{j(\alpha)}^{(x, \xi)} \mod S_0$$

$$= p_1(x, \xi) + \frac{i}{2} \sum_{\nu} \frac{\partial^2}{\partial x_\nu \partial \xi_\nu} \Big(\sum_j a_j(x, \xi)^2 \Big)$$

$$= p_1(x, \xi) + \frac{i}{2} \sum_{\nu} \frac{\partial^2}{\partial x_\nu \partial \xi_\nu}\, p_2(x, \xi)$$

The symbol $\sigma(x, \xi)$ is called the subprincipal symbol of P, and it plays the role of a second order characteristic form, being significant in cases of higher multiplicity. On the double characteristics of $p(x, \xi)$, the value of σ is invariant under canonical transformations. In the case above P will be hypoelliptic provided that $\operatorname{Re} \sigma \ge 0$ on Σ (93, p. 374).

Now a vector subspace of a symplectic space which contains its orthogonal complement with respect to the symplectic form is called involutive. This is not the case above when single Poisson brackets generate all of S! Hence if $T_{(x_0\xi_0)}\Sigma$ is not involutive for $(x_0, \xi_0) \in \Sigma$ and $p_2(x, \xi) \geq 0$ vanishes to second order on Σ then P is hypoelliptic with loss of 1 derivative.

In a forthcoming study by Beals and Greiner (6) of "model" equations of this form with linear coefficients, appropriate to the Heisenberg group, it is shown that the operator P is invertible provided the coefficient of X_0 does not belong to a certain singular set, which in various cases may be a set of eigenvalues, or the complement (in R) of an interval. Similar results have been obtained for first order systems with linear coefficients, of the Lewy type.

For certain operators with double characteristics the symbol $\sigma(x, \xi)$ determines a class of microlocally equivalent operators. Taylor (93) shows by a detailed discussion that if $P, P^2 \in S^m$, both have principal part $p(x, \xi) \geq 0$ which vanishes to second order on a symplectic variety Σ_2, then P and P^2 are microlocally equivalent mod S^{m-2} if and only if $\sigma(P) = \bar{\sigma}(\tilde{P})$ on Σ_2. If also X is three-dimensional, then for any $(x_0, \xi_0) \in \Sigma_2$ there is a conic neighbourhood Γ an elliptic operator B, and a Fourier integral operator J such that

$$BJPJ^{-1} = -\frac{1}{2}(Z_1\bar{Z}_1 + \bar{Z}_1 Z_1) + i\alpha T \bmod S_0 , \quad Z_1 = \frac{\partial}{\partial z_1} + i\bar{z}_1\frac{\partial}{\partial t}, \quad T = \frac{\partial}{\partial t}$$

Consider next the case of an operator $p = ab$, where a and b vanish to first order on Σ_1 and Σ_2 respectively, and we assume $\{a, b\} = 0$ on $\Sigma_1 \cap \Sigma_2$, i.e. the involutive case. Assume also that da, db and $\xi_j dx_j$ are linearly independent. Then after multiplication by an elliptic pseudo differential operator and conjugation by a Fourier integral operator, P can be transformed to

$$D_{x_1}D_{x_2} + A(x, D_x)$$

where $A(x, D_x) \in S'$. With $v_1 = u$, $v_2 = D_1 u$ we obtain a system

$$D_1 v_1 = v_2 \; ; \; D_2 v_2 = -Av_1 + f \; .$$

To construct a parametrix following (93) we assume that $\sigma(P) = 0$ (the "Levi condition") and hence $-A(x, \xi) = \xi_2 B_0(x, \xi')$. Hence $B_0 D_2 v_1 + D_2 v_2 = f$ and by a change of basis

$$v = \begin{bmatrix} 1 & 0 \\ B_0 & 1 \end{bmatrix}^{-1} w$$

we find the system now has the form

$$\begin{bmatrix} D_1 & 0 \\ 0 & D_2 \end{bmatrix} w = B_2 W + g \; .$$

By setting $x_1 = t$, $x_2 = t + y$, this changes to

$$\begin{bmatrix} D_t & 0 \\ 0 & D_t + D_{y_1} \end{bmatrix} w = B_2 W + g$$

where $B_2 \in S^0$ in a conic neighbourhood of the set Σ_2 of double characteristics $\{\tau = \tau + y_1 = 0\}$. With a series expansion similar to that used for the first order hyperbolic equation above, and a suitable elliptic operator inversion one can make a further reduction to a form where B_2 involves only D_y and D_t. Now B_2 is still, in general, a pseudo differential operator, but the behaviour in respect of the two characteristic surfaces has been isolated and displayed explicitly. If B_2 is zero, constant, or independent of D_y then the explicit solution with suitable characteristic initial conditions of the Goursat problem (25) can be found by elementary or classical methods.

In more general cases of this type one would expect a reduction to a characteristic system of the general type studied long ago by Riquier (82) and Janet (50), and more recently by Spencer (91) and Pommaret (78); see also (94). However the complications attendent upon the nature of pseudodifferential operators appear to have so far stood in the way of very general results of this kind.

We remark that R. Lascar (56, Chapter 1) has given a result for a single pseudodifferential operator P of an essentially hyperbolic type with characteristics locally of higher multiplicity: Let $D_0 = \Gamma_0 \cap L_0^+$ and $C_0 = \Gamma_0 \cap L_0$, where Γ_0 is a certain conic neighbourhood, L_0 a spacelike hypersurface transverse to the characteristic Hamiltonian flows, and L_0^+ a half-space next this surface defined with a certain orientation. Then if $Pu = f$, if $WF(Pu)$ does not meet D_0 and $WF(u)$ does not meet C_0, then $WF(u)$ does not meet D_0.

In a particular case of hyperbolic type, (56, Chapter 2)

$$P(x, \xi) = -(\xi_0 - \Lambda_1(x, \xi'))(\xi_0 - \Lambda_2(x, \xi')) + M(x, \xi')$$

where Λ_i and M are C^∞ classical symbols, R. Lascar constructs a parametrix in the case of "glancing" characteristic surfaces, and gives an estimate for the wave front set. He requires $\sigma(P) = 0$ on the characteristic set. The detailed construction and discussion occupies a carefully prepared typescript of some seventy pages.

In another vein, a study of coincident characteristic surfaces has been made by Melrose (65, 67) as part of an investigation of general boundary value problems for second order linear partial differential equations. This in turn is a stage in his general program of study of boundary value problems with emphasis on singular cases of diffraction problems, glancing and gliding rays, and their corresponding involutory differential geometry (65, 66, 67). See also (52).

1.8 Characteristics with non-involutory intersections.

Considering to start the simplest case of this kind, let P have principal symbol of the form

$$p_m(x, \xi) = ab$$

where the characteristic surfaces $a = 0$ and $b = 0$ intersect transversally with $\{a, b\} \neq 0$ on their intersection Σ. We may suppose da, db and $\Sigma_j \xi_j dx_j$ are independent on Σ. By elementary transformations involving elliptic operators we can assume $m = 1$, $a \in S^0$, $b \in S^1$, and $\{a, b\} = 1$ on a conic neighbourhood of a given point (x_0, ξ_0) of Σ. Hence, by conjugation with a suitable Fourier integral operator, we can suppose

$$P = y_1 D_{y_1} + A(x, D_x)$$

where the distinguished x-coordinate has been denoted y, and $A \in S^0$ is independent of y and D_y. Parametrices for this operator have been constructed by Ivrii (49) Hanges (38, 39, 40), Melrose (67) and Taylor (93): Define an operator-valued distribution with values in $S^0_{1,0}$ by

$$F(y, y') = H(y' - y)(y' + i0)^{-1-A}(y + i0)^A$$

then

$$(y\frac{\partial}{\partial y} - A)F = \delta(y - y')I .$$

Taylor (93) obtains four equivalent versions depending on $\pm$ signs, of the following result. Let $L_{ij}(x_0, \xi_0) = \{(y, x_0, \zeta, \xi_0) \in T^*(R^n)\backslash 0 : y\zeta = 0$ and either $(-1)^i y > 0$ or $(-1)^j \zeta > 0\}$. Suppose $u \in D'(R^n)$ with $(0, x_0, 0, \xi_0) \notin WF((yD_y - B)u)$, and $L_{ij}(x_0, \xi_0) \cap WF(u)$ is empty. Then $(0, x_0, 0, \xi_0) \notin WF(u)$. In effect, the Green's function of this operator is $G(y, y') = |y|^A |y'|^{-1-A} H(y - y')$, at least for the interval $y > 0$.

Properties of the principal part $A_0(x, \xi)$ of $A(x, D)$ play a part in the next result we shall describe in which integer values must be avoided. Thus let $A_0(x, \xi) \notin \{0, 1, 2, \ldots\}$. Then there is a right parametrix K for $y\frac{\partial}{\partial y} - A$, satisfying $(y\frac{\partial}{\partial y} - A)K = \delta(y - y')I + R$, where $Ru \in C^\infty$ when $\{\xi = 0\} \notin WF(u)$.

To construct K, note that $A - jI$ has principal part $A_0(x,\xi) - j \neq 0$ $(\xi \neq 0)$ and so is elliptic. Hence there is a parametrix $S_j \in S^0$ such that $S_j(A - j) = I + R_j$ with $R_j \in S^{-\infty}$. We repeat this construction for $j = 0, 1, 2, \ldots$ until an integer k is reached with $A_0(x, \xi) - k < 0$ and we recall there is a Green's function G_k with $(y\frac{\partial}{\partial y} - (A-k))G_k = \delta(y-y')I$. Now for $j = 1, \ldots k-1$ let $M_k: C^\infty(R, D'(M)) \to C^\infty(R, D'(M))$ be defined by

$$u = \sum_{j=0}^{k-1} \frac{\partial^j}{\partial y^j} u(x, 0) + y^k M_k u.$$

We can now set

$$Hu = \sum_{j=1}^{k-1} \frac{y^j}{j!} S_j \frac{\partial^j u}{\partial y^j}(x, 0) + y^k G_k M_k u.$$

and it can be verified that

$$(y\frac{\partial}{\partial y} - A)\,H = I + \sum_{j=1}^{k-1} \frac{y^j}{j!} R_j \,[D_{y^j}]_{y=0}.$$

This right hand side series is of the required $S^{-\infty}$ type. Consideration of the adjoint operator then yields the following result: Let $(y\frac{\partial}{\partial y} - B)u = f$ where $B_0(x, \xi)$ avoids the set $Z = \{-1, -2, \ldots\}$. Suppose $(0, x_0, 0, \xi_0) \notin WF(f)$ and that

$$\{(0, x_0, \eta, \xi_0) : \eta \neq 0\} \cap WF(u) = \emptyset .$$

Then $(0, x_0, 0, \xi_0) \notin WF(u)$. The avoiding condition can be expressed in invariant terms related to the original operator P with $p_m(x, \xi) = a\,b$: Then for $j = 1, 2$, if $(x_0, \xi_0) \notin WF(Pu)$ and

$$\frac{(-1)^{j+1} i\sigma(P)}{\{a, b\}} + \frac{1}{2} \notin \{0, -1, -2, \ldots\}$$

and $(\gamma_{j1} \cap \gamma_{j2}) \cap WF(u) = \emptyset$, where γ_{j1} and γ_{j2} are certain half rays, then $(x_0, \xi_0) \notin WF(u)$. Results of this kind have been discussed by Hanges (38,39), Ivrii (49), Alinhac (1, 2, 3) and Taylor (93).

An investigation of the class of operators $R = PQ^2 + AQ + B$, which is in general non-involutive, with P of order 0, Q of order 1, A and B of order 0, has been made by Nagaraj (72) who shows that this operator is microlocally equivalent (under a suitable canonical transformation) to the form

$$R' = tD_t^2 + A_0(x, t, D_x, D_t)D_t + B_0(x, t, D_x, D_t)$$

where λ and β have order 0; i.e. $\in S^0$. To simplify matters we restrict attention to the case $\lambda, \beta \in C$ and consider first the case $\beta = 1$, $\lambda \neq 1$. Then an elementary solution of $R'u = f$ is given by $K_{(\lambda,0)}(x, t, y, s) = [H(t-s)F^+_{(\lambda,0)}(t, s)] \otimes \delta(x-y)$ where

$$F^+_{(\lambda,0)}(t, s) = \frac{1}{1-\lambda}\,[(t+i0)^{1-\lambda} \otimes (s+i0)^{\lambda-1} - 1]$$

It is shown that if $(x_0, 0, \xi_0, 0) \notin WF(M_{(\lambda,0)}u)$, for real $\lambda \neq 1$, and neither of the two null bicharacteristics of t and τ issuing from the triple characteristic $(x_0, 0, \xi_0, 0)$ meets $WF(u)$, then $(x_0, 0, \xi_0, 0) \notin WF(u)$.

If $\lambda = 1$ the parametrix involves logarithms but a similar result holds.

If $\beta \neq 0$ one makes a substitution $z^2 = 4\beta t$ and a form of Bessel's equation is obtained, $(z^2D_z^2 + zD_z + z^2 - k^2)u = f$, $k = 1 - \lambda$. The analysis breaks into two slightly different cases, according as to whether λ is an integer or not. If λ is a (numerical) integer, the result may be stated: Let $u \in D'(R^{\lambda+1})$ with $(x_0, 0, \xi_0, 0) \notin WF(Ru)$ where $\lambda \in \{\pm\eta\}$. Then if $\{(x_0, 0, \xi_0, \tau) : \tau \neq 0\} \cup \{(x_0, t, \xi_0, 0)\} \cup WF(u) = \emptyset$, then $(x_0, 0, \xi_0, 0) \notin WF(u)$. However, the extension of this result to pseudo differential operators λ, β is not clear.

For the general complex non-involutive case Duistermaat and Sjöstrand (20) have given an existence proof and construction of a fundamental solution of $Pu = f$, where P is a properly supported pseudodifferential operator of type (1,0) on a manifold X with principal symbol p positively homogeneous of degree m. Thus the non-involutory condition $\{p,\bar{p}\} \neq 0$ holds, and in this case, by an application of the Darboux theorem (45,65), a reduction to the operator $D_n + ix_nD_{n-1}$ is possible; this was found in the analytic case by Sato, Kashiwara and Kawai (51,85) who also treated the higher order cases $D_n + ix_n^k D_{n-1}$ corresponding to higher order characteristic zeros of $\operatorname{Im} p$ when p is suitably standardized; see also (45,104) for the C^∞ case. The kernel and cokernel of P are treated in these papers, and solvability is resolved in a very general framework. A related result by Godin (26) is that if a pseudodifferential operator with double characteristics does not propagate singularities, then its adjoint is not microlocally solvable. The connection between propagation of singularities along characteristics, and solvability can be expressed in various ways and, for example, Rodino and Parenti (75,83) have developed a theory based on a concept of ψ-filters and ψ-solvability.

1.9 Characteristics of higher order, and variable order.

When a zero of higher order is traversed along a characteristic, then phenomena first encountered in conical refraction can appear. The locus of singularities is in general widened by new convex closure components, tangential in a dual sense of algebraic geometry. The reduction of algebraic surface singularities by perturbations also plays a role in the hyperbolic case. We draw attention here to the calculus of localizations of Atiyah, Gårding and Bott (5), and the examples and particular cases shown by Ludwig and Granoff (58). Generalized conical refraction has recently been studied by B. Lascar (54), B. Lascar and Sjöstrand (55), Laubin (58), and Melrose and Uhlmann (70).

With the advent of Fourier integral operators and simplifying canonical transformations, differential operators appeared to regain much of their earlier leading role. But the complexities of micro-local analysis, together with all the earlier problems of linear partial differential operators, seem to have brought about a pause on the threshold of even more general linear pseudodifferential system theory for higher order characteristics. Thus the present occasion is timely to review the search for new and hopefully both simple and general perspectives regarding characteristics and the propagation of singularities.

CHAPTER 2 THE UNCERTAINTY PRINCIPLE

2.1 Background

The study of linear partial differential equations with constant coefficients owes much of its success and generality to the technique of Fourier analysis. Thus the Fourier transform

$$\hat{f}(\xi) = \frac{1}{(2\pi)^{n/2}} \int_{R^n} e^{-ix\cdot\xi} f(x)dx$$

provides an analysis by superposition of the exponential translation group characters $e^{\pm ix\cdot\xi}$ where $x\cdot\xi = \sum_{j=1}^{n} x_j \xi_j$. Because the transform of $D^\alpha\hat{f}$ is $\xi^\alpha\hat{f}$, this can be used to convert the linear constant coefficient equation $P(D)u = f$ into the division problem $P(\xi)\hat{u} = \hat{f}$. Though the details of inversion by the reciprocal formula

$$f(x) = \frac{1}{(2\pi)^{n/2}} \int_{R^n} e^{ix\cdot\xi} \hat{f}(\xi)d\xi$$

or one of its complex analogues, and the subsequent interpretation, are substantial, they can be successfully executed in very general cases. Thus Malgrange (see 47, Chaps. 7,10) gave in 1955 a proof of the existence of a fundamental solution in R^n for any such operator $P(D)$.

The difficulties of extending such analysis to equations with variable coefficients were brought home in the most startling way by H. Lewy's discovery about 1956 of the example of a linear first order system with linear coefficients (related to the Cauchy Riemann equations in two complex variables) that had no solutions, even in distributions, if the nonhomogeneous term was C^∞ and not analytic. Thus the study of linear partial differential equations with variable coefficients developed along lines that were related to Lewy's example and the subsequent generalizations of it by Hörmander. While this had led to a great wealth of analysis including pseudodifferential operators, Fourier integral operators wave front sets and much other microlocal analysis, it has represented a return to concepts associated with classical Hamiltonian mechanics rather than a pursuit of quantum mechanical analogies.

However the division problem of Fourier transforms is closely analogous to the diagonalization process for Hermitian operators in quantum mechanics. It is therefore natural that a stage should be reached when a further investigation of the analogy with quantum mechanics, and the associated viewpoints, should appear worthwhile. Here will be presented a description of such a development, the work of C. Fefferman as described in (23) and certain earlier papers (16,24). We have followed Fefferman's choice of title: "the uncertainty principle" for his major descriptive article (23).

Quantum mechanics, like microlocal analysis, is characterized by its use of linear function spaces, and the simultaneous representation of both position operators x and momentum operators

$\xi = D_x = -i\partial/\partial x$. Simultaneous numerical valuations can be assigned only to operators that are mutually commuting, which is not true for x and D_x, or x and ξ. Indeed, since $u = u(x) = \partial_x(xu) - x\partial u$, we have

$$\begin{aligned}\int |u|^2 dx &= \int u\bar{u}dx \\ &= \int \left(\partial_x(xu) - x\partial_x u\right)\bar{u}\,dx \\ &= -\int xu\partial_x\bar{u}\,dx - \int \partial_x u\,\overline{xu}\,dx \\ &= -2\mathrm{Re}\int xu\partial_x\bar{u}\,dx \le 2\|xu\|_2\,\|\partial_x u\|_2 .\end{aligned}$$

With Parseval's formula $\|u\|_2 = \|\hat{u}\|_2$ we obtain then

$$\|u\|_2^2 = \|\hat{u}\|_2^2 \le 2\|xu\|_2\,\|\xi\hat{u}\|_2$$

for any complex valued $u(x) \in L^2(-\infty,\infty)$. This elementary version of the uncertainty principle shows at once that the goal of simultaneous diagonalization of operators x and ξ is impossible. For if u becomes concentrated near $x = 0$, with $\|u\|_2$ remaining fixed, then $\|xu\|_2$ will tend to zero so $\|\xi\hat{u}\|_2$ must become large. Hence $\hat{u} = \hat{u}(\xi)$ cannot be concentrated near the origin $\xi = 0$ but must have large "mass" or large values for some large values of ξ . We may recall, for example, that $\hat{\delta}(\xi)$ = const. Thus the pointwise diagonalization possible for constant coefficients in the x-space is impossible in phase space. The whole thrust of Fefferman's study is to overcome the difficulties thus encountered, which is done by considering finite suitably shaped boxes as support elements for diagonalization, rather than individual points.

To what extent can a function and its Fourier transform both be concentrated on short intervals - say about the origin in each variable? The following result (Titchmarsh 95, p77-81) shows that this is to a considerable extent possible: the Hermite type functions $\phi_n(x) = (-1)^n e^{\frac{1}{2}x^2}\partial_x^n(e^{-x^2})$ form an orthogonal set on $(-\infty,\infty)$, and $\hat{\phi}_n(\xi) = i^n\phi_n(\xi)$. Thus $e^{-\frac{1}{2}x^2}$, which is self reciprocal under the Fourier transform, is the most concentrated in this double or dual sense; the $\phi_n(x)$ form a complete set with $\phi_n(x) = e^{-\frac{1}{2}x^2}H_n(x)$ where $H_n(x)$ is the <u>nth</u> Hermite polynomial.

2.2 The sharp form of the uncertainty principle

Following Fefferman, we may state the uncertainty principle as follows: a function $u(x)$, mostly concentrated in $|x - x_0| = \delta_x$, cannot have its Fourier transform $\hat{u}(\xi)$ mostly concentrated in $|\xi - \xi_0| \le \delta_\xi$, unless $\delta_x \cdot \delta_\xi \ge 1$. For counting eigenvalues of an operator $a(x,\xi) = \sum_{|\alpha|\le m} a_\alpha(x)\xi^\alpha$, each box B of the form $\{(x,\xi);|x-x_0| < \delta, |\xi-\xi_0| < \delta^{-1}\}$ should count for one eigenvalue, so the number of eigenvalues less than K should at first sight be approximately the volume of the set $S(A,K) = \{(x,\xi);A(x,\xi) < K\}$. While this is asymptotically correct for elliptic A when $K \to \infty$, errors can occur unless the uncertainty principle is interpreted more carefully, using instead the number of disjoint distorted unit boxes B that can be packed into the volume. This sharper version of uncertainty is called by Fefferman the SAK principle.

This sharp version of the uncertainty principle can also be used in the study of existence and regularity for solutions of partial differential equations, which can usually be reduced to a priori estimates of the form

$$c\|P(x,D)u\| \le \|Q(x,D)u\| + \text{small error}$$

where P and Q are differential or pseudodifferential operators. When does such an estimate hold for $u \in L^2$? It holds if, for example, $|P(x,\xi)| \le |Q(x,\xi)|$ + small error, though the proof is difficult even in this case. However, from the viewpoint of the uncertainty principle we should imagine that functions can be localized in (x,ξ) space no further than to a suitable box B; so the appropriate (indeed necessary and sufficient) condition should be

$$c \max_B |p| \le \max_B |q| + (\text{small error})_B$$

and this is a weaker condition.

To carry out our approximate diagonalization into boxes B of unit volume, we write $u = \sum_\nu u_\nu$ for $u \in L^2$, in such a way that u_ν and $\hat{u}_\nu$ are each localized in the sense of being mostly concentrated in a single box. We can use overlapping boxes with ϕ_ν supported in B_ν, and a partition of unity $1 = \Sigma_\nu \phi_\nu^2(x,\xi)$ in phase space. Then the given linear operator $L(x,D)$ will act on each piece approximately by multiplying it by a scalar Λ_ν , where $\Lambda_\nu \approx L(x_{0\nu},\xi_{0\nu})$. For if the Fourier transform $\hat{u}_\nu$ is concentrated near ξ_0 , then Du has Fourier transform approximately $\xi_0\hat{u}(\xi)$, while multiplication by x is approximately multiplication by x_0 in the same box B_ν . The uncertainty principle gives us the limit beyond which we cannot reconcile the conflicting goals of concentration upon point supports in both x and ξ simultaneously.

The best type of function of (x,ξ) concentrated on a box for us to use is of the type $\phi\big((x-x_0)/\delta\big)e^{i\xi_0\cdot x}$ where ϕ is a fixed test function of compact support. Such a function ϕ_B is microlocalized in the box $B : \{|x-x_0| < \delta , |\xi-\xi_0| < \delta^{-1}\}$.

Thus to diagonalize $L(x,D)$ approximately, we divide phase space $R^n \times R^n$ into boxes $B_\nu = \{(x,\xi)|x-x_\nu| < \delta_\nu , |\xi-\xi_\nu| < \delta_\nu^{-1}\}$ and associate a ϕ_ν microlocalized to each B_ν . Then, approximately, the ϕ_ν are orthogonal and form a basis for L^2 , while

$$L(x,D)\phi_\nu = L(x_\nu,\xi_\nu)\phi_\nu + E_\nu$$

with the error E_ν satisfying an estimate of the type

$$\|E_\nu\| \le O(|\xi_\nu|^{m-s}\|\phi_{B_\nu}\|) .$$

With a good partition into boxes, we can achieve a value $s = 1/2$ and have thus diagonalized L with this magnitude of error.

If L is elliptic, the quantity $L(x_\nu,\xi_\nu)$ is of magnitude $|\xi_\nu|^m$ for $\xi_\nu \ne 0$, so that the above error is indeed comparatively small for large $|\xi|$. This makes possible an alternative proof of the standard elliptic regularity theorems. However it is necessary to improve this level of approximation to deal with cases of real characteristics when $L(x,\xi) = 0$. We must therefore adapt the shape of the boxes B_ν to the characteristic variety $V = \{(x,\xi)|L(x,\xi)\} = 0$, so that the error terms remain small compared to the principal terms.

The resulting technique for cutting and bending symbols $L(x,\xi)$ can be understood on three levels, of which the simplest is the following:

2.3 Level I. Microlocal analysis: cutting all operators into large pieces modulo lower order errors. Fefferman refers to microlocal analysis of this kind as the "algorithm of the 70's" for proving theorems on partial differential equations. For a nondegenerate vector field X , we straighten the field locally by a smooth coordinate transformation and thus reduce the problem to the elementary case $X = \partial/\partial x_1$. This is possible, by the technique of pseudodifferential operators we have already described.

When $p(x,\xi)$ is an elliptic symbol, we have $|p(x,\xi)| \geq c(1+|\xi|)^m$ so that $p^{-1}(x,\xi)$ is a symbol in S^{-m} with $pp^{-1} = p^{-1}p = I$ modulo S^{-1}, and by a successive approximation technique this can be improved step by step so that $p(x,\xi)$ is inverted modulo symbols of arbitrarily large negative order. As is well known, this leads to the solution of $p(x,D)u = f$, modulo smooth functions.

If we subdivide phase space into boxes B_ν centred at (x_ν,ξ_ν) , with unit sides in the x directions and sides $\sim \frac{1}{4}(1+|\xi|)$ in the ξ directions, the ϕ_ν belong uniformly to S^0 as a vector valued symbol. Hence the operator $U : f \to (f_\nu)_{\nu\in Z} = \big(\phi_\nu(x,\xi)f\big)_{\nu\in Z}$ is a vector valued pseudodifferential operator. It can be shown to be unitary: $U^*U = I$, and $UL(x,D) = L(x,D)U$ modulo lower order errors.

With Fourier integral operators one can also bend the boxes and the symbols, with canonical transformations of coordinates (x,ξ) in phase space. As with coordinate changes, canonical transformations Φ preserve Poisson brackets:

$$\{F,G\} \circ \Phi = \{F \circ \Phi , G \circ \Phi\} ,$$

and they also preserve volume in phase space. It was observed by Egoroff that a pseudodifferential operator will be canonically transformed by a unitary operator, as follows:

Egorov's Theorem: Let Φ be a suitable canonical transformation that carries B into the double B*. Let $p(x,\xi) \in S^m$ be supported in $\Phi(B)$ and let $p(y,\eta) = P \circ \Phi(y,\eta)$. Then the operators $p(z,D)$ and $\tilde{p}(y,D)$ are related by $\tilde{p}(y,D) = Up(z,D)U^{-1}$ + lower order terms, where U is a suitable unitary transformation.

In general, the operator U is given as a Fourier integral operator

$$U\big(f(y)\big) = \int e(y,\zeta)e^{iS(y,\xi)}\hat{f}(\zeta)d\zeta ,$$

where $e \in S^0$, $S \subseteq S^1$. The phase function $S(y,\zeta)$, which plays the role of a generating function of a canonical transformation, is related to $\Phi(y,\eta)$ as follows:

$$\Phi(y,\eta) = (z,\zeta)\ ,\ \eta_k = \frac{\partial S(y,\zeta)}{\partial y_k}\ ,\ \text{and}\quad z_k = \frac{\partial S(y,\zeta)}{\partial \zeta_k}\ .$$

The proof of Egoroff's theorem, as in the calculus of pseudodifferential operators, is by stationary phase methods and it will be omitted here. We refer to canonical transformations applied to equations such as $p(x,D)u = f$, as "bending" the equation to obtain $\tilde{p}(\tilde{x},\tilde{D})\tilde{u} = \tilde{f}$. The pseudodifferential calculus, together with Egoroff's theorem, gives the finest cutting and bending technique that will work for all symbols in S^m .

By a canonical transformation and multiplication by an elliptic symbol, a real pseudodifferential symbol with simple zeros only can be reduced to the form $\frac{\partial}{\partial x}$, and hence can be solved by integration, together with inversion of the transformations used and reconstitution of the global solution by combining local solutions through a partition of unity. Though not effectively explicit, this process clearly leads to existence and regularity theorems of considerable scope.

These standard methods of microlocal analysis have a close analogy with quantum mechanics, through the use of the Fourier transform and related integral operators. In quantum mechanics, an observable quantity is a self adjoint operator A on a Hilbert space, for example position corresponds to multiplication by x , momentum to the operator $-i\partial/\partial x$, both operating on state functions $\psi \in L^2(R^n)$. The expected value or average observed value of A is given by $(A\psi,\psi)$, where the ψ evolves in time according to the equation $\frac{d\psi}{dt} = iH\ .\ \psi$ with A the Hamiltonian operator representing the total energy observable of the system. In the equivalent "Heisenberg picture", the ψ remain constant in time, but A varies according to the Heisenberg equation of motion $\frac{dA}{dt} = i[H,A]$. These theories are invariant under unitary transformations $U : L^2 \to L^2$; thus with $\tilde{\psi} = U\psi$, $\tilde{A} = UAU^{-1}$ the equations are all preserved and the expectation predictions for the outcome of any experiment are unchanged. Fefferman gives the following table of comparison:

	Classical	Quantum
State of System	(x,ξ)	$\psi \in L^2$
Observable	F function	A operator
Result of measurement	Deterministic; always $F(x,\xi)$	Probabilistic on average $(A\psi,\psi)$
Object controlling dynamics	Poisson bracket $\{A,B\}$	Commutator $[A,B]$
Change to equivalent viewpoint	Canonical transformation	Unitary operator

Standard microlocal analysis, or, more precisely, the introduction of Fourier integral operators, is equivalent to quantization: one passes from Poisson brackets to commutators, and from canonical transformations to unitary operators. The analogue of the principle of uncertainty is that we cannot localize to points in phase space, but only to boxes B_ν of unit volume, which are obtained by canonical transformations as images of a unit cube.

2.4 Level II. Cutting and bending adapted to a particular operator.

Further subdivisions of blocks, related to a particular operator P, will be performed by repeated bisection of the blocks, until a family of blocks is reached, on which the symbol $p(x,\xi)$ is in a suitable sense "non-degenerate". This is known as a Calderon-Zygmund decomposition; it is very fine where the symbol $p(x,\xi)$ is small, so that $P(x,D)$ is well represented by these microlocalized pieces. The error terms will be of lower order then the main terms, as in Level I. This process is strong enough to yield useful results such as the Nirenberg-Trèves conjecture (P) and Hörmander's theorem on squares of vector fields, but it is still not the best approximation to diagonalizing that can be done for an operator $P(x,\xi)$. Hence we shall pass directly on to

2.5 Level III. Cutting an operator into small enough pieces modulo a one-percent error.

When the boxes B_ν are further subdivided, the operator $P(x,D)$ is more accurately diagonalized, but the canonical transformations needed become wilder, so that eventually Egorov's theorem cannot be used, and the errors cannot be kept to a lower order of magnitude than the main terms. However, the diagonalization method can still work as the basis of a successive approximation process, as long as the errors are no larger than a fixed small constant times the main term - say 0.01 or 1%. Approximate inverses can still be constructed, if sharp a priori estimates are available, and eigenvalues can be calculated. In fact the eigenvalue corresponding to an appropriately curved box B_α is of magnitude $\max_{(x,\xi)\in B_\alpha} |p(x,\xi)|$, in conformity with the SAK principle stated at the outset above.

Our SAK principle and diagonalization technique have application to the local existence problem of linear partial differential operators: which operators L are solvable in the sense that $Lu = f \in C^\infty$ has local solutions. Recall that Lewy's example

$$\left[\left(\frac{\partial}{\partial x} + i\frac{\partial}{\partial y}\right) + (x-iy)\frac{\partial}{\partial t}\right] u = f \quad , \quad f \in C^\infty$$

has no solutions in any small neighbourhood. The work of Hörmander, Nirenberg and Trèves (47, chap26,74) showed that solvability can be related to the local geometry of the symbol $L(x,\xi)$. For example, if $L = \frac{\partial}{\partial t} + \sum_{k=1}^{n} a_k(t)\frac{\partial}{\partial x_k}$, then $Lu = f$ can be solved by taking a Fourier transform in the x variables. This yields an ordinary differential equation $\left[\frac{\partial}{\partial t} + \sum_{k=1}^{n} a_k(t)\xi_k\right]\hat{u}(t,\xi) = \hat{f}(t,\xi)$ which is easily solved by means of the integrating factor $\exp\left(\int^t \sum_{k=1}^{n} a_k(s)\xi_k dS\right)$. It turns out that exponential growth of this integrating factor must be avoided, or else there will be no solutions of the original problem. The necessary and sufficient condition for this is found to be the following condition: (P) For $\xi \neq 0$, the function $t \to \sum_{k=1}^{n} a_k(t)\xi_k$ never changes sign. A more general invariant formulation of this Nirenberg-Trèves condition is the well known statement: The operator p+iq of principal type is locally solvable iff q never changes sign on the bicharacteristic curves of p. The proof involves microlocal analysis of the Level II type, and will not be repeated here; see the discusson below.

Hypoelliptic operators L are those for which u is always C^∞ except where Lu = f is not C^∞. A basic example of a hypoelliptic operator is given by Hörmander's remarkable theorem, that the operator $L = X_0 + \sum_{j=1}^{N} X_j^2$ is hypoelliptic if the vector fields $X_0,\ldots,X_N$ and their repeated commutators span the tangent space at each point. However the inversion of these operators can be reduced to the geometric study of certain related non-Euclidean balls $B_L(\chi, \;)$, as discovered by Stein using nilpotent groups. This work has been developed by several authors, leading to a calculus on Heisenberg manifolds (6). If the X_j are taken as the translation invariant vector fields of the Lie algebra on a nilpotent group, then $L = X_0 + \Sigma X_k^2$ is also translation invariant, so that L^{-1} is a convolution operator on the group, with kernel homogeneous with respect to the natural dilatations acting on the group. The kernel or fundamental solution is also clearly related to the shape of the non-Euclidean balls, or spheres, natural to the nilpotent group, and can be constructed by a special lifting process (84).

For vector fields with variable coefficients, where the nilpotent group techniques are not available, an analogous sufficient condition for hypoellipticity has been given by Oleinik and Radkevich(73). These problems can all be studied by the SAK diagonalization methods.

Boundary value problems for a partial differential equation on a domain Ω can often be reduced to pseudodifferential operator equations on the boundary Ω, as shown for instance by Hörmander (42). Such a reduction can often be made using classical layer potentials of Newtonian or other suitable type. Hence the conditions for local or global solvability of pseudodifferential operator equations take on increased significance. The analogue of the Nirenberg-Trèves condition (P) for pseudodifferential operators is this apparently more relaxed condition: $(\overline{\Psi})$ Along the null bicharacteristics of p, the symbol q can only change sign from negative to positive.

It has been shown by Moyer (71) and Hörmander (47, vol.4,p.96) that the operator p+iq is locally solvable only if condition $(\overline{\Psi})$ holds.

A full converse result giving a necessary and sufficient condition is not yet established, in the most general cases such as that of principal type operators. However for first order operators p+iq that satisfy condition ($\overline{\Psi}$) the following theorem has been established by Egorov (21, 23): Let $p_1,\ldots,p_N$ denote p,q and their repeated Poisson brackets up to order m . If

$$\sum_1^N |p_k(x,\xi)| \geq c|\xi|$$

for large ξ , then L = p+iq is hypoelliptic and satisfies a sharp estimate $\|u\| + \|Lu\| \geq c\|u\|_{\frac{1}{m+1}}$.

In proving his result, Egorov was the first to use curved boxes B_ν, which however were of large volume. The combining of the local results to give the theorem in the large is carried out in Hörmander (46).

With this introduction we shall move on to two topics considered in (24). First the techniques will be applied to the spectra of Schrödinger operators, and then to pseudodifferential operators of certain general classes.

2.6 Eigenvalues of Schrödinger operators.

We shall study discrete eigenvalues of Schrödinger operators $L = -\Delta + V(x)$ on R^n , where Δ is the Laplacian in R^n and V(x) is an assigned potential function. The ultimate goal is to relate the eigenvalues and eigenfunctions of L to the growth of the symbol $L = |\xi|^2 + V(x)$ on boxes B_ν as described above. For polynomial potentials such estimates are elementary, in the sense that the number of eigenvalues of L which are less than a given (energy) bound E is essentially the same as the number of disjoint boxes B_ν that fit inside the region $|\xi|^2 + V(x) < E$. For more general potentials there are analogous estimates but the proofs are too difficult to describe here. A case of especial interest is the 2N body problem of Coulomb forces among N electrons and N nuclei in R^3, for which estimates independent of N are desirable. Fefferman succeeds in sharpening existing results and showing, in a sense, that matter is made of atoms which bind together to form molecules. Here we shall consider just the polynomial case, for reasons of space.

The classical results on the asymptotics of eignvalues show that the number $N(\lambda,L)$ of eigenvalues less than λ is approximately equal, to the phase space volume $V(\lambda,L) = |\{(x,\xi)\,|\,|\xi|^2 + V(x) < \lambda\}|$. Asymptotically, then, the nth eigenvalue λ_n should be the smallest λ for which $V(\lambda,L) = N$. In general the following result, which is closely related to Sobolev's inequality, holds:

Theorem 1. In R^n $(n \geq 3)$, $N(\lambda,L) \leq C_n V(\lambda,L)$ for a positive constant C_n .

Example 1 (Two uncoupled harmonic oscillators), $L = -\Delta + u_1x^2 + u_2y^2$ on R^2 . Let $u_1 = 1$, u_2 be small, then $\lambda_1 = 1 + u_2^{\frac{1}{2}} > 1$. But given $\varepsilon > 0$, $N > 1$ we can take u_2 so small that $V(\varepsilon,L) > N$. Thus volume counting predicts many very small eigenvalues, although the least eigenvalue in fact exceeds 1.

Example V. (Particle in a box). Let $I = I_1 \times I_2 \times \ldots \times I_n$ where the intervals on sides I_j have lengths $\delta_1 \le \delta_2 \le \ldots \le \delta_n$. For $E > 0$ sufficiently small, the operator $L = -\Delta - E\chi_I \ge 0$, while for $E > E_{critical}(\delta_1, \delta_2, \ldots, \delta_n)$ one finds negative eigenvalues corresponding to capture by the potential box well. How does $E_{critical}$ depend on the box sides? Volume counting suggests that $E_{critical} \sim \sim (\delta_1\delta_2 \ldots \delta_n)^{-2/n}$ but in fact E depends strongly only on δ_1 and δ_2 and weakly on δ_3, for estimates

$$\frac{C}{\delta_1\delta_2} \ge E_{critical} \ge \frac{C_\varepsilon}{\delta_1\delta_2(\delta_3/\delta_1)^t}$$

can be established by our methods.

These and other examples show that the classical methods, which one should remember are primarily asymptotic for large n, can produce great discrepancies for the initial eigenvalues.

Let us repeat such estimates using instead the number $N_{UP}(\lambda, L)$ of disjoint boxes $B_\nu = \{|x-x_0| < \delta, |\xi-\xi_0| < \delta^{-1}\}$ which fit inside $\{|\xi|^2 + V(x) < \lambda\}$. This gives for the lowest eigenvalue $\lambda_1(L)$ the estimate $\lambda_{UP}(L)$ of the least number for which $N_{UP}(\lambda, L) \ge 1$, namely

$$\lambda_{UP}(L) = \inf_B \max_{(x,\xi)\in B} \left(|\xi|^2 + V(x)\right)$$

$$\sim \inf_{x_0,\delta} \{\delta^{-2} + \max_{|x-x_0|<\delta} V(x)\} .$$

Theorem 2. If $V \ge 0$ is a polynomial of degree d, then $c\lambda_{UP}(L) \le \le \lambda_1(L) \le C\lambda_{UP}(L)$ where c depends only on n and d, C only on n.

Divide R^n into a grid of cubes $\{Q_\nu\}$ of side $\lambda^{-1/2}$ and redefine $N_{UP}(\lambda, L)$ as the number of Q_ν on which $\max_{Q_\nu} V \le \lambda$; this is consistent with the earlier definition.

To estimate the higher eigenvalues of L, we have the converse

Theorem 3. If $V \ge 0$ is a polynomial of degree d, then $N_{UP}(c\lambda, L) \le \le N(\lambda, L) \le N_{UP}(C\lambda, L)$ where C depends only on n and d, while c depends only on n.

This result will yield the order of magnitude of λ_N, uniformly in N.

If the polynomial V(x) is not restricted to be positive, there is a sharper version of Theorem 2:

Theorem 4. If V is any polynomial of degree d, then

$$\inf_{x_0,\delta} \{c\delta^{-2} + \max_{|x-x_0|<\delta} V(x)\} \le \lambda_1(L) \le \inf_{x_0,\delta} \{C\delta^{-2} + \max_{|x-x_0|<\delta} V(x)\} ,$$

where C depends only on n, c depends only on n and d.

2.7 Proofs of the eigenvalue properties.

The proofs of Theorems 2, 3, and 4 depend on the following:

Main Lemma. Assume $V(x) \ge 0$ is a polynomial of degree $\le d$ on a cube Q in R^n. Suppose $Av_Q V \ge (\text{diam}Q)^{-2}$. Then for any function $u(x)$ in Q we have

$$\int_Q \{|\nabla u|^2 + V|u|^2\}dx \geq c(\mathrm{diam}Q)^{-2}\int_Q |u|^2 dx ,$$

where c depends only on n and d.

To prove this main lemma we begin by noting that for polynomials $P(x)$ of degree $\leq d$, we can easily show that

(a) $\mathrm{Av}_Q|P| \leq \max_Q|P| \leq C\mathrm{Av}_Q|P|$,

(b) $\max_Q|\nabla P| \leq C(\mathrm{diam}Q)^{-1}\max_Q|P|$,

(c) If $P \geq 0$ on Q, there is a subcube $Q' \subseteq Q$ with $\mathrm{diam}\,Q' \geq c\,\mathrm{diam}\,Q$ on which $\min_{Q'}P \geq \frac{1}{2}\max_Q P$.

Now let $u(x)$ be any function on Q, and observe that

$$\int_Q |\nabla u|^2 dx \geq \frac{c(\mathrm{diam}Q)^{-2}}{|Q|}\iint_{Q\times Q}|u(x) - u(y)|^2 dx\,dy$$

while

$$\int_Q V(x)|u(x)|^2 dx = \frac{1}{|Q|}\iint_{Q\times Q} V(y)|u(y)|^2 dx\,dy$$

so that

$$\int_Q \{|\nabla u(x)|^2 + V(x)|u(x)|^2\}\,dx$$

$$\geq \frac{1}{|Q|}\iint_{Q\times Q}[c(\mathrm{diam}Q)^{-2}|u(x)-u(y)|^2 + V(y)|u(y)|^2]\,dx\,dy$$

$$\geq \frac{1}{|Q|}\iint_{Q\times Q}\left\{\min\left(V(y),c(\mathrm{diam}Q)^{-2}\right)(|u(x)-u(y)|^2 + |u(y)|^2\right\}dx\,dy$$

$$\geq \frac{1}{|Q|}\iint_{Q\times Q}\left(\min V(y),c(\mathrm{diam}Q)^{-2}\right)\frac{|u(x)|^2}{2}dx\,dy$$

$$\geq \frac{1}{|Q|}\int_Q \left(\min V(y),c(\mathrm{diam}Q)^{-2}\right)dy\int_Q |u(x)|^2 dx .$$

From property (c) and the hypothesis $\mathrm{Av}_Q V \geq (\mathrm{diam}Q)^{-2}$, we have

$$\tfrac{1}{2}\min\left(V(y),c(\mathrm{diam}Q)^{-2}\right) \geq c(\mathrm{diam}Q)^{-2}$$

for a fixed proportion of the measure of Q. Thus

$$\frac{1}{|Q|}\int_Q \tfrac{1}{2}\min\left(V(y),c(\mathrm{diam}Q)^{-2}\right)dy \geq c'(\mathrm{diam}Q)^{-2}$$

and the Main Lemma now follows.

Let us use this result to establish

Theorem 2: $c\lambda_{UP}(L) \leq \lambda_1(L) \leq C\lambda_{UP}(L)$. Here the upper bound for $\lambda_1(L)$ is easy, since $\lambda_1(L) = \inf\limits_{\psi\neq 0}\frac{(L\psi,\psi)}{\|\psi\|^2} \leq \frac{(L\psi_0,\psi_0)}{\|\psi_0\|^2}$ for any given $\psi_0 \in L$, $\psi_0 \not\equiv 0$. Let ψ_0 run over all translations and dilatations of a fixed smooth function supported in the unit ball, and the upper bound follows at once from its definition.

The lower bound for $\lambda_1(L)$ is equivalent to the estimate

$$(Lu,u) = \|\nabla u\|^2 + (Vu,u) \geq c\lambda_{UP}(L)\|u\|^2$$

for $u \in C^\infty(R^n)$. To establish this, we first cut R^n into a grid of cubes Q each with side $C_1|\lambda_{UP}(L)|^{-\frac{1}{2}}$. If C_1 is large enough, each Q will contain a ball $B(x_0,\delta)$ with $\delta = 2[\lambda_{UP}(L)]^{-\frac{1}{2}}$ so by the definition of λ_{UP} we have

$$\lambda_{UP} \leq \delta^{-2} + \max_{B(x_0\delta)} V(x) \leq \frac{\lambda_{UP}}{4} + \max_Q V(x) .$$

Hence

$$\max_Q V(x) \geq \frac{3}{4}\lambda_{UP} \geq \frac{3}{4}C\ (\mathrm{diam}Q)^{-2} .$$

Now $\mathrm{Av}_Q V \sim \max_Q V$ so by taking C_1 large enough we can arrange that

$$\mathrm{Av}_Q V(x) \geq (\mathrm{diam}Q)^{-2} ,$$

and

$$(\mathrm{diam}Q)^{-2} = c'\lambda_{UP}(L) .$$

These two relations together with the Main Lemma yield

$$\int_Q \{|\nabla(x)|^2 + V(x)|u(x)|^2\}dx \geq c\lambda_{UP}(L)\int_Q |u(x)|^2 dx ,$$

for every Q of the grid, and now summation over all the cubes leads to the estimate above which we needed to prove. This completes the proof of Theorem 2.

To prove Theorem 3 we recall from elementary variational theory of eigenvalues that

(1) $N(\lambda,L) \geq N$ if there is an N-dimensional subspace $H \subseteq L$ such that $(Lu,u) \leq \lambda\|u\|^2$ whenever $u \in H$.

(2) $N(\lambda,L) \leq N$ if there is a codimension N subspace $H' \subseteq L^2$ such that $(L,u) \geq \|u\|^2$ whenever $u \in H'$.

Let us first show that $N(\lambda,L) \geq N = N_{UP}(c\lambda)$. Given $c\lambda$ there is a grid $\{Q_\nu\}$ of cubes with side $(c\lambda)^{-\frac{1}{2}}$. Let $Q_1,\ldots,Q_N$ be those cubes of this grid on which $\max_Q V \leq c\lambda$. Take a fixed function ψ supported in the unit cube and translate and dilate it to obtain $\psi_1,\ldots,\psi_N$ supported similarly in $Q_1,\ldots,Q_N$. Now define H as the span of $\psi_1,\ldots,\psi_N$; H is evidently N-dimensional, while for $u = \sum_{j=1}^N \alpha_j\psi_j \in H$ we find

$$\begin{aligned}(Lu,u) &= \|\nabla u\|^2 + (Vu,u)\\ &= \sum_{j=1}^N \{|\alpha_j|^2\|\nabla\psi_j\|^2 + |\alpha_j|^2(V\psi_j,\psi_j)\}\\ &\leq \sum_{j=1}^N C(\mathrm{diam}Q)^{-2}|\alpha_j|^2\|\psi_j\|^2 \leq \lambda\|u\|^2\end{aligned}$$

since the ψ_j are translates of a fixed function ψ_0, and $\max_{Q_j} V(x) \leq c\lambda$. So by relation (1) above, we find $N(\lambda,L) \geq N_{UP}(c\lambda,L)$.

To establish the other part of Theorem 3, we must show $N(\lambda,L) \le$ $\le N_{UP}(C\lambda,L) = N$. Corresponding to $C\lambda$ is a new grid of cubes $\{Q'_\nu\}$ of side $(C\lambda)^{-\frac{1}{2}}$, and we choose $Q_1,\dots,Q_N$ as those cubes on which $\max V \le C\lambda$. Let H' be the space of all $u \in L^2$ with integrals zero over $Q_1,\dots,Q_N$. H' has codimension N , and we must show that $(Lu,u) = \|\nabla u\|^2 + V(u,u) \ge \lambda\|u\|^2$ for $u \in H'$. But for $\nu \ne 1,2,\dots,N$ we have $\max_{Q_\nu} V(x) \ge C\lambda$, by definition, so by the main lemma

$$\int_{Q_\nu} \{|\nabla u|^2 + V|u|^2\}dx \ge \lambda \int_{Q_\nu} |u|^2 dx \ , \ \nu > N \ .$$

holds for any u in Q_ν . We must now find a similar result for $\nu = 1,2,\dots,N$. But any $u \in L^2(Q)$ satisfies

$$\int_{Q_\nu} |\nabla u|^2 dx \ge c(\mathrm{diam}Q_\nu)^{-2} \int_{Q_\nu} |u(x) - \mathrm{Av}_{Q_\nu} u|^2 dx$$

$$\ge \lambda \int_{Q_\nu} |u(x) - \mathrm{AV}_{Q_\nu} u|^2 dx \ .$$

With $u \in H'$, the average term is zero, and if we note that $V \ge 0$ the corresponding inequality clearly holds for $\gamma = 1,2,\dots,N$, $u \in H'$. The final result establishing Theorem 3 now follows by summation over all cubes.

Finally let us prove Theorem 4. The upper bound, as in Theorem 2, is again straightforward so we shall concentrate on the lower bound, which amounts to the following estimate. Let V be a polynomial of degree $\le d$ on R^n which satisfies the basic condition $\max_Q V \ge$ $\ge -c_1(\mathrm{diam}Q)^{-2}$ for every cube Q . Then $\|\nabla u\|^2 + (Vu,u) \ge 0$.

To establish this estimate suppose u is supported in a very large cube Q^0 . Unless $V = \text{const.}$, we have

$$\max_{Q^0} V - \min_{Q^0} V \ge C(\mathrm{diam}Q)^{-2}$$

provided Q^0 is large enough.

Now perform a Calderon-Zygmund decomposition of Q^0 by bisection into 2^n subcubes, and repeated bisections. We stop the bisections for any cube Q that satisfies $\max_Q V - \min_Q V \le C_1(\mathrm{diam}Q)^{-2}$. This is bound to happen eventually, since at each bisection into 2^n subcubes, the left side diminishes, while the right side increases by a factor of 4. At the end of this bisection process we have the big cube Q^0 partitioned into subcubes Q_ν each satisfying the sandwich inequality

$$cC_1(\mathrm{diam}Q_\nu)^{-2} \le \max_{Q_\nu} V - \min_{Q_\nu} V \le C_1(\mathrm{diam}Q_\nu)^{-2} \ .$$

Here c depends only on n and d , while the left hand inequality holds because the bisection stopping condition was not satisfied at the next to last stage.

Now take C_1 large enough (depending on n and d) and apply the main lemma to $\overline{V}(x) = V(x) - \min_{Q_\nu} V$. We obtain

$$\int_{Q_\nu} \{|\nabla u|^2 + \overline{V}|u|^2\}dx \geq \left(\min_{Q_\nu} V + \overline{c}(\operatorname{diam} Q_\nu)^{-2}\right) \times \int_{Q_\nu} |u|^2 \, dx \ .$$

But if our basic condition on V, given above, holds with c_1 small enough, then we have $\min_{Q_\nu} V \geq -\overline{c}(\operatorname{diam}Q_\nu)^{-2}$. To show this, we suppose $\min_{Q_\nu} V$ is taken at $x^0 \in Q_\nu$, and let Q be a subcube of Q with $\operatorname{diam}Q = \beta \operatorname{diam}Q_\nu$ and $x^0 \in Q$. By observation (b) in the proof of the main lemma, we have

$$\max_Q V \leq \min_Q V + C\beta(\max_{Q_\nu} V - \min_{Q_\nu} V) \ .$$

Employing the basic condition on the left, and the above sandwich inequality on the right, we find

$$-c_1(\operatorname{diam}Q)^{-2} \leq \min_{Q_\nu} V + C\,C_1(\operatorname{diam}Q_\nu)^{-2} \ .$$

Replacing $\operatorname{diam}Q$ by $\beta\operatorname{diam}Q_\nu$ on the left, we find

$$\min_{Q_\nu} V \geq -(c_1\beta^{-2} + CC_1\beta)(\operatorname{diam}Q)^{-2} \ .$$

Now choose $\beta \ll C_1C$ and then $c_1 \ll \beta^2$. We find that $\min_{Q_\nu} V \geq$ $\geq -\overline{c}(\operatorname{diam}Q_\nu)^{-2}$ as claimed; this shows that

$$\int_{Q_\nu} \{|\nabla u|^2 + V|u|^2\}dx \geq 0 \ .$$

Summing over all cubes we obtain the final result : $\|\nabla u\|^2 + (Vu,u) \geq 0$ which completes the proof of Theorem 4.

Space does not permit us to describe the further work in which Fefferman extends these results to more general potentials, using the technique of subdivision into cubes, and the Thomas-Fermi approximation of the N-body problem, which leads to a theory of the stability of matter and of chemical bonding. For this we must simply refer to his review paper (23) and the further references contained in it. It is a sign of the vigour of modern mathematical analysis that these techniques of Fourier analysis should make renewed contact at this deeper level with basic problems of mathematical physics and statistical mechanics.

2.8 Decompositions of phase space

The process of cutting phase space into blocks, and of subdividing these blocks by a Calderon-Zygmund bisection sequence, leads to approximate diagonalization of differential and pseudodifferential operators.

The accuracy of these approximations is of course limited by the Uncertainty Principle, but is nonetheless enough to yield many important results. Here we shall discuss the lowest eigenvalue of a pseudodifferential operator, the sufficiency of the Nirenberg-Trèves condition (P) for local solvability, and hypoellipticity with a sharp estimate for sums of squares of vector fields.

To find the lowest eigenvalue, we consider a non-negative symbol $A(x,\xi)$ which satisfies second order estimates of the form $|\partial^\alpha \partial^\beta A| \le C_{\alpha\beta} M^{2-|\beta|}$ which we abbreviate as $A \in S^2(1\times M)$. Such estimates hold for a classical second order symbol on a block $|x| \le 1$, $|\xi| \le M$ as in our introductory sections. By the SAK principle, the number of eigenvalues of $A(x,D)$ less than K will be essentially the same as the number of distorted unit boxes B_ν contained in $S(A,K) = \{(x,\xi) \in R^{2n} | A(x,\xi) < K\}$. Thus we would predict that $\lambda_1(A)$ is comparable to $\mu_1(A) = \min_\nu \max_{(x,\xi)\in B_\nu} A(x,\xi)$.

The boxes B_ν should be "testing boxes", defined as images of the unit cube under a canonical transformation $\Phi(z,\zeta) \to (x,\xi)$ which maps $|z|, |\zeta| < M^\varepsilon$ into R^{2n} with $|\partial^\alpha_{z,\zeta} x| \le M^{-\varepsilon|\alpha|}, |\partial^\alpha_{z,\zeta}\xi| \le M^{1-\varepsilon|\alpha|}$ for $|\alpha| \ge 1$. Then we have

Theorem 1 To a symbol $A \in S^2(1\times M)$, $A \ge 0$ can be associated a family of testing boxes $\{B_\nu\}$ such that

$$\lambda_1(A) \le C_\varepsilon \mu_1(A) + C_\varepsilon M^{2\varepsilon}, \ \mu_1(A) \le C_\varepsilon \lambda_1(A) + C_\varepsilon M^{2\varepsilon}.$$

In particular $\lambda_1(A)$ and $\mu_1(A)$ are comparable unless both are $\lesssim M^\varepsilon$. From this result follows

Theorem 2 Suppose $X_1, X_2, \ldots, X_N$ and their commutators of order $\le m$ span the tangent space at every point. Then for $L = \sum_{j=1}^{N} X_j^2$ we have

$$C\|u\|^2 + C\,\mathrm{Re}\,(Lu,u) \ge \|u\|^2_{\frac{1}{m+1}}.$$

This result implies directly that $Lu \in H^s_{loc}(\Omega)$ implies $u \in H^{s+\frac{1}{m+1}}_{loc}(\Omega)$, so L must be hypoelliptic, by a bootstrap process (84).

To study local solvability, we consider differential operators $L(x,D)$ such that their principal symbol $p+iq$ is of principal type, that is, with $p+iq$ and $\mathrm{grad}(p+iq)$ not vanishing together at any point in $T^*(\Omega)\backslash 0$. From the null bicharacteristic of p through (x_0,ξ_0), which is the solution of Hamilton's equations $\frac{dX_p}{dt} = \frac{\partial p}{\partial \xi_k}$, $\frac{d\xi_k}{dt} = \frac{\partial p}{\partial x_k}$. Condition (P) states that q has a constant sign along this null curve.

Theorem 3 A partial differential equation of principal type is locally solvable if and only if its symbol satisfies condition (P).

2.9 Proofs of the three theorems

In each block of a decomposition we make coordinate changes to simplify the operator, as in the following.

Lemma 1. Let $X = \sum_{k=1}^{n} a_k(x)\partial/\partial x_k$ be a real C^∞ vector field on the cube $|x| \leq 1$. Then there is a Calderon-Zygmund decomposition of $|x| \leq 1$ into cubes $\{Q_\nu\}$ such that

(a) In each Q_ν there is a coordinate change satisfying natural estimates such that X becomes $\partial/\partial y_1$.

(b) If $1 = \Sigma\theta_\nu$ is a partition of unity with θ_ν supported essentially in Q_ν, then X commutes with the decomposition operator $u \to (u_\nu) = (\theta_\nu u)$ modulo an error bound on L^2.

For the proof we bisect $|x| \leq 1$ repeatedly, stopping at a cube Q only if $\max_{x\in Q}\max_k |a_k(x)| \geq 2C(\mathrm{diam}Q)$, with the large constant C chosen so that $|\nabla a_k| \leq C$ for $|x| \leq 1$. On each cube Q_ν we now have an $x_0 \in Q_\nu$ with $|a_{k_0}(x_0)| \geq 2C(\mathrm{diam}Q_\nu)$ for some k_0, $1 \leq k_0 \leq n$. As well, $|a_k(x)| \leq 4C(\mathrm{diam}Q_\nu)$ for $x \in Q$, because the stopping condition did not hold at the previous stage. In each Q_ν let $y = (X-X_0)/(\mathrm{diam}Q_\nu)$ so Q_ν goes over into a unit cube and X becomes $\Sigma_k\tilde{a}_k(y)\partial/\partial y_k$ with $\tilde{a}_k(y) = (\mathrm{diam}Q_\nu)^{-1}a_k(x_0+(\mathrm{diam}Q_\nu)y)$, and $|\partial_y^\alpha\tilde{a}_k| = (\mathrm{diam}Q_\nu)^{|\alpha|-1}|\partial^\alpha a_k|$ for $|\alpha| \geq 1$. Thus we have bounds, as above, for $\tilde{a}_k(y)$ and its derivatives, and X is a "nice" vector field on the new unit cube Q_ν, while $|\tilde{a}_{k_0}| \geq C$ throughout the unit cube now follows easily from the known bounds on $\tilde{a}_k(y)$ and its gradient. This means that X can be brought to the form $\partial/\partial z_1$ on Q_ν by a further coordinate change, so part (a) of the Lemma is established. To establish part (b) we have to show that $X\theta_\nu$ is uniformly bounded, which is elementary, so we shall omit details. The point of this proof is that the decomposition must be very fine where X is nearly degenerate.

To understand second order operators $L = \Sigma_{k\ell}a_{k\ell}(x)\partial^2/\partial x_k\partial_{k\ell}$ + + lower terms in the same way, we call L non-degenerate if the coefficient matrix $(a_{k\ell}(x))$ is non-singular.

Lemma 2. By a change of variables, a nondegenerate second order operator can be expressed in the normal form

$$L = -\left(\frac{\partial}{\partial y_1}\right)^2 - \sum_{j,k\geq 2}^{N} b_{jk}(y_1,y')\frac{\partial^2}{\partial y_j\partial y_k} + \text{lower order terms} .$$

The proof is an exercise in Riemann geometry and is omitted. This normal form can be regarded as a Schrödinger operator in y_1 where the potential $V(y_1)$ is an operator in n-1 variables. As in Lemma 1, we can show that the unit cube has a Calderon-Zygmund decomposition with the above normal form in each subcube, such that a partition of unity conjugated with L introduces only a bounded error term. Again, the fineness of the decomposition into subcubes has to be related to the degeneracy of the operator.

For pseudodifferential operators in $S^m(1\times M)$, there are similar lemmas:

Lemma 4. Let $A(x,\xi) \in S^1(1\times M)$ be a real symbol with

$$\max_{|\alpha|+|\beta|=1} \max_{|x|, M^{-1}|\xi|\le 1} \frac{\partial_x^\alpha \partial_\xi^\beta A}{M^{1-|\beta|}} \ge C .$$

Then either A is first order elliptic, or else as in Egorov's theorem, A can be put in the form $A\circ\Phi^{-1}(y,\zeta) = \zeta_1$, for a suitable canonical transformation $\Phi : (x,\xi) \to (y,\zeta)$.

Lemma 5. Let $A(x,\xi) \ge 0$ belong to $S^2(1\times M)$ and assume $\max_{|\alpha|+|\beta|=1} \max_{|x|, M^{-1}|\xi|\le 1} \frac{|\partial_x^\alpha \partial_\xi^\beta A|}{M^{2-|\beta|}} \ge C$. Then either A is second order elliptic, or if not, by a canonical transformation Φ it can be brought to the form

$$A\circ\Phi^{-1}(y,D) = -\left(\frac{\partial}{\partial y_1}\right)^2 + V(y_1)$$

where $V(y_1)$ is a second order pseudodifferential operator in n-1 variables.

Fefferman then describes the detailed process of microlocalization that is needed to obtain similar results for localized cubes Q_ν, using pseudodifferential operator theory with the Calderon-Vaillancourt theorem and Beals calculus. Here we shall assume these results in order to pass on to the proofs of the three theorems. A further proposition we shall assume is that if $A \ge 0$ is a second order symbol, then $A(x,D) \ge -C$; this is readily proved by induction on the dimension.

Main Lemma. Let $L = \left(\frac{\partial}{\partial y_1}\right)^2 + V(y_1)$ where $V(y_1) = V(y_1,y',Dy')$ and $0 \le V(y_1,y',\zeta') \in S^2(1\times M)$. Then $L > cK > M^\varepsilon$ if and only if $\mathrm{Av}_{y\in I} V(y_1) \ge cK$ for every interval I of length $K^{-\frac{1}{2}}$.

This holds if $V(y_1)$ is a scalar polynomial, and the condition $K \ge M^\varepsilon$ permits us to treat $V(y_1)$ as if it were a polynomial. For a Taylor expansion of V about $y_1 \in I$ up to order d leads to an error $O(M^2K^{-d/2})$ which is bounded if $K > M^\varepsilon$ and $d > 2/\varepsilon$. However the full proof is too difficult to attempt in detail here.

For Theorem 1 we shall prove only the following "corollary": If $A(x,\xi) \ge 0$ is a symbol in $S^2(1\times M)$ then $\Lambda_1(A) \ge c_\varepsilon K - C_\varepsilon M^{2\varepsilon}$ where B runs through all testing boxes, and $K = \inf_B \max_{(x,\xi)\in B} A(x,\xi)$. To show $A(x,D) \ge cK$ we can now assume $K < M^\varepsilon$ as otherwise the proposition $A(x,D) \ge -C$ gives the desired result. The microlocal decomposition methods reduce the problem to three cases: $A(x,D)$ bounded or $A(x,D)$ elliptic, which are immediate; and the third, nontrivial case $A(x,D) = -(\partial/\partial x_1)^2 + V(x_1)$ with V a pseudodifferential potential. To show $A(x,D) \geq cK$, it is sufficient to show $Av_{y_1 \in I} V(y_1) \geq cK$ for any y_1-interval of length $> K^{-\frac{1}{2}}$. So assume the corollary holds in n-1 variables, and show that $\bar{V}(y',\zeta') = Av_{y_1 \in I} V(y_1,y',\zeta)$ satisfies $\max_{(y',\zeta') \in B'} \bar{V}(y',\zeta') \geq c'K$ for each testing box $B' \in R^{2(n-1)}$. This would establish an induction proof on the dimension. But if this average property does not hold, i.e. if $\max_{(y',\zeta') \in B'} V(y',\zeta') < c'K$ for some testing box B, then since V is essentially a polynomial of bounded degree in y_1, we must have $V(y_1,y',\zeta') \leq K/4$ for $y_1 \in I^*$ where I^* is the double of I, with $|I^*| < 2K^{-\frac{1}{2}}$. Now construct a testing box $B = \{(y_1,y',\zeta_1,\zeta') \in R^{2n} | y_1 \in I^*, |\zeta_1| < K^{\frac{1}{2}}/2, (y',\zeta') \in B\}$. Clearly

$$\max_B A(y,\zeta) = \max_B\{\zeta_1^2 + V(y_1,y',\zeta')\}$$

$$= \frac{1}{4}K + \max_{y_1 \in I^*, (y'\zeta') \in B} V(y_1,y',\zeta')$$

$$\leq \frac{1}{4}K + \frac{1}{4}K = \frac{1}{2}K .$$

But this contradicts the definition of $K = \inf_B \max_B A$. This contradiction establishes the induction property and completes the proof of the corollary.

To establish Theorem 2 on squares of vector fields, we microlocalize to a standard box $|x| \leq 1$, $|\xi| \leq M$. The theorem asserts, in effect, that $L \geq cM^{2/m+1}$, and the corollary just proved reduces this to an estimate $\max_{(x,\xi) \in B} L(x,\xi) \geq cM^{\frac{2}{m+1}}$ for any testing box B. Since $L = \Sigma_j p^2{}_j$ with $p_j(x,\xi)$ the symbol of vector field X_j, the hypothesis on commutators is that some repeated Poisson bracket is elliptic: $|\{p_{j_1}\{p_{j_2},\dots\{p_{j_{m-1}}, p_{j_{m'}}\}\dots\} \geq cM$ on $|x| \leq 1$, $|\xi| \sim M$, $m' \leq m$. Now the testing box is the image of the unit cube under a "good" canonical transformation Φ, so by comparing with this transformation we can rewrite the condition in the same form on the unit cube. Since the iterated Poisson bracket is an $m'+1$ degree polynomial in derivatives of order at most m', we now have

$$\max_j \|\bar{p}_j\|_{C^{m'}(Q^0)} \geq c'M^{1/m+1}$$

where $\bar{p}_j = p_j \circ \Phi$. But the "good estimates" for Φ and the $S^1(1 \times M)$ property for p_j show that the higher derivatives of $\bar{p}_j$ are very small. Hence $\|\bar{p}_j - P_j\|_{C^m(Q^0)} \leq 1$ for polynomials P_j of bounded degree. Since $\|P_j\|_{C^m(Q^0)} \leq C\max_{Q^0}|P_j|$ holds for such polynomials, we find the apparently stronger estimate

$$\max_j \max_{(x,\xi) \in Q^0} |\bar{p}_j(z,\zeta)| \geq c''M^{1/m+1} .$$

Observing that $\bar{p}_j = p_j \circ \Phi$ we can now obtain

$$\max_j \max_B |p_j| \geq cM^{1/m+1} .$$

This completes the proof of Theorem 2.

Finally, we outline the proof of Theorem 3 on the sufficiency of condition (P) for local solvability. On a microlocal block $|x| \leq 1$, $|\xi| \leq M$ we can straighten out the imaginary part of the symbol, and so write $L = i\tau + a(t,x,\xi)$ with $a \in S^1(1 \times M)$, where, by condition (P), $t \to a(t,x,\xi)$ does not change sign. Let us regard $(\partial/\partial t + a(t,x,D_x))u = f$, $|t| \leq T$, as an evolution equation, for which there are three standard cases: First, if $a \geq 0$ we can solve with t increasing, and $u = 0$ at $t = -T$; we can omit details and observe only that the solution is well behaved. Similarly, if $a \leq 0$ we can solve with $u = 0$ at $t = +T$. Second, if $a(t,x,\xi)$ changes sign with, say, x_1, then the equation breaks up into two uncoupled evolution problems, which were shown by Nirenberg and Trèves to be solvable as in the first case (74). Thirdly, if $a(t,x,\xi)$ is bounded, the equation is certainly solvable with good properties for the solution.

Now make a Calderon-Zygmund decomposition of the initial block $|x| \leq 1$, $|\xi| \leq M$ by repeated bisection, stopping at a Q with sides $\delta \times M\delta$ if either $\mathrm{vol}Q \sim 1$ or if

$$\max_{|\alpha|+|\beta|\leq 1} \frac{|\partial_x^\alpha \partial_\xi^\beta a(t,x,\xi)|}{(M\delta)^{1-|\beta|}\delta^{-|\alpha|}} \geq C$$

for some t . The property $a(t,x,\xi) \in S'(Q)$ still holds, and the evolution equation can be partitioned into a family of microlocal problems of the same form, with essential supports in each cube. For cubes of vol ~ 1 we are in the third case, while if $\mathrm{vol}Q >> 1$, a canonical transformation will reduce the problem to either (a) $a(t_0,x,\xi)$ is first order elliptic for some $t_0 \in [-T,T]$, and hence does not change sign, or (b) $a(t_0,x,\xi) = M\delta_\nu^2 x_1$ for some t_0 so the sign changes at $x_1 = 0$. Since condition (P) is assumed to hold, we are in the first case above for (a) and the second for (b), and these microlocalized problems are all solvable. This completes the proof of Theorem 3.

2.10. Approximate diagonalization.

Let us return to the general problem of diagonalizing an operator $L(x,D)$ by decomposing phase space into suitable blocks, and briefly review Fefferman's approach. One writes $L = A + \varepsilon$, where A is strictly diagonalized to the blocks B_α , and ε is small compared to A , in the sense that $\|\varepsilon u\|_2^2 \leq \delta \Sigma_\alpha \Lambda_\alpha^2 \|u_\alpha\|^2$, with $\delta << 1$ while u_α is localized by a partition of unity to the block B_α where $A \approx \Lambda_\alpha = A(x_\alpha,\xi_\alpha)$. By comparison, the main term A satisfies $\|Au\|^2 \leq C\Sigma_\alpha \Lambda_\alpha^2 \|u_\alpha\|^2$. Supposing $\delta < |\Lambda_\alpha|$ for all α , then we have $\|\varepsilon u\| << \|Au\|$ so that a converging successive approximation based on A can be established. The inverse operator A^{-1} is represented by the same block decomposition with the inverse block eigenvalues Λ_α^{-1} , at least in the first approximation.

Each box B_α can be "straightened out" by a canonical transformation $\Phi_\alpha : I \to B_\alpha$ with $I = \{|x| \leq 1, |\xi| \leq B\}$ where B is a large constant. Then, by Egorov's theorem

$$L(x,D) \approx \Sigma U_\alpha^*(L_\alpha \circ \Phi_\alpha)U_\alpha$$

where U_α is a suitable Fourier integral operator associated with Φ_α. However $\overline{L}_\alpha = L_\alpha \circ \Phi_\alpha$, although localized to a straightened block B_α, is too big to belong to a good symbol class $S^m(I)$. It is necessary to divide by Λ_α to control this operator; then we obtain

$$L(x,D) \approx \Sigma_\alpha \Lambda_\alpha U_\alpha^* L_\alpha^+(x,D)U_\alpha$$

where $L^+(x,D) \in S^m(I)$ and I has volume $\sim B^n$. In effect, therefore, L^+ acts on a finite dimensional or compact vector space of functions microlocalized to I and the error term will satisfy

$$\|\varepsilon u\|^2 \leq \delta \Sigma_\alpha \Lambda_\alpha \|u_\alpha\|^2$$

with δ a negative power of B.

As the Φ_α do not belong to the standard classes, a special construction of the U_α is required. To make the diagonalization work a subelliptic estimate

$$\|L_\alpha^+(x,D)u\| \geq cB^{m-2+\varepsilon}\|u\|$$

is necessary in $|x| \leq 1$, $|\xi| \leq B$, $L_\alpha^+ \in S^m(1 \times B)$, since then $\|\Lambda_\alpha u_\alpha\| \geq cB^{m-2+\varepsilon}\|u_\alpha\|$ will follow as required.

For a second order operator with the non-negative symbol $L(x,\xi) = \sum_{j,k} a_{jk}(x)\xi_j\xi_k + \sum_k b_k(x)\xi_k + V(x)$, microlocalized to $|x| < 1$, $|\xi| < B$, with polynomial coefficients of degree $\leq d$, and $|a_{j,k}(x)| \leq C$, $|b_k(x)| \leq CB$, $|V(x)| \leq CB^2$, so that $L \in S^2(I)$, a suitable normal form can be defined by induction on the dimension n. For $n = 0$ the block is a point and $L(x,\xi)$ a number V, in normal form if $V > cB^2$. In n dimensions L is in normal form if either (a) L is elliptic, or, (b) $L(x,\xi) = \xi_1^2 + \overline{L}(x_1,x',\xi')$ with $\overline{L}(x',\xi') = \int_{|x_1| \leq 1} \overline{L}(x_1,x',\xi')dx_1$ in normal form in n-1 dimensions.

Then $L \in S^2(1 \times B)$ while the main lemma of the preceding section shows that $L(x,D) \geq cB^\varepsilon$ microlocally in $|x| \leq 1$, $|\xi| \leq B$. So the necessary subelliptic estimate can be proved in this case.

In the first order case $L = i\tau + a(t)\xi + V(t,x)$, with suitable conditions, a relatively rough form of subelliptic estimate, namely,

$$\|L(t,x,D_t,D_x)u\|^2 \geq B^{-2}\|u_t\|^2 + B^{-2}\|iau_x + Vu\|^2$$

for $u \in C_0^\infty$ is sufficient to initiate the machinery, and to prove an estimate $\|p(x,D)u\| + \|q(x,D)u < C\|(p+iq)(x,D)u\| + C\|u\|_{(\varepsilon)}$ for $p+iq$ satisfying condition (ψ). For reasons of space, discussion of this proof is omitted.

Despite its simplicity in principle, the actual construction of solutions is very elaborate, and much simplification, clarification and explicit working out of cases remains to be done if useful direct results are to be reached in a wide range of cases including higher order differential operators. However the method suggests the possibility of combining two hitherto distinct approaches, the method of a priori estimates, and the classical approximation methods, to reach new ground in the theory of linear partial differential equations with variable coefficients.

Chapter 3. Huygens' Principle and Hadamard's Conjecture.

3.1 Historical background.

The first hyperbolic partial differential equations to be studied in detail were the wave equations in 1 , 2 and 3 space dimensions. The one-dimensional wave equation, which describes the vibrations of an elastic string, was solved by d'Alembert in the 18th century, and the concept of a travelling wave emerged. At the turn of the 19th century Poisson using spherical means found the solution of the initial value problem for the wave equation in 3 space dimensions, the case of constant coefficients being understood. Later the Maxwell theory of electromagnetic waves showed how the wave equation plays a role in the propagation of light in 3 dimensional space, and brought the physical theory of wave propagation into relation with the mathematical theory of linear partial differential equations.

In the seventeenth century Christian Huygens had developed a theory of the propagation of light by primary and secondary waves, the clean cut nature of observed light propagation being explained by the perfect cancellation of certain combinations of secondary waves, namely those which occur off the leading wave front. Corresponding to a well known property of the time-evolving semi-group property of the solutions of the wave equation in 3 space dimensions, the clean-cut property of these solutions was said to satisfy Huygens' Principle of wave propagation.

By contrast, the wave equation with constant coefficients in two space dimensions does not satisfy Huygens' Principle because the support of the elementary solution includes the interior of the wave cone, as well as its surface.

Later, through the work of Herglotz, it became known that Huygens' principle holds for wave equations with constant coefficients (and no lower order terms) in an odd number of space dimensions. In 1945 Petrowsky published a study of hyperbolic polynomials and their algebraic geometry and gave the name "lacuna" to this phenomenon - the vanishing of the solution in a component region between sheets of the wave surface. His work was later extended by Atiyah, Bott and Gårding (5) to whose paper we refer for further detail of the general topic of lacunas.

In his 1923 volume of Lectures on Cauchy's Problem, Hadamard (35) gave the construction by a series of a fundamental solution for a wave equation with variable coefficients, and then formulated his conjecture on Huygens' Principle. This was to the effect that only the wave equations with constant coefficients, and no lower order terms, or equations reducible to this by elementary transformations, will have clean cut wave propagation in 3 , 5 , 7 , ... space dimensions. That is, only for the constant coefficient case is the interior of the wave cone effectively a lacuna for odd space dimensions. This chapter is devoted to the subsequent history and the present standing of this famous conjecture by Hadamard, which has been shown to be not altogether correct and which is still only partially resolved, even in 3 space dimensions.

Hadamard himself made one important contribution to the solution of the problem, by showing that the vanishing of the logarithmic term in the formal series for the fundamental solution is a necessary and sufficient condition for clean cut wave propagation in the Huygens sense. This has made the problem accessible to local methods based on series expansions and the use of tensors and differential geometry. Recently, methods related to modern relativity theory have been used in the 3 space dimensional, or 4 dimensional space - time case, and the Hadamard conjecture has been resolved for space - times of Petrov class N (11). Even in this case the conjecture is not universally true for there is a counter - example (32) but this is now known to be the only such counter example of Petrov type N in 4 dimensional space time.

The methods used in pursuing Hadamard's conjecture have an interesting relationship to those of microlocal analysis, which they antedate. They involve expansion in series of function values, rather than orders of singularities, and so are in a sense more precise, and also less general.

3.2 Formulation of the problem.

The general second order homogeneous linear partial differential equation in n independent variables can be written in the coordinate-invariant form

$$L(u) = g^{ab}u_{;ab} + A^a u_{,a} + Cu = 0,$$

where the Einstein summation convention is understood. Here $g_{ab} = g_{ba}$ $(a , b = 1 , \ldots , n)$ are the contravariant components of the symmetric metric tensor of a Riemannian space V_n of signature $2 - n$. The subscript comma denotes differentiation and the semi-colon covariant differentiation with respect to this metric and its accompanying connection. The coefficients g^{ab}, A^a and C are assumed to be C^∞ in a given set of coordinates $x_1 = t$, $x_2, \ldots x_n$.

For a non singular Riemannian metric, we may introduce the related affine connection

$$\Gamma^i_{jk} = \frac{1}{2} g^{i\ell} \left[\frac{\partial g_{ij}}{\partial x_k} + \frac{\partial g_{kl}}{\partial x_j} - \frac{\partial g_{jk}}{\partial x_\ell}\right]$$

which are also known as Christoffel symbols. The covariant derivative of a covector field $w = (w_i)$ is then a covariant tensor field

$$(\nabla w)_{ij} = \nabla_i w_j \equiv \frac{\partial w_j}{\partial x_i} - \Gamma^k_{ij} w_k.$$

The covariant derivative of a contravariant vector field $v = (v^i)$ is likewise given by

$$(\nabla v)^j_i = \nabla_i v^j = \frac{\partial v^i}{\partial x_j} + \Gamma^j_{ik} v^k .$$

Covariant derivatives of higher rank tensors will contain a connection type term for each index with signs as above in the covariant and contravariant cases. We recall that covariant derivatives do not commute; they transform as tensors of the appropriate type and the Riemann curvature tensor R_{abcd} can be defined by the relation

$$(\nabla_j \nabla_k - \nabla_k \nabla_j) w_i = R^{\ell}{}_{ijk} w_\ell$$

where

$$R^{\ell}{}_{ijk} = g^{\ell m} R_{mijk}$$

and indices generally are raised and lowered by contraction with the metric tensor g^{ab} or its inverse or reciprocal tensor g_{ab}.

For a contravariant vector field v, we define the divergence as

$$\operatorname{div} v = \nabla_a v^a = |g|^{-\frac{1}{2}} \frac{\partial}{\partial x_a} (|g|^{\frac{1}{2}} v^a) ,$$

where $g = \det(g_{ab})$; the metric factor when differentiated provides the additional connection term in the covariant derivative, which makes the divergence a scalar invariant under smooth coordinate transformations. We define the gradient $\nabla_a \phi$ of a scalar invariant $\phi = \phi(x)$ by the partial derivatives:

$$\phi_{,a} \equiv \nabla_a \phi \equiv \frac{\partial \phi}{\partial x_a}$$

for in this case there is no tensor index in ϕ and no connection term.

The d'Alembertian operator is

$$\Box \phi = \operatorname{div} \operatorname{grad} \phi = |g|^{-\frac{1}{2}} \frac{\partial}{\partial x_a} (|g|^{\frac{1}{2}} g^{ab} \frac{\partial \phi}{\partial x_b})$$

where the Einstein summation convention applies for $a , b = 1 , \ldots , n$. When a timelike variable $x_1 = t$ is selected and Riemannian normal coordinates constructed (25 , p. 16) , so that $g_{11} = 1$ and $g_{1b} = 0$ near P , then we may write the line element in the form

$$ds^2 = g_{ab} dx_a dx_b = dt^2 + g_{\alpha\beta} dx_\alpha dx_\beta$$

where $\alpha , \beta = 2 , \ldots , n$. The restricted spacelike metric tensor $g_{\alpha\beta}$ then is nonsingular and negative definite. We may then construct the spacelike Laplacian operator

$$\begin{aligned} \Delta u &= \operatorname{div} \operatorname{grad} u \\ &= -|g|^{-\frac{1}{2}} \frac{\partial}{\partial x_\alpha} (|g|^{\frac{1}{2}} g^{\alpha\beta} \frac{\partial u}{\partial x_\beta}) \end{aligned}$$

where $\alpha, \beta = 2, \dots, n$ and $|g| = |\det g|$ keeps its earlier value. Then we have $\Box u = u_{tt} - \Delta u$ in the region of Riemannian normal coordinates and globally for constant coefficients.

Cauchy's problem is to determine a solution of this hyperbolic equation with given values and normal derivatives on a spacelike initial surface (that is, a surface whose normal is timelike and whose square has the same sign as g^{11}). However we shall concentrate attention on the retrograde elementary solution E which by superposition can be used to build up solutions of Cauchy's problem. We take the vertex P or field point of this solution to be the origin, and observe that its singular support is the retrograde wave cone surface with vertex at P, on which E behaves like a distribution homogeneous of order $1-n$ in all the variables, at least asymptotically near P. The support of E is the interior of the retrograde part of the wave cone with vertex at P, in general, and certainly whenever the total number n of variables is odd. The equation is of Huygens' type precisely when E vanishes identically within the retrograde cone. In this case $S(E) = SS(E)$ is the retrograde conical characteristic surface with vertex at P. As noted above, the wave equations with constant coefficients

$$L(u) = u_{tt} - \Delta u = 0$$

are of Huygens' type when $n = 2m \geq 4$ is even.

Hadamard first attempted to show that every Huygens' equation could be shown equivalent to one of these constant coefficient cases, n even, under the following transformations:

a) a suitable transformation of coordinates

b) multiplication of both sides of the equation by a non-vanishing function of position, inducing a conformal transformation of the metric.

c) replacement of the unknown u by $\lambda(x)u$ where $\lambda(x)$ is a nonvanishing function of position.

These three types of transformation will be used throughout. Observe that a combination (bc): $\frac{1}{\lambda}L(\lambda u)$ has the same metric coefficients g^{ab} as L, so that the characteristic form $g^{ab}\phi_{,a}\phi_{,b}$ is unchanged.

Applying these transformations, Mathisson (59, 60), Hadamard (35, 36) and Asgeirsson (4) verified the Hadamard conjecture in the particular case $n = 4$, $g^{ab} =$ constant. However a tentative proof for $n = 4$ of the case of g^{ab} variable by Mathisson, referred to by Hadamard in his 1942 paper (36) dedicated to Mathisson's memory, never appeared. As we shall see, such a proof could not have been correct. The first counter-examples to Hadamard's conjecture were given by Stellmacher (92 a, b) as follows for $n = 6, 8, 10, \dots$

$$L(u) = u_{tt} - \Delta u + \left(\frac{\lambda_1}{t^2} - \sum_{i=2}^{n} \frac{\lambda_i}{x_i^2} \right) u = 0$$

where $-\lambda_i = \nu_i(\nu_i + 1)$, $\nu_i = 0, 1, 2, \dots$ and $\Sigma_{i=1}^{n} \nu_i = \frac{n}{2} - 2$. These cases are not equivalent to any constant coefficient wave equation if any $\lambda_i \neq 0$.

Significant counterexamples have also been discovered by Günther (30, 31, 32, 33) for $n = 4$; they arise from the wave equation on the Lorentz spaces of "maximum mobility" studied by Petrov (77) and having metric $ds^2 = 2dx_1dx_2 - a_{\alpha\beta}(x_1)dx^\alpha dx^\beta$, where $\alpha, \beta = 3, 4$ and $a_{\alpha\beta}(x_1)$ is positive definite with components depending only on x_1. In General Relativity this yields an exact plane wave solution for the vacuum or Einstein-Maxwell field equations. In a study by Ehlers and Kundt (22) using a different coordinate system the metric has the plane wave form $ds^2 = 2dv[du + (Dz^2 + \bar{D}\bar{z}^2 + ez\bar{z})dv - 2dzd\bar{z}$ where $D = D(v)$ and $e = e(v) = e$.

For this plane wave metric the Maxwell field equations in the differential form $dF = 0$, $\delta F = 0$ have been shown to satisfy Huygens' principle as well (32, 33, 53, 86, 87). Also for this metric, Wünsch (102) has verified Huygens' Principle for the Weyl equation $\nabla_A^B \phi_B = 0$ and the wave equation $\Delta\phi_A = 0$ where ϕ_A is a one-index two-spinor.

For the scalar wave equation $\Box u = u_{tt} - \Delta u$ on a conformally empty four dimensional space-time, McLenaghan (61) has shown that the only cases satisfying Huygens' principle are the flat and the above plane wave space-times. In effect, these are the only conformally empty spacetimes for which the cancellation of secondary waves is exact off the wave cone.

3.3 The elementary solution.

When the field point P is fixed, the fundamental solution or elementary solution then satisfies the adjoint differential equation

$$L^*v = g^{ab}v_{;ab} - (A^a v)_{,a} + Cv = \delta(x - x_0).$$

From this point on we shall consider only the case $n = 4$ for which the form of the elementary solution is given by Friedlander (25, Chap. 4). For the retarded solution we have in the interior of the retarded light cone,

$$E_P(x) = V_0(x_0, x)\,\delta(\Gamma(x_0, x)) + V_1(x_0, x)$$

where x_0 gives the coordinates of P and x of the "source point" Q. Here also

$$V_0(x_0, x) = \frac{1}{2\pi} \exp\left\{-\frac{1}{4}\int_0^{s(x)} (g^{ab}\Gamma_{;ab} - 8 - A^a\Gamma_{,a})\frac{dt}{t}\right\}$$

where integration runs over the geodesic joining P to Q. Throughout, $\Gamma(x_0, x) \equiv s^2(x_0, x)$ where s is the goedesic distance from P to Q. Assuming normal coordinates with origin at P, we have (25, p. 17) $\Gamma = g_{ij}(0)x^i x^j = g_{ij}(x)x^i x^j$ while $\Gamma_{,i}\Gamma_{,i} = 4\Gamma$ and $g_{ij}(x)x^j = g_{ij}(0)x^j$, $i = 1, \ldots, n$.

The additional "logarithmic" term $V_1(x_0, x)$ satisfies $L_x^*(V_1(x_0, x)) = 0$ in the interior of the light cone and

$$V_1(x_0\,,x) = \frac{V_0(x_0\,,x)}{s(x_0\,,x)} \int_0^{s(x)} \frac{G(V)}{V}\,dt$$

when x is on the retrograde cone with vertex P. As Hadamard has shown (36) the Huygens condition is the vanishing of $V_1(x_0\,,x)$ for x in the retrograde cone of vertex P at x_0, this to hold for all x_0 in the region of interest. From the expression for V_1 it can be seen that Hadamard's condition implies $L^*(V_0(x_0\,,x)) = C$, for all x_0, when x is on the retrograde cone with vertex x_0. By further calculation the formula above for V_0 yields

$$V_0(x_0\,,x) = \frac{1}{2\pi\rho^{\frac{1}{2}}(x_0\,,x)} \exp\left\{\frac{1}{4}\int_0^{s(x)} A^a\Gamma_{,a}\,\frac{dt}{t}\right\}$$

where

$$\rho(x_0\,,x) = 8\,(\,g(x)\,g(x_0))^{\frac{1}{2}}\left[\det\frac{\partial^2\Gamma}{\partial x_0^a\partial x_0^b}\right]^{-1}$$

is called the discriminant function, while as above $g(x) = \det(g_{ab}(x))$.

To obtain further information from Hadamard's condition, one must simplify as far as possible by means of the elementary transformations (a), (b) and (c), and then develop a Taylor series expansion of the condition around the vertex of the wave cone (63). Let us look first at the transformations. Applying (b) : $L(u) \to e^{-2\phi}L(u)$ and (bc) : $L(u) \to \frac{1}{\lambda}\,L(\lambda u) = \bar{L}(u)$, we find

$$\bar{L}(u) = \bar{g}^{ab}u_{;ab} + \bar{A}^a u_{,a} + \bar{C}u = \lambda^{-1}e^{-2\phi}L(\lambda u)$$

where $\bar{g}^{ab} = e^{-2\phi}g^{ab}$, $\bar{g}_{ab} = e^{2\phi}g_{ab}$ and $\bar{A}_a = A_a + 2(\log\lambda)_{,a} - (n-2)\phi_{,a}$, $\bar{A}^a = \bar{g}^{ab}\bar{A}_b$ with

$$\bar{C} = e^{-2\phi}(C + \lambda^{-1}\Box\,\lambda + A^a(\log\lambda)_{,a})\ .$$

Under coordinate transformations (a) the various tensors involved will transform as indicated by their indices. To this information we can now add that under transformations (b) and (bc) the conformal curvature tensor will be unchanged : $\bar{C}^a{}_{bcd} = C^a{}_{bcd}$, since the metric has undergone a conformal transformation only. Since coefficients A_a are changed only by a gradient, it follows that $\bar{H}_{ab} = \bar{A}_{a,b} - \bar{A}_{b,a} = A_{a,b} - A_{b,a} = H_{ab}$. With square brackets to indicate an alternating sum, we can write $H_{ab} = A_{[a,b]}$.

The scalar C_1 defined by

$$C_1 = C - \frac{1}{2}A^a{}_{;a} - \frac{1}{4}A_aA^a - \frac{(n-2)}{4(n-1)}R$$

is now easily shown to satisfy the transformation law $\bar{C}_1 = e^{-2\phi}C_1$. For the adjoint differential operator we obtain the transformation $\overline{L^*}(v) = \lambda e^{-4\phi}\times L^*(\lambda^{-1}e^{-2\phi}v)$ while $E_{x_0}(x) = \lambda\lambda_C^{-1}\,e^{-2\phi}E_{x_0}(x)$ with

$\lambda_0 \equiv \lambda(x_0)$. Hence also, at least on the light cone, $\bar{V}_0 = \lambda_0^{-1} a_1 \lambda e^{-2\phi} V_0$, and $V_1 = \lambda_0^{-1} \lambda e^{-2\phi} V_1$, where

$$a_1 = \frac{1}{s} \int_0^{s(x)} e^{2\phi} dt .$$

Then it follows that $L^*(V_0) = \lambda_0^{-1} a_1 \times \lambda e^{-4\phi} L^*(V_0)$ at least on the wave cone, so that Hadamard's condition is verified to be invariant under the transformations of types (b) and (c).

The Riemann curvature tensor of our V_4 is defined by $R^a{}_{bcd} = 2 \times (\partial_{[c}\Gamma^a_{d]b} + \Gamma^f_{b[d}\Gamma^a_{c]f}$ where the square brackets indicate the anti symmetric combination over the indices bracketed. We have $R_{abcd} = R_{cdab} = -R_{bacd}$. The Ricci tensor is $R_{ab} = R^c{}_{acb}$ which is contracted over two indices and $R_{ab} = R_{ba}$. We also define $R = g^{ab}R_{ab}$ and $L_{ab} = -R_{ab} + \frac{1}{6} g_{ab} R$, $S_{abc} = L_{a[b;c]}$ and finally the conformal curvature tensor of Weyl, which is $C_{abcd} = R_{abcd} - 2g_{[a[d}L_{b]c]}$.

3.4 The Series Expansion.

The sequence of necessary conditions can now be derived as follows. Let P with coordinates x_0 be the field point and make a transformation of type (b) so that $L^0{}_{ab} = L^0{}_{(ab;c)} = L^0{}_{(ab;cd)} = \dots = 0$ where we have dropped the bar over L , and the superscript 0 denotes evaluation at P . We then specify a (bc) type transformation with

$$\lambda = \exp\left[-\frac{1}{4} \int_0^{s(x)} A^a \Gamma_{,a} \frac{dt}{t} \right],$$

for which $\lambda_0 = 1$ and $V_0(x_0 , x) = \frac{1}{2\pi} \rho^{\frac{1}{2}}$. Also we choose normal coordinates x^a with coordinate x_0 ; then V_0 reduces to the simple form $V = (2\pi)^{-1}(g^0/g)^{\frac{1}{4}}$.

Hadamard's condition $L^*(V_0(x_0 , x)) = 0$ on the cone can then be expressed as $\sigma(x_0 , x) = 0$ on the cone, where

$$\sigma = \gamma + A^a g^{bc} g_{bc;a} + 4A^a{}_{;a} - 4C$$

and

$$\gamma = (g^{ab} g^{cd} g_{cd;a})_{;b} + \frac{1}{4} g^{ab} g_{ab;c} g^{cd} g^{ef} g_{ef;d} .$$

Expanding σ around x_0 we obtain the following sequence of conditions, where TS( ) denotes the trace-free symmetric part of the tensor enclosed in the parentheses:

$$\overset{0}{\sigma} = 0 , \overset{0}{\sigma}_{,a} = 0 , TS(\overset{0}{\sigma}_{;ab}) = 0 , TS(\overset{0}{\sigma}_{;abc}) = 0 , \dots .$$

The derivatives of σ must be systematically calculated from the Taylor expansions about x_0 of g_{ab} , g^{ab} , A^a and C . Here, following (63), we give these expansions to the second order only, which will suffice for the first condition in the sequence that has been derived. We have

$$g_{ab} = \overset{0}{g}_{ab} + \frac{1}{3}\overset{0}{R}_{acdb}x^c x^d$$

$$A_a = \overset{0}{H}_{ab}x^b + \frac{2}{3}\overset{0}{H}_{ab;c}x^b x^c$$

and

$$C = \overset{0}{C} + \overset{0}{C}_{;a}x^a + \overset{0}{C}_{;ab}x^a x^b$$

From the definitions of σ and γ_0 we find these conditions imply $\overset{0}{\sigma} = 4\overset{0}{C}$ so that the first condition is $\overset{0}{C} = 0$, with the choice already made of the transformations (a), (b), (c). To express this condition in an invariant form we must use the Cotton invariant defined as above by

$$C_I = C - \frac{1}{2}A^a{}_{;a} - \frac{1}{4}A^a A_a - \frac{1}{6}R .$$

This invariant under transformations (a), (b), and (c) must therefore vanish at every point P for a Huygens equation. This gives the first main condition.

Further, higher order, conditions of this type have been derived by increasingly laborious calculation. For a description of this process we refer to (62). Each of these necessary conditions is expressed by the vanishing of a trace-free symmetric tensor which is invariant under the three types of transformation.

The first five conditions found in this way are as follows (30, 41, 61, 62, 98, 99)

$$\text{I} \qquad C = \frac{1}{2}A^k{}_{;k} + \frac{1}{4}A^k A_k + \frac{1}{6}R$$

$$\text{II} \qquad H_{ab;}{}^b = 0$$

$$\text{III} \qquad S_{abk;}{}^k - \frac{1}{2}C^k{}_{ab}{}^\ell L_{k\ell} = -5(h_a{}^k H_{bk} - \frac{1}{4}g_{ab}H_{k\ell}H^{k\ell})$$

$$\text{IV} \qquad TS(S_{abk}H^k{}_c + C^k{}_{ab}{}^\ell H_{ck;\ell}) = 0$$

$$\begin{aligned}\text{V} \qquad & TS(3C_{kab\ell;m}C^{k}{}_{cd}{}^{\ell m}{}_{;} + 8\,C^k{}_{ab}{}^\ell{}_{;c}S_{k\ell d} + 40\,S_{ab}{}^k S_{cdk} - \\ & - 8\,C^k{}_{ab}{}^\ell S_{k\ell c;d} - 24\,C^k{}_{ab}{}^\ell S_{cdk;\ell} + 4\,C^k{}_{ab}{}^\ell C_\ell{}^m{}_{ck}L_{dm} + \\ & + 12\,C^k{}_{ab}{}^\ell C^m{}_{cd\ell}L_{km} + 12\,H_{ka;bc}H^k{}_d - 16\,H_{ka;b}H^k{}_{c;d} - \\ & - 84\,H^k{}_a C_{kbc\ell}H^\ell{}_f - 18\,H_{ka}H^k{}_b L_{cd}) = 0\end{aligned}$$

where, as above, $H_{ab} = A_{[a,b]}$, $L_{ab} = -R_{ab} + \frac{1}{6}g_{ab}R$ and $S_{abc} = L_{a[b;c]}$.

3.5 The case of empty space-time

When the Ricci tensor $R_{ab} = 0$ the space-time is empty in the relativistic interpretation. The five conditions above are enough to resolve the Hadamard conjecture in this case (61 , 63). The Einstein field equations are $R_{ab} - \frac{1}{2} g_{ab}R = -\lambda T_{ab}$ where the energy momentum tensor in our interpretation is

$$T_{ab} = \frac{1}{4} g_{ab} H_{k\ell}H^{k\ell} - H_{ak} H_b{}^k$$

When $T_{ab} = 0$, the general equation is reducible to the self-adjoint form $g^{ab}u_{;ab} + \frac{1}{6} Ru = 0$, as follows. Indeed $T_{ab} = 0$ is known to imply $H_{k\ell} = 0$ (59) , so that $A_{a;b} = A_{b;a}$ and it follows that the differential form $A = A_a dx^a$ is closed and so locally equal to a derived form dg. The transformation of type (bc) with $\lambda = \exp(-\frac{1}{2}g)$ then leads to an equation with $\bar{A}_a = 0$ and it will follow from condition I that $\bar{C} = \frac{R}{6}$. Thus if we assume that H_{ab} is the Maxwell field, we need consider for empty space time only the self-adjoint wave equation $g^{ab}u_{;ab} + \frac{1}{6} Ru = 0$. McLenaghan (61) has shown that this equation satisfies Huygens' principle on an empty space time if and only if the space time is flat or is a plane wave space time as described in Section 2 above but with $e = 0$. This result follows from Condition V above which in this case now reduces to

$$TS(C_{kab\ell;m}C^k{}_{cd}{}^{\ell}{}_;{}^m) = 0$$

while the other conditions hold identically. We shall omit the proof.

A somewhat stronger result in the same genre is that an equation of the general second order hyperbolic type will satisfy Huygens' Principle on a conformally empty space time if and only if it is equivalent to the above self adjoint wave equation on a plane wave space-time with $e = 0$ in the plane wave metric. A necessary condition that a space time be conformal to an empty space time is the vanishing of the Bach tensor $B_{ab} = S_{abk;}{}^k - \frac{1}{2} C^k{}_{ab}{}^{\ell}L_{k\ell}$ (30). Thus it follows from Condition III that the "energy-momentum tensor" of the H_{ab} field must be zero, so that H_{ab} itself can be shown to vanish and the reduction to self-adjoint form can proceed as above. The result then follows from the preceding theorem of McLenaghan.

This theorem has also been extended by Wünsch (101) to the case of an Einstein space-time with metric satisfying $R_{ab} = \lambda g_{ab}$. When $\lambda \neq 0$ Huygens' principle will hold only if the space time is of constant curvature. More generally, for symmetric spaces where $R_{abcd;e} = 0$ the problem has been resolved only for the self-adjoint wave equation $g^{ab}u_{;ab} + \frac{1}{6} Ru = 0$ which is Huygens', only if space time is conformally flat or a symmetric plane wave space with e and D constants in the plane wave metric.

Another related result of Wünsch (101) concerns 2×2 decomposable space times with metric

$$ds^2 = g_{\alpha\beta}(x^1, x^2)dx^{\alpha}dx^{\beta} + g_{\mu\nu}(x^3, x^4)dx^{\mu}dx^{\nu}$$

where $\alpha, \beta = 1, 2$ and $\mu, \nu = 3, 4$. In this case the self adjoint wave equation can satisfy Huygens' principle if and only if the space-time is conformally flat. There are a number of similar results for other special cases which are of interest in relativity, but are not very general. (33), (34), (48 a,b), (53), (86 a,b), (97), (102), (103). These results are for the most part in 4 variables, with a few in 6 and very few for higher dimensions.

3.6 The Seventh Condition.

The derivation of further conditions based on higher terms in the series expansion about the vertex of the wave cone is the only obvious way to further narrow the set of equations that might be of Huygens' type. However the task becomes increasingly arduous as the order of the terms increases, and the highest condition yet derived is the seventh. It also appears that the odd order conditions are the most significant, and that the sixth condition does not yield significant new limitations.

A derivation of the seventh condition was published by Rinke and Wünsch in 1981 (81) and in the notation already used its form is as follows.

$$
\begin{aligned}
\text{VII}\quad & TS[3C^{k}{}_{ab}{}^{\ell m}{}_{;c}\, C_{kde\ell;mf} + C^{k}{}_{ab}{}^{\ell}{}_{;cd}(10\, S_{k\ell e;f} + 6\, S_{efk;\ell}) + \\
& + 64\, S_{abk;c}\, S_{de}{}^{k}{}_{;f} - C^{k}{}_{ab}{}^{\ell}(3C^{m}{}_{cdk;ef}L_{\ell m} + 5C_{kcd\ell;me}L^{m}{}_{f} + \\
& + 7\, C^{m}{}_{cdk;\ell e}L_{mf} + 13\, S_{k\ell c;d}L_{ef} + 12 S_{cdk;\ell}L_{ef} + 71 S_{cdk;e}L_{\ell f}) - \\
& - 10\, C^{k}{}_{ab}{}^{\ell}{}_{;c}(S_{k\ell d;ef} + 3S_{dek;\ell f}) - 20S_{abk;cd}S_{ef}{}^{k} + \\
& + 50\, S_{abk}\, S_{cd}{}^{k}\, L_{ef} + 5C^{k}{}_{ab}{}^{\ell}{}_{;c}(2C^{m}{}_{k\ell d;e}\, L_{mf} + 3C^{m}{}_{dek;\ell}L_{mf} + \\
& + S_{k\ell d}\, L_{ef} + 3C_{kde}{}^{m}{}_{;f}\, L_{\ell m} + 15S_{dek}L_{ef}) + 10C^{k}{}_{ab}{}^{\ell}(C_{kcd}{}^{m}{}_{;e}L_{(\ell m;f)} + \\
& + S_{cdk}L_{(\ell e;f)} - \frac{1}{12} R_{;c}C_{kde\ell;f}) - 4C^{k}{}_{ab}{}^{\ell}(2C_{k}{}^{mn}{}_{c}C_{\ell nmd;ef} - \\
& - 10\, C^{m}{}_{cd}{}^{n}C_{kef\ell;mn} + 20C_{\ell cd}{}^{m}S_{kme;f}) - 20C_{k}{}^{mn}{}_{a}C_{\ell mnb}C^{k}{}_{cd}{}^{\ell}{}_{;ef} + \\
& + 4\, C^{k}{}_{ab}{}^{\ell}(7C_{k}{}^{mn}{}_{c;d}C_{\ell mnd}\, L_{ef} - 10\, C_{kef\ell}\, C^{m}{}_{cd}{}^{n}\, L_{mn}) - \\
& - 5C^{k}{}_{ab}{}^{\ell}(3C_{k}{}^{mn}{}_{c;d}C_{\ell mne;f} + 54C_{\ell cd}{}^{m}{}_{;e}S_{kmf} + 74C_{\ell cd}{}^{m}{}_{;k}S_{efm} - \\
& - \frac{76}{3}\, C_{ck\ell}{}^{m}{}_{;d}S_{efm} - \frac{404}{3}\, S_{cdk}\, S_{ef\ell}) + 30C_{k}{}^{mn}{}_{a}C^{k}{}_{bc}{}^{\ell}{}_{;d}C_{\ell efm;n} + \\
& + 25C^{k}{}_{ab}{}^{\ell}C_{\ell cd}{}^{m}L_{km}L_{ef} + \frac{1}{6}\, C^{k}{}_{ab}{}^{\ell}C_{kcd\ell}(87L^{m}{}_{e}L_{mf} + 19RL_{ef})] = 0.
\end{aligned}
$$

3.7 Spaces of Petrov class N.

For the Hadamard problem in four dimensional space-time, the available results suggest that the only wave equations $g^{ab}u_{;ab} + \frac{1}{6}Ru=0$ with the Huygens property are on conformally flat or conformally plane wave space-times. One approach to the proof of this revised conjecture is being followed by Carminati and McLenaghan who divide the problem into five cases according to their Petrov type (77 , 87). This is a natural approach to the problem as Petrov type is invariant under conformal transformations.

A proof for Petrov type N has been found and will shortly appear (11). Condition VII is sufficient to yield the solution of Hadamard's problem in this case, at least for the self-adjoint scalar wave equation. However for Maxwell's and Weyl's equations the question is still open and higher conditions may yet be required before they can be settled.

The formal result obtained in (11) is as follows: THEOREM (McLenaghan and Carminati) The wave equation $\Box u + \frac{1}{6} Ru = 0$ on a Petrov type N space time satisfies Huygens' Principle if and only if the space-time is conformally related to a plane wave space-time with a coordinate system $(u , v , z , \bar{z})$ and a function ϕ such that the metric has the plane wave form

$$ds^2 = e^{-2\phi}\{2dv[du + (D(v)z^2 + \bar{D}(v)\bar{z}^2 + e(v)z\bar{z})dv] - 2dzd\bar{z}\}$$

where $D(v)$ and $e(v)$ are arbitrary functions of v .

A sketch of the proof will now be given. Assume that the space-time manifold V_4 is of Petrov type N (77 , 87). This amounts to the assumption that there exists a necessarily null vector field ℓ which is a null vector field for the conformal curvature tensor: $C_{abcd}\ell^d = 0$, at every point. Note that the above plane wave metric is of this type with $\ell = \partial/\partial u$.

We first show , using Conditions III and V that there is a coordinate system with metric

$$ds^2 = e^{-2\phi}\{2dv[du + (\frac{1}{2}(p_z(v , z) + \bar{p}_{\bar{z}}(v , z))u + m(v , z , \bar{z}))dv] - 2(dz + p(v , z)dv)(d\bar{z} + p(v , \bar{z})dv)\}$$

where the functions p and m satisfy $p(v,z) = p_2(v)z^2 + p_1(v)_z + p_0(v)$ and $m(v , z , \bar{z}) = \bar{z}G(v , z) + zG(v , \bar{z}) + H(v , z) + \bar{H}(\bar{v} , \bar{z})$. Here also the functions of two variables G and H are either explicitly $G(v,z) = g_1(v)z + g_0(v)$, $H(v,z) = h_2(v)z^2$; or else they satisfy certain differential equations, viz

$$G_{zz}(v,z) = a_2(v)[a_1(v)z]^{\frac{1}{a_1(v)}}$$

$$H_{zz}(v,z) = \frac{a_2(v)b_1(v)}{1 + a_1(v)}[d(v)z]^{\frac{1}{d(v)}} z .$$

Here $a_2(v)$ is an arbitrary non-vanishing function while a_1 and b_1 satisfy either $a_1(v) = -\frac{1}{5}$, $(b_1(v))^2 = \frac{2}{25}(\frac{k_2}{k_1} - 1)$, or $(17 - 2k_2/k_1)|a_1(v)|^2 + 4(a_1(v) + \bar{a}_1(v)) + 1 = 0$, $b_1 = 0$. In the latter case the functions $p_1(v)$ above are arbitrary. The parameters k_1 and k_2 are respectively 3 and 4 for the scalar wave equation problem, but can be 5 and 16 for Maxwell's equations, and 8 and 13, respectively, for the case of Weyl's equations. These metrics all satisfy Conditions III and V.

We now introduce a two-component spinor formalism of Penrose (76). Here tensors and spinors are related by complex connection quantities $\sigma_a{}^{A\dot{A}}$ where $a = 1, \ldots, 4$; $A = 0, 1$. These quantities are Hermitian in the spinor indices $A, \dot{A}$ and satisfy conditions $\sigma_a{}^{A\dot{A}}\sigma^b{}_{B\dot{B}} = \delta^A_B\delta^{\dot{A}}_{\dot{B}}$. The spinor indices $B, \dot{B}$ have been lowered in this equation by the skew symmetric spinors ϵ_{AB}, $\epsilon_{\dot{A}\dot{B}}$ defined by $\epsilon_{01} = \epsilon_{\dot{0}\dot{1}} = 1$, and using the convention $\xi_A = \xi^B\epsilon_{BA}$. Likewise spinor indices can be raised by contraction with the inverse skew symmetric spinors ϵ^{AB}, $\epsilon^{\dot{A}\dot{B}}$.

The conformal curvature tensor C_{abcd} may now be represented by a four index Weyl spinor ψ_{ABCD} as follows:

$$C_{abcd}\sigma^a_{A\dot{A}}\sigma^b_{B\dot{B}}\sigma^c_{C\dot{C}}\sigma^d_{D\dot{D}} = \psi_{ABCD}\epsilon_{\dot{A}\dot{B}}\epsilon_{\dot{D}\dot{C}} + \bar{\psi}_{\dot{A}\dot{B}\dot{C}\dot{D}}\epsilon_{AB}\epsilon_{DC}$$

while the tensor L_{ab} defined above is represented by a trace-free Ricci spinor $\Phi_{AB\dot{A}\dot{B}}$, where

$$L_{ab}\sigma^a_{A\dot{A}}\sigma^b_{B\dot{B}} = 2(\Phi_{AB\dot{A}\dot{B}} - \Lambda\epsilon_{AB}\epsilon_{\dot{A}\dot{B}})$$

and $\Lambda = R/24$.

The covariant derivative of a spinor ξ_A is defined as $\xi_{A;a} = \xi_{A,a} - \xi_B\Gamma^B_{Aa}$ where Γ^B_{Aa} denote the spinor affine connection determined by requiring the covariant derivative to be real, linear, to satisfy Leibniz product rule, and to satisfy $\sigma_a{}^{A\dot{A}}{}_{;b} = \epsilon_{AB;b} = 0$.

There is also need for a basis (o_A, ι_A) for the space of valence one spinors satisfying $o_A\iota^A = 1$; these may also be used to define a spinor dyad $T_a{}^A$ by $T_0{}^A = o^A$; $T_1{}^A = \iota^A$, and an associated null tetrad $(\ell, n, m, \bar{m})$ defined by

$$\ell^a = \sigma^a_{A\dot{A}}o^A\bar{o}^{\dot{A}}, \quad n^a = \sigma^a_{A\dot{A}}\iota^A\bar{\iota}^{\dot{A}}, \quad \text{and} \quad m^a = \sigma^a_{A\dot{A}}o^A\bar{\iota}^{\dot{A}}.$$

Their only non zero inner products are $\ell_a n^a = -m_a\bar{m}^a = 1$. Then the metric tensor can be expressed as $g_{ab} = 2\ell_{(a}n_{b)} - 2m_{(a}\bar{m}_{b)}$ where parentheses indicate symmetric parts of the indices enclosed.

We must also use certain Newman-Penrose components of the Weyl tensor and the trace-free Ricci tensor, namely

$$\psi_0 = \psi_{ABCD}o^{ABCD}, \quad \psi_1 = \psi_{ABCD}o^{ABC}\iota^D$$

$$\psi_2 = \psi_{ABCD}o^{AB}\iota^{CD}, \quad \psi_3 = \psi_{ABCD}o^A\iota^{BCD}$$

$$\psi_4 = \psi_{ABCD}\iota^{ABCD}$$

and

$$\Phi_{00} = \Phi_{AB\dot{A}\dot{B}}o^{AB}\bar{o}^{\dot{A}\dot{B}}; \quad \Phi_{01} = \Phi_{AB\dot{A}\dot{B}}o^{AB}\bar{o}^{\dot{A}}\bar{\iota}^{\dot{B}}$$

$$\Phi_{02} = \Phi_{AB\dot{A}\dot{B}}o^{AB}\bar{\iota}^{\dot{A}\dot{B}}; \quad \Phi_{11} = \Phi_{AB\dot{A}\dot{B}}o^A\iota^B\bar{o}^{\dot{A}}\bar{\iota}^{\dot{B}}$$

$$\Phi_{12} = \Phi_{AB\dot{A}\dot{B}}o^A\iota^B\bar{\iota}^{\dot{A}}\bar{\iota}^{\dot{B}}; \quad \Phi_{22} = \Phi_{AB\dot{A}\dot{B}}\iota^{AB}\bar{\iota}^{\dot{A}\dot{B}}$$

where o^{AB} denotes $o^{A}o^{B}$, and so on. The Newman-Penrose differential operators are defined as $D = e^{a}\partial/\partial x_{a}$ $\Delta = n^{a}\partial/\partial x_{a}$, $\delta = m^{a}\partial/\partial x_{a}$; $\bar{\delta} = \bar{m}^{a}\partial/\partial x_{a}$.

The null tetrad preserving the direction of ℓ is determined up to a subgroup G_4 of proper orthochronous Lorentz transformations L_{+}^{+} as defined by $\ell' = e^{a}\ell$, $m' = e^{ib}(m + \bar{q}\ell)$, $n' = e^{-a}(n + qm + \bar{q}\bar{m}) + q\bar{q}\ell$ where a, b are real valued functions and q is complex. Then also $o' = e^{w/2}o$, $\iota' = e^{-w/2}(\iota + qo)$ where $w = a + ib$. We may note that the conformal transformation $\tilde{g}_{ab} = e^{2\phi}g_{ab}$ is induced by the transformation $\tilde{\ell}_{a} = \ell_{a}$, $\tilde{n}_{a} = e^{2\phi}n_{a}$, $\tilde{m}_{a} = e^{\phi}m_{a}$.

The Newman-Penrose spin coefficients associated to the dyad T_{a}^{A} and the null tetrad $(\ell, n, m, \bar{m})$ are

$$o_{A;B\dot{B}} = \gamma o_{A}o_{B}\bar{o}_{\dot{B}} - \alpha o_{A}o_{B}\bar{\iota}_{\dot{B}} - \beta o_{A}\iota_{B}\bar{o}_{\dot{B}} + \epsilon o_{A}\iota_{B}\bar{\iota}_{\dot{B}} - \tau o_{A}o_{B}\bar{o}_{\dot{B}} + \\ + \rho\iota_{A}o_{B}\bar{\iota}_{\dot{B}} + \sigma\iota_{A}\iota_{B}\bar{o}_{\dot{B}} - \kappa\iota_{A}\iota_{B}\bar{\iota}_{\dot{B}}$$

and

$$\iota_{A;B\dot{B}} = \nu o_{A}o_{B}\bar{o}_{\dot{B}} - \lambda o_{A}o_{B}\bar{\iota}_{\dot{B}} - \mu o_{A}\iota_{B}\bar{o}_{\dot{B}} + \pi o_{A}\iota_{B}\bar{\iota}_{\dot{B}} - \gamma\iota_{A}o_{B}\bar{o}_{\dot{B}} + \\ + \alpha\iota_{A}o_{B}\bar{\iota}_{\dot{B}} + \beta\iota_{A}\iota_{B}\bar{o}_{\dot{B}} - \epsilon\iota_{A}\iota_{B}\bar{\iota}_{\dot{B}}$$

and among the transformations of these coefficients induced by the above subgroup G_4 are these: $\tilde{\kappa} = e^{-3\phi}\kappa$; $\tilde{\rho} = e^{-2\phi}(\rho - D\phi)$; $\tilde{\sigma} = e^{-2\phi}\sigma$, $\tilde{\tau} = e^{-\phi}(\tau - \delta\phi)$.

3.8 Application of Conditions III and V.

The third and fifth main conditions can be expressed in spinor form by contracting with an appropriate number of the σ's and observing that a trace-free symmetric tensor goes over into a Hermitian spinor symmetric in its dotted and undotted indices. Thus are obtained the conditions

$$\text{III} \quad \psi_{ABKL;}{}^{K}{}_{\dot{A}}{}^{L}{}_{\dot{B}} + \psi_{\dot{A}\dot{B}\dot{K}\dot{L};}{}^{\dot{K}}{}_{A}{}^{\dot{L}}{}_{B} + \psi_{AB}{}^{KL}\Phi_{KL\dot{A}\dot{B}} + \psi_{\dot{A}\dot{B}}{}^{\dot{K}\dot{L}}\Phi_{KL AB} = 0,$$

and

$$\text{V} \quad k_1\psi_{ABCD;K\dot{K}}\psi_{\dot{A}\dot{B}\dot{C}\dot{D};}{}^{K\dot{K}} + k_2\psi^{K}{}_{(ABC;D)}{}_{(\dot{A}}\bar{\psi}_{\dot{B}\dot{C}\dot{D})\dot{L};}{}^{\dot{L}}{}_{K} + \\ + k_2\psi^{\dot{K}}{}_{(\dot{A}\dot{B}\dot{C};\dot{D})}{}_{(A}\bar{\psi}_{BCD)L;\dot{K}}{}^{L} - 2(8k_1 - k_2)\psi_{(ABC|K|;}{}^{K}{}_{(\dot{A}}\psi_{\dot{B}\dot{C}\dot{D})\dot{K};}{}^{\dot{K}}{}_{D)} - \\ - k_2\psi^{K}{}_{(ABC}\bar{\psi}_{(\dot{A}\dot{B}\dot{C}|\dot{L}|;}{}^{L}{}_{|K|D)\dot{D})} - k_2\psi^{\dot{K}}{}_{(\dot{A}\dot{B}\dot{C}}\bar{\psi}_{(ABC|L|;}{}^{\dot{L}}{}_{|\dot{K}|D\dot{D})} + \\ + 4k_1\psi^{K}{}_{(ABC}\bar{\psi}_{\dot{A}\dot{B}\dot{C}|\dot{L}|;}{}^{\dot{L}}{}_{D)K\dot{D})} + 4k_1\psi^{\dot{K}}{}_{(\dot{A}\dot{B}\dot{C}}\bar{\psi}_{(ABC|L|;}{}^{L}{}_{\dot{D})D)\dot{K}} + \\ + 2(k_2 - 4k_1)\psi^{K}{}_{(ABC}\Phi_{D)K\dot{K}(\dot{A}}\bar{\psi}_{\dot{B}\dot{C}\dot{D})}{}^{\dot{K}} - 2(4k_1 + k_2)\Lambda\psi_{ABCD}\bar{\psi}_{\dot{A}\dot{B}\dot{C}\dot{D}} = 0$$

Next we suppose the space time is of Petrov type N which in spinor form amounts to the existence of a principal spinor o_A such that the Weyl spinor has the form $\psi_{ABCD} = \psi o_{A}o_{B}o_{C}o_{D}$ where $\psi = \psi_4$. We can select o_A to be the first spinor in a dyad which in view of the preceding formulas implies that ψ_0, ψ_1, ψ_2 and ψ_3 are all zero. In fact we can choose $\psi_4 = 1$ in view of the previous dyad transformation formulas. We then proceed by substituting in the two conditions, eliminating covariant derivatives by using the spin conditions listed above, contracted with suitable products of o^A and ι^A. Each dyad equation thus obtained will be a conformal invariant.

For example the first contraction to consider in V is with $o^{ABC}\iota^{D}\bar{o}^{\dot A}\bar\iota^{\dot B\dot C\dot D}$ and this yields $(k_2-4k_1)\kappa^2 = 0$. Since $k_2 \neq 4k_1$ in any of the possible applications, we find $\kappa = 0$. Hence the principal null congruence of C_{abcd} defined by the principal null vector field $\ell^a = \sigma^a{}_{A\dot A}o^A\bar{o}^{\dot A}$ will be geodesic.

Next we note that it is always possible to set $\bar{\tilde\delta} = -\tilde\delta$, in view of the transformation formula $\tilde\delta = e^{-2\phi}(\rho - D\phi)$ because this implies $D\phi = \frac{1}{2}(\rho + \bar\rho)$ which always has a solution ϕ that defines an appropriate conformal transformation in which $\bar{\tilde\rho} = -\tilde\rho$. Hence, dropping the tildes; we can assume $\bar\rho = -\rho$.

Continuing with further contractions, $o^{AB}\iota^{CD}\bar\iota^{\dot A\dot B\dot C\dot D}$ with V yields $2k_1 D\sigma + \sigma[(12k_1 + k_2)\rho + 2(k_1 - k_2)\epsilon + 2(17k_1 - k_2)\bar\epsilon] = 0$ while $o^{A}\iota^{BCD}o^{\dot A}\iota^{\dot B\dot C\dot D}$ with V yields $(k_2 - k_1)[2D(\varepsilon + \bar\varepsilon) - 7\rho^2 + 12\rho(\varepsilon - \bar\varepsilon) + 6\varepsilon^2 + 6\bar\varepsilon^2 + \Phi_{\infty}] + (15k_2 - 92k_1)\sigma\bar\sigma - 4(28k_1 + k_2)\varepsilon\bar\varepsilon = 0$ and $o^{AB}\bar\iota^{\dot A\dot B}$ with III yields $D(4\varepsilon - \rho) - 4\varepsilon\bar\varepsilon - 7\varepsilon\rho + \rho^2 - \sigma\bar\sigma + \bar\varepsilon\rho + 12\varepsilon^2 + 2\sigma^2 + \Phi_{\infty} = 0$. From the Newman-Penrose calculus we have $D\rho = \rho^2 + \sigma\bar\sigma + \rho(\varepsilon + \bar\varepsilon) + \Phi_{\infty}$ and $D\sigma = \sigma(3\epsilon - \bar\epsilon)$ so eliminating $D\sigma$ we find

$$\sigma[(12k_1 + k_2)\rho + 2(4k_1 - k_2)\varepsilon + 2(16k_1 - k_2)\bar\varepsilon] = 0$$

Assuming $\sigma = 0$ the square bracket is zero. Taking its real part we have $(10k_1 - k_2)(\varepsilon + \bar\varepsilon) = 0$ so that $\bar\varepsilon = -\varepsilon$, since $10k_1 \neq k_2$ for any applicable case. Solving back for ϵ we obtain $\varepsilon = (12k_1 + k_2)\rho/24k_1$.

We now require equations (4.2a), (4.2b) of (76 b). In our case they reduce to $D\rho = \rho^2 + \sigma\bar\sigma + \rho(\varepsilon + \bar\varepsilon) + \Phi_{\infty}$ and $D\sigma = \sigma(3\epsilon - \bar\epsilon)$. We can now eliminate $D\sigma$ from the contraction above, obtaining $\sigma[(12k_1 + k_2)\rho + 2(4k_1 - k_2)\varepsilon + 2(16k_1 - k_2)\bar\varepsilon] = 0$. Under the assumptions made it can be shown that the expression in the square bracket does not vanish. Hence $\sigma = 0$, and this condition remains invariant under the conformal transformations discussed above. The vanishing of σ is said to make the principal null congruence of C_{abcd} defined by ℓ^a shear-free.

We can also eliminate $D(\varepsilon + \bar\varepsilon)$ between the last two of the contractions listed above, and obtain $(4k_1 - k_2)[\rho^2 + 2\rho(\varepsilon - \bar\varepsilon)] - 16k_1\varepsilon\bar\varepsilon = 0$. Completing the square yields $(4k_1 - k_2)(\rho + \varepsilon - \bar\varepsilon)^2 - (4k_1 - k_2)(\varepsilon + \bar\varepsilon)^2 - 4k_2\varepsilon\bar\varepsilon = 0$. Since $\bar\rho = -\rho$ and $4k_1 > k_2$ in all cases, the first term is negative. It follows that $\varepsilon = 0$ and also $\rho = 0$. These conditions are not invariant under the conformal transformations studied above, but their invariant form can be found: it is $\rho = \bar\rho$ and $D\psi + 4\varepsilon\psi = 0$.

The vanishing of κ, ρ, ε and σ in one coordinate system also implies the vanishing of certain components of the trace-free Ricci tensor: these are $\Phi_{00} = 0$, $\Phi_{01} = 0$ as may be seen from (76 b, Eqns. 4. 2a, 4.2k).

In the conformal transformation formulas we can set $\tilde\tau = 0$ while maintaining $\tilde{\rho} = 0$, provided ϕ satisfies the differential system $D\phi = 0$, $\delta\phi = \tau$, $\delta\phi = \bar{\tau}$. McLenaghan and Carminati show that the conditions of integrability of this system are satisfied, so a solution exists. Henceforth it is thus possible to assume $\tau = 0$, dropping the tilde.

The conditions obtained so far can be summarized: Conditions III and V imply that for any null tetrad $(\ell, u, m, \bar{m})$, ℓ a principal null vector of the type N Weyl tensor, there is a conformal transformation ϕ for which $\kappa = \sigma = \rho = \tau = 0$, $D\psi + 4\epsilon\psi = 0$, $\psi_0 = \psi_1 = \psi_2 = \psi_3 = 0$; $\Phi_{00} = \Phi_{01} = \Phi_{02} = \Lambda = 0$. To put these conditions into a more useful form we differentiate the Weyl spinor $\psi_{ABCD} = \psi o_A o_B o_C o_D$, obtaining $\psi_{ABCD;E\dot{E}} = \psi_{ABCD} K_{E\dot{E}}$ where, by the spin conditions, the quantity $K_{E\dot{E}} = \psi^{-1}\psi_{;E\dot{E}} + 4\gamma o_E \bar{o}_{\dot{E}} - 4\alpha o_E \bar{\iota}_E - 4\beta \iota_E o_{\dot{E}} + 4\epsilon \bar{\iota}_E \bar{\iota}_{\dot{E}}$.

This spinor equation is equivalent to the tensor form $\overset{+}{C}_{abcd;e} = \overset{+}{C}_{abcd} K_e$ where $\overset{+}{C}_{abcd} = \frac{1}{2}(C_{abcd} - i{*C}_{abcd})$, $K_e = o_e{}^{E\dot{E}} K_{E\dot{E}}$, and ${*C}_{abcd} = \frac{1}{2}\epsilon_{abef} C^{ef}{}_{cd}$ with ε_{abef} the Levi Civita tensor. This means V_4 is a complex recurrent space time (64). Thus any Petrov type N space time satisfying III and V is conformally related to a complex recurrent space time, with conformal factor $e^{-2\phi}$, and ϕ satisfying the conditions of the proof above.

3.9 Reduction of the complex recurrent metric.

Now drop the assumption $\psi = 1$ and work with a general spinor dyad with o_A a principal null spinor of ψ_{ABCD} . We then find

$$\psi_{ABCD;E\dot{E}} = o_{ABCD}(\theta_1 o_E \bar{o}_{\dot{E}} + \theta_2 o_E \bar{\iota}_{\dot{E}} + \theta_3 \iota_E \bar{o}_{\dot{E}})$$

where $\theta_1 = \Delta\psi + 4\gamma\psi$, $\theta_2 = -(\bar{\delta}\psi + 4\alpha\psi)$ and $\theta_3 = -(\delta\psi + 4\beta\psi)$. Contracting with ε^{DC} we have $\psi_{ABCL;}{}^{L}{}_{\dot{E}} = \theta_3 o_{ABC} \bar{o}_{\dot{E}}$, and we shall need the transformation formula $\theta_3' = e^{-2a-ib}\theta_3$. Covariant differentiation now yields

$$\psi_{ABCL;}{}^{L}{}_{EF\dot{F}} = o_{ABC}\bar{o}_{\dot{E}}\,(\Sigma_8 o_F \bar{o}_F + \Sigma_9 o_F \bar{\iota}_F + \Sigma_{10} \iota_F \bar{o}_F)$$

where

$$\Sigma_8 = \Delta\theta_3 + (3\gamma + \bar{\gamma})\theta_3 \ , \ \Sigma_9 = -[\bar{\delta}\theta_3 + (3\alpha + \beta)\theta_3]$$

and

$$\Sigma_{10} = -[\delta\theta_3 + (3\beta + \alpha)\theta_3] \ .$$

The coefficient of $\iota_F \bar{\iota}_F$ in the above formula turns out to be identically zero. We also require the formula

$$\Phi_{AB\dot{A}\dot{B}} = \Phi_{22} o_{AB} \bar{o}_{\dot{A}\dot{B}} - 2\Phi_{21} o_{AB} \bar{o}_{(\dot{A}} \bar{\iota}_{\dot{B})} - 2\Phi_{12} o_{(A} \iota_{B)} \bar{o}_{\dot{A}\dot{B}} + $$
$$+ 4\Phi_{11} o_{(A} \iota_{B)} \bar{o}_{(\dot{A}} \bar{\iota}_{\dot{B})}$$

which follows from the conditions listed above.

These conditions now imply that the remaining information in III and V is comprised by the equations $\Sigma_{10} + \bar{\Sigma}_{10} = 0$, $4k_1\bar{\psi}\Sigma_9 + 4k_1\psi\bar{\Sigma}_9 + k_1\theta_2\bar{\theta}_2 + (17k_1 - 2k_2)\theta_3\bar{\theta}_3 + 2(4k_1 - k_2)\psi\bar{\psi}\Phi_{11} = 0$.

To solve these equations we use canonical coordinates $(u, v, z, \bar{z})$ for a type N complex recurrent space-time (64), wherein

$$ds^2 = 2dv[du + (ek^2(v)u^2 + \ell(v, z, \bar{z})u + m(v, z, \bar{z}))dv] - $$
$$- 2k^{-2}(v)(1 + ez\bar{z})^{-2}(dz + p(v, z)dv)(d\bar{z} + \bar{p}(v, z)dv)$$

with $e = -1, 0$ or 1; $k^2(v) = 1 + e^2K^2(v)$ and $\ell(v, z, \bar{z}) = \frac{1}{2}(p_z(v,z) + \bar{p}_{\bar{z}}(v,z)) - e(1+ez\bar{z})^{-1}(\bar{z}p(v,z) + z\bar{p}(v,z))$.

Now a canonical null tetrad for this metric, in which ℓ^a is a principal null vector of the type N Weyl conformal curvature tensor, can be defined by the following choice of Newman-Penrose operators (64): $D = \partial/\partial u$, $\delta = -k(v)(1+ez\bar{z})\partial/\partial z$, $\Delta = \partial/\partial v - p(v,z)\partial/\partial z - \bar{p}(v,z)\partial/\partial\bar{z} - (ek^2(v)u^2 + \ell(v, z, \bar{z})u + m(v, z, \bar{z}))\partial/\partial u$. McLenaghan and Carminati then show that the condition $D\psi + 4\epsilon\psi = 0$ now becomes $p_{zzz}(v,z) = 0$, so that $p(v,z) = p_2(v)z^2 + p_1(v)z + p_0(v)$. Examining next the condition $\Sigma_{10} + \bar{\Sigma}_{10} = 0$ they calculate

$$\Sigma_{10} = k^2[(1+ez\bar{z})^2\psi_{z\bar{z}} - 2ez(1+ez\bar{z})\psi_{\bar{z}} + 2e^2z^2\psi]$$

and then, by eliminating ψ that

$$\Sigma_{10} = k^4(1+ez\bar{z})^3[(1+ez\bar{z})m_{zz\bar{z}\bar{z}} + 2e(zm_{zz\bar{z}} + \bar{z}m_{z\bar{z}\bar{z}}) + 4em_{z\bar{z}}]$$

so that $\bar{\Sigma}_{10} = \Sigma_{10}$ and consequently $\Sigma_{10} = 0$, an invariant condition. The first form above for Σ_{10} then gives an equation for ψ with general solution $\psi(v, z, \bar{z}) = (1+ez\bar{z})(\bar{z}A(v,z) + B(v,z))$, where $A(v,z)$ and $B(v,z)$ are arbitrary functions. In the case $e = 0$, an alternate solution is given by $m(v,z,\bar{z}) = \bar{z}G(v,z) + z\bar{G}(v,z) + H(v,z) + \bar{H}(v,z)$ with arbitrary functions $G(v,z)$ and $H(v,z)$. When $e = 0$, then $\psi(v,z,\bar{z}) = m_{zz}(v,z,\bar{z})$ so that $G_{zz}(v,z) = A(v,z)$ and $H_{zz}(v,z) = B(v,z)$.

The last condition above on Σ_9 must now be solved, and in the case $e = 0$ it takes the form $4(z\bar{A}+\bar{B})A' + 4(\bar{z}A+B)\bar{A}' + (\bar{z}A'+B') \times (z\bar{A}'+\bar{B}') + (17 - 2k_2/k_1)A\bar{A} = 0$.

When $A' = 0$ this becomes $B'\bar{B}' + (17 - 2k_2/k_1)A\bar{A} = 0$ and since $2k_2 < 17k_1$ in all applicable cases this implies $A = 0$ and $B' = 0$. Hence the functions G and H have the forms $G(v,z) = g_1(v)z + g_0(v)$ and $H(v,z) = h_2(v)z^2 + h_1(v)z + h_0(v)$, where the g_i and h_i are arbitrary functions of v. Hence we obtain directly that

$$m(v,z,\bar{z}) = D(v)z^2 + \bar{D}(v)z^{-2} + e(v)z\bar{z} + F(v)z + \bar{F}(v)\bar{z} + g(v).$$

This now gives the first form of the metric as stated at the outset of the first stage of this proof.

Considering now the case $A' \neq 0$ we divide our condition above by $A'\bar{A}'$ and apply $\partial^4/\partial z^2\partial\bar{z}^2$ to it to obtain

$$\left(\frac{B'}{A'}\right)''\left(\frac{\bar{B}'}{\bar{A}'}\right)'' + \left(17 - \frac{2k_2}{k_1}\right)\left(\frac{A}{A'}\right)''\left(\frac{\bar{A}}{\bar{A}'}\right)'' = 0 .$$

Since $2k_2 < 17k_1$ in all applicable cases, this clearly implies $(A/A')'' = 0$ and $(B'/A')'' = 0$. Hence $A/A' = a_1(v)z + a_0(v)$ and $B'/A' = b_1(v)z + b_0(v)$ where the new coefficients $a_i(v)$, $b_i(v)$ satisfy in particular a subsidiary relation found by the differentiation $\partial^2/\partial z\partial\bar{z}$ only, namely

$$\left(17 - \frac{2k_2}{k_1}\right)|a_1(v)|^2 + 4(a_1(v) + \bar{a}_1(v)) + |b_1(v)|^2 + 1 = 0 .$$

By a certain transformation among u, v, and z it can also be shown that $a_0(v)$ can be made zero. Then it will readily follow that $A(v,z) = a_2(v)[a_1(v)z]^{1/a_1(v)}$, where $a_2(v) \neq 0$ is arbitrary. The equation for B' can now be integrated yielding

$$B(v,z) = b_2(v) + a_2(v)(1 + a_1(v))^{-1}[b_1(v)z + b_0(v)(1 + a_1(v))] \times \times (a_1(v)z)^{\frac{1}{a_1(v)}} ,$$

where $b_2(v)$ is yet to be determined.

To find $b_2(v)$ substitute for A and B into the general condition when $e = 0$, and obtain a relation of the form

$$b_2(v)a_1^{-1}(v)(a_1(v)z)^{1 - \frac{1}{a_1(v)}} + P(v,z,\bar{z}) = 0 ,$$

where P is a polynomial of degree less than 3 in z and $\bar{z}$. But from the above subsidiary relation for $a_1(v)$ and $b_1(v)$ it is easily shown that $1 - a_1(v)^{-1}$ is never equal to $0, 1$ or 2. Now by taking $\partial^3/\partial z^3$ of the polynomial relation above, we obtain

$$b_2(v)a_2^{-1}(v)(a_1(v) - 1)(a_1(v) + 1)(a_1(v)z)^{-2 - \frac{1}{a_1(v)}} = 0$$

and now $b_2(v) = 0$ follows at once. Furthermore $P(v,z,\bar{z})$ must now vanish, so we obtain also that $b_1(v)(5a_1(v) + 1) = 0$, $b_0(v)(4a_1(v) + 1) + \bar{b}_0(v)b_1(v) = 0$, and $|b_0(v)|^2 = 0$. Thus $b_0(v) \equiv 0$ and either $b_1(v) = 0$ or $a_1(v) = -1/5$. In the first case $b = 0$ and $a_1(v)$ satisfies $(17 - 2k_2/k_1)|a_1(v)|^2 + 4(a_1(v) + \bar{a}_1(v)) + 1 = 0$ which does have solutions for each of the three applicable cases. This establishes the second of the cases listed at the beginning of Stage 1. In the other case $a_1(v) = -1/5$ we find $A(v,z) = a_2(v)(-z/5)^{-5}$ and $B(v,z) = 5/4a_2(v)b_1(v)(-z/5)^{-5}z$ where $|b_1(v)|^2 = 2/25(k_2/k_1 - 1) > 0$. This leads to the first case and Stage 1 is thus completed, at least in the case $e = 0$.

However, by a further lengthy calculation, McLenaghan and Carminati show that the case $e \neq 0$ does not lead to any solutions. Hence Conditions III and V lead to the form of the metric given at the outset of the proof.

3.10 Application of Condition VII

To complete the main proof we must now apply Condition VII to the solutions of the first stage. As the description of the proof up to this point has shown its specific, detailed, and arduous character, we omit all details of the spinor calculations. We shall simply report that one particular spinorial contraction of Condition VII for the wave equation $\square u + 1/6Ru = 0$ yields $\bar{A}A_{zz} = 0$, so that $A_{zz} = 0$ and hence $a_2(v)(1 - a_1(v))[a_1(v)z]^{-2 - 1/a_1(v)} = 0$. Unless $a_2(v) = 0$, $a_1(v)$ must be equal to unity, and this is not consistent with the solutions of the second alternative listed at the outset. Hence any possible solutions must derive from the first alternative given there. In this case it is also found that Condition VII becomes in effect $p_2(v) = 0$, so that $p(z,v) = p_1(v)z + p_0(v)$. But in this case the metric can be transformed (64) to the plane wave form stated in the theorem. Moreover it is known from the work of Günther (32) that these plane wave metrics do have the Huygens property. This completes the outline of the proof of the theorem for Petrov type N metrics for the wave equation $\square u + \frac{1}{6} Ru = 0$

3.11 Conclusion.

As the Petrov type N metrics include the plane wave counter-example cases, it is possible that the other Petrov types will not be as intricate as type N , for the purposes of deciding the Hadamard conjecture in four space-time dimensions. However it is evident that somewhat different approaches in detail will be required. The general case of wave equations with first derivative coefficients would then still remain, but could be expected to be less difficult than the basic problem of the metric, or second derivative coefficients. For higher dimensions, the Hadamard conjecture remains far from any complete resolution.

This problem also has potential astronomical, or cosmological, interest, in view of the following question: can inferences on the curvature of space-time be made from observations of radiation from distant objects, or novae, or other sources? As yet, this aspect seems very little developed.

REFERENCES

1. Alinhac, S.: Problemes de Cauchy pour des operateurs singuliers, Bull. Soc. Mat. France, 102 (1974), pp 289-315.
2. __________: Parametrix et propagation des singularités pour un problème de Cauchy a multiplicité variable, Asterisque, 34-35 (1976), pp 3-26
3. __________: Parametrix pour une system hyperbolique à multiplicité variable, Comm. PDE, 2(3) (1977), pp 251-296.
4. Asgeirsson, L.: Some hints on Huygens' Principle and Hadamard's conjecture, Comm. Pure and App. Math., 9 (1956), pp 307-326.
5. Atiyah, M., Bott, R., Gårding, L.: Lacunas for hyperbolic differential operators with constant coefficients, Acta Math., I 124 (1970), pp 109-189, II 131 (1973), pp 145-206.
6. Beals, R. W., and Greiner, P. C.: Calculus on Heisenberg manifolds, to appear, Princeton U. P., 1986.

7. Beals, M., and Reed, M.: Propagation of singularities for hyperbolic pseudo differential operators with non-smooth coefficients, Comm. P. A. M., 35 (1982), pp 169-184.
8. Bony, J. M., and Schapira, P.: Propagation des singularités analytiques pour les solutions des equations aux derivées partielles, Ann. Inst. Fourier, 26 (1976), pp 81-140.
9. Bruhat, Y.: Theorème d'existence pour certaines systemes d'equations aux derivées partielles non lineaires, Acta Math., 88 (1952), pp 141-225.
10. Cahen, M., and McLenaghan, R.: Metrique des espaces lorentziens symetriques a quatre dimensions, C. R. Acad Sci. Paris, 266 (1968), pp 1125-1128.
11. Carminati, J., and McLenaghan, R. G.: Some new results on the validity of Huygens' principle for the scalar wave equation on a curved space-time, in Gravitation, Geometry and Relativistic Physics, Proceedings of the Journées Relativistés 1984, Aussois, France, edited by Laboratoire, "Gravitation et Cosmologie Relativistes," Inst. H. Poincaré, Lecture Notes in Physics 212, Springer, Berlin, 1984.
12. ______________________: An explicit determination of the Petrov type N space times on which the conformally invariant scalar wave equation satisfies Huygen's principle, Physics Letters, 105 A, no. 7 (1984), pp 351-4, and to appear.
13. Chazarain, J.: Operateurs hyperboliques à caracteristiques de multiplicité constante, Ann. Inst. Fourier, 24 (1974), pp 173-202.
14. __________: Propagation des singularités pour une classe d'operateurs à caracteristiques multiples et resolubilité locale, Ann. Inst. Fourier, 24 (1974), pp 209-233.
15. Chevalier, M.: Sur le noyau de diffusion de l'operateur laplacien, C. R. Acad. Sci. Paris, 264 (1967), pp 380-382.
16. Cordoba, A., and Fefferman, C.: Wave packets and Fourier integral operators, Comm. P. D. E., 3(11) (1978), pp 979-1005.
17. Debever, R.: Le rayonnement gravitationnel et tenseur de Riemann en relativité générale, Cahs. Physique, 168-9 (1964), pp 303-349.
18.a) Douglis, A.: The problem of Cauchy for linear hyperbolic equations of second order, Comm. P. A. M., 7 (1954), pp 271-295.
18.b) __________: A criterion for the validity of Huygens' principle, Comm. P. A. M., 9 (1956), pp 391-402.
19. Duistermaat, J., and Hörmander, L.: Fourier Integral Operators, II, Acta Math., 128 (1972), pp 183-269.
20. Duistermaat, J. J., and Sjostrand, J.: A Global Construction for Pseudo differential operators with non-involutive characteristics, Inventiones Math., 20 (1973), pp 209-225.
21. Egorov, Yu. V.: Subelliptic operators, Uspeckhi Mat. Nauk, 30 (1975), pp 57-104; English translation, Russian Math. Surveys, 30 (1975), (2), pp 57-114; (3), pp 57-104.
22. Ehlers, J., and Kundt, K.: Exact solutions of the gravitational field equations, article in Gravitation an introduction to current research, ed. L. Witten, Wiley, New York (1964).
23. Fefferman, C. L.: The Uncertainty Principle, Bull. A. M. S., 9(2) (1983), pp 129-206.

24. Fefferman, C. L., and Phong, D. H.: The uncertainty principle and sharp Gårding inequalities, Comm. P. A. Math., 34 (1981), pp 285-331.
25. Friedlander, F. G.: The wave equation in a curved space-time, Cambridge U. P. (1975), x + 282 p.
26. Godin, P.: A class of pseudo-differential operators which do not propagate singularities, Comm. PDE, 5(7) (1980), pp 683-781.
27. Goldschmidt, H.: Existence theorems for analytic linear partial differential equations, Ann. Math., 86 (1967), 246-270.
28. ______________: Integrability criteria for systems of non-linear partial differential equations, J. Diff. Geom., 1 (1967), 269-307.
29. Greiner, P., and Stein, E. M.: Estimates for the $\bar{\partial}$ - Neumann problem, Math Notes, 19 Princeton U. P. (1977).
30. Günther, P.: Zur Gultigkeit des Huygensschen Princips bei partiellen Differential-gleichungen von normalen hyperbolischen Typus, S. - B. Sächs Akad Wiss Leipzig Math - Natur Kl, 100 (1952), 1-43.
31. __________: Uber einige spezielle Probleme aus des Theorie der linearen partiellen Differentialgleichungen Zweiter Ordnung S. - B. Sachs. Akad. Wiss. Leipzig, Math. Natur. Kl, 102 (1957), pp 1-50.
32. __________: Ein Beispeil einer nichttrivalen huygensschen Differentialgleichungen mit 4 unabhängingen Veranderlichen, Archive for Rat. Mech. and Analysis, 18 (1965), pp 103-106.
33. __________: Einige Sätze uber huygenssche Differential-gleichungen Wiss. Zeit. Karl Marx Univ. Leipzig Math.-Naturwiss. Reihe, 3 (1965), pp 497-507.
34. Günther, P., and Wünsch, V.: Maxwellsche Gleichungen und Huygenssches Prinzip, I, Math. Nachr., 63 (1974), 97-121.
35. Hadamard, J.: Lectures on Cauchy's problem in linear partial differential equations, Silliman Lectures, Yale U.P. 1923.
36. __________: The problem of diffusion of waves, Ann. Math. 43 (1942), pp 510 - 522.
37. Hamada, Y., Leray, J., and Wagschal, C.: Systems d'equations aux derivées partielles à caracteristiques multiples: problème de Cauchy ramifié; hyperbolicité partielle, J. Math. pures appl., 55 (1976), pp 297-352.
38. Hanges, N.: Parametrices and propagation of singularities for operators with non-involutive characteristics, Indiana Univ. Math. J., 28 (1) (1979), pp 87-97.
39. __________: Propagation of Singularities for a class of operators with double characteristics, in Seminar on Singularities of Solutions of linear partial differential equations, Princeton U. P. (1979), 113-126.
40. __________: Propagation of analyticity along real bicharacteristics, Duke Math. J., 48 (1981), pp 269-277.
41. Hölder, E.: Poissonsche Wellenformel in nicht euclidiscken Raumen, Ber. Verh. Sachs. Akad. Wiss. Leipzig, 99 (1938), pp 53-66.

42. Hörmander, L.: Pseudodifferential operators and non-elliptic boundary problems, Ann. Math., 83 (1966), pp 129-209.
43. ____________: Hypoelliptic second order differential equations, Acta Math., 119 (1967), pp 147-171.
44. ____________: Fourier Integral Operators, I, Acta Math., 127 (1971), pp 79-183.
45. ____________: Spectral analysis of singularities, in Seminar on Singularities of Solutions of linear partial differential equations, Princeton U. P. (1979), pp 3-49.
46. ____________: Subelliptic operators, ibid, Princeton U. P. (1979), pp 127-208.
47. ____________: The Analysis of Linear Partial Differential Operators, 4 vols., Springer, 1983, 1985.
48.a) Ibragimov, N. H., and Mamontov, E. V.: Sur le problème de J. Hadamard relatif à la diffusion des ondes, C. R. Acad. Sci. Paris, 270 (1970), pp 456-8.
48.b) ____________________: On the Cauchy problem for the equation $u_{tt} - u_{xx} - \Sigma_{i,j=1}^{n-1} a_{ij}(x-t) U_{y_i y_j} = 0$, Math. Sbornik, 102 (144) (1977), pp 347-363.
49. Ivrii, V. Ja.: Wave fronts of solutions of some microlocally hyperbolic pseudodifferential equations, Soviet Math. Dokl., 17 (1976), pp 233-6.
50. Janet, M.: Les systèmes d'equations aux derivées partielles, J. de Math. (8), vol. 3 (1920), pp 65-151.
51. Kashiwara, M., Kawai, T.: Second microlocalization and asymptotic expansions, Springer Lecture Notes in Physics, 126 (1980), pp 21-76.
52.a) Kataoka, K.: Microlocal theory of boundary value problems, I, J. Fac. Sci. Univ. of Tokyo, Sect. 1 A, 27 (1980), pp 355-399.
52.b) __________: II Theorems on regularity up to the boundary for reflective and diffractive operators, J. Fac. Sci. Univ. Tokyo, Sect. 1 A, 28 (1981), pp 31-56.
53. Künzle, H. P.: Maxwell fields satisfying Huygens' Principle, Proc. Camb. Phil. Soc., 64 (1968), pp 779-785.
54. Lascar, B.: Propagation des singularités pour des equations hyperboliques a caracteristique de multiplicité au plus double et singularites Masloviennes. Am. J. Math., 104 (1982), pp 227-286.
55. Lascar, B., and Sjöstrand, J.: Equation de Schrödinger et propagation des singularités pour des operateurs pseudo differentials à caracteristique reelles de multiplicité variable, I, Asterisque, 95 (1982), pp 167-207, II, Comm. in P. D. E., 10 (5) (1985), pp 467-523.
56. Lascar, R.: Propagation des singularités des Solutions d'Equations Pseudo-Differentielles a caracteristiques de Multiplicités Variables, Springer, Lecture Notes in Mathematics, no. 856 (1981), pp 237.
57. Laubin, P.: Refraction conique et propagation des singularités analytiques, J. Math. pure et appl., 63 (1984), pp 149-168.

58. Ludwig, D., and Granoff, B.: Propagation of singularities along characteristics with non-uniform multiplicity, J. Math. Anal. Appl., 21 (1968), pp 566-574.
59. Mathisson, M.: Le problème de M. Hadamard relatif à la diffusion des ondes, Acta Math., 71 (1939), pp 249-282.
60. ____________: Eine Lösungsmethode for Differential gleichungen vom normalen hyperbolischen Typus, Math. Ann., 107 (1932, pp 400-419.
61. McLenaghan, R. G.: An explicit determination of the empty space times on which the wave equation satisfies Huygens' principle, Proc. Camb. Phil. Soc., 65 (1969), pp 139-155.
62. ________________: On the validity of Huygens' Principle for second order partial differential equations with four independent variables, Part I: Derivation of necessary conditions , Ann. Inst. H. Poincaré, 20 (1974), pp 153-188.
63. ________________: Huygen's Principle, Ann. Inst. H. Poincaré, Section A, 37 (1982), pp 211-236.
64. McLenaghan, R. G., and Leroy, J.: Complex recurrent space-times, Proc. Roy. Soc. London, A 327 (1972), pp 229-249.
65. Melrose, R. B.: Equivalence of glancing hypersurfaces, I, Inventiones Math., 37 (1976), pp 165-191; II, Math. Ann., 255 (1981), pp 159-198.
66. _____________: Differential Boundary Value Problems of Principal Type, in Seminar on Singularities of Solutions of linear partial differential equations, Princeton U. P., (1979), pp 81-112.
67. _____________: Transformation of boundary problems, Acta Math., 147 (1981), pp 149-236.
68. Melrose, R. B., and Sjöstrand, J.: Singularities of boundary value problems, I, Comm. P. A. M., 31 (1978), pp 593-617.
69. ______________________________: Singularities of boundary value problems, II, Comm. P. A. M., 35 (1982), pp 129-168.
70. Melrose, R., and Uhlmann, G.: Microlocal structure of involutive conical refraction, Duke Math. J., 46 (1979), pp 571-582.
71. Moyer, R.: On the Nirenberg-Trèves condition for local solvability, J. Differential Equations, 26 (1977), pp 223-239.
72. Nagaraj, B. R.: Microlocal analysis of Operators with non-involutive characteristics, manuscript.
73. Oleinik, O., and Radkevitch, E.: Second order equations with non-negative characteristic form (translated from Russian), Plenum Press, New York (1973), vii + 259 p.
74. Nirenberg, L., and Trèves, F.: On local solvability of linear partial differential equations, Comm. P. A. M., 23 (1970), I Necessary conditions, pp 1-38; II Sufficient conditions, pp 459-509.
75. Parenti, C., and Rodino, L.: A pseudo differential operator which shifts the wave front set, Proc. Amer. Math. Soc., 72 (1978), pp 251-257.
76.a) Penrose, R.: A spinor approach to general relativity, Ann. Physics, 10 (1960), pp 171-201.
76.b) Penrose, R., and Newman, E. T.: An approach to gravitational radiation by a method of spin coefficients, J. Math. Phys., 3 (1962), pp. 566-578.

77. Petrov, A. Z.: Einstein-Raume, Akademic Verlag, Berlin, (1964).
78. Pommaret, J. F.: Systems of partial differential equations and Lie pseudogroups, Paris, 1978, ix + 407p.
79. Rauch, J., and Reed, M. C.: Propagation of singularities in non strictly hyperbolic semi linear systems: Examples, Comm. P. A. Math., 35 (1982), pp 555-565.
80. Riesz, M.: L'intégrale de Riemann-Liouville et le problème de Cauchy, Acta Math., 81 (1949), pp 1-223.
81. Rinke, B., and Wünsch, V.: Zum Huygensschen Prinzip bei der skalaren Wellengleichung, Beitr. Zur Analysis, 18 (1981), pp 43-75.
82. Riquier, C.: Les Systemes d'Equations aux derivées partielles, Paris, 1910.
83. Rodino, L.: Microlocal Analysis for spatially inhomogeneous pseudodifferential operators, Ann. Scuola. Norm. Sup. Pisa. Cl. Sci., (4) 9 (1982), no. 2, pp 211-253.
84. Rothschild, L., and Stein, E. M.: Hypoelliptic differential operators and nilpotent groups, Acta Math., 137 (1976), pp 247-320.
85. Sato, M., Kawai, T., and Kashiwara, M.: Microfunctions and pseudo differential equations, Springer Lecture Notes in Mathematics, 287.
86.a) Schimming, R.: Zur Gultigkeit des huygenssehen Prinzips bei einer speziellen Metrik, Z. A. M. M., 51 (1971), pp 201-208.
86.b) ____________: Spektrale Geometrie und Huygenssches Prinzip für Tensorfelder und Differentialformen, I, Z.A.A., 1 (1982), pp 71-95.
87. Schmützer, E., Kramer, D., Stephani, H.: et al, Exact solutions of Einstein's Field Equations, Cambridge U. P. - VEB Deutscher Verlag der Wissenschaften, Berlin, (1980), p 425.
88. Sjöstrand, J.: Parametrices for pseudodifferential operators with multiple characteristics, Ark für Math., 12 (1974), pp 85-130.
89. ____________: Propagation of singularities for operators with multiple involutive characteristics, Ann. Inst. Fourier, 26 (1976), pp 141-155.
90. ____________: Singularités analytiques microlocales, Asterisque, Paris, 95 (1982), p 207.
91. Spencer, D. C.: Overdetermined systems of linear partial differential equations, Bull. A. M. S., 75 (1965), pp 1-114.
92.a) Stellmacher, K. L.: Ein Beispeil einer Huygensschen Differentialgleichungen, Nachr. Akad. Wiss. Gottingen - Math. Phys. Kl II, 10 (1953), pp 133-138.
92.b) ________________: Eine Klasse huygenscher Differentialgleichungen und ihre Integration, Math. Ann., 130 (1955), pp 219-233.
93. Taylor, M. E.: Pseudo differential operators, Princeton, 1981.
94.a) Thomas, J. M.: Riquier's Existence Theorems, Annals of Math., 30 (1929), pp 285-310 and 35 (1934), pp 306-311.
94.b) ____________: Differential Systems, A. M. S. Colloquium Pub., vol. 21 (1937), p 118.
95. Titchmarsh, E. C.: Introduction to the theory of Fourier integrals, Oxford U. P., (1937), viii + p 391.
96. Trèves, F.: Introduction to pseudodifferential and Fourier integral operators, vols. 1 and 2, New York and London, 1980.

97. Vandercapellen, G.: Contributions a l'etude du principle d'Huygens en espace temps courbe, Memoire de Licence, Université de l'Etat a Mons, (1980).

98. Wünsch, V.: Uber selbstadjungierte Huygenssche Differentialgleichungen mit vier unabhängigen Variablen, Math. Nachr., 47 (1970), pp 131-154.

99. _________: Maxwellsche Gleichungen und Huyghenssches Prinzip II, Math. Nachr., 73 (1976), pp 19-36.

100. _________: Uber eine Klasse Konforminvarianter Tensoren, Math. Nachr., 73 (1976), pp 37-58.

101. _________: Cauchy-Problem und Huygenssches Prinzip bei einigen Klassen spinorieller Feldgleichungen I, Beitr. zur Analysis, 12 (1978), pp 47-76.

102. _________: Cauchy-Problem und Huygenssches Prinzip bei einigen Klassen spinorieller Feldgleichungen II, Beitr. zur Analysis, 13 (1979), pp 147-177.

103. _________: Conformally invariant variational problems and Huygens' principle, Math. Nachrichten, 120 (1985), pp 175-193.

104. Yamamoto, K.: On the reduction of certain pseudo-differential operators with non-involutive characteristics, J. Diff. Eq., 26 (1977), pp 435-442.

ON THE WAVE EQUATION IN PLANE REGIONS WITH POLYGONAL BOUNDARY

F. G. Friedlander
Department of Mathematics
University College London

Abstract: We first discuss the wave equation in a plane sector, using a fundamental solution due to Sommerfeld and some elementary estimates. We then make two applications, the first to the diffraction of singularities of solutions of the wave equation at a corner, and the second one to the construction of solutions of the wave equation in a region with polygonal boundary.

1. Let X be the plane sector

(1.1) $$X = \{x \in R^2 : x_1 = r\cos\theta,\ x_2 = r\sin\theta,\ r > 0,\ 0 < \theta < \alpha\}$$

where $0 < \alpha < 2\pi$. Consider the following boundary value problem for the wave equation on $X \times R$:

(1.2) $$\square u = f \text{ on } X \times R,\ u = 0 \text{ on } \partial X \times R,\ u = 0 \text{ if } t \ll 0.$$

Here

(1.3) $$\square = (\partial/\partial t)^2 - \Delta$$

is the wave operator, Δ being the laplacian on R^2. As it stands, this problem is indeterminate, and some additional hypothesis on the behaviour of u at the vertex of X is needed to ensure uniqueness. In fact, one can prove:

Theorem 1.1. Suppose that $f \in C_0^\infty(X \times R)$. Then there is a unique $u \in C^\infty((\bar{X} \setminus 0) \times R)$ which satisfies (1.2) and for which, with $v(r,\theta,t) = u(r\cos\theta, r\sin\theta, t)$, one has

(1.4) $$(r\partial/\partial r)^i (\partial/\partial\theta)^j (\partial/\partial t)^k v = O(r^{\frac{1}{2}}) \text{ as } r \to 0$$

for all $i \geq 0$, $j \geq 0$, $k \geq 0$, uniformly in t when t is in a bounded subset of R.

The proof of this is outlined at the end of Section 2. Here we

H. G. Garnir (ed.), Advances in Microlocal Analysis, 135–150.

only remark that, by a routine argument, it follows from (1.4) with $i=j=0$, $k=1$, and $i = 1$, $j = k = 0$, respectively, that for every $s \in R$ there is a $C(s) > 0$ such that

$$(1.5) \qquad \int_X \int_{-\infty}^{s} (|\partial u|\partial t)^2 + |\text{grad } u|^2) dxdt \leq C(s) \int_X \int_{-\infty}^{s} |f|^2 dxdt.$$

Thus u, being a solution of (1.2) 'with finite energy', is necessarily unique.

By a device due to Sommerfeld, the mixed problem (1.2) can be replaced by an initial value problem on a covering manifold. If one goes over to polar coordinates (r,θ) as in (1.1), then X is mapped to $R^+ \times (0,\alpha)$, and the wave operator (1.3) becomes

$$(1.6) \qquad P = (\partial/\partial t)^2 - (\partial/\partial r)^2 - r^{-1}(\partial/\partial r) - r^{-2}(\partial/\partial\theta)^2.$$

One can now consider P as a differential operator on the manifold $\tilde{M} = R^+ \times R \times R$. Put

$$(1.7) \qquad g(r,\theta,t) = f(r \cos \theta, r \sin \theta, t), \ (r,\theta,t) \in R^+ \times [0,\alpha] \times R,$$

and extend g to a function on $\tilde{M}$ by putting

$$(1.8) \qquad g(r,\theta,t) = - g(r,-\theta,t) \quad , \quad - \alpha \leqq \theta \leqq 0,$$

$$g(r,\theta,t) = g(r,\theta - 2\alpha,t) \ , \quad \theta \in R.$$

Thus, $g \in C_0^\infty(\tilde{M})$ and, as a function of θ, g is odd and 2α-periodic.

Suppose now that one can determine $v \in C^\infty(\tilde{M})$ such that

$$(1.9) \qquad Pv = g,$$

and that $\theta \to v$ is again odd and 2α-periodic. Then it is clear that the pullback

$$(1.10) \qquad u(x,t) = v(r,\theta,t), \qquad (x,t) \in X \times R,$$

where x and (r,θ) are related as in (1.1), is a solution of (1.2) which is in $C^\infty((\bar{X} \setminus 0) \times R)$. So one is led to consider 'many-valued solutions of the wave equation', functions or distributions which satisfy (1.8) on $\tilde{M}$, and are 2α- periodic in θ. Alternatively, one can take them to be defined on the manifold

$$(1.11) \qquad M = R^+ \times R/2\alpha Z \times R.$$

If $0 < \alpha < \pi$, then (1.8) on M can also be considered as the wave equation on a right circular cylinder embedded in R^3, and equipped with the Riemannian metric induced by the standard metric on R^3.

The literature on this subject is extensive, and has recently been augmented by a substantial paper [CT]. (For references up to 1958, see [F1].) In effect, Sommerfeld obtained a fundamental solution of (1.8), and thence of (1.2), in 1901. We shall show here that this yields solutions of (1.2) that satisfy (1.4). By transposition, one can then deduce a result on the propagation of singularities of solutions of the homogeneous wave equation in X x R with Dirichlet boundary conditions on X x R. A slightly extended version of the problem (1.2) is then shown to give, by virtue of (1.4), a satisfactory basis for the construction of solutions of the wave equation in a polygonal region.

The differential operator r^2P on M is totally characteristic, in the sense defined in [M]. Although the spaces of distributions associated there with such operators are not used here, the hypotheses on g in Propositions 2.4 and 2.5 below are in the same spirit.

2. We shall be working with functions and distributions on $\tilde{M}$ which are 2α-periodic on θ; considering them as living on M, we shall use notations such as $C_o^\infty(M)$, $\mathcal{D}'(M)$. We use the measure

$$dm = r\,dr\,d\theta\,dt \tag{2.1}$$

which is the pullback of Lebesgue measure dx dt under the projection $\tilde{M} \to (R^2 \setminus 0) \times R$. The pairing of $\mathcal{D}'(M)$ and of $C_o^\infty(M)$ will be taken to be the continuous extension of the bilinear form

$$L^\infty_{loc}(M) \times C_o^\infty(M) \to \mathbb{C}:\ (v,\phi) \to \langle v,\phi\rangle = \int v\,\phi\,dm. \tag{2.2}$$

The Dirac kernel $\delta_M \in \mathcal{D}'(M \times M)$ is then

$$\delta_\mu(m,m') = r'^{-1}\,\delta(r-r')\delta(t-t')\sum_{n=-\infty}^{\infty}\delta(\theta-\theta'-2n\alpha), \tag{2.3}$$

where $m = (r,\theta,t)$, $m' = (r',\theta',t')$. By definition, a forward fundamental solution of P on M is a distribution $E \in \mathcal{D}'(M \times M)$ such that

$$\begin{aligned} P_{(m)}E = P_{(m')}E &= \delta_M(m,m') \\ \operatorname{supp} E &\subset \{(m,m'):\ t \geq t'\}. \end{aligned} \tag{2.4}$$

We first dispose of an elementary (and uninteresting) case:

<u>Proposition 2.1.</u> If $\alpha = \pi/N$, where N is a positive integer, then

$$E = (2\pi)^{-1}H(t-t')\sum_{n=0}^{N-1}\left((t-t')^2 - r^2 - r'^2 + 2rr'\cos(\theta-\theta'-2n\alpha)\right)_+^{-\frac{1}{2}} \tag{2.5}$$

satisfies (2.4); here H(.) is the Heaviside function.

<u>Proof.</u> In this case, the second member of (2.3) is 2π-periodic in both θ and θ', and pulls back to

$$\sum_{n=1}^{N-1} \delta(x-x'_n)\delta(t-t') \in \mathcal{D}'(R^3 \times R^3),$$

where

$$x'_{n,1} + n'_{n,2} = (x'_1 + i\, x'_2)\exp(2i\pi n/N),\ n = 0,1,\ldots,N-1.$$

The second member of (2.5) is the sum of the corresponding forward fundamental solutions of the wave operator on R^3, in terms of polar coordinates; so the Proposition is proved.

Next, we state Sommerfeld's fundamental solution. It is obviously sufficient to do this when $\theta' = t' = 0$, and we write

$$E(r,\theta,t,r',0,0) = F(r,\theta,t,r');$$

here F is a function of r' with values in $\mathcal{D}'(M)$ such that

(2.6) $$P\, F = r'^{-1}\delta(r-r')\delta(t) \sum_{n=-\infty}^{\infty} \delta(\theta - 2n\alpha).$$

<u>Proposition 2.2.</u> The following locally integrable function satisfies (2.6) in $\mathcal{D}'(M)$:

(2.7a) $$F = (2\pi)^{-1}H(t) \sum_{n=-\infty}^{\infty} \chi(\theta - 2n) \times$$
$$\times\ (t^2-r^2-r'^2+ 2rr'\cos(\theta - 2n\alpha))_+^{-\frac{1}{2}}\ ,\ \text{if } t < r + r'\ ,$$

(2.7b) $$F = (2\pi)^{-1} \int_0^{\infty} K(\eta,\theta)(2rr'\cosh\eta + r^2+r'^2-t^2)_+^{-\frac{1}{2}}\, d\eta$$
$$\text{if } t > r + r'\ ,$$

where

(2.8) $$\chi(\theta) = 1 \text{ if } |\theta| < \pi\ ,\ \chi(\theta) = 0 \text{ if } |\theta| \geqq \pi$$

and, for $\eta > 0$,

(2.9) $$K(\eta,\theta) = \frac{1}{2\alpha}\left(\frac{\sinh a\eta}{\cosh a\eta - \cos a(\pi,\theta)} + \frac{\sinh a\eta}{\cosh a\eta - \cos a(\pi,\theta)}\right),$$

$$= \tfrac{1}{2}(1 + 2 \sum_{k=0}^{\infty} e^{-ka\eta} \cos ka\pi \cos ka\theta),$$

with

$$a = \pi/\alpha. \tag{2.10}$$

Remark. It is easy to obtain (2.7a) by exploiting the relation between the wave operators on M and on R^3, and a dependence domain argument. One can then seek to extend F to $\{t > r+r'\}$ by setting the restriction of $(2rr')^{\frac{1}{2}}F$ to $\{t > 0\}$ equal to the pullback of some $F^*(Y,\theta) \in \mathcal{D}'((-1,\infty) \times R/2\alpha Z)$ under the map

$$(r,\theta,t) \to ((t^2-r^2-r'^2)/2rr',\theta).$$

Technically, this is simpler to carry through on R x M, where it reduces the extension problem to an elementary exercise on Laplace's equation in the upper half plane. One can then obtain (2.7b) by Hadamard's method of descent and a contour integral manipulation; see [F3] for this approach on $\tilde{M}$, that is to say without the 2α- periodicity.

It is an immediate consequence of (2.7a,b) that

$$\operatorname{supp} F \subset \{t \geq |r-r'|\} \tag{2.11}$$

Furthermore, one has

$$\int_0^\infty \int_{-\alpha}^{\alpha} |F| r'dr'd\theta = \int_0^\infty \int_{-\alpha}^{\alpha} F\, r'dr'd\theta \leq 2t_+. \tag{2.12}$$

Here the first equality is trivial, as $F \geq 0$. The inequality is then proved by straightforward estimates, splitting the integral into the sum of one over $\{r' > (t-r)_+\}$ and one over $\{r' < (t-r)_+\}$, and noting that K in (2.7b) is nonnegative; we omit the details, which will be published elsewhere.

Proposition 2.3. Let $m = (r,\theta,t)$, $m' = (r',\theta',t')$ and put

$$E(m,m') = F(r,\theta-\theta',t-t',r'). \tag{2.13}$$

The E is a forward fundamental solution of P on M. One has $E \geq 0$,

$$\operatorname{supp} E \subset \{t-t' \geq |r-r'|\} \tag{2.14}$$

and, for any $t_o \in R$,

$$\int_{\{t' > t_o\}} E\, dm \leq (t-t_o)_+^2 . \tag{2.15}$$

Proof. Clear, from Proposition 2.2 and the equations (2.11), (2.12).

As a distribution kernel, E is just the kernel of an integral operator,

$$(Eg)(m) = \int E(m,m')g(m')dm', \qquad g \in C_0^\infty(M).$$

Evidently, Eg is well defined for a larger class of functions. In fact, one has:

Proposition 2.2. Suppose that $g \in C^\infty(M)$, and that there is a $t_0 \in \mathbb{R}$ such that $g = 0$ for $t < t_0$. Assume in addition that

$$(2.16)\qquad (r\partial/\partial r)^i(\partial/\partial\theta)^j(\partial/\partial t)^k g \in L^\infty_{loc}(M)$$

for all $i \geqq 0$, $j \geqq 0$, $k \geqq 0$. Define Eg by

$$(2.17)\qquad (Eg)(m) = \int E(m,m')g(m')dm'.$$

Then (i) $Eg \in C^\infty(M)$, and

$$(2.18)\qquad [\partial/\partial\theta,E]g = [\partial/\partial t,E]g = 0,\ [r\partial/\partial r + t\partial/\partial t,E]g = 2g;$$

(ii) on has

$$(2.19)\qquad PEg = g, \quad Eg = 0 \text{ for } t < t_0;$$

(iii) if r_0 and T are positive real numbers and

$$(2.20)\qquad D = D(r_0,T) = \{m \in M : r+t < r_0+t_0+T,\ t_0 < t < t_0+T\},$$

then, for N = 0, 1, ... there are constants $C_N = C_N(D) > 0$ such that

$$(2.21)\qquad \sum_{i+j+k\leqq N} \sup_D |(r\partial/\partial r)^i(\partial/\partial\theta)^j(\partial/\partial t)^k Eg|$$
$$\leqq C_N \sum_{i+j+k\leqq N} \sup_D |(r/\partial r)^i(\partial/\partial\theta)^j(\partial/\partial t)^k g|.$$

Proof. That Eg is well defined follows from (2.16) with $i = j = k = 0$, since (2.14) implies that the domain of integration in (2.17) is a subset of

$$(2.22)\qquad \{t_0 < t' < t - |r-r'|\}.$$

By (2.13), one can put (2.17) in the form

$$(2.23)\qquad (Eg)(m) = \int_0^\infty \int_{-\alpha}^{\alpha} \int_{-\infty}^{\infty} F(r,\theta',t',r')g(r',\theta-\theta',t-t')r'dr'd\theta'dt'.$$

By (2.7ab), $(r,t',r') \to r'F(r,\theta',t',r')$ is homogeneous of degree zero. So one can put $r' = rr''$, $t = rt''$, to obtain

$$(Eg)(m) = r^2 \int_0^\infty \int_{-\alpha}^{\alpha} \int_{-\infty}^{\infty} F(1,\theta',t'',r'')g(rr'',\theta-\theta',t-rt'')r''dr''d\theta'dt''$$

In view of (2.16), it is now clear that one can differentiate repeatedly under the integral sign with respect to r, θ and t. This shows that $Eg \in C^\infty(M)$, and also yields the identities (2.18).

To prove (ii), it is sufficient to observe that, as E is a fundamental solution of the differential operator P, Fubini's theorem implies that (2.19) holds in $\mathcal{D}'(M)$; by part (i), which has already been proved, it therefore holds in the usual sense as well.

As to (iii), one first notes that (2.15) and (2.17) give

$$|(Eg)(m)| \leqq (t-t_o)_+^2 \sup\{|g(m')| : t' < t\}.$$

It has already been observed that the domain of integration in (2.17) is a subset of (2.22), hence a subset of D when $m \in D$. So

$$\sup_D |Eg| \leqq T^2 \sup_D |g| ,$$

which is (2.21) when $N = 0$. The other estimates now follow from this and the identities (2.18), applied repeatedly; so we are done.

The estimates (2.21) can be sharpened, for instance if g vanishes for small r. As our main objective here is the boundary value problem (1.2), which leads to functions g on M that are odd in θ, the following result will be sufficient for our purpose.

Proposition 2.2. Let $g \in C^\infty(M)$. Suppose that $g = 0$ for $t < t_o$, for some real number t_o, and that, furthermore,

$$(2.24)\qquad \int_{-\alpha}^{\alpha} g(r,\theta,t)\, d\theta = 0 , \qquad (r,t) \in \mathbb{R}^+ \times \mathbb{R}.$$

Assume in addition that, for every set $D \subset M$ of the form(2.20) there are constants $C_{ijk} > 0$ such that

$$(2.25)\qquad |(r\partial/\partial r)^i(\partial/\partial\theta)^j(\partial/\partial t)^k g| \leqq C_{ijk} r^{\frac{1}{2}} , \qquad m \in D,$$

and all $i \geqq 0$, $j \geqq 0$, $k \geqq 0$. Set $Eg = v$. Then

$$(2.26) \qquad \int_{-\alpha}^{\alpha} v(r,\theta,t)\, d\theta = 0, \qquad (r,t) \in R^{+} \times R,$$

and there are positive constants C'_{ijk}, depending on the C_{ijk} and on D, such that

$$(2.27) \qquad |(r\partial/\partial r)^{i}(\partial/\partial\theta)^{j}(\partial/\partial t)^{k}\, v| \leqq C'_{ijk} r^{\frac{1}{2}}\ , \quad m \in D\ ,$$

for all $i \geqq 0$, $j \geqq 0$, $k \geqq 0$.

The proof is omitted, as it is too long to be given here. It is a straightforward exercise based on (2.7a,b) and the Fourier series (2.9) for the function K in the integrand of (2.7b). The equation (2.26) is of course immediate, from (2.23) and (2.24), and Fubini's theorem.

Proof of Theorem 1.1. Consider the boundary value problem (1.2). If $f \in C_o^\infty(X \times R)$, and g is defined by (1.7) and (1.8), then the hypotheses of Proposition 2.5, and hence also those of Proposition 2.4, hold. It is clear from (2.23) that $v = Eg$ is an odd function of θ then, since F is even in θ. So it follows from Proposition 2.4 that, if u is defined as the pullback of v by means of (1.10), one obtains a solution of (1.2) with the regularity properties asserted in the theorem. Furthermore, the estimates (2.27) imply that (1.4) holds, as the limitation on r implied by the hypothesis $m \;\; D(r_o,T)$ can easily be removed by a dependence domain argument.

It is not difficult to extend Proposition 2.5 to functions g which do not satisfy (2.24). For example, if $g \in C_o^\infty(M)$, one can show that Eg tends to a limit as $r \to 0$, which is a C^∞ function of t only, and easily computed. Moreover, one has

$$Eg - Eg|_{r=0} = O(r^{\frac{1}{2}}),\ (r\partial/\partial r)Eg = O(r^{\frac{1}{2}})\ .$$

The Fourier coefficients v_n of $v = Eg$ satisfy the Euler-Darboux equations

$$((\partial/\partial t)^2 - (\partial/\partial r)^2 - r^{-1}(\partial/\partial r) + n^2a^2r^{-2})v_n = g_n\ , \ n = 0, \pm 1, \ldots;$$

the Fourier coefficients g_n of g are in $C_o^\infty(R^{+} \times R)$. It can be shown [F2] that one has

$$r^{-|n|a} v_n \quad C^\infty(\bar{R}^{+} \times R)\ ;$$

the limits $r^{-|n|a} v_n|_{r\,=\,0}$ can be computed, and are functions of t only.

There can be little doubt that the Fourier series (in θ) of v, which of course converges in $C^\infty(M)$, also gives an asymptotic expansion valid as $r \to 0$, but our simple estimates are not strong enough to prove this.

For elliptic equations, asymptotic expansions of similar type can be established in much greater generality [MM].

3. The transpose of E is

$$(3.1) \qquad {}^tE(m,m') = E(m',m) = F(r,\theta-\theta',t'-t,r'),$$

where the second equality arises from the fact that F is even in θ, and a symmetric function of r and r'. So ${}^tE$ is obtained from E by the 'time-reversal map' $(t,t') \to (-t,-t')$. Instead of (2.14), one has

$$(3.2) \qquad \operatorname{supp} {}^tE \subset \{t'-t \geq |r-r'|\}$$

Propositions 2.4 and 2.5 have time-reversed counterparts, valid for functions g that vanish for t sufficiently large, with Eg replaced by ${}^tEg$. Minor changes are required in the statements; these are left to the reader.

It follows that, as distribution kernels, both E and ${}^tE$ give rise to (continuous) maps $C_0^\infty \to C^\infty$. These in turn extend, by continuity (or by transposition) to maps $\mathcal{E}' \to \mathcal{D}'$. Because of (2.14) and (3.2), the maps $v \to Ev$ and $v \to {}^tEv$ are also well defined if v = 0 for t << 0 or for t >> 0, respectively.

Both E and ${}^tE$ are left inverses of P on $\mathcal{E}'(M)$. In order to obtain results on the propagation of singularities of solutions of the wave equation in X, one must introduce a larger class of distributions for which this is the case. Let $\mathcal{D}'_\partial(M)$ be the subspace of $\mathcal{D}'(M)$ consisting of distributions v for which there is an $r_o > 0$ and a C^∞ function

$$v(r,.): (0,r_o) \to \mathcal{D}'(R/2\alpha Z \times R)$$

such that

$$\langle v,\emptyset\rangle = \int_0^\infty \langle v(r,.),\emptyset(r,.)\rangle \, r\, dr$$

for all $\emptyset$ $C_0^\infty(M)$ supported in $\{0 < r < r_0\}$. Note that, by the partial hypoellipticity properties of P, any v $\mathcal{D}'(M)$ such that Pv = 0 on $\{0 < r < r_0\}$ for some $r_0 > 0$ is in $\mathcal{D}'_\partial(M)$. By a routine argument (see [F2] for a similar case) one can deduce the following from Proposition 2.5:

Lemma 3.1. Suppose that v $\mathcal{D}'_\partial(M)$, and that (i) v = 0 for t << 0 (respectively, t >> 0), and (ii) v is odd in θ, and both v(r,.) and $(r\partial/\partial r)v(r,.)$ are continuous at r = 0. Then EPv = v (respectively, ${}^tEPv = v$).

The wave front set of E can be analyzed by means of (2.13), (2.7a,b), and an alternative form of F which can for example be found in [F1, p.118] (The derivation given there is incorrect, but the result is valid; see also [F3] for the non-periodic case.) By combining this information with Lemma 3.1 and Sommerfeld's device, standard properties of wave front sets allow one to analyze the propagation of singularities in $X \times R$. The novel feature here is diffraction at the vertex, so one must exclude the purely reflective cases corresponding to Proposition 2.2. One then has the following, which is here stated without proof; for a similar result, see [CT] and also [V].

Proposition 3.2. Suppose that π/α is not a positive integer. Let $u \in \mathcal{D}'(X \times R)$ be such that $\Box u = 0$, and that its pullback under $(r,\theta,t) \to (x,t) = (r\cos\theta, r\sin\theta, t)$ is the restriction to $R^+ \times (0,\alpha) \times R$ of some $v \in \mathcal{D}'_\partial(M)$ which satisfies hypothesis (ii) of Lemma 3.1. (In particular, one has $u = 0$ on $(\partial X \setminus 0) \times R$.) Suppose that WF(u) contains the bicharacteristic

$$\gamma_o = \{(x,t,\xi,\tau) \in T^*(X \times R) : x = \zeta_o t,\ t < 0,\ \xi = -\tau_o\zeta_o,$$

$$\tau = \tau_o\ ,\ \text{where } |\zeta_o| = 1 \text{ and } \tau_o \neq 0\}.$$

Then WF(u) contains at least one bicharacteristic of the form

$$\gamma = \{(x,t,\xi,\tau) \in T^*(X \times R) : x = \zeta t,\ t > 0,\ \xi = -\tau_o\zeta,$$

$$\tau = \tau_o\ ,\ \text{where } |\zeta| = 1\}.$$

Similarly, if $\gamma \in$ WF(u), then WF(u) must contain a bicharacteristic γ_o for some ζ_o.

Note. If $u \in \mathcal{D}'(X \times R)$ satisfies the wave equation in $X \times R$, with Dirichlet conditions on the boundary, and is extendible across $(\partial X \setminus 0) \times R$, then it follows from well known results on partial hypo-ellipticity in the interior and at the boundary that u^* can be extended to a $v \in \mathcal{D}'_\partial(M)$ which is odd in θ. So the hypotheses essentially only bear upon the behaviour of u at the vertex.

It would be of some interest to determine whether, say, $\gamma_o \in$ WF(u) implies that all admissible γ are present in WF(u), or whether in some special circumstances one can have just one 'incoming' and one 'outgoing' singularity-carrying bicharacteristic.

4. Propositions 2.4 and 2.5 can be used to construct a solution of the boundary value problem (1.2) when the sector X is replaced by a plane region whose boundary is a (not necessarily connected) polygon, with the solution satisfying estimates of type (1.4) at each vertex. We shall only consider triangular regions here. It is evident that, with minor modifications, the method can be adapted to the general case.

Let $\Omega \subset R^2$ be the interior of a triangle. Denote the vertices of this triangle by A_μ, $\mu = 1, 2, 3$, and the angle set at A_μ by α_μ. Each A_μ determines a sector X_μ, bounded by the two rays from A_μ obtained by prolonging the sides of Ω that meet at A_μ, and containing Ω.

For each $\mu = 1, 2, 3$, one can introduce polar coordinates (r_μ, θ_μ) such that

$$(4.1) \qquad X_\mu = \{x \in R^2 \colon x_1 = A_{\mu,1} + r_\mu \cos(\beta_\mu + \theta_\mu),\ x_2 = A_{\mu,2} + r_\mu \sin(\beta_\mu + \theta_\mu)$$

$$r_\mu > 0,\ 0 < \theta_\mu < \alpha_\mu\},$$

where the β_μ are certain constants. Thus one has three diffeomorphisms

$$h_\mu \colon R^+ \times (0, \alpha_\mu) \times R \to X_\mu \times R,\ \mu = 1, 2, 3.$$

Write $C^\infty(\dot{\Omega} \times R)$ for functions of class $C^\infty(\Omega \times R)$ which are smooth up to $\partial\Omega \setminus \{A_1, A_2, A_3\}$.

Theorem 4.1. If $f \in C_o^\infty(\Omega \times R)$, then there is a unique $u \in C^\infty(\dot{\Omega} \times R)$ such that

$$(4.2) \qquad \Box u = f \text{ on } \Omega \times R, \quad u = 0 \text{ on } \partial\Omega \times R, \quad u = 0 \text{ if } t << 0,$$

and that, for $\mu = 1, 2, 3$ and all nonnegative integers i, j, k

$$(4.3) \qquad (r_\mu \partial/\partial r_\mu)^i (\partial/\partial\theta_\mu)^j (\partial/\partial t)^k\, h_\mu^* u = O(r_\mu^{\frac{1}{2}})$$

as $r_\mu \to 0$, uniformly in t for bounded t.

Remark 1. The estimates (4.3) with $i=j=0$, $k=1$ and with $i = 1$, $j = k = 0$ already imply that the solution u of the Theorem has finite energy, in the sense that, for every real number s, one has an estimate of the type (1.5) with X replaced by Ω. A fortiori, the estimates (4.3) imply uniqueness.

Remark 2. If $\alpha_\mu = \pi/N_\mu$ for $\mu = 1, 2, 3$, where the N_μ are integers, then Proposition 2.2 applies to each sector X_μ. It is well known, and easy to prove, that the only possibilities are (i) the equilateral triangle and (ii) the (right-angled) trianges with $\alpha_1 = \pi/2$, $\alpha_2 = \pi_3$, $\alpha_3 = \pi/6$ and with $\alpha_1 = \pi/2$, $\alpha_2 = \alpha_3 = \pi/4$. The proof of Theorem 4.1 given below applies, but is redundant, because the solution of (4.2) in question can be constructed by the elementary 'image' method'. The simple details are left to the reader. Incidentally, the only other 'purely reflective' polygon is the rectangle.

The proof of Theorem 4.1 requires several steps. The first is a dependence domain argument. By definition, the (backward) dependence domain of a point (x,t) R^3 is the interior of the backward characteristic cone with vertex (x,t),

$$(4.4) \qquad I(x,t) = \{(x',t') \colon t' < t - |x-x'|\}.$$

From now on, the subscript μ will always take the values 1, 2 and 3. We denote the side of our triangle opposite to A_μ by Σ_μ, and let

$$(4.5) \qquad d_\mu = \inf \{|A_\mu - x| : x \in \Sigma_\mu\}$$

be the distance from A_μ to Σ_μ. For any $t_o \in R$, put

$$(4.6) \qquad I_\mu(t_o) = I(A_\mu, d_\mu + t_o) \cap (\Omega \times (t_o, \infty)).$$

Lemma 4.2. If $(x,t) \in I_\mu(t_o)$, then $\overline{I(x,t)} \cap (\Omega \times (t_o,\infty))$ is disjoint from $\Sigma_\mu \times (t_o,\infty)$.

Proof. Clear, from (4.5), (4.6), and the inclusion properties of dependence domains.

The lemma shows that if f is supported in $\Omega \times (t_o,\infty)$, then the restriction of the solutions of (4.2), (4.3) to $I_\mu(t_o)$ is the solution of the Sommerfeld problem in the sector X_μ. However, it is better to work with truncated characteristic cones, so we set, for $T \in (0,d_\mu)$,

$$(4.7) \qquad D_\mu^T(t_o) = \{(x,t) \in I_\mu(t_o) : t_o < t < t_o + T\}.$$

Lemma 4.3. If $T > 0$ is sufficiently small then

$$(4.8) \qquad D_1^T(t_o) \cup D_2^T(t_o) \cup D_3^T(t_o) = \Omega \times (t_o, t_o+T).$$

Proof. For sufficiently small positive s, the union of the sets

$$\{x \in \Omega : |x - A_\mu| < d_\mu - s\}, \quad \mu = 1, 2, 3$$

is a covering of Ω. This is proved by a simple geometric argument, left to the reader. (For example, one can consider the cases of an acute triangle, and of a right-angled or obtuse triangle, separately.)
As

$$D_\mu^T(t_o) = \{(x,t) \in \Omega \times R : |x - A_\mu| < d_\mu + t_o - t,\ t_o < t < t_o + T\},$$

the lemma follows.

Before stating the next lemma, we note that the solution of the sector problem (1.2) that satisfies (1.4) is

$$(4.9) \qquad u(x,t) = \int G(x,t,x',t') f(x',t') dx'dt'$$

where the integral is over $X \times R$, and the Sommerfeld Green's function is defined by

(4.10) $G(x,t,x',t') = F(r,\theta-\theta',t-t',r') - F(r,\theta+\theta',t-t',r')$;

here x and x' are points of X with the polar coordinates (r,θ) and (r',θ'), respectively. This follows from (1.7), (1.8), (2.13) and Propositions 2.4 and 2.5. Furthermore, it is not difficult to see that

(4.11) $\text{supp } G \subset \{(x,t,x',t') \in X \times R \times X \times R: (x',t') \in \overline{I(x,t)}\}$.

Returning to the problem in hand, we introduce the three Sommerfeld Green's functions G_1, G_2 and G_3 for the sectors X_1, X_2 and X_3, respectively. We then have

Lemma 4.4. Let T be as in Lemma 4.3. Suppose that $f \in C_0^\infty(\Omega \times R^+)$, and put, for $\mu = 1, 2, 3$,

(4.12) $u_\mu(x,t) = \int G_\mu(x,t,x',t')f(x',t')dx'dt'$, $(x,t) \in D_\mu^T(0)$

Then there is a unique $u \in C^\infty(\dot\Omega \times (0,T))$ satisfying (4.2) and (4.3) for $0 < t < T$, such that

(4.13) $u|D_\mu^T(0) = u_\mu$, $\mu = 1, 2, 3$.

Proof. It follows from (4.11) and Lemma 4.2 that the domain of integration in (4.12) is a subset of $D_\mu^T(0)$, so that u_μ is the solution of Sommerfeld's problem for the sector X_μ, restricted to $D^T(0)$. We now claim that

(4.14) $u_\mu = u_\lambda$ on $D_\mu^T(0) \cap D_\lambda^T(0)$, $\mu,\lambda = 1, 2, 3$.

Indeed, if (x,t) is in the intersection of all three $D_\mu^T(0)$, then it is clear from Lemma 4.2 that $I(x,t) \cap (\Omega \times R^+)$ does not meet any $\Sigma_\mu \times R$. It is not difficult to infer from (2.7a,b) and (4.10) that all the G_μ are then equal to the free space forward fundamental solution of the wave operator, so that $u_1 = u_2 = u_3 = u_o$, say, the free space solution of the inhomogeneous wave equation vanishing for $t \ll 0$. Again, if for example $(x,t) \in (D_1^T(0) \cap D_2^T(0)) \setminus D_3^T(0)$, then $I(x,t) \cap (\Omega \times R^+)$ meets $\partial\Omega \times R^+$ only at points $\Sigma_3 \times R^+$. In this case, it follows from (2.7a,b) and (4.10) that $u_1 = u_2 = u_o - u_o^*$, where u_o^* is derived from u_o by reflection in the plane containing $\Sigma_3 \times R$. Finally, (4.14) is trivial when $\mu = \lambda$.

By Lemma 4.3, one can now construct u so as to satisfy (4.13), say by a partition of unity. Applying Theorem 1.1 to each u_μ, one concludes that (4.2) and (4.3) hold, and so the lemma is proved.

For the final step in the proof of Theorem 4.1, we note that one can assume without loss of generality that

(4.15) $\text{supp } f \subset \Omega \times (0,\delta)$

where δ is any pre-assigned positive real number, as the general case can be reduced to this by a finite partition of unity and translation in t. Then Lemma 4.4 gives the solution of (4.2), (4.3) for $t < T$, and one has to extend it to larger values of t. One way to do this is as follows.

With T as in Lemma 4.3, chose real numbers δ_1, δ_2 and δ such that

(4.16) $$0 < \delta_1 < \delta_2 < \delta < \tfrac{1}{2}T,$$

and then chose $p(t) \in C^\infty(R)$ such that

(4.17) $$p = 0 \text{ if } t < \delta_1, \quad p = 1 \text{ if } t > \delta_2.$$

Now suppose that, for any $s \in R$, one is given $v \in C^\infty(\dot{\Omega} \times (s,s+T))$ such that

(4.18) $$\Box v = 0 \text{ on } \Omega \times (s,s+\delta), \quad v = 0 \text{ on } \partial\Omega \times (s,s+\delta)\ ,$$

and that, for all μ and all nonnegative integers i,j and k there are $C_{ijk} > 0$ such that

(4.19) $$|(r_\mu \partial/\partial r_\mu)^i (\partial/\partial\theta_\mu)^j (\partial/\partial t)^k\ h_\mu^* v| \leq C_{ijk}\ r_\mu^{+\frac{1}{2}}, \quad (x,t) \in D_\mu^\delta(s).$$

Put

(4.20) $$g = \Box(p(t-s)v(x,y))\ , \quad s < t < s+\delta$$

and set $g = 0$ for all other t; note that, by (4.16), (4.17) and (4.18), one has $g = 0$ for $t < \delta_1$ and for $t > \delta_2$, in fact. Set, for all μ,

(4.21) $$w_\mu(x,t) = \int G_\mu(x,t,x',t')g(x',t')dx'dt'\ , \ (x,t) \in D_\mu^T(s).$$

Then one has

<u>Lemma 4.5.</u> There is a unique $w \in C^\infty(\dot{\Omega} \times (s,s+T))$ such that

(4.22) $$w|D_\mu^T(s) = w_\mu\ , \quad \mu = 1,\ 2,\ 3$$

and

(4.23) $$\Box w = g \text{ on } \Omega \times (s,s+T),\ w = 0 \text{ on } \partial\Omega \times (s,s+T).$$

Furthermore, for $\mu = 1, 2, 3$ and all nonnegative integers i,j,k there are constants $B_{ijk} > 0$ such that

(4.24) $$|r_\mu \partial/\partial r_\mu)^i (\partial/\partial\theta_\mu)^j (\partial/\partial t)^k\ h_\mu^* w| \leq B_{ijk}\ r_\mu^{\frac{1}{2}}\ , \ (x,t) \in D_\mu^T(s).$$

Finally, one has

$$w(x,t) = p(t-s)v(x,t) \quad \text{if} \quad s < t < s+\delta. \tag{4.25}$$

Proof. It follows from (4.18) and the regularity hypotheses on v that, for all μ and all nonnegative integers k,

$$(\partial/\partial\theta_\mu)^{2k}(h^*v_\mu) = 0 \quad \text{if} \quad \theta_\mu = 0 \ \text{or}\ \theta_\mu = \alpha_\mu$$

Hence one can extend $h_\mu^* v$ to a C^∞ function that is odd and $2\alpha_\mu$-periodic in θ_μ, on a domain of the form (2.20). This clearly carries over to $h_\mu^* g$, and gives inhomogeneities which satisfy the hypotheses of Proposition 2.5. One can thus set $w_\mu^* = E_\mu(h_\mu^* g)$, and pull this back to $D_\mu^T(s)$ under h_μ^{-1}, thus obtaining the w_μ given by (4.21). The construction of w now parallels the proof of Lemma 4.4, and Propositions 2.4 and 2.5 ensure that (4.23) and (4.24) hold. Finally, w-p(t-s)v satisfies the homogeneous wave equation for $t < s+T$ and vanishes for $t \leq s+\delta_1$; as it also satisfies Dirichlet boundary conditions, one can appeal to (4.19) and (4.24), and invoke the uniqueness theorem for the wave equation, and this gives (4.25). So the lemma is proved

Proof of Theorem 4.1. Assume that (4.15) holds. Suppose that a solution of (4.2) satisfying (4.3) is already known for $t < s+\delta$ where $s > \delta$; call this $\tilde{u}$. Take v in Lemma 4.5 to be the restriction of $\tilde{u}$ to $\Omega \times (s,s+\delta)$, determine w as in the lemma, and put

$$u = (1 - p(t-s))\tilde{u} + p(t-s)w, \tag{4.26}$$

with the convention that the first term in the second member is zero for $t > s+\delta$. It is then clear from Lemma 4.5 that u extends $\tilde{u}$ to $\Omega \times (-\infty,s+T)$. As Lemma 4.4 gives u for $t < T$, and we have chosen $\delta < 2T$, it follows that one can determine the solution of (4.2) satisfying (4.3) on $\Omega \times (-\infty,s)$ for any finite s in a finite number of steps, and so the theorem is proved.

References

[CT] J. Cheeger and M. E. Taylor, Diffraction of waves by conical singularities, I, II, Comm.Pure Appl.Math.25, 275-331, 487-529 (1982).

[F1] F. G. Friedlander, Sound Pulses, Cambridge University Press, 1958.

[F2] F. G. Friedlander, A singular initial-boundary value problem for a generalized Euler-Darboux equation, J. Diff. Equ.40, 121-154 (1981).

(F3] F. G. Friedlander, Multivalued solutions of the wave equation, Math.Proc.Camb.Phil.Soc. 90, 335-341 (1981); Corrigenda, ibid., 95, 187 (1984).

[M] R. B. Melrose, Transformation of boundary value problems, Acta Math. 147, 149-236 (1981).

[MM] R. B. Melrose and G. A. Mendoza, Elliptic boundary value problems on spaces with conic points, Journées 'Equations aux derivées partielles', St-Jean-de-Monts, 1981.

[V] J. P. Varenne, Diffraction par un angle ou un dièdre, C.R.Acad.Sc. Série A, t.290, 175-178 (1980).

THE NECESSITY OF THE IRREGULARITY CONDITION FOR SOLVABILITY IN GEVREY CLASSES (s) AND {s}

Hikosaburo Komatsu
Department of Mathematics
Faculty of Science
University of Tokyo
Hongo, Tokyo, 113 Japan

ABSTRACT. The author reviews briefly the classical theory of homogeneous solutions of linear ordinary differential equations near an irregular singular point and its application to the existence of ultradistribution solutions of Gevrey classes. Then he develops an analogous theory for formal solutions of linear partial differential equations near a characteristic surface of constant multiplicity. As a consequence he shows that the irregularity condition he introduced earlier in [13] and [14] is necessary in general in order that a formally hyperbolic equation with real analytic coefficients be well posed in a corresponding Gevrey class of functions and ultradistributions.

1. IRREGULARITY OF ORDINARY DIFFERENTIAL EQUATIONS

Let

$$P(z, d/dz) = \sum_{i=0}^{m} a_i(z)(d/dz)^i$$

be a linear ordinary differential operator with holomorphic coefficients $a_i(z)$ defined near the origin. If the origin is a singular point of multiplicity d, then the operator is decomposed as

$$P(z, d/dz) = \sum_{i=0}^{m} q_i(z)\, z^{d_i}\, (d/dz)^i$$

with $d_m = d$ and $q_i(0) \neq 0$ unless $q_i(z) \equiv 0$. In that case we set $q_i(z) \equiv 0$ and $d_i = \infty$. Then the irregularity σ of the singular point 0 is defined by

$$\sigma = \max\{1, \max\{(d - d_i)/(m - i);\ 0 \leqq i < m\}\}. \tag{1}$$

We always have $1 \leqq \sigma \leqq d$. By Fuchs's theory 0 is a regular singular point if and only if $\sigma = 1$. In this case the homogeneous equation

H. G. Garnir (ed.), Advances in Microlocal Analysis, 151–164.

$$P(z, d/dz)\, U(z) = 0 \tag{2}$$

has m linearly independent solutions of the form

$$U(z) = z^{\alpha} (\log z)^{k} (u_0 + u_1 z + \dots + u_j z^j + \dots),$$

where α is a complex number, k is a nonnegative integer and the series converges.

Let 0 be an irregular singular point, i. e. $\sigma > 1$. Then in each sector $\Delta = \{z;\ 0 < |z| < \varepsilon,\ \vartheta_1 < \arg z < \vartheta_2\}$ of opening $\vartheta_2 - \vartheta_1$ less than $\pi/(\sigma - 1)$ equation (2) has m linearly independent solutions of the form

$$U(z) = e^{\psi(z)} z^{\gamma} \sum_{k=0}^{\mu} (\log z)^{k} u_k(z), \tag{3}$$

where

$$\psi(z) = \frac{-\alpha}{(\sigma-1)z^{\sigma-1}} + \dots + \frac{\beta}{z^{1/q}}$$

is a polynomial in $z^{-1/q}$ for an integer $q > 0$, γ is a complex number and u_k is a holomorphic function on Δ with the asymptotic expansion

$$u_k(z) \sim u_{k,0} + u_{k,1} z^{1/q} + \dots + u_{k,j} z^{j/q} + \dots$$

as z tends to 0. For each root α of the algebraic equation

$$\sum_{d-d_i=\sigma(m-i)} q_i(0)\, \alpha^{i} = 0, \tag{4}$$

there is a solution $U(z)$ as above.

When all roots of (4) are simple and non-zero, the result is very old (H. Poincaré (1886), G. D. Birkhoff (1909)). But a complete proof in the general case was obtained relatively recently (W. J. Trjitzinsky (1934), M. Hukuhara (1937, 42), H. L. Turrittin (1955), W. Wasow [29], B. Malgrange [21], J.-P. Ramis (1980 -)). Formal solutions are rather easy to find but they do not converge. Therefore one had to construct genuine solutions with given asymptotic expansions by other methods such as Laplace transforms, integral equations and factorial series.

We employed this fact to prove the necessity of the irregularity condition for existence of solutions of the equation

$$P(x, d/dx)\, u(x) = f(x) \tag{5}$$

in the Gevrey class (s) or $\{s\}$ of ultradistributions on the real domain.

Let $s > 1$ and Ω be an open set in R^n. We denote by $E^{(s)}(\Omega)$ (resp. $E^{\{s\}}(\Omega)$) the space of all infinitely differentiable functions f on Ω such that for each compact set K in Ω and $h > 0$ there is a constant C (resp. there are constants h and C) satisfying

$$\sup_{x\in K} |\partial^\alpha f(x)| \leq C\, h^{|\alpha|} |\alpha|!^s.$$

Let $*$ be either (s) or $\{s\}$ and denote by $D^*(\Omega)$ the space of all f in $E^*(\Omega)$ with compact support. Then the space $D^{*\prime}(\Omega)$ of ultradistributions of class $*$ on Ω is by definition the dual of the space $D^*(\Omega)$ equipped with a natural locally convex topology (cf. [16 - 18]). It is often convenient to admit $* = (\infty)$ and $\{1\}$ so that $E^{(\infty)}$ and $D^{(\infty)\prime}$ are Schwartz's E and D' and $E^{\{1\}}$ and $D^{\{1\}\prime}$ are Sato's A and B.

We mean by the irregularity condition the following condition for irregularity σ at every singular point:

$$\begin{aligned} &\sigma \leq s/(s-1) \quad \text{if} \quad * = (s), \\ &\sigma < s/(s-1) \quad \text{if} \quad * = \{s\}. \end{aligned} \tag{6}$$

Under this condition we have a very good theory of equation (5) in $D^{*\prime}$ (see [9, 10] (resp. [11]) when the coefficients are in A (resp. E^*)). For example, there are exactly $m + d$ linearly independent homogeneous solutions of (5) in $D^{*\prime}(\Omega)$ in a neighborhood Ω of 0. However, let α be a non-zero root of (4) and let $U(z)$ be an analytic continuation of the solution (3) corresponding to α. Then we have the estimates

$$c \exp(c|y|^{-(\sigma-1)}) \leq \sup_{x\in\Omega} |U(x + iy)| \leq C \exp(C|y|^{-(\sigma-1)})$$

for positive constants c and C if $|y|$ is small and $y > 0$ or $y < 0$. By a theorem in [16] these inequalities show that the boundary value $u(x) = U(x + i0)$ or $U(x - i0)$ is in $D^{(s)\prime}(\Omega)$ but not in $D^{\{s\}\prime}(\Omega)$ for $s = \sigma/(\sigma - 1)$. Therefore at least one solution is lost in the class $\{s\}$ for which the irregularity condition does not hold.

2. IRREGULARITY OF PARTIAL DIFFERENTIAL EQUATIONS

Let

$$P(x, \partial) = \sum_{|\alpha|\leq m} a_\alpha(x)\, \partial^\alpha$$

be a linear partial differential operator defined in an open set Ω in R^{n+1} or C^{n+1}. We denote by the corresponding small letter the characteristic polynomial

$$p(x, \xi) = \sum_{|\alpha|=m} a_\alpha(x)\, \xi^\alpha.$$

The characteristic variety $Ch(P) = \{(x, \xi) \in T^*\Omega \setminus 0;\ p(x, \xi) = 0\}$ plays the same role as the singular points for ordinary differential operators. Let $(\overset{o}{x}, \overset{o}{\xi})$ be a non-singular characteristic element, that is, a point in the non-singular part of $Ch(P)$ at which $\sum \xi_j dx_j$ does not vanish on $Ch(P)$. Assume either that the coefficients of $P(x, \partial)$ are real analytic [12] or that $P(x, \partial)$ is a formally hyperbolic operator of constant multiplicity [14]. Then there is a partial differential operator $K(x, \partial)$ which is simple characteristic at $(\overset{o}{x}, \overset{o}{\xi})$ and such that $k(x, \xi)$ is a factor of $p(x, \xi)$ and there are partial differential operators $Q_i(x, \partial)$ defined near $\overset{o}{x}$ such that

$$P(x, \partial) = \sum_{i=0}^{m} Q_i(x, \partial) K(x, \partial)^{d_i}, \tag{7}$$

where either $Q_i \equiv 0$ and $d_i = \infty$ or $q_i(x, \xi)$ does not vanish identically on the characteristic variety $Ch(K)$ near $(\overset{o}{x}, \overset{o}{\xi})$ in $R^{n+1} \times C^{n+1}$ and the order of $Q_i K^{d_i}$ is equal to i. m is the order of P and $d = d_m$ is the multiplicity of the characteristic element $(\overset{o}{x}, \overset{o}{\xi})$. The assumption of non-singularity implies $q_m(\overset{o}{x}, \overset{o}{\xi}) \neq 0$. Then we define the irregularity σ of $P(x, \partial)$ at $(\overset{o}{x}, \overset{o}{\xi})$ by (1). Clearly we have $1 \leqq \sigma \leqq d$. We call (7) the De Paris decomposition after De Paris [2]. The De Paris decomposition is not unique and depends on the coordinate system but the irregularity is uniquely determined by P and $(\overset{o}{x}, \overset{o}{\xi})$.

Actually the irregularity is a microlocal invariant. T. Aoki [1] defined the irregularity of a microdifferential operator (= analytic pseudo-differential operator) $P(x, \partial)$ relative to a microdifferential operator $K(x, \partial)$ of simple characteristic at $(\overset{o}{x}, \overset{o}{\xi})$ and proved the compatibility of two definitions of irregularity when P and K are differential operators.

3. HYPERBOLIC EQUATIONS AND IRREGULARITY

When $\sigma = 1$, $P(x, \partial)$ is said to satisfy Levi's condition. Originally this was introduced in various forms as a condition under which a hyperbolic equation of constant multiplicity is well posed in $E^{(\infty)}$ (E. E. Levi (1909), A. Lax (1956), Mizohata - Ohya (1968), De Paris [2], J. Chazarain (1974)). More generally, let $* = (s)$, $1 < s \leqq \infty$, or $\{s\}$, $1 \leqq s < \infty$. We say that $P(x, \partial)$ satisfies the irregularity condition for the Gevrey class $*$ if (6) holds at any charactereristic elememt $(\overset{o}{x}, \overset{o}{\xi})$. As we reported at the Nato Conference in Liège [13], then the equation becomes well posed in $E*$ and $D*'$ (Ohya (1964), Leray - Ohya (1964, 67), Hamada - Leray - Wagschal (1976), Ivrii [7], De Paris - Wagschal (1978), H. Komatsu [13, 14]). For example, we have the following [14]:

Theorem 1. Let $\Omega = (-T, T) \times R^n$ and let $P(x, \partial)$ be a linear partial differential operator of order m and with coefficients in $E*(\Omega)$. We assume that $P(x, \partial)$ is formally hyperbolic, that is,

(i) The hypersurfaces $\{x_0 = \text{const}\}$ are non-characteristic, or

$p(x; 1, 0, \ldots, 0) \neq 0;$

(ii) The characteristic equation

$$p(x; \xi_0, \xi') = 0 \tag{8}$$

has only real roots ξ_0 for any $x \in \Omega$ and $\xi' \in R^n$.

Further assume that every characteristic element is non-singular and satisfies irregularity condition (6) for the class $*$ and that the roots ξ_0 of characteristic equation (8) are bounded on $\Omega \times S^{n-1}$.

Then for any data

$$f \in E((-T, T), D^{*\prime}(R^n)),$$

$$g_j \in D^{*\prime}(R^n), \quad j = 0, 1, \ldots, m-1,$$

the Cauchy problem

$$\begin{cases} P(x, \partial)u(x) = f(x), \\ \partial_0^j u(0, x') = g_j(x'), \quad j = 0, 1, \ldots, m-1, \end{cases}$$

has a unique solution

$$u \in E((-T, T), D^{*\prime}(R^n)).$$

If

$$f \in E((-T, T), E^*(R^n)) \quad (\text{resp. } E^*(\Omega)),$$

$$g_j \in E^*(R^n), \; j = 1, \ldots, m-1,$$

then

$$u \in E((-T, T), E^*(R^n)) \quad (\text{resp. } E^*(\Omega)).$$

If

$$f \in D^{*\prime}(\Omega) \quad \text{and} \quad \operatorname{supp} f \subset \{x_0 \geqq t\},$$

then there is a unique solution

$$u \in D^{*\prime}(\Omega) \quad \text{with} \quad \operatorname{supp} u \subset \{x_0 \geqq t\}$$

of $P(x, \partial)u(x) = f(x)$.

The necessity of Levi's condition for solvability in $E^{(\infty)}$ has been proved by Mizohata - Ohya [24], Flaschka - Strang [4] and Ivrii -

Petkov [4]. The following theorem shows that the irregularity condition (6) is necessary in general in order that the conclusions of Theorem 1 hold.

Theorem 2. Let $P(x, \partial)$ be a formally hyperbolic operator with real analytic coefficients in a neighborhood of $\mathring{x} \in R^{n+1}$. Suppose that $(\mathring{x}, \mathring{\xi})$ is a non-singular characteristic element at which P has irregularity $\sigma > 1$ and such that the equation

$$\sum_{d-d_i=\sigma(m-i)} q_i(\mathring{x}, \mathring{\xi})\, \alpha^{d_i} = 0 \tag{9}$$

has only simple roots $\alpha_1, \dots, \alpha_d$. Let $s = \sigma/(\sigma - 1)$.

Then there is a neighborhood Ω_0 of $\mathring{x}$ such that for any neighborhood Ω of $\mathring{x}$ in Ω_0 there is a solution $u \in D^{(s)}{}'(\Omega)$ of $P(x, \partial)\, u(x) = f(x)$ which is not in $D^{\{s\}}{}'$ on $\Omega_+ = \{x \in \Omega;\ x_0 > \mathring{x}_0\}$ and is in $E^{\{s\}}$ on a neighborhood $\tilde{\Omega}$ of $\Omega \cap \{x_0 = \mathring{x}_0\}$. In particular, all the Cauchy data

$$\partial_0^j u(\mathring{x}_0, x') \in E^{\{s\}}(\Omega'), \quad j = 0, 1, \dots,$$

where $\Omega' = \{x' \in R^n;\ (\mathring{x}_0, x') \in \Omega\}$.

We gave a proof in [15] under the stronger assumption that all roots of (9) are simple and non-zero. We will sketch a proof later.

Ivrii [7] and Mizohata [23] have also obtained necessary conditions for solvability in the Gevrey class $\{s\}$ in other formulations. Ivrii starts with the assumption that for every $f \in D^{\{s\}}(\Omega)$ with $\operatorname{supp} f \subset \{x_0 \geq t\}$ there is a solution $u \in D^{\{s\}}{}'(\Omega)$ with $\operatorname{supp} u \subset \{x_0 \geq t\}$. This is inconsistent with the conclusion of Therem 2.

In fact, let u be the solution of Theorem 2. We may assume that Ω is included in the dependence domain of Ω'. We take a cut-off function $k(x)$ in $E^{(s)}(\Omega)$ which vanishes on a neighborhood of $\Omega \cap \{x_0 \leq \mathring{x}_0\}$ and is equal to 1 on a neighborhood of $\Omega_+ \setminus \tilde{\Omega}$. Then $u_1(x) = k(x)u(x)$ is in $D^{(s)}{}'(\Omega) \setminus D^{\{s\}}{}'(\Omega_+)$ and $f_1(x) = P(x, \partial)u_1(x)$ is in $E^{\{s\}}(\Omega)$ and has support in $\Omega_+ \cap \tilde{\Omega}$. Suppose that $u_1(x)$ is not in $D^{\{s\}}{}'(\omega)$ for a relatively compact open set ω in Ω_+. Let $h(x)$ be a cut-off function in $E^{(s)}(\Omega)$ which is equal to 1 on a neighborhood of the influence domain $\tilde{\omega}$ of ω and vanishes outside a neighborhood of $\tilde{\omega}$ and let $f_2(x) = h(x)f_1(x)$. Then f_2 is a function in $D^{\{s\}}(\Omega)$ with support in Ω_+ and the solution $u_2(x)$ of $P(x, \partial)u_2(x)$

$= f_2(x)$ with $\operatorname{supp} u_2 \subset \Omega_+$ is not in $\mathcal{D}^{\{s\}}{}'(\Omega)$ because it coincides with $u_1(x)$ on ω by the Holmgren theorem.

4. FORMAL SOLUTIONS

To prove Theore 2 we employ the method of Y. Hamada [5, 6] and S. Ouchi [25, 26] as they discussed the Cauchy problem with meromorphic data. Its origin may be traced back to P. D. Lax [20] and S. Mizohata [22].

Let $P(z, \partial)$ be an analytic continuation of $P(x, \partial)$ or, more generally, a linear partial differential operator with holomorphic coefficients defined near $\overset{\circ}{x}$ and with the De Paris decomposition (6). Since $K(z, \partial)$ is simple characteritic at $(\overset{\circ}{x}, \overset{\circ}{\xi})$, there is a holomorphic function $\varphi(z)$ satisfying

$$k(z, \operatorname{grad} \varphi(z)) = 0,$$
$$\varphi(\overset{\circ}{x}) = 0, \quad \operatorname{grad} \varphi(\overset{\circ}{x}) = \overset{\circ}{\xi}.$$

We call it a characteristic phase function. For application to Theorem 2 we choose $\varphi(z)$ so that it takes real values on the real domain.

Let $w(\chi)$ be an arbitrary wave form which is a (generalized) function of one variable χ. By Leibniz's rule we can find partial differential operators $P_\varphi^i(z, \partial)$ of order at most i such that

$$P(z, \partial)(w(\varphi(z))u(z)) = \sum_{i=0}^{m} w^{(i)}(\varphi)\, P_\varphi^{m-i}(z, \partial)u(z).$$

The operators P_φ^i depend only on φ and do not on w, so that, if we take $w(\chi) = e^{\lambda\chi}$ with a parameter λ, we have

$$P(z, \partial)(e^{\lambda\varphi(z)}u(z)) = e^{\lambda\varphi(z)}\, P_\varphi(z, \partial, \lambda)u(z),$$

where

$$P_\varphi(z, \partial, \lambda) = \sum_{i=0}^{m} P_\varphi^{m-i}(z, \partial)\, \lambda^i.$$

We call a formal power series

$$U(z, \lambda) = \sum_{j=-\infty}^{\infty} u_j(z)\, \lambda^j$$

in λ a formal operator solution associated with the phase function φ if it satisfies

$$P_\varphi(z, \partial, \lambda)U(z, \lambda) = 0.$$

Then for any sequence $w^{(j)}(\chi)$ of (generalized) functions satisfying

$$dw^{(j)}(\chi)/d\chi = w^{(j+1)}(\chi), \quad j = 0, \pm 1, \pm 2, \ldots,$$

the (generalized) function

$$u(z) = U(z, \partial_\varphi)w^{(0)}(\varphi(z)) = \sum_{j=-\infty}^{\infty} u_j(z)w^{(j)}(\varphi(z)) \tag{11}$$

satisfies the equation

$$P(z, \partial)u(z) = 0 \tag{12}$$

formally in the sense that all the coefficients of $w^{(j)}(\varphi(z))$ vanish.

If we take $w^{(j)}(\chi) = \lambda^j e^{\lambda\chi}$, we obtain an asymptotic solution in the sense of P. D. Lax [20]. In this case series (11) does not converge but it enabled him to prove the necessity of condition (ii) of Theorem 1 and the propagation of singularities along bicharacteristic strips for hyperbolic equations of simple characteristics. This work seems to be one of the origins of microlocal analysis.

Later Mizohata [22] and Hamada [5, 6] estimated the coefficients $u_j(z)$ of (10) and proved the convergence of (11) for $w^{(j)}(\chi) = f^{(j+k)}(\chi)$, where

$$f^{(j)}(\chi) = \begin{cases} \dfrac{(-1)^j j!}{\chi^{j+1}}, & j \geqq 0, \\ \dfrac{\chi^{-j-1}}{(-j-1)!}\left(\log \chi - 1 - \dfrac{1}{2} - \ldots - \dfrac{1}{-j-1}\right), & j < 0, \end{cases} \tag{13}$$

We improved the estimates by Hamada [6] in the following way [12].

Theorem 3. Let $P(z, \partial)$ be a linear partial differential operator with holomorphic coefficients and of irregularity σ at the non-singular characteristic element $(\overset{\circ}{x}, \overset{\circ}{\xi})$ of multiplicity d. If the hypersurface $z_0 = \overset{\circ}{x}_0$ is transversal to the bicharacteristic curve of $K(z, \partial)$ through $(\overset{\circ}{x}, \overset{\circ}{\xi})$, then for any holomorphic functions $h_0(z'), \ldots, h_{d-1}(z')$ defined in a neighborhood of $\overset{\circ}{x}$, there is a formal operator solution (10) on a complex neighborhood Ω_0 of $\overset{\circ}{x}$ satisfying the initial condition

$$\partial_0^k u_j(\overset{\circ}{x}_0, z') = \delta_{j,0}\, h_k(z'), \quad 0 \leqq k < d, \tag{14}$$

and the estimates

$$|u_j(z)| \leqq C^{-j+1}(-j)!, \qquad j \leqq 0, \tag{15}$$

$$|u_j(z)| \leqq \begin{cases} C^{j+1}\left(\dfrac{|z_0 - \overset{o}{x}_0|^j}{j!}\right)^{\sigma/(\sigma-1)}, & \sigma > 1 \text{ and } j > 0, \\ 0, & \sigma = 1 \text{ and } j > 0, \end{cases} \tag{16}$$

with a constant C.

Moreover, there are no other solutions satisfying (14), (15) and

$$|u_j(z)| \leqq C_\varepsilon\, \varepsilon^j/j!, \quad j > 0, \tag{17}$$

for any $\varepsilon > 0$ with a constant C_ε.

The function $w_1(\chi)$ defined by

$$w_1(\chi) = \begin{cases} \exp(-\chi^{-1/(s-1)}), & \chi > 0, \\ 0, & \chi \leqq 0, \end{cases} \tag{18}$$

belongs to the Gevrey class $E^{\{s\}}(R)$ but does not to $E^{(s)}(R)$. If $1 < s \leqq \sigma/(\sigma - 1)$, then it follows from estimates (15) and (16) that (11) converges in $E^{\{s\}}(\Omega)$ for $w^{(0)}(\chi) = w_1(\chi)$ and its derivatives and primitives with support in $\{\chi \geqq 0\}$. Hence it represents a null solution in a neighborhood Ω of $\overset{o}{x}$ [12].

To prove Theorem 2 we consider the case $s = \sigma/(\sigma - 1)$ and take $w^{(0)}(\chi) = w_1(c\chi)$ for a real $c \neq 0$. Then, as in [12], it is easily proved that (11) converges in $D^{(s)'}(\Omega_0)$ and represents a solution u of (12). Moreover, it converges in $E^{\{s\}}(\tilde{\Omega})$ if $\tilde{\Omega}$ is a sufficiently small neighborhood of $\overset{o}{\Omega}'$. In particular, the initial values $\partial_0^j u(x_0, x')$ are all in $E^{\{s\}}(\Omega')$. The proof will be completed if we show that there are initial data $h_j(z')$ such that for any neighborhood Ω of $\overset{o}{x}$ the solution u is not in $D^{\{s\}'}(\Omega)$ for $|c|$ sufficiently large. To do so, we need estimates from below.

5. ASYMPTOTIC BEHAVIOR OF FORMAL OPERATOR SOLUTIONS

The following theorem is essentially due to Ōuchi [25, 26].

Theorem 4. In addition to the assumptions of Theorem 3 assume that $\sigma > 1$ and that the roots α_i of (9) are all simple. Let

$$1/s = (\sigma - 1)/\sigma = r/q,$$

where q and r are relatively prime natural numbers. Then the formal operator solution $U(z, \lambda)$ of Theorem 3 is decomposed as

$$U(z, \lambda) = U_{I+II}(z, \lambda) + U_{III}(z, \lambda),$$

where

$$U_{III}(z, \lambda) = \sum_{j=-\infty}^{-1} u_{III,j}(z) \lambda^{j/q}$$

is a formal power series in $\lambda^{-1/q}$ whose coefficients have the estimates

$$|u_{III,j}(z)| \leqq M^{1-j} \Gamma(1 - j/q) \tag{19}$$

with a constant M. The first term $U_{I+II}(z, \lambda)$ is a power series in $\lambda^{1/q}$ and $\lambda^{-1/q}$ and converges for $|\lambda| > \Lambda_1$ for a Λ_1. On each sector Σ of opening less than $s\pi$ it is decomposed into the sum $U_I(z, \lambda) + U_{II}(z, \lambda)$ which have the asymptotic expansions

$$\begin{aligned} U_I(z, \lambda) &\sim \sum_{i=1}^{d} e^{\lambda^{1/s}\psi_i(z, \lambda)} \sum_{j=0}^{\infty} a_{i,j}(z, \lambda) \lambda^{-j/s}, \\ U_{II}(z, \lambda) &\sim \sum_{j=0}^{\infty} b_j(z, \lambda) \lambda^{-j/s} \end{aligned} \tag{20}$$

as $|\lambda| \longrightarrow \infty$ in Σ provided that $|z - \overset{o}{x}|$ is sufficiently small. Here $\psi_i(z, \lambda)$, $a_{i,j}(z, \lambda)$ and $b_j(z, \lambda)$ are holomorphic functions of z and $\lambda^{-1/q}$ on $\Omega_0 \times \{\lambda; |\lambda| > \Lambda_1\}$ with the asymptotic behavior

$$\begin{aligned} \psi_i(z, \lambda) &= \tilde{\alpha}_i(z_0 - \overset{o}{x}_0) + O((|z - \overset{o}{x}| + |\lambda|^{-1/q})^2), \\ a_{i,j}(z, \lambda) &= a_{i,j} + O(|z - \overset{o}{x}| + |\lambda|^{-1/q}), \\ b_j(z, \lambda) &= b_j + O(|z - \overset{o}{x}| + |\lambda|^{-1/q}) \end{aligned} \tag{21}$$

as (z, λ) tends to $(\overset{o}{x}, \infty)$, where $\tilde{\alpha}_i$, $a_{i,j}$ and b_j are constants. Actually $\tilde{\alpha}_i$ is a constant times the root α_i of (9) and for each $\alpha_i \neq 0$ there are initial data $h_k(z')$ for which $a_{i,0} \neq 0$.

Since $\partial_\varphi^{j/q} w(\varphi)$ for $j < 0$ is represented by the Riemann-Liouville integral, estimates (19) proves that $U_{III}(z, \partial_\varphi)$ is an integral operator with bounded holomorphic kernel for sufficiently small Ω_0.

Hence $U_{III}(x, \partial_\varphi)w^{(0)}(\varphi(x))$ belongs to $E^{\{s\}}(\Omega)$ for $w^{(0)}(\chi) = w_1(c\chi)$.

To prove that $U_{I+II}(x, \partial_\varphi)w^{(0)}(\varphi(x))$ is not in $D^{\{s\}}{}'(\Omega)$, we employ the Heaviside calculus based on the Laplace transforms. Let $w^{(0)}(\chi)$ be a hyperfunction of exponential type and with support in $[a, \infty)$ (resp. $(-\infty, b]$) and let

$$\hat{w}^{(0)}(\lambda) = \int_{-\infty}^{\infty} e^{-\lambda\chi}\, w^{(0)}(\chi)\, d\chi$$

be its Laplace transform. If the formal power series

$$v(\chi, \lambda) = \sum_{j=-\infty}^{\infty} v_j(\chi)\, \lambda^{j/q}$$

converges for $|\lambda| > \Lambda_1$ and is bounded by $C_\varepsilon e^{\varepsilon|\lambda|}$ for any $\varepsilon > 0$, then $v(\chi) = v(\chi, \partial_\chi)w^{(0)}(\chi)$ with support in $[a, \infty)$ (resp. $(-\infty, b]$) is represented as the boundary value

$$v(\chi) = V(\chi + i0) - V(\chi - i0)$$

of the holomorphic function

$$V(\zeta) = \frac{1}{2\pi i}\int_{\Lambda}^{\infty} e^{\lambda\zeta}\, v(\zeta, \lambda)\, \hat{w}^{(0)}(\lambda)\, d\lambda, \tag{22}$$

where Λ (resp. $-\Lambda$) is suficiently large.

Ōuchi considered the case where $\hat{w}^{(0)}(\lambda) = \lambda^k$ and discussed the asymptotic behavior of integral (22). In our case, let $w^{(0)}(\chi) = w_c(\chi) = w_1(c\chi)$. If $c > 0$, then the Laplace transform $\hat{w}_c(\lambda)$ is holomorphic in the sector $|\arg\lambda| < \pi s/2$ and in each subsector $|\arg\lambda| \leqq \pi s'/2$, $s' < s$, it has the uniform asymptotic expansion

$$\hat{w}_c(\lambda) = (c_0/c)(\lambda/c)^{-1+1/(2s)}\, e^{-c_1(\lambda/c)^{1/s}}\, (1+O((\lambda/c)^{-1/s}))$$

as λ tends to ∞, where c_0 and c_1 are positive constants depending only on s.

In the integral

$$U_I(z, \zeta) = \frac{1}{2\pi i}\int_{\Lambda}^{\infty} e^{\lambda\zeta}\, u_I(z, \lambda)\, \hat{w}_c(\lambda)\, d\lambda,$$

the exponent

$$\lambda^{1/s}\psi_i(z, \lambda) - c_1(\lambda/c)^{1/s} = \lambda^{1/s}(\psi_i(z, \lambda) - c_1 c^{-1/s})$$

of each term of the asymptotic expansion of the product $u_I \hat{w}_c$ is expanded into the convergent series

$$t_r(z)\lambda^{r/q} + t_{r-1}(z)\lambda^{(r-1)/q} + \dots + t_0(z) + t_{-1}(z)\lambda^{-1/q} + \dots$$

and the leading coefficient $t_r(z)$ has the asymptotic expansion

$$t_r(z) \sim \tilde{\tau}_i(z_0 - \overset{o}{x}_0) - c_1 c^{-1/s} + O(|z - \overset{o}{x}|^2) \tag{23}$$

as z tends to $\overset{o}{x}$. If

$$|\arg t_r| < (1 + 1/s)\pi/2, \tag{24}$$

then the integral

$$U(\zeta) = \frac{1}{2\pi i}\int_\Lambda^\infty e^{\lambda\zeta} e^{t_r\lambda^{r/q}+\dots+t_1\lambda^{1/q}} k(\lambda)\, d\lambda,$$

where $k(\lambda)$ is a holomorphic function on the domain $\Sigma = \{\lambda \in C;\ |\lambda| > \Lambda_1,\ |\arg\lambda| < s'\pi/2\}$, can be evaluated by the method of steepest descent.

Suppose that at least one $\tilde{\tau}_i$ in Theorem 4 has non-negative real part. Take such a $\tilde{\tau}_i$. Then for any point $x \in \Omega$ sufficiently close to $\overset{o}{x}$ and satisfying $\varphi(x) = 0$ and $x_0 - \overset{o}{x}_0 > 0$ we can find a large $c > 0$ such that $t_r(z)$ of (23) is away from 0 and satisfies (24) uniformly for all z in a complex neighborhood Ω_1 of x in Ω_0. Thus on a sector Z we can evaluate the integral with that exponent. If we subdivide Z, if necessary, only one integral becomes dominant and we can find positive constants C_1 and C_2 such that

$$|U_I(z, \zeta)| \geq C_1 \exp(C_2 |\zeta|^{-1/(s-1)})$$

for $z \in \Omega_1$ and $\zeta \in Z$ with sufficiently small $|\zeta|$.

On the other hand, it is easily proved that $U_{II}(z, \zeta)$ is bounded on $\Omega_0 \times \{\zeta;\ |\arg(-\zeta)| < \pi\}$. Therefore setting $\omega = \Omega_1 \cap R^{n+1}$, we have the estimate

$$\sup_{x\in\omega} |U_{I+II}(x+iy, \varphi(x+iy))| \geq C_3 \exp(C_4|y|^{-1/(s-1)})$$

with positive constants C_3 and C_4 for the defining function $U_{I+II}(z, \varphi(z))$ of the hyperfunction $u_{I+II}(x, \partial_\varphi)w_c(\varphi(x))$. This proves that the hyperfunction is not in $D^{\{s\}'}(\Omega)$ by the characterization theorem of ultradistributions of class $\{s\}$ ([16], Petzsche [28] and de Roever [3]). When there are only $\tilde{\tau}_i$ with negative real part, we take $c < 0$.

REFERENCES

[1] T. Aoki, An invariant measuring the irregularity of a differential operator and a microdifferential operator, J. Math. Pures Appl., 61(1982), 131 - 148.
[2] J.-C. De Paris, Problème de Cauchy oscillatoire pour un opérateur différentiel à caractéristiques multiples; lien avec l'hyperbolicité, J. Math. Pures Appl., 51(1972), 231 - 256.
[3] J. W. de Roever, Hyperfunctional singular support of ultradistributions, J. Fac. Sci. Univ. Tokyo, Sec. IA, 31(1985), 585 - 631.
[4] H. Flaschka - G. Strang, The correctness of the Cauchy problem, Advances in Math., 6(1971), 347 - 379.
[5] Y. Hamada, The singularities of the solutions of the Cauchy problem, Publ. RIMS, Kyoto Univ., 5(1969), 21 - 40.
[6] Y. Hamada, Problème analytique de Cauchy à caractéristiques multiples dont les données de Cauchy ont des singularités polaires, C. R. Acad. Sci. Paris, Sér. A, 276(1973), 1681 - 1684.
[7] V. Ya. Ivrii, Conditions for correctness in Gevrey classes of the Cauchy problem for weakly hyperbolic equations, Siberian Math. J., 17(1976), 422 - 435 (Original Russian: Sibirsk. Mat. Z., 17(1976), 547 - 563).
[8] V. Ya. Ivrii - V. M. Petkov, Necessary conditions for the Cauchy problem for non-strictly hyperbolic equations to be well-posed, Russian Math. Surveys, 29(1974), no. 5, 1 - 70 (Original Russian: Uspehi Mat. Nauk, 29(1974), no. 5, 3 - 70).
[9] H. Komatsu, On the index of ordinary differential operators, J. Fac. Sci. Univ. Tokyo, Sec. IA, 18(1971), 379 - 398.
[10] H. Komatsu, On the regularity of hyperfunction solutions of linear ordinary differential equations with real analytic coefficients, J. Fac. Sci. Univ. Tokyo, Sec. IA, 20(1973), 107 - 119.
[11] H. Komatsu, Linear ordinary differential equations with Gevrey coefficients, J. Diff. Equations, 45(1982), 272 - 306.
[12] H. Komatsu, Irregularity of characteristic elements and construction of null-solutions, J. Fac. Sci. Univ. Tokyo, Sec. IA, 23(1976), 297 - 342.
[13] H. Komatsu, Ultradistributions and hyperbolicity, Boundary Value Problems for Linear Evolution Partial Differential Equations, Reidel, 1977, pp. 157 - 173.
[14] H. Komatsu, Linear hyperbolic equations with Gevrey coefficients, J. Math. Pures Appl., 59(1980), 145 - 185.
[15] H. Komatsu, Irregularity of hyperbolic operators, Proc. Workshop on Hyperbolic Equations and Related Topics to appear.
[16] H. Komatsu, Ultradistributions, I, Structure theorems and a characterization, J. Fac. Sci. Univ. Tokyo, Sec. IA, 20(1973), 25 - 105.
[17] H. Komatsu, Ultradistributions, II, The kernel theorem and ultradistributions with support in a submanifold, J. Fac. Sci. Univ. Tokyo, Sec. IA, 24(1977), 607 - 628.
[18] H. Komatsu, Ultradistributions, III, Vector valued ultradistributions and the theory of kernels, J. Fac. Sci. Univ.Tokyo, Sec. IA, 29(1982), 653 - 718.
[19] Y. Laurent, Théorie de la Deuxième Microlocalisation dans le Domain

Complexe, Birkhäuser, 1985.
[20] P. D. Lax, Asymptotic solutions of oscillatory initial value problems, Duke Math. J., 24(1957), 627 - 646.
[21] B. Malgrange, Sur les points singuliers des équations différentiel les linéaires, Enseign. Math., 20(1974), 147 - 176.
[22] S. Mizohata, Solutions nulles et solutions non analytiques, J. Math. Kyoto Univ., 1(1962), 272 - 302.
[23] S. Mizohata, Sur l'indice de Gevrey, Propagation des Singularités et Opérateurs Différentiels, Séminaire Vaillant 1984 - 1985, Hermann, 1985, pp. 106 - 120.
[24] S. Mizohata - Y. Ohya, Sur la condition d'hyperbolicité pour les équations à caractéristiques multiples, II, Japan. J. Math., 40(1971), 63 - 104.
[25] S. Ōuchi, Asymptotic behaviour of singular solutions of linear partial differential equations in the complex domain, J. Fac. Sci. Univ. Tokyo, Sec. IA, 27(1980), 1 - 36.
[26] S. Ōuchi, An integral representation of singular solutions of linear partial differential equations in the complex domain, J. Fac. Sci. Univ. Tokyo, Sec. IA, 27(1980), 37 - 85.
[27] S. Ōuchi, Existence of singular solutions and null solutions for linear partial differential equations, to appear in J. Fac. Sci. Univ. Tokyo, Sec. IA.
[28] H.-J. Petzsche, Generalized functions and the boundary values of holomorphic functions, J. Fac. Sci. Univ. Tokyo, Sec. IA, 31(1984), 391 - 431.
[29] W. Wasow, Asymptotic Expansions for Ordinary Differential Equations, Interscience, 1965.

ASYMPTOTIC SOLUTIONS OF HYPERBOLIC BOUNDARY VALUE PROBLEMS WITH DIFFRACTION

Pascal Laubin
Department of Mathematics
University of Liège
15, Av. des Tilleuls, 4000 LIEGE
Belgique

1. INTRODUCTION

This paper concerns the propagation of analytic singularities in boundary value problems. We are mainly interested in the problem

$$(1.1) \qquad (-\Delta+D_t^2)u \in A(\mathbb{R}\times\Omega)$$

$$(1.2) \qquad u_{|\mathbb{R}\times\partial\Omega} \in A(\mathbb{R}\times\partial\Omega)$$

where Ω is an open subset of $\mathbb{R}^{n-1}$ with analytic boundary. The Dirichlet boundary condition (1.2) can be replaced by many others, in particular by the Neumann boundary condition

$$(1.3) \qquad D_\nu u_{|\mathbb{R}\times\partial\Omega} \in A(\mathbb{R}\times\partial\Omega).$$

It is wellknown that the problem (1.1)-(1.2) can be reduced locally to

$$(1.4) \qquad Pu \in A(M)$$

$$(1.5) \qquad u_{|\partial M} \in A(\partial M)$$

where

$$P(x,D) = D_{x_n}^2 + R(x,D_{x'}),$$

$M = U' \times [0,a[$, $a > 0$ and U' is an open neighbourhood of 0 in $\mathbb{R}^{n-1}$. The "tangential" operator $R(x,D_{x'})$ is a second order operator of real principal type with analytic coefficients in $U \times]-a,a[$. Moreover, if $p(x,\xi) = \xi_n^2 + r(x,\xi')$ is the principal symbol of R and $r_o(x',\xi') = r(x',0,\xi')$ we have

$$\partial_{\xi'}r \neq 0 \text{ if } r = 0.$$

This problem has been studied by many authors, [3], [6], [8], [9],

H. G. Garnir (ed.), Advances in Microlocal Analysis, 165–202.

[10]. The reduced analytic wave front set of a solution u of (1.4)-(1.5) is defined by

$$WF_{ba}(u) = WF_a(u|_{\overset{\circ}{M}}) \cup WF_a(u|_{\partial M}) \cup WF_a(D_{x_n} u|_{\partial M}).$$

The elliptic, hyperbolic and glancing regions are

$$\left\{\begin{matrix} E \\ G \\ H \end{matrix}\right\} = \{(x',\xi') \in \dot{T}^{*}(\partial M) : r_o(x',\xi') \left\{\begin{matrix} > \\ = \\ < \end{matrix}\right\} 0\}.$$

In E no propagation occurs. In H we have the phenomenon of transversal reflexion. These results are due to Schapira. In G the situation is more intricated. We consider here $G_+ = \{(x',\xi') \in G : \partial_{x_n} r(x',0,\xi') < 0\}$, the so-called diffractive region. For the wave equation it corresponds to the points where the obstacle K = Ω is strictly convex in the ray direction.

Through a given point $\rho_o' = (x_o',\xi_o') \in G_+$ we have four half bicharacteristic curves in M. Indeed we can define

$$\gamma_{\{\begin{smallmatrix}1\\3\end{smallmatrix}\}}(s) = \exp(sH_{r_o})(\rho_o') \ , \ 0 < \pm\, s < \delta$$

in $\dot{T}^{*}(\partial M)$ and

$$\gamma_{\{\begin{smallmatrix}2\\4\end{smallmatrix}\}}(s) = \exp(sH_p)(x_o',0,\xi_o',0) \ , \ 0 < \pm\, s < \delta$$

in $\dot{T}^{*}(M)$. We have the following results of Sjöstrand, [9], [10].

Theorem 1.1. *If* u *solves* (1.4)-(1.5) *and* γ_3,γ_4 *do not meet* $WF_{ab}u$ *for some* $\delta > 0$ *then* $\rho_o \notin WF_{ab}u$.

Theorem 1.2. *If* u *solves* (1.4)-(1.5) *and* γ_2,γ_3 *do not meet* $WF_{ab}u$ *for some* $\delta > 0$ *then* $\rho_o \notin WF_{ab}u$.

There is also a more general result than Theorem 1.1. which says that $WF_{ab}u$ is a union of maximally extended analytic rays, [10].

A third result concerning G_+ was first stated by Kataoka, [3].

Theorem 1.3. *If* u *satisfies* (1.4) *and* γ_2,γ_4 *do not meet* $WF_a u$ *then* $\rho_o \notin WF_{ab}u$.

The surprising fact here is that no boundary condition is needed. Kataoka's proof uses the theory of mild hyperfunctions. G. Lebeau has also proved this theorem. He uses the second analytic wave front set along an isotropic variety. Thanks to a theorem which allows to reduce the problem to a flat boundary in the original coordinates, he has shown that theorem 1.3 is also valid for higher orders of tangency.

Here we give a proof which follows an idea of Sjöstrand. It consists of an explicit construction of asymptotic solutions to (1.4) which are singular on the bicharacteristic of p but don't propagate singularities at the boundary. It is more close to the proofs of Theorem 1.1 and 1.2.

In section 2 we construct the asymptotic solutions. The main problem is that the eiconal equation has only multivalued solutions. We override this problem thanks to an expression of the solution ϕ as critical value of an univalued function having degenerated critical points. In section 3 we prove theorem 1.3. The Appendix contains a careful study of the structure of the curves of steepest descend near a degenerated critical point.

I am gratefull to J. Sjöstrand for helpfull discussions on the following constructions.

2. ASYMPTOTIC SOLUTIONS WITHOUT BOUNDARY PROPAGATION

In this section we construct asymptotic solutions to the diffractive problem whose singularities do not propagate at the boundary. We then use them to prove theorem 1.3.

2.1. Phase functions

We work microlocally near a point $(0,\xi_o') \in \dot{T}^{\star}(\partial M)$ satisfying

$$r_o(0,\xi_o') = 0 \ , \ \partial_{\xi'} r_o(0,\xi_o') \neq 0, \text{ and } \partial_{x_n} r(0,\xi_o') < 0.$$

There is a real analytic function $\psi(x',\theta')$ in a neighbourhood of a point $(x_o',\theta_o') \in \mathbb{R}^{n-1} \times \mathbb{R}^{n-1}$ such that

$$(2.1) \qquad \begin{cases} r_o(x',\partial_{x'}\psi(x',\theta')) + \theta_1 = 0 \\ \partial_{x'}\psi(0,\theta_o') = \xi_o' \\ \det \partial_{x'}\partial_{\theta'}\psi(0,\theta_o') \neq 0. \end{cases}$$

It follows that $\theta_{o,1}' = 0$. If for example $\partial_{\xi_1} r_o(0,\xi_o') \neq 0$, ψ can be chosen as the solution of

$$\begin{cases} r_o(x',\partial_{x'}\psi) + \theta_1 = 0 \\ \psi(0,x'',\theta') = x''.\theta'' \ , \ \partial_{x'}\psi(0,\theta_o') = \xi_o' \end{cases}$$

with $x' = (x_1,x'')$.

Our purpose is to solve the eiconal equation

$$(2.2) \qquad \begin{cases} (\partial_{x_n}\phi)^2 + r(x,\partial_{x'}\phi) = 0 \\ \phi(x',0,\theta') = \psi(x',\theta'). \end{cases}$$

This equation cannot be solved directly since the Hamiltonian field H_p is not transversal to $x_n = 0$. We search ϕ as the critical value of another phase function. This will lead to multivalued solutions of (2.2).

Let $H(x',\xi_n,\theta',t)$ be the holomorphic solution of

$$(2.3) \qquad \begin{cases} \partial_t H + \xi_n^2 + r(x',-\partial_{\xi_n}H,\partial_{x'}H) = 0 \\ H(x',\xi_n,\theta',0) = \psi(x',\theta'), \end{cases}$$

defined in a neighbourhood of $(0,0,\theta_o',0)$. From (2.3) and its derivatives with respect to t, it follows easily that

$$(2.4) \quad H(x',\xi_n,\theta',t)=\psi(x',\theta') + \theta_1 t + \frac{\xi_n^3 - (\xi_n+t\partial_{x_n}r(x',0,\partial_{x'}\psi))^3}{3\,\partial_{x_n}r(x',0,\partial_{x'}\psi)}$$

$$- \frac{\xi_n t^3}{3}(\partial_{\xi'} r(x',0,\partial_{x'}\psi).\partial_{x'}[\partial_{x_n} r(x',0,\partial_{x'}\psi)]+2\xi_n\partial^2_{x_n} r(x',0,\partial_{x'}\psi))+\mathcal{O}(t^4).$$

Hence

$$\partial_t^2 H(x',0,\theta',0) = 0 \text{ and } \partial_t^3 H(x',0,\theta'_o,0) = -2(\partial_{x_n} r(x',0,\partial_{x'}\psi))^2.$$

The implicit functions' theorem gives an holomorphic function $\tilde{t}(x',\xi_n,\theta')$ such that

(2.5) $$\partial_t^2 H(x',\xi_n,\theta',\tilde{t}(x',\xi_n,\theta')) = 0 , \quad \tilde{t}(x',0,\theta') = 0.$$

The expansion (2.4) leads to

$$\partial_{\xi_n}\tilde{t}(x',0,\theta') = -(\partial_{x_n} r(x',0,\partial_{x'}\psi))^{-1}.$$

For a fixed θ' and t small, let Λ_t be the lagrangian variety that consists of the points $(x',-\partial_{\xi_n} H(x',\xi_n,\theta',t),\partial_{x'}H(x',\xi_n,\theta',t),\xi_n)$. By the Hamilton-Jacobi theory of first order equations, we have

$$\exp (sH_p)(\Lambda_t) = \Lambda_{t+s}.$$

Let Γ_t be the set of points in Λ_t where H_p is tangent to Λ_t. Clearly

$$\exp (sH_p)(\Gamma_t) = \Gamma_{t+s}.$$

From the equations (2.1) and (2.3), we see that Γ_o consists of the points $(x',0,\partial_{x'}\psi(x',\theta'),0)$. Hence Γ_t is a subvariety of Λ_t with codimension 1 for every t. If $(x',-\partial_{\xi_n} H,\partial_{x'}H,\xi_n) \in \Gamma_t$ we have

$$\partial_{(x',\xi_n)}\partial_t H(x',\xi_n,\theta',t) = \partial_{(x',\xi_n)}[p(x',-\partial_{\xi_n} H,\partial_{x'}H,\xi_n)] = 0$$

for $dp = 0$ on $T_\rho(\Lambda_t)$ when $\rho \in \Gamma_t$. It follows that

$$\partial_t^2 H(x',\xi_n,\theta',t) = - \partial_t[p(x',-\partial_{\xi_n} H,\partial_{x'}H,\xi_n)] = 0.$$

Since

$$\partial_{\xi_n}\partial_t^2 H(x',\xi_n,\theta',0) = -2\,\partial_{x_n} r(x',0,\partial_{x'}\psi) \neq 0,$$

this proves that

$$\partial_t^2 H(x',\xi_n,\theta',t) = 0$$

is the equation of Γ_t in Λ_t.

On one hand $-\partial_t H(x',\xi_n,\theta',t)$ is the value of p at the point of Λ_t which corresponds to (x',ξ_n). On the other hand $\partial_t H(x',0,\theta',0) = \theta_1$. Since p is constant on its bicaracteristic curves we get

$$\partial_t H(x',\xi_n,\theta',t) = \theta_1 \text{ if } (x',-\partial_{\xi_n} H,\partial_{x'}H,\xi_n) \in \Gamma_t.$$

Therefore

$$\partial_t H(x',\xi_n,\theta',\tilde{t}(x',\xi_n,\theta')) = \theta_1. \tag{2.6}$$

By the division theorem we have

$$\partial_t H(x',\xi_n,\theta',t)=F(x',\xi_n,\theta',t)[(t-\tilde{t}(x',\xi_n,\theta')-\mu(x',\xi_n,\theta'))^2-\lambda(x',\xi_n,\theta')]$$

in a neighbourhood of $(0,0,\theta'_o,0)$ with

$$F(0,0,\theta'_o,0) \neq 0 \text{ and } \mu(0,0,\theta'_o) = \lambda(0,0,\theta'_o) = 0.$$

Identifying the two first derivatives with respect to t at $t=\tilde{t}(x',\xi_n,\theta')$ we get

$$\mu(x',\xi_n,0,\theta'') = \lambda(x',\xi_n,0,\theta'') = 0$$

and

$$\partial_{\theta_1}\lambda(x',0,0,\theta'') = (\partial_{x_n} r(x',0,\partial_{x'}\psi))^{-2}.$$

It follows that

$$\partial_t H(x',\xi_n,\theta',t) = F(x',\xi_n,\theta',t)(t-t_+(x',\xi_n,\sqrt{\theta}_1,\theta''))(t-t_-(x',\xi_n,\sqrt{\theta}_1,\theta'')) \tag{2.7}$$

with

$$t_\pm(x',\xi_n,\sqrt{\theta}_1,\theta'')=\tilde{t}(x',\xi_n,\theta')\mp\sqrt{\theta}_1 b(x',\xi_n,\theta')+\theta_1 c(x',\xi_n,\theta'). \tag{2.8}$$

Here b,c are holomorphic functions near $(0,0,\theta'_o)$ and

$$b(x',0,0,\theta'') = -(\partial_{x_n} r(x',0,\partial_{x'}\psi))^{-1}.$$

In these formulas, $\sqrt{\theta}_1$ is one of the square roots of θ_1, the same at each place. Moreover (2.4) gives

$$0 = \partial_t H(x',\xi_n,\theta',0) = F(x',\sqrt{\theta}_1,\theta',0)t_+(x',\sqrt{\theta}_1,\sqrt{\theta}_1,\theta'')t_-(x',\sqrt{\theta}_1,\sqrt{\theta}_1,\theta'').$$

Formula (2.8) shows that the last factor cannot vanishe if $\theta_1 \neq 0$, hence

$$t_+(x',\sqrt{\theta}_1,\sqrt{\theta}_1,\theta'') = 0. \tag{2.9}$$

Now introduce the holomorphic function G by

$$G(x',\xi_n,\sqrt{\theta}_1,\theta'') = H(x',\xi_n,\theta',t_+(x',\xi_n,\sqrt{\theta}_1,\theta''))$$

with the same convention as above. It follows from (2.3) and (2.9) that

$$\text{(2.10)} \qquad \begin{cases} \xi_n^2 + r(x',-\partial_{\xi_n} G, \partial_{x'} G) = 0 \\ G(x',\sqrt{\theta}_1,\sqrt{\theta}_1,\theta'') = \psi(x',\theta'). \end{cases}$$

Moreover (2.3) and (2.7) lead to

$$\text{(2.11)} \quad G(x',\xi_n,\sqrt{\theta}_1,\theta'') = \psi(x',\theta') + \theta_1\tilde{t}(x',\xi_n,\theta') + \frac{\xi_n^3}{3\partial_{x_n} r(x',0,\partial_{x'}\psi)}$$

$$- \frac{2}{3}\, b(x',\xi_n,\theta')\theta_1^{3/2} + \mathcal{O}(|\theta_1|^2+|\xi_n|^4).$$

Taking $\xi_n = \sqrt{\theta}_1$ in (2.10) and using (2.1) we get

$$\text{(2.12)} \qquad \partial_{\xi_n} G(x',\sqrt{\theta}_1,\sqrt{\theta}_1,\theta'') = 0$$

by the implicit functions theorem. This relation and the two first derivatives of (2.10) with respect to ξ_n at $\xi_n = \sqrt{\theta}_1$ give

$$\partial^2_{\xi_n} G(x',\sqrt{\theta}_1,\sqrt{\theta}_1,\theta'') = \frac{2\sqrt{\theta}_1}{\partial_{x_n} r(x',0,\partial_{x'}\psi)}$$

and

$$\partial^3_{\xi_n} G(x',0,0,\theta'') = 2(\partial_{x_n} r(x',0,\partial_{x'}\psi(x',0,\theta'')))^{-1}.$$

There exists an holomorphic function $\tilde{\xi}_n$ satisfying

$$\partial^2_{\xi_n} G(x',\tilde{\xi}_n(x',\sqrt{\theta}_1,\theta''),\sqrt{\theta}_1,\theta'') = 0.$$

The expansion (2.11) shows that

$$\tilde{\xi}_n(x',\sqrt{\theta}_1,\theta'') = \mathcal{O}(\theta_1)$$

and

$$\partial_{\xi_n} G(x',\tilde{\xi}_n(x',\sqrt{\theta}_1,\theta''),\sqrt{\theta}_1,\theta'') = \theta_1\partial_{\xi_n}\tilde{t}(x',\tilde{\xi}_n(x',\sqrt{\theta}_1,\theta''),\theta') + \mathcal{O}(\theta_1^{3/2}).$$

Therefore

$$\text{(2.13)} \qquad \partial_{\xi_n} G(x',\tilde{\xi}_n(x',\sqrt{\theta}_1,\theta''),\sqrt{\theta}_1,\theta'') = a(x',\sqrt{\theta}_1,\theta'')\theta_1$$

where a is an holomorphic function such that

$$a(x',0,\theta'') = -(\partial_{x_n} r(x',0,\partial_{x'}\psi))^{-1}.$$

The solutions of (2.2) are the critical values of $\xi_n \to x_n\xi_n + G(x',\xi_n,\sqrt{\theta_1},\theta'')$. Let us study the critical points of this function. As above the division theorem gives holomorphic functions A,α,β such that

$$x_n + \partial_{\xi_n} G(x',\xi_n,\sqrt{\theta_1},\theta'')$$
$$= A(x',\xi_n,\sqrt{\theta_1},\theta'')((\xi_n-\tilde{\xi}_n(x',\sqrt{\theta_1},\theta'') - \alpha(x',\sqrt{\theta_1},\theta''))^2 - \beta(x,\sqrt{\theta_1},\theta'')).$$

Moreover A doesn't vanishes and α,β are equal to 0 if $x_n = \theta_1 = 0$. Identifying the two first derivatives with respect to ξ_n at $\xi_n = \tilde{\xi}_n(x',\sqrt{\theta_1},\theta'')$ we obtain

$$\alpha(x,\sqrt{\theta_1},\theta'') = X(x,\sqrt{\theta_1},\theta'')(x_n + a(x',\sqrt{\theta_1},\theta'')\theta_1)$$
$$\beta(x,\sqrt{\theta_1},\theta'') = Y(x,\sqrt{\theta_1},\theta'')(x_n + a(x',\sqrt{\theta_1},\theta'')\theta_1)$$

where X, Y are holomorphic in a neighbourhood of $(0,\theta'_o)$ and

$$Y(x',0,0,\theta'') = -\partial_{x_n} r(x',0,\partial_{x'}\psi(x',0,\theta'')).$$

It follows that there exists an holomorphic function ξ_n^+ in a neighbourhood of $(0,0,\theta'_o)$ such that

$$(2.14)\quad x_n + \partial_{\xi_n} G(x',\xi_n,\sqrt{\theta_1},\theta'')$$
$$= A(x,\xi_n,\sqrt{\theta_1},\theta'')(\xi_n-\xi_n^+(x',\sqrt{x_n+a(x',\sqrt{\theta_1},\theta'')\theta_1},\sqrt{\theta_1},\theta''))$$
$$(\xi_n-\xi_n^+(x',-\sqrt{x_n+a(x',\sqrt{\theta_1},\theta'')\theta_1},\sqrt{\theta_1},\theta''))$$

and

$$(2.15)\quad \xi_n^+(x',\sqrt{x_n+a(x',\sqrt{\theta_1},\theta'')\theta_1},\sqrt{\theta_1},\theta'') = \tilde{\xi}_n(x',\sqrt{\theta_1},\theta'')$$
$$+ \sqrt{x_n+a(x',\sqrt{\theta_1},\theta'')\theta_1}\,\sqrt{Y(x,\sqrt{\theta_1},\theta'')} + (x_n+a(x',\sqrt{\theta_1},\theta'')\theta_1)X(x,\sqrt{\theta_1},\theta'').$$

Here again the only restriction on the square roots is that they take the same value at each places. Clearly (2.12), (2.14)-(2.15) give

$$(2.16)\quad \xi_n^+(x',\sqrt{a(x',\sqrt{\theta_1},\theta'')\theta_1},\sqrt{\theta_1},\theta'') = \sqrt{\theta_1}$$

if the same determination is chosen at the three places for $\sqrt{\theta_1}$ and the argument of $a(x',\sqrt{\theta_1},\theta'')$ is close to 0.

Define the holomorphic function ϕ in a neighbourhood of $(0,\theta'_o)$ by

$$\phi(x', \sqrt{x_n+a(x',\sqrt{\theta}_1,\theta'')\theta_1},\sqrt{\theta}_1,\theta'')=x_n\xi_n^+(x', \sqrt{x_n+a(x',\sqrt{\theta}_1,\theta'')\theta_1},\sqrt{\theta}_1,\theta'') + G(x',\xi_n^+(x', \sqrt{x_n+a(x',\sqrt{\theta}_1,\theta'')\theta_1},\sqrt{\theta}_1,\theta''),\sqrt{\theta}_1,\theta'').$$

This is equivalent to

$$\phi(x',y_n,\sqrt{\theta}_1,\theta'') = (y_n^2-a(x',\sqrt{\theta}_1,\theta'')\theta_1)\xi_n^+(x',y_n,\sqrt{\theta}_1,\theta'')+G(x',\xi_n^+(x',y_n,\sqrt{\theta}_1,\theta''),\sqrt{\theta}_1,\theta'').$$

We have

$$\partial_{y_n}\phi(x',y_n,\sqrt{\theta}_1,\theta'') = 2y_n\xi_n^+(x',y_n,\sqrt{\theta}_1,\theta'')+(y_n^2-a(x',\sqrt{\theta}_1,\theta'')\theta_1)\partial_{y_n}\xi_n^+(x',y_n,\sqrt{\theta}_1,\theta'') + \partial_{\xi_n}G(x',\xi_n^+(x',y_n,\sqrt{\theta}_1,\theta''),\sqrt{\theta}_1,\theta'')\,\partial_{y_n}\xi_n^+(x',y_n,\sqrt{\theta}_1,\theta'').$$

On one hand it follows from (2.13), (2.15) and the definition of $\tilde{\xi}_n$ that

$$\partial_{y_n}\phi(x',0,\sqrt{\theta}_1,\theta'') = 0$$

and

$$\partial_{y_n}^2\phi(x',0,\sqrt{\theta}_1,\theta'') = 2\tilde{\xi}_n(x',\sqrt{\theta}_1,\theta'').$$

On the other hand we have

$$\partial_{y_n}^3\phi(x',0,0,\phi'') = 6\partial_{y_n}\xi_n^+(x',0,0,\theta'')+\partial_{\xi_n}^3 G(x',0,0,\theta'')(\partial_y\xi_n^+(x',0,0,\theta''))^3$$

and

$$\partial_{y_n}\xi_n^+(x',0,0,\theta'') = (-\partial_{x_n} r(x',0,\partial_{x'}\psi(x',0,\theta'')))^{1/2}$$

by definition of ξ_n^+. It follows that

$$\partial_{y_n}^3\phi(x',0,0,\theta'') = 4(-\partial_{x_n} r(x',0,\partial_{x'}\psi(x',0,\theta'')))^{1/2}.$$

Remark that by definition of G and ϕ we have

$$\phi(x',y_n,\sqrt{\theta}_1,\theta'') = \phi(x',y_n,0,\theta'') + \mathcal{O}(\theta_1). \tag{2.17}$$

Let

$$\phi_1(x,\sqrt{\theta_1},\theta'') = \frac{1}{2}(\phi(x',\sqrt{x_n+a(x',\sqrt{\theta}_1,\theta'')\theta_1},\sqrt{\theta}_1,\theta'')+\phi(x',-\sqrt{x_n+a(x',\sqrt{\theta}_1,\theta'')\theta_1},\sqrt{\theta}_1,\theta''))$$

and

$$\rho_1(x',y_n,\sqrt{\theta}_1,\theta'') = \frac{1}{2}(\phi(x',y_n,\sqrt{\theta}_1,\theta'')-\phi(x',-y_n,\sqrt{\theta}_1,\theta'')).$$

Of course we have

(2.18) $$\phi_1(x,\sqrt{\theta_1},\theta'') = \phi_1(x,0,\theta'') + O(\theta_1).$$

Since ρ_1 is an odd function of y_n and $\partial^k_{y_n}\rho_1 = 0$ at $y_n=0$ if $k < 3$, we may write

$$\rho_1(x',y_n,\sqrt{\theta_1},\theta'') = \frac{2}{3}y_n^3\,\rho_2(x',y_n^2-a(x',\sqrt{\theta_1},\theta'')\theta_1,\sqrt{\theta_1},\theta'')$$

with

$$\rho_2(x',0,0,\theta'') = (-\partial_{x_n} r(x',0,\partial_{x'}\psi(x',0,\theta'')))^{1/2}.$$

As above we have

(2.19) $$\rho_2(x,\sqrt{\theta_1},\theta'') = \rho_2(x,0,\theta'') + O(\theta_1).$$

If we introduce

(2.20) $$\phi_2(x,\sqrt{\theta_1},\theta'') = (x_n+a(x',\sqrt{\theta_1},\theta'')\theta_1)(\rho_2(x,\sqrt{\theta_1},\theta''))^{2/3},$$

we obtain the following decomposition of ϕ

$$\phi(x',\sqrt{x_n+a(x',\sqrt{\theta_1},\theta'')\theta_1},\sqrt{\theta_1},\ \theta'') = \phi_1(x,\sqrt{\theta_1},\theta'')+\frac{2}{3}(\phi_2(x,\sqrt{\theta_1},\theta''))^{3/2}.$$

Of course using the definition of ϕ and (2.14) we see that ϕ is a solution to (2.2) in every open set where the square roots are welldefined. The four "branches" of ϕ satisfy the eiconal equation but the initial condition is fullfilled only if the two square roots are consistent (see (2.16)).

2.2. Properties of the phase functions

We prove some evaluations of the phase functions that will be needed later. First we expand $x_n\xi_n + H$ at $t = \tilde{t}(x',\xi_n,\theta')$.

Lemma 2.1. *We have*

$$x_n\xi_n + H(x',\xi_n,\theta',t) = \psi(x',\theta') + \theta_1(t-\tilde{t}(x',\xi_n,\theta')) - \frac{(t-\tilde{t}(x',\xi_n,\theta'))^3}{3\alpha^2(x',\theta')}$$
$$+(x_n+\alpha(x',\theta')\theta_1)\xi_n - \alpha(x',\theta')\frac{\xi_n^3}{3} + O(|\theta_1|^2+|\xi_n|^4+|t-\tilde{t}(x',\xi_n,\theta')|^4)$$

where

$$\alpha(x',\theta') = -1/\,\partial_{x_n} r(x',0,\partial_{x'}\psi(x',\theta')).$$

Proof. Apply Taylor's formula to H at $t = \tilde{t}(x',\xi_n,\theta')$. Using (2.5) and (2.6) it follows that

$$H(x',\xi_n,\theta',t) = H(x',\xi_n,\theta',\tilde{t}(x',\xi_n,\theta')) + \theta_1(t-\tilde{t}(x',\xi_n,\theta'))$$
$$+\frac{1}{6}\partial_t^3 H(x',\xi_n,\theta',\ \tilde{t}(x',\xi_n,\theta'))(t-\tilde{t}(x',\xi_n,\theta'))^3 + O(|t-\tilde{t}(x',\xi_n,\theta')|^4).$$

Since $\tilde{t} = O(\xi_n)$, (2.4) gives

$$\partial_t^3 H(x',\xi_n,\theta',\tilde{t}(x',\xi_n,\theta')) = -2(\partial_{x_n} r(x',0,\partial_{x'}\psi(x',\theta')))^2 + O(\xi_n).$$

Moreover (2.3), (2.4) and (2.6) lead to

$$\partial_{\xi_n}[H(x',\xi_n,\theta',\tilde{t}(x',\xi_n,\theta'))]_{\xi_n=0} = \theta_1\alpha(x',\theta')$$

$$\partial^2_{\xi_n}[H(x',\xi_n,\theta',\tilde{t}(x',\xi_n,\theta'))]_{\xi_n=0} = O(\theta_1)$$

$$\partial^3_{\xi_n}[H(x',\xi_n,\theta',\tilde{t}(x',\xi_n,\theta'))]_{\xi_n=0} = -2\alpha(x',\theta') + O(\theta_1)$$

The lemma follows easily.

Next we need the link between ϕ and the bicharacteristic curves of p.

Lemma 2.2. *Denote by* $v'(x',\xi_n,\theta',t)$ *the holomorphic function which satisfies*

$$\partial_{\theta'}H(x',\xi_n,\theta',t) = \partial_{\theta'}\psi(v'(x',\xi_n,\theta',t),\theta')$$

and let

$$y'(x',\sqrt{x_n+a(x',\sqrt{\theta_1},\theta'')\theta_1},\sqrt{\theta_1},\theta'')$$
$$= v'(x',\xi_n^+(x',\sqrt{x_n+a(x',\sqrt{\theta_1},\theta'')\theta_1},\sqrt{\theta_1},\theta''),\theta',$$
$$t_+(x',\xi_n^+(x',\sqrt{x_n+a(x',\sqrt{\theta_1},\theta'')\theta_1},\sqrt{\theta_1},\theta''),\sqrt{\theta_1},\theta'')).$$

Then we have

(2.21) $$\partial_{\theta'}\phi(x',\sqrt{x_n+a(x',\sqrt{\theta_1},\theta'')\theta_1},\sqrt{\theta_1},\theta'')$$
$$= \partial_{\theta'}\psi(y'(x',\sqrt{x_n+a(x',\sqrt{\theta_1},\theta'')\theta_1},\sqrt{\theta_1},\theta''),\theta')$$

and

(2.22) $$\exp(tH_p)(y',0,\partial_y\psi(y',\theta'),\sqrt{\theta_1})$$
$$=(x,\partial_x\phi(x',\sqrt{x_n+a(x',\sqrt{\theta_1},\theta'')\theta_1},\sqrt{\theta_1},\theta''))$$

if

$$t = t_+(x',\xi_n^+(x',\sqrt{x_n+a(x',\sqrt{\theta_1},\theta'')\theta_1},\sqrt{\theta_1},\theta''),\sqrt{\theta_1},\theta'')$$
$$y' = y'(x',\sqrt{x_n+a(x',\sqrt{\theta_1},\theta'')\theta_1},\sqrt{\theta_1},\theta'').$$

Particularly

$$\partial_x\phi(x',\sqrt{x_n+a(x',\sqrt{\theta_1},\theta'')\theta_1},\sqrt{\theta_1},\theta'') = (\partial_{x'}\psi(x',\theta'),0) + O(|x_n|^{1/2}+|\theta_1|^{1/2})$$
$$\partial_{\theta'}\phi(x',\sqrt{x_n+a(x',\sqrt{\theta_1},\theta'')\theta_1},\sqrt{\theta_1},\theta'') = \partial_{\theta'}\psi(x',\theta') + O(|x_n|^{1/2}+|\theta_1|^{1/2}).$$

Proof. By the Hamilton-Jacobi theory of first order equations applied to (2.3), we know that

(2.23) $$\partial_{\theta'}H(x',\xi_n,\theta',t) = \partial_{\theta'}\psi(y',\theta')$$

(2.24) $$\exp(tH_p)(y',0,\partial_{y'}\psi(y',\theta'),\eta_n) = (x',-\partial_{\xi_n}H(x',\xi_n,\theta',t),\partial_{x'}H(x',\xi_n,\theta',t),\xi_n)$$

if

$$x' = x'(y',\eta_n,\theta',t) \text{ and } \xi_n = \xi_n(y',\eta_n,\theta',t)$$

for some functions $(x',\xi_n)(y',\eta_n,\theta',t)$. Denote by $\eta_n(y',\xi_n,\theta',t)$ the inverse map of $\eta_n \to \xi_n(y',\eta_n,\theta',t)$. Then (2.23) shows that $y' \to x'(y', \eta_n(y',\xi_n,\theta',t),\theta',t)$ is the inverse of $x' \to v'(x',\xi_n,\theta',t)$.

In (2.23), (2.24) introduce successively

$$\eta_n = \eta_n(y',\xi_n,\theta',t)$$

$$y' = v'(x',\xi_n,\theta',t)$$

$$t = t_+(x',\xi_n,\sqrt{\theta_1},\theta'')$$

$$\xi_n = \xi_n^+(x',\sqrt{x_n+a(x',\sqrt{\theta_1},\theta'')\theta_1},\sqrt{\theta_1},\theta'').$$

Using the definitions of G and ϕ as critical values we get (2.21) and (2.22). In fact the only thing to prove is that η_n becomes exactly $\sqrt{\theta_1}$. Since the equation of H gives

$$\partial_t H(x',\xi_n,\theta',t) = \theta_1-\eta_n^2$$

in the conditions of validity of (2.21)-(2.22), it is clear that η_n^2 becomes θ_1 by the definition of t_+. Moreover if $x_n = 0$ then $\xi_n^+ = \sqrt{\theta_1}$ and $t_+ = 0$ by (2.16) and (2.9). Therefore $\eta_n = \xi_n^+ = \sqrt{\theta_1}$. This shows that the good sign is plus.

Because of the square roots, ϕ is not always real when x,θ' are real. However we can give a good bound for its imaginary part.

Lemma 2.3. *If* $\Re z > 0$ *and* $\Re w > 0$ *then*

$$|z^{3/2} - w^{3/2}| \leq \frac{3}{4}\,|z-w|(|z|^{1/2}+ |w|^{1/2})$$

provided that we choose arg z *and* arg w *in* $]-\frac{\pi}{2}, \frac{\pi}{2}[$.

<u>Proof</u>. We may assume $|z| \neq |w|$. Then we have

$$|z^{3/2}-w^{3/2}|=\left|\int_0^1 \partial_t[(1-t)z+tw]^{3/2}dt\right|= \frac{3}{2}|z-w|\left|\int_0^1((1-t)z+tw)^{1/2}dt\right|$$

$$\leq \frac{3}{2}|z-w|\int_0^1((1-t)|z|+t|w|)^{1/2}dt = |z-w|(|z|^{3/2}-|w|^{3/2})/(|z|-|w|)$$

$$\leq \frac{3}{4}\,|z-w|(|z|^{1/2} + |w|^{1/2}).$$

Lemma 2.4. *There exists a constant* C > 0 *such that*

$$|I\phi(x',\sqrt{x_n+a(x',\sqrt{\theta_1},\theta'')\theta_1},\sqrt{\theta_1},\theta'')| \leq C|\theta_1|^{3/2}$$

for every choice of the square roots if x',θ' *are real and* $x_n \geq 0$.

Proof. From (2.18) and (2.19) we know that

$$|I\left\{\begin{matrix}\phi_1\\ \phi_2\end{matrix}\right\}(x,\sqrt{\theta_1},\theta'')| \leq C|\theta_1|^{3/2}$$

Take $c_o > 0$ such that

$$|a(x',\sqrt{\theta_1},\theta'') \leq c_o$$

in a neighbourhood of $(0,\theta_o')$. If $0 \leq x_n \leq c_o|\theta_1|$ it follows that

$$|(x_n + a(x',\sqrt{\theta_1},\theta'')\theta_1)^{3/2}| \leq (2c_o|\theta_1|)^{3/2}.$$

If $x_n > c_o|\theta_1|$ we use lemma 2.3. It shows that

$$\begin{aligned}&|I(x_n+a(x',\sqrt{\theta_1},\theta'')\theta_1)^{3/2}|\\ &\leq |(x_n+a(x',\sqrt{\theta_1},\theta'')\theta_1)^{3/2} - (x_n+\Re(a(x',\sqrt{\theta_1},\theta'')\theta_1))^{3/2}|\\ &\leq C|I(a(x',\sqrt{\theta_1},\theta'')\theta_1)| \leq C'|\theta_1|^{3/2}.\end{aligned}$$

Summing up we get

$$\begin{aligned}&|\phi(x',\sqrt{x_n+a(x',\sqrt{\theta_1},\theta'')\theta_1},\sqrt{\theta_1},\theta'')|\\ &\leq|I\phi_1(x,\sqrt{\theta_1},\theta'')|+\tfrac{2}{3}|I((x_n+a(x',\sqrt{\theta_1},\theta'')\theta_1)^{3/2}\rho_2(x,\sqrt{\theta_1},\theta''))|\leq C''|\theta_1|^{3/2}.\end{aligned}$$

We also need an estimation of $I\phi$ when $I(x',\theta')$ is small.

Lemma 2.5. *There exists a constant* C > 0 *such that*

$$|(a+s)^{3/2} - (a^{3/2} + \tfrac{3}{2}a^{1/2}s)| \leq C|s|^{3/2}$$

if $\Re a > |\Re s|$ *and the real parts of the square roots are positive.*

Proof. Of course we may assume $|s| < |a|$. Then we have

$$\begin{aligned}|(a+s)^{3/2}-(a^{3/2} + \tfrac{3}{2}a^{1/2}s)| &= \tfrac{3}{4}|s^2\int_0^1 \frac{1-t}{\sqrt{a+st}}\,dt|\\ &\leq \tfrac{3}{4}|s|^{3/2}\int_0^1 \sqrt{1-t}\,dt = \frac{|s|^{3/2}}{2}\end{aligned}$$

since $|a+st| \geq |a| - t|s| \geq (1-t)\ |s|$.

Lemma 2.6. *Let* $c_o > 0$ *and assume that*

$$|a(x',\sqrt{\theta_1},\theta'')| \leqslant c_o$$

in a complex neighbourhood of $(0,\theta'_o)$. *Then there are constants* $\varepsilon, C_o > 0$ *such that*

$$|\phi(x'+ish',\sqrt{x_n+a(x'+ish',\sqrt{\theta_1+isk_1},\theta''+isk'')(\theta_1+isk_1)},\sqrt{\theta_1+isk_1},\theta''+isk'')$$

$$-\phi(x',\sqrt{x_n+a(x',\sqrt{\theta_1},\theta'')\theta_1},\sqrt{\theta_1},\theta'')-is\partial_{x'}\phi(x',\sqrt{x_n+a(x',\sqrt{\theta_1},\theta'')\theta_1},\sqrt{\theta_1},\theta'').h'$$

$$-is\ \partial_{\theta'}\phi(x',\sqrt{x_n+a(x',\sqrt{\theta_1},\theta'')\theta_1},\sqrt{\theta_1},\theta'').k'| \leqslant C_o(s^{3/2}+|\theta_1|^{3/2})$$

if $0 \leqslant s < \varepsilon$, x,θ' *are real,* $|(x,\theta')| < \varepsilon$, $x_n \geqslant 0$, $h',k' \in \mathbb{R}^{n-1}$, $|h'|$ *and* $|k'| < 1$. *This is true for every choices of the square roots with the only restriction that*

$$\mathcal{R}\sqrt{x_n+a(x'+ish',\sqrt{\theta_1+isk_1},\theta''+isk'')(\theta_1+isk_1)} \text{ and } \mathcal{R}\sqrt{x_n+a(x',\sqrt{\theta_1},\theta'')\theta_1}$$

have the same sign if $x_n > c_o(s+|\theta_1|)$.

Proof. We use the following representation of ϕ

$$\phi = \phi_1(x,\sqrt{\theta_1},\theta'') + \frac{2}{3}\,(x_n+a(x',\sqrt{\theta_1},\theta'')\theta_1)^{3/2}\ \rho_2(x,\sqrt{\theta_1},\theta'').$$

By (2.18) we know that

$$\phi_1(x,\sqrt{\theta_1},\theta'') = \phi_{11}(x,\theta') + \theta_1^{3/2}\ \phi_{12}(x,\theta'), \tag{2.25}$$

where ϕ_{11} and ϕ_{12} are holomorphic functions. It follows easily that

$$\phi_1(x'+ish',x_n,\sqrt{\theta_1+isk_1},\theta''+isk'')$$

$$= \phi_{11}(x,\theta')+is\ \partial_{x'}\phi_{11}(x,\theta').h'+is\partial_{\theta'}\phi_{11}(x,\theta').k'+O(s^{3/2}+|\theta_1|^{3/2})$$

$$=\phi_1(x,\sqrt{\theta_1},\theta'')+is\partial_{x'}\phi_1(x,\sqrt{\theta_1},\theta'').h'+is\partial_{\theta'}\phi_1(x,\sqrt{\theta_1},\theta'').k'+O(x^{3/2}+|\theta_1|^{3/2}).$$

The same estimate is also valid for the functions $\rho_2(x,\sqrt{\theta_1},\theta'')$ and $x_n+a(x',\sqrt{\theta_1},\theta'')\theta_1$ since we have for them a decomposition in the form (2.25). This proves the Lemma if $x_n \leqslant c_o(s+|\theta_1|)$. Indeed, in this case all the terms including ρ_2 are smaller than the right hand side of the inequality.

From now on we assume $x_n > c_o(s+|\theta_1|)$. By the Lemma 2.5 and the expansion (2.25) applied to $x_n + a\theta_1$, we have

$$(x_n+a(x'+ish',\sqrt{\theta_1+isk_1},\theta''+isk')(\theta_1+isk_1))^{3/2}=(x_n+a(x',\sqrt{\theta_1},\theta'')\theta_1)^{3/2}$$

$$+\frac{3is}{2}(x_n+a(x',\sqrt{\theta_1},\theta'')\theta_1)^{1/2}[\partial_{x'}(\theta_1 a(x',\sqrt{\theta_1},\theta'').h'+\partial_{\theta'}(\theta_1 a(x',\sqrt{\theta_1},\theta'')).k']$$

modulo the same error as above. Using this equality and the expressions of ϕ_1, ρ_2 obtained from (2.25) we get

$$\phi(x'+ish',\sqrt{x_n+a(x'+ish',\sqrt{\theta_1+isk_1},\theta''+isk'')(\theta_1+isk_1)},\sqrt{\theta_1+isk_1},\theta''+isk'')$$

$$= \phi(x',\sqrt{x_n+a(x',\sqrt{\theta_1},\theta'')\theta_1},\sqrt{\theta_1},\theta'')$$

$$+ is\,\partial_{x'}\phi_1(x,\sqrt{\theta_1},\theta'').h' + is\partial_{\theta'}\phi_1(x,\sqrt{\theta_1},\theta'').k''$$

$$+\frac{3is}{2}(x_n+a(x',\sqrt{\theta_1},\theta'')\theta_1)^{1/2}\rho_2(x,\sqrt{\theta_1},\theta'')[\partial_{x'}(\theta_1 a(x',\sqrt{\theta_1},\theta'')).h'$$

$$+\partial_{\theta'}(\theta_1 a(x',\sqrt{\theta_1},\theta'')).k']$$

$$+ is(x_n+a(x',\sqrt{\theta_1},\theta'')\theta_1)^{3/2}(\partial_{x'}\rho_2(x,\sqrt{\theta_1},\theta'').h'+\partial_{\theta'}\rho_2(x,\sqrt{\theta_1},\theta'').k')$$

$$+ \mathcal{O}(s^{3/2}+|\theta_1|^{3/2}).$$

This completes the proof.

2.3. Asymptotic solutions

First of all we construct a formal analytic symbol $A = (a_k), k \in N$, in a neighbourhood of $(0,0,\theta_o',0)$ satisfying

$$(2.26)\quad \begin{cases} e^{-i\lambda H(x',\xi_n,\theta',t)}(\tilde{D}_t+P(x',-\tilde{D}_{\xi_n},\tilde{D}_{x'},\xi_n,\lambda)) \\ \qquad (e^{i\lambda H(x',\xi_n,\theta',t)}A(x',\xi_n,\theta',t,\lambda)) = 0 \\ A(x',\xi_n,\theta',0,\lambda) = 1 \text{ or } \xi_n. \end{cases}$$

The two initial conditions 1 and ξ_n are necessary to prove theorem 1.3. We treat them simultaneously. The equality (2.26) means that after expansion of the action of the pseudo-differential operator we have annulation at each order in λ. We refer to [11] for the notations and the existence of the formal analytic symbol A. We use the notation $\tilde{D}$ for $\partial/i\lambda$. In fact since H satisfies the eiconal equation of $\tilde{D}_t + P$, each function a_k is the solution of a first order equation along the bicharacteristic curves of p.

In the level of formal analytic symbols, write

$$\eta_n^2 + R(x,\xi',\lambda) = \eta_n^2 + R(x',x_n-\sigma_n,\xi',\lambda) + \sigma_n \circ Q(x,\xi',\sigma_n,\lambda)$$

where the base variables are (x',x_n,ξ_n), the dual variables are (ξ',η_n,σ_n) and

$$Q(x,\xi',\sigma_n,\lambda) = \int_0^1 \partial_{x_n} R(x',x_n-t\sigma_n,\xi',\lambda)dt.$$

Since A and H are independent of x_n, it follows from the rules of computation of pseudo-differential operators that

$$\begin{aligned}(2.27)\quad & e^{-i\lambda(x_n\xi_n+H)} P(x,\tilde{D}_x,\lambda)(e^{i\lambda(x_n\xi_n+H)}A) \\ & = e^{-i\lambda H}P(x',-\tilde{D}_{\xi_n},\tilde{D}_{x'},\xi_n,\lambda)(e^{i\lambda H}A)+e^{-i\lambda(x_n\xi_n+H)}\tilde{D}_{\xi_n}(e^{i\lambda(x_n\xi_n+H)}C)\end{aligned}$$

where

$$C(x,\xi_n,\theta',t,\lambda) = e^{-i\lambda(x_n\xi_n+H)} Q(x,\tilde{D}_{x'},\tilde{D}_{\xi_n},\lambda)(e^{i\lambda(x_n\xi_n+H)}A).$$

Introduce

$$a(x',\xi_n,\theta',t,\lambda) = \sum_{0\leq k<\lambda/C_1} \lambda^{-k} a_k(x',\xi_n,\theta',t) \ ,\lambda > 1,$$

a realization of the formal symbol A. If the constant C_1 is large enough and c is the associated realization of C then we have

$$(2.28)\quad (\tilde{D}_t+P(x,\tilde{D}_x,\lambda))(e^{i\lambda(x_n\xi_n+H)}a) = \tilde{D}_{\xi_n}(e^{i\lambda(x_n\xi_n+H)}c) + \mathcal{O}(e^{i\lambda(x_n\xi_n+H)-\varepsilon\lambda})$$

in a complex neighbourhood of $(0,0,\theta_o',0)$ for some $\varepsilon > 0$.

Let σ,τ be positive constants,

$$A_\tau = e^{i\pi/2}\tau \ , \ B_\tau = e^{7i\pi/6}\tau, \ C_\tau = e^{-i\pi/6}\tau$$

and define A_σ, B_σ, C_σ similarly. Define

$$u_+(x,\theta',\lambda) = \frac{\lambda}{2\pi}\int_{A_\sigma}^{C_\sigma} e^{i\lambda x_n\xi_n}d\xi_n \int_{B_\tau}^{A_\tau} e^{i\lambda H(x',\xi_n,\theta',t)} a(x',\xi_n,\theta',t,\lambda)dt$$

$$u_-(x,\theta',\lambda) = \frac{\lambda}{2\pi}\int_{B_\sigma}^{A_\sigma} e^{i\lambda x_n\xi_n}d\xi_n \int_{C_\tau}^{A_\tau} e^{i\lambda H(x',\xi_n,\theta',t)} a(x',\xi_n,\theta',t,\lambda)dt.$$

The constants σ and τ are chosen such that $0 < C_o\tau^{4/3} < \sigma < \tau/C_o$ with C_o large enough and to be determined below. Of course σ,τ are so small that (2.28) is valid in the integrals. We also introduce $u = u_+ + u_-$.

Proposition 2.7. *There exists* $\varepsilon, r > 0$ *such that*

$$P(x,D)u_{\pm}(x,\theta',\lambda) = O(e^{-\varepsilon\lambda}) \tag{2.29}$$

and

$$(u_{+}+u_{-})(x',0,\theta',\lambda) = \{{}^{0}_{1}\}\, e^{i\lambda\psi(x',\theta')} + O(e^{-\varepsilon\lambda}) \tag{2.30}$$

$$\tilde{D}_{x_n}(u_{+}+u_{-})(x',0,\theta',\lambda) = \{{}^{1}_{0}\}\, e^{i\lambda\psi(x',\theta')} + O(e^{-\varepsilon\lambda}) \tag{2.31}$$

uniformly when $|x| < r$, $|\theta'-\theta_0'| < r$. *The upper value in the brackets corresponds to the initial value* 1 *in* (2.26) *and the lower one to* ξ_n.

<u>Proof</u>. First we prove (2.29). To fix the ideas we prove it for u_+. In the definition of u_+ take the following paths of integration

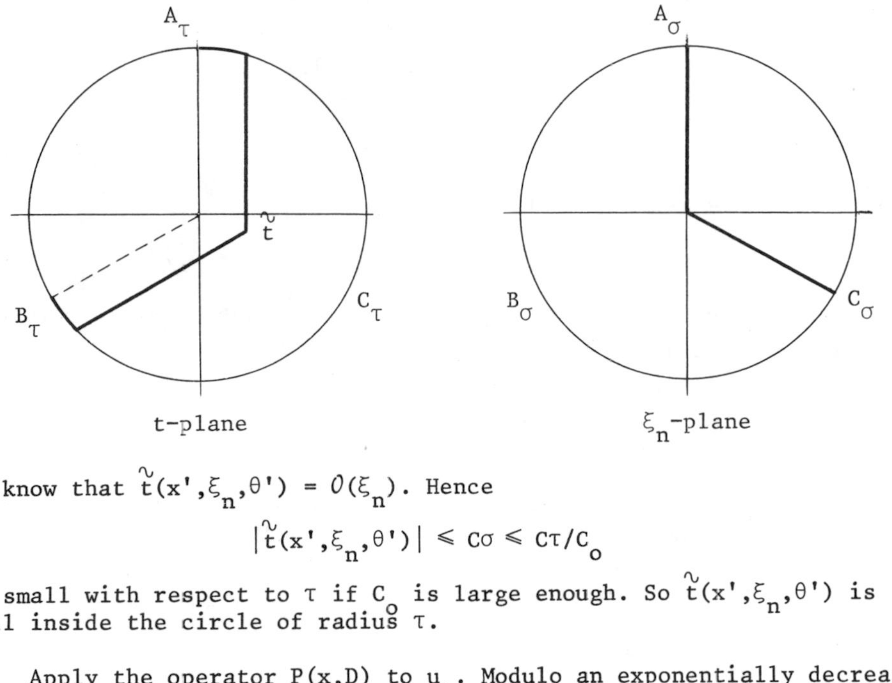

We know that $\tilde{t}(x',\xi_n,\theta') = O(\xi_n)$. Hence

$$|\tilde{t}(x',\xi_n,\theta')| \leqslant C\sigma \leqslant C\tau/C_o$$

is small with respect to τ if C_o is large enough. So $\tilde{t}(x',\xi_n,\theta')$ is well inside the circle of radius τ.

Apply the operator $P(x,D)$ to u_+. Modulo an exponentially decreasing term, we may transform the derivatives with respect to x into derivatives with respect to t and ξ_n using (2.28). On the chosen integration paths the arguments of ξ_n and $t - \tilde{t}$ always remain close to $\pi/2$, $7\pi/6$ or $-\pi/6$. Thus the cubic terms in lemma 2.1 have a good imaginary part. Integration by parts gives rise to terms where $|t| = \tau$ or $|\xi_n| = \sigma$. One of the cubic terms is thus always strictly positive. Consider for example the case of $|\xi_n| = \sigma$. Using lemma 2.1 we obtain that the imaginary part of $x_n\xi_n + H$ on the path is greater than

$$\psi(x',\theta') + c\sigma^3 - C(|\theta_1|\tau + |x_n|\sigma + |\theta_1|^2 + \sigma^4 + \tau^4)$$

with $c > 0$. If r is small and C_o large enough this is strictly positive. We have proved (2.29).

Now, in the definition of u_+ and u_- write

$$\int_{B_\tau}^{A_\tau} dt = \int_0^{A_\tau} dt - \int_0^{B_\tau} dt \text{ and } \int_{C_\tau}^{A_\tau} dt = \int_0^{A_\tau} dt - \int_0^{C_\tau} dt.$$

We obtain

(2.32)
$$u_+ + u_- = \frac{\lambda}{2\pi}\left(\int_{B_\sigma}^{C_\sigma} d\xi_n \int_0^{A_\tau} + \int_{C_\sigma}^{A_\sigma} d\xi_n \int_0^{B_\tau} + \int_{A_\sigma}^{B_\sigma} d\xi_n \int_0^{C_\tau}\right) e^{i\lambda(x_n\xi_n + H)}\, a\, dt.$$

Here we choose contours in ξ_n consisting of a third of circle of radius σ joining B_σ to C_σ, C_σ^n to A_σ and A_σ to B_σ.

We also modify the t-contours in the following way. If $|\xi_n| = \sigma$, $\partial_t H$ is not equal to zero. Indeed we have

$$\partial_t H(x',\xi_n,\theta',0) = \theta_1 - \xi_n^2,$$

hence one can find r_1 such that $0 < r_1 < \tau$ and

$$|\partial_t H(x',\xi_n,\theta',t)| \geq |\xi_n|^2 - |\theta_1| - C|t| \geq \sigma^2/8$$

if $\sigma/2 < |\xi_n| < 3\sigma/2$, $|t| < r_1$ and $|x'| < r, |\theta'-\theta_o'| < r$. Let $c_1 = r_1\sigma^2/16$. The differential equation

$$\partial_s \gamma(s) = \frac{i}{\partial_t H(x',\xi_n,\theta',\gamma(s))}$$

$$\gamma(0) = 0$$

has a solution defined for $s \in [0,c_1]$ if $\sigma/2 < |\xi_n| < 3\sigma/2$, $|x'| < r$ and $|\theta'-\theta_o'| < r$. Moreover

$$|\gamma(s)| \leq \frac{8}{\sigma^2} s \leq \frac{r_1}{2}.$$

By construction we have

$$H(x',\xi_n,\theta',\gamma(s)) = \psi(x',\theta') + is$$

hence $\mathcal{R}H$ is constant along γ and γ is a path of steepest descend. Let

(2.33) $$\hat{u}(x',\xi_n,\theta',\lambda) = i\lambda \int_\gamma e^{i\lambda H(x',\xi_n,\theta',t)} a(x',\xi_n,\theta',t,\lambda)dt.$$

If $|\xi_n| = \sigma$ and $\arg \xi_n \in [-5\pi/6, -\pi/6]$ then $\arg \tilde{t}$ is also close to the interval $[-5\pi/6, -\pi/6]$. If θ_1 is small enough it follows from the structure of the curves of steepest descend of $\mathcal{I}H$ (see Appendix A) that 0 is in the valley of A_τ. Therefore we can choose a path on which

$\mathcal{I}H$ increases, joining 0 to A_τ and containing γ. It follows that

(2.34)

$$\left| \int_0^{A_\tau} e^{i\lambda H(x',\xi_n,\theta',t)} a(x',\xi_n,\theta',t,\lambda)dt - \frac{1}{i\lambda} \hat{u}(x',\xi_n,\theta',\lambda) \right| \leqslant Ce^{-c_1\lambda}$$

if r is small enough. In the same way we can identify the integral from 0 to B_τ and from 0 to C_τ with $\hat{u}$ when $\arg \xi_n \in [-\pi/6,\pi/2]$ and $\arg \xi_n \in [\pi/2,7\pi/6]$ respectively. Using (2.32) and (2.33) we get

$$u(x,\theta',\lambda) = \frac{1}{2i\pi} \int_{|\xi_n|=\sigma} e^{i\lambda x_n\xi_n} \hat{u}(x',\xi_n,\theta',\lambda)d\xi_n \tag{2.35}$$

modulo an exponentially decreasing term if $|x| < r$ and $|\theta'-\theta'_o| < r$.

The definition of $\hat{u}$ can be written

$$\hat{u}(x',\xi_n,\theta',\lambda) = e^{i\lambda\psi(x',\theta')}b(x',\xi_n,\theta',\lambda) \tag{2.36}$$

with

$$b(x',\xi_n,\theta',\lambda) = i\lambda \int_0^{c_1} e^{-s\lambda} a(x',\xi_n,\theta',\gamma(s),\lambda)\partial_s\gamma(s)ds.$$

Let us prove that b is a classical analytic symbol near $|\xi_n| = \sigma$. Introduce

$$a_{j\ell}(x',\xi_n,\theta') = i\partial_s^\ell \, [a_j(x',\xi_n,\theta',\gamma(s))\partial_s\gamma(s)]_{s=o}$$

and

$$b_k(x',\xi_n,\theta') = \sum_{j+\ell=k} a_{j\ell}(x',\xi_n,\theta').$$

Since a is a analytic symbol there is a $M > 0$ such that

$$|a_{j\ell}(x',\xi_n,\theta')| \leqslant M^{1+j+\ell} \, j!\ell!$$

Hence (b_k), $k \in \mathbb{N}$, is a formal analytic symbol. Furthermore we have

$$b(x',\xi_n,\theta',\lambda) - \sum_{0\leqslant k<\lambda/C_1} \lambda^{-k}b_k(x',\xi_n,\theta')$$

$$= \lambda \sum_{o\leqslant j<\lambda/C_1} \lambda^{-j} \int_0^{c_1} e^{-s\lambda}[ia_j(x',\xi_n,\theta',\gamma(s))\partial_s\gamma(s)$$

$$- \sum_{o\leqslant j+\ell<\lambda/C_1} \frac{s^\ell}{\ell!} a_{j\ell}(x',\xi_n,\theta')]ds$$

$$- i\lambda \sum_{0\leqslant j+\ell<\lambda/C_1} \lambda^{-j}a_{j\ell}(x',\xi_n,\theta') \int_{c_1}^{+\infty} e^{-s\lambda} \frac{s^\ell}{\ell!} ds.$$

If the integer p satisfies $j+p-1 < \lambda/C_1 \leqslant j+p$, we have

$$|ia_j(x',\xi_n,\theta',\gamma(s))\partial_s\gamma(s) - \sum_{0\leqslant j+\ell<\lambda/C_1} \frac{s^\ell}{\ell!} a_{j\ell}(x',\xi_n,\theta')|$$

$$= |s^p \int_0^1 \frac{(1-v)^{p-1}}{(p-1)!} \partial_t^p [ia_j(x',\xi_n,\theta',\gamma(t))\partial_t\gamma(t)]_{t=vs}ds| \leqslant M^{1+j+p}j!\, s^p$$

by Taylor's formula. Hence

$$|b(x',\xi_n,\theta',\lambda) - \sum_{0\leqslant k<\lambda/C_1} \lambda^{-k}b_k(x',\xi_n,\theta')|$$

$$\leqslant \sum_{\lambda/C_1\leqslant j+p<1+\lambda/C_1} \lambda^{-(j+p)}M^{1+j+p}j!p! + e^{-c_1\lambda/2} \sum_{0\leqslant j+\ell<\lambda/C_1} \lambda^{-(j+\ell)}M^{1+j+\ell}\ell!$$

since

$$\lambda\int_{c_1}^{+\infty} e^{-s\lambda}\frac{s^p}{p!}\,ds = \lambda e^{-c_1\lambda}\int_0^{+\infty} e^{-s\lambda}\frac{(s+c_1)^p}{p!}ds = \frac{e^{-c_1\lambda}}{\lambda^p}\sum_{j=o}^{p}\frac{(c_1\lambda)^j}{j!}$$

$$\leqslant (\tfrac{2}{\lambda})^p e^{-c_1\lambda}\sum_{j=o}^{p}\frac{(c_1\lambda/2)^j}{j!} \leqslant (\tfrac{2}{\lambda})^p\, e^{-c_1\lambda/2}.$$

Now choose $\lambda > C_1 > 2eM$. We get the bound

$$M(1+\frac{\lambda}{C_1})e^{-\lambda/C_1} + Me^{-c_1\lambda/2}\sum_{0\leqslant k<\lambda/C_1}(k+1)e^{-k}.$$

We have proved that

$$(2.37)\qquad |b(x',\xi_n,\theta',\lambda)-\sum_{0\leqslant k<\lambda/C_1}\lambda^{-k}\, b_k(x',\xi_n,\theta')| \leqslant Ce^{-\varepsilon\lambda}$$

for some constants C,ε if $\sigma/2 < |\xi_n| < 3\sigma/2$ and $|x'| < r$, $|\theta'-\theta_o'| < r$.

To study b more closely we use (2.28), (2.33) and (2.36). We have

$$(2.38)\qquad P(x,\tilde{D},\lambda)(e^{i\lambda(x_n\xi_n+\psi(x',\theta'))}\, b(x',\xi_n,\theta',\lambda))$$

$$= i\lambda\int_\gamma(-\tilde{D}_t(e^{i\lambda(x_n\xi_n+H)}a)+\tilde{D}_{\xi_n}(e^{i\lambda(x_n\xi_n+H)}c))dt+\mathcal{O}(e^{i\lambda(x_n\xi_n+\psi)-\varepsilon\lambda})$$

$$= \left\{ \frac{1}{\xi_n} \right\} e^{i\lambda(x_n\xi_n+\psi)} + \tilde{D}_{\xi_n}\left(e^{i\lambda(x_n\xi_n+\psi)} d(x,\xi_n,\theta',\lambda)\right) + O\left(e^{i\lambda(x_n\xi_n+\psi)-\varepsilon\lambda}\right)$$

if we introduce

$$d(x,\xi_n,\theta',\lambda) = i\lambda e^{i\lambda\psi(x',\theta')} \int_\gamma e^{i\lambda H(x',\xi_n,\theta',t)} c(x,\xi_n,\theta',t,\lambda)dt.$$

Of course the estimations which prove (2.37) can also be applied to d so d is a classical analytic symbol. Denote B and D the formal symbols defined by b and d. It follows from (2.38) that, in the level of formal symbols

$$(2.39) \qquad e^{-i\lambda(x_n\xi_n+\psi)} P(x,\tilde{D},\lambda)\left(e^{i\lambda(x_n\xi_n+\psi)} B\right)$$

$$= e^{-i\lambda\psi}(\xi_n^2+R(x',-\tilde{D}_{\xi_n},\tilde{D}_{x'},\lambda))(e^{i\lambda\psi}B)+e^{-i\lambda(x_n\xi_n+\psi)}\tilde{D}_{\xi_n}\left(e^{i\lambda(x_n\xi_n+\psi)} E\right)$$

$$= \left\{ \frac{1}{\xi_n} \right\} + e^{-i\lambda(x_n\xi_n+\psi)}\tilde{D}_{\xi_n}\left(e^{i\lambda(x_n\xi_n+\psi)} D\right)$$

where

$$E = e^{-i\lambda(x_n\xi_n+\psi)} Q(x,\tilde{D}_{x'},\tilde{D}_{\xi_n},\lambda)\left(e^{i\lambda(x_n\xi_n+\psi)} B\right).$$

The last equality in (2.39) shows that

$$(x_n+\tilde{D}_{\xi_n})(E-D)$$

does not depend on x_n. It implies E = D. Hence

$$(2.40) \qquad e^{-i\lambda\psi}P(x',-\tilde{D}_{\xi_n},\tilde{D}_{x'},\xi_n,\lambda)(e^{i\lambda\psi}B) = \left\{\frac{1}{\xi_n}\right\}.$$

The pseudo-differential operator acting on B in (2.40) is of the form

$$\xi_n^2 - \theta_1 + \sum_{k=1}^{\infty} \lambda^{-k} L_k(x',\theta',D_{x'},D_{\xi_n}).$$

This proves that

$$(2.41) \quad b_o(x',\xi_n,\theta') = \frac{\left\{\frac{1}{\xi_n}\right\}}{\xi_n^2-\theta_1} \text{ and } b_k(x',\xi_n,\theta') = \frac{\alpha_k(x',\xi_n,\theta')}{(\xi_n^2-\theta_1)^{2k+1}}$$

where α_k is a polynomial with respect to ξ_n of degree less or equal to $2(2k-1)+\{0,1\}$ if $k \geqslant 1$. This is proved easily by induction since L_j is a differential operator of order less or equal to j.

Now we are in position to prove (2.30) and (2.31). We know from (2.41) that b_k is a rational function of ξ_n which decreases at least as far as ξ_n^{-3k} at infinity if $k \geqslant 1$. Moreover the integral of a rational function on a path containing all its singular points is null if it behaves as ξ_n^{-2} at infinity. It follows that only b_o gives a contribution to (2.35) and its derivative with respect to x_n at $x_n = 0$. Hence

$$u(x',0,\theta',\lambda) = \frac{1}{2i\pi} e^{i\lambda\psi(x',\theta')} \int_{|\xi_n|=\sigma} \frac{\begin{Bmatrix}1\\ \xi_n\end{Bmatrix}}{\xi_n^2-\theta_1} d\xi_n + O(e^{-\varepsilon\lambda})$$

$$= \begin{Bmatrix}0\\ 1\end{Bmatrix} e^{i\lambda\psi(x',\theta')} + O(e^{-\varepsilon\lambda})$$

and

$$\tilde{D}_{x_n} u(x',0,\theta',\lambda) = \frac{1}{2i\pi} e^{i\lambda\psi(x',\theta')} \int_{|\xi_n|=\sigma} \frac{\begin{Bmatrix}\xi_n\\ \xi_n^2\end{Bmatrix}}{\xi_n^2-\theta_1} d\xi_n + O(e^{-\varepsilon\lambda})$$

$$= \begin{Bmatrix}1\\ 0\end{Bmatrix} e^{i\lambda\psi(x',\theta')} + O(e^{-\varepsilon\lambda}).$$

This completes the proof of theorem 2.7.

Proposition 2.8. *There exists* $C, r > 0$ *such that*

$$(2.42) \qquad |u_\pm(x,\theta',\lambda)| + |\tilde{D}_{x_n} u_\pm(x,\theta',\lambda)| \leqslant C\lambda\, e^{-\lambda S(x,\theta')}$$

if $|x| < r$, $|\theta' - \theta_o'| < r$, $\lambda > 1$ *and* $S(x,\theta')$ *is the lower bound of the four branches of*

$$I\phi(x',\sqrt{x_n + a(x',\sqrt{\theta_1},\theta'')\theta_1},\sqrt{\theta_1},\theta'').$$

If x_n *is real and greater than* $c_o|\theta_1|$ *with the same* c_o *as in Lemma 2.6, we have only to consider the branches with*

$$\mathcal{R}\sqrt{x_n + a(x',\sqrt{\theta_1},\theta'')\theta_1} > 0$$

for u_+ *and the other ones for* u_-. *Moreover there exists* $c > 0$ *and for each* $\alpha \in \mathbb{N}^n$ *a constant* C_α *such that*

$$(2.43) \qquad |D_x^\alpha u_\pm(x,\theta',\lambda)| \leqslant C_\alpha \lambda^{1+|\alpha|} e^{c\lambda|\theta_1|^{3/2}}$$

if x,θ' *are real,* $x_n \geqslant 0$, $|x| < r$, $|\theta'-\theta_o'| < r$ *and* $\lambda > 1$.

Proof. To fix the ideas we only consider u_+. By Lemma 2.1 and the choice of the points A_τ, B_τ, the imaginary part of H at the points $t = A_\tau$ or B_τ is greater than

$$I\psi + c\tau^3 - C(|\theta_1|\tau+|\theta_1|^2+\tau^4+\sigma^4)$$

with $c > 0$. If C_o is large enough and r is small this is greater than $c\tau^3/2$. Since r is small with respect to τ it follows that, at A_τ or B_τ, IH is greater than at the critical points $t_\pm$. Using the results of Appendix A we can find a curve joining B_τ to A_τ on which H is everywhere greater than its smallest value at the critical points. It follows that

$$\left|\int_{B_\tau}^{A_\tau} e^{i\lambda H(x',\xi_n,\theta',t)} a(x',\xi_n,\theta',t,\lambda)dt\right| \leqslant Ce^{-\lambda IG(x',\xi_n,\theta',t_\pm(x',\xi_n,\sqrt{\theta_1},\theta''))}$$

where we choose the sign that gives the smallest value of IG. The second integral in the definition of u_+ can be estimated in the same way. Indeed, formula (2.11) shows that $I(x_n\xi_n+G)$ is greater at $\xi_n=A_\sigma$ or c_σ than at the critical points $\xi_n^\pm$ if r is small with respect to σ. Thus we can choose a path joining A_σ to C_σ with the same property as above. If x_n is real and greater than $c_o|\theta_1|$, the argument of $x_n + a(x',\sqrt{\theta_1},\theta'')\theta_1$ belongs to $]-\pi/3, \pi/3[$. Using the result of Appendix A on the paths of steepest descent containing the critical points, we see that the one which joins A_σ to C_σ contains the critical point whose real part is positive.

To prove (2.43) we perform the derivatives under the integral signs, choose the same paths as above and use Lemma 2.4.

2.4. Superposition solutions

The asymptotic solutions $u_\pm$ cannot be used unless they are modified to satisfy boundary conditions which are exponentially decreasing outside some neighbourhood of $x' = 0$. From now on, we work in a neighbourhood of $(0,\theta_o')$ where the conclusions of Proposition 2.7 and 2.8 are valid.

Write

$$\psi(x',\theta') - \psi(y',\theta') = (x'-y').\xi'(x',y',\theta').$$

Of course

$$\xi'(x',y',\theta') = \int_0^1 \partial_{x'}\psi((1-t)x'+ty',\theta')dt.$$

Hence, by (2.1), the equation $\xi' = \xi'(x',y',\theta')$ can be solved in the form $\theta'=\theta'(x',y',\xi')$. Introduce

$$J(x',y',\xi') = \det \partial_{\xi'}\theta'(x',y',\xi').$$

It is known, see [11], that there exist formal analytic symbols F and $G_1,\ldots,G_{n-1}$ such that

$$(2.43)\quad J(x',y',\xi')F(y',\theta'(x',y',\xi'),\lambda)=1+\sum_{j=1}^{n-1}(x_j-y_j)G_j(x',y',\xi',\lambda)$$

$$+\sum_{j=1}^{n-1}\tilde{D}_{\xi_j'}G_j(x',y',\xi',\lambda)$$

i.e.

$$\int e^{i\lambda(\psi(x',\theta')-\psi(y',\theta'))}F(y',\theta',\lambda)d\theta'$$

is formally the kernel of the identity operator.

Now we introduce the function of $x' \in \mathbb{R}^{n-1}$

$$w(x',\lambda) = \frac{\lambda^{(n-1)/4}}{c_{n-1}}\int_{|u'|<s} e^{-\frac{\lambda}{4}|x'-u'|^4}du',$$

with

$$c_{n-1} = \int_{\mathbb{R}^{n-1}} e^{-|x'|^4/4}dx' \text{ and } s > 0.$$

Clearly w extends as an holomorphic function to $\mathbb{C}^{n-1}$. They are constants δ, $c > 0$ such that

$$(2.44)\quad \mathcal{R}(z'-u')^4 \geq \delta|\mathcal{R}z'-u'|^4-c|\mathcal{I}z'|^4 \text{ if } z' \in \mathbb{C}^{n-1},\ u' \in \mathbb{R}^{n-1}.$$

Hence

$$|w(x',\lambda)| \leq Ce^{c\lambda|\mathcal{I}x'|^4},\quad x' \in \mathbb{C}^{n-1}.$$

Moreover, if K is a compact subset of $\{u' \in \mathbb{R}^{n-1} : |u'| < s\}$ and G is a closed subset of $\mathbb{R}^{n-1}$ which does not meet $\{u' \in \mathbb{R}^{n-1} : |u'| \leq s\}$, one can find $C,\varepsilon,\delta > 0$ such that

$$|1-w(x',\lambda)| \leq Ce^{-\varepsilon\lambda} \text{ if } \mathcal{R}x' \in K,\ |\mathcal{I}x'| < \delta$$

$$|w(x',\lambda)| \leq Ce^{-\varepsilon\lambda} \text{ if } \mathcal{R}x' \in G,\ |\mathcal{I}x'| < \delta.$$

This follows easily from the definition of w and (2.44).

Denote by f a realization of the formal symbol F. Let

$$U_{\pm}(x,\theta',\lambda) = \left(\frac{\lambda}{2\pi}\right)^{n-1} \iint_{\substack{|y'|<\rho \\ |\sigma'-\theta'_o|<s}} u_{\pm}(x,\sigma',\lambda)e^{-i\lambda\psi(y',\sigma')}f(y',\sigma',\lambda)$$

$$e^{i\lambda\psi(y',\theta')}w(y',\lambda)dy'd\sigma'$$

and $U = U_+ + U_-$. The constants s,ρ are chosen such that $0<C_2 s<\rho<1/C_2$ where C_2 is a large constant to be fixed below.

The properties (2.29)-(2.31) of $u_{\pm}$ give rise to the following result.

Proposition 2.9. *There is a* $\varepsilon > 0$ *such that*

$$P(x,D)U_{\pm}(x,\theta',\lambda) = O(e^{-\varepsilon\lambda}) \tag{2.45}$$

uniformly if $|x'| < \rho$, $|x_n| < \rho$ *and* $|\theta'-\theta'_o| < s$. *Moreover*

$$U(x',0,\theta',\lambda) = \left\{\begin{matrix}0\\1\end{matrix}\right\} w(x',\lambda)e^{i\lambda\psi(x',\theta')} + O(e^{-\varepsilon\lambda}) \tag{2.46}$$

$$\tilde{D}_{x_n} U(x',0,\theta',\lambda) = \left\{\begin{matrix}1\\0\end{matrix}\right\} w(x',\lambda)e^{i\lambda\psi(x',\theta')} + O(e^{-\varepsilon\lambda}) \tag{2.47}$$

uniformly if x',θ' *are real and* $|x'| < \rho$, $|\theta'-\theta'_o| < s/2$.

Proof. Formula (2.45) follows immediately from (2.29). Using (2.30)-(2.31) we see that to prove (2.46)-(2.47) we have to estimate

$$I(x',\theta',\lambda) = \left(\frac{\lambda}{2\pi}\right)^{n-1} \iint_{\substack{|y'|<\rho \\ |\sigma'-\theta'_o|<s}} e^{i\lambda(\psi(x',\sigma')-\psi(y',\sigma'))}f(y',\sigma',\lambda)$$

$$e^{i\lambda\psi(y',\theta')}w(y',\lambda)dy'd\sigma'.$$

Let

$$g(\sigma',\lambda) = \frac{\lambda^{(n-1)/4}}{c_{n-1}} \int_{|u'-\theta'_o|<2s/3} e^{-\frac{\lambda}{4}|\sigma'-u'|^4} du'.$$

If we introduce $g(\sigma',\lambda)$ in the definition of I, the error will be exponentially decreasing. Indeed consider $\chi_1 \in D(\{y' \in \mathbb{R}^{n-1} : |y'| < \rho\})$ equal to 1 in a neighbourhood of $\{y' \in \mathbb{R}^{n-1} : |y'| \leqslant \rho/2\}$. In the error term we switch to the contour

$$y' \to y'+i\mu(\partial_{y'}\psi(y',\theta') - \partial_{y'}\psi(y',\sigma'))\ \chi_1(y'),\ \mu > 0.$$

The imaginary part of the phase function then becomes greater than

$$c\mu|\partial_{y'}\psi(y',\theta') - \partial_{y'}\psi(y',\sigma')|^2\ \chi_1(y') \geq c'\mu|\theta'-\sigma'|^2\chi_1(y').$$

In the domain where $\chi_1(y') = 1$ we can conclude using this estimation if $|\sigma'-\theta'_o| > s_1 > s/2$ since we assume $|\theta'-\theta'_o| < s/2$. If $|\sigma'-\theta'_o|<s_1<2s/3$ the function $1-g$ is exponentially decreasing. If $\chi_1(y') \neq 1$, w is exponentially decreasing if C_2 is large.

Now we have an exponential decrease everywhere on the boundary of the domain of integration. We perform the change of variables $\sigma' = \theta'(x',y',\xi')$. Using (2.43) we get

(2.48)

$$I(x',\theta',\lambda)=(\frac{\lambda}{2\pi})^{n-1}(\iint e^{i\lambda((x'-y').\xi'+\psi(y',\theta'))}w(y',\lambda)g(\theta'(x',y',\xi'),\lambda)\, dy'd\xi'$$

$$- \sum_{j=1}^{n-1} \iint e^{i\lambda((x'-y').\xi'+\psi(y',\theta'))}G_j(x',y',\xi',\lambda)w(y',\lambda)\, \tilde{D}_{\xi'_j}[g(\theta'(x',y',\xi'),\lambda)]dy'd\xi')$$

modulo an exponentially decreasing term. The domain of integration is

$$|y'| < \rho\ ,\ |\theta'(x',y',\xi')-\theta'_o| < s.$$

To prove that the last sum is negligible,take again $s_1 \in]s/2,\ 2s/3[$. Since

$$|\partial_{y'}\psi(y',\theta')-\xi'|=|\partial_{y'}\psi(y',\theta')-\partial_{y'}\psi(y',\theta'(y',y',\xi'))|$$
$$\geq c|\theta'-\theta'(y',y',\xi')|$$

it follows that

$$|x'-y'| + |\partial_{y'}\psi(y',\theta')-\xi'| \neq 0$$

if

$$s_1 \leq |\theta'(x',y',\xi')-\theta'_o| \leq s,\ |x'| \leq \rho,\ |y'| \leq \rho \text{ and } |\theta'-\theta'_o| \leq s/2.$$

Take χ_2 in $D(\{\theta' \in \mathbb{R}^{n-1} : |\theta'-\theta'_o| < s\})$ equal to 1 in a neighbourhood of $\{\theta' \in \mathbb{R}^{n-1} : |\theta'-\theta'_o| < 2s/3\}$ and consider the complex shift

$$y' \to y' + i\mu(\partial_{y'}\psi(y',\theta')-\xi')\chi_1(y')$$

$$\xi' \to \xi' + i\mu(x'-y')\chi_2(\theta'(x',y',\xi')).$$

The imaginary part of the phase is then greater than

$$c\mu(|x'-y'|^2\chi_2(\theta'(x',y',\xi')) + |\partial_{y'}\psi(y',\theta')-\xi'|^2\chi_1(y')).$$

Using the properties of w and g as above, we obtain that the sum is exponentially decreasing.

Now we can expand the first term in (2.48) by the stationnary phase formula. Hence

$$I(x',\theta',\lambda) = w(x',\lambda)e^{i\lambda\psi(x',\theta')}+O(e^{-\varepsilon\lambda}).$$

Proposition 2.10. *There exists constants* $C,\varepsilon,\delta>0$ *such that*

(2.49) $$|U_\pm(x,\theta',\lambda)| + |\tilde{D}_{x_n} U_\pm(x,\theta',\lambda)| \leqslant Ce^{-\varepsilon\lambda}$$

if x,θ' *are real,* $\rho/2 < |x'| < \rho$, $|\theta'-\theta'_o| < \delta$ *and* $0 \leqslant x_n < \delta$

Proof. As in the proof of Proposition 2.9, choose cutoff functions $\chi_1 \in D(\{y':|y'|<\rho\})$ equal to 1 in a neighbourhood of $\{y':|y'|\leqslant\rho/2\}$ and $\chi_2 \in D(\{\theta':|\theta'-\theta'_o|<s\})$ equal to 1 in a neighbourhood of $\{\theta':|\theta'-\theta'_o| \leqslant 2s/3\}$. In the definition of $U_\pm$, switch the real integration domain into

$$y' \to y' + i\mu\chi_1(y')(\partial_{y'}\psi(y',\theta')-\partial_{y'}\psi(y',\sigma'))|\theta'-\sigma'|^{-2/3}$$

$$\sigma' \to \sigma' + i\mu\chi_2(\sigma')(\partial_{\sigma'}\psi(x',\sigma')-\partial_{\sigma'}\psi(y',\sigma'))|x'-y'|^2,$$

with $\mu > 0$. To fix the ideas we consider U_+. On the new integration domain we can estimate u_+ by the Proposition 2.8. Afterwards we use Lemma 2.6. The imaginary part of the phase function in u_+ becomes

$$\mu\chi_2(\sigma)\mathcal{R}(\partial_\sigma,\phi)(x',\sqrt{x_n+a(x',\sqrt{\sigma_1},\sigma'')\sigma_1},\sqrt{\sigma_1},\sigma'').(\partial_{\sigma'}\psi(x',\sigma')-\partial_{\sigma'}\psi(y,\sigma'))$$

$$|x'-y'|^2+O(|\sigma_1|^{3/2}+\mu^{3/2}\chi_2(\sigma')|x'-y'|^{9/2})$$

where we can choose all the real part of the square roots positive. It follows from the Lemma 2.2 that

(2.50)
$$|\mathcal{R}(\partial_\sigma,\phi)(x',\sqrt{x_n+a(x',\sqrt{\sigma_1},\sigma'')\sigma_1},\sqrt{\sigma_1},\sigma'')-\partial_{\sigma'}\psi(x',\sigma')| \leqslant C(x_n^{1/2}+|\sigma_1|^{1/2}).$$

Hence we may replace $\mathcal{R}(\partial_\sigma,\phi)$ by $\partial_{\sigma'}\psi$ without adding new error term other than $x_n^{3/2}$. Summing up, the imaginary part of the phase function in the definition of U_+ is

$$\mu\chi_1(y')|\partial_{y'}\psi(y',\theta')-\partial_{y'}\psi(y',\sigma')|^2|\theta'-\sigma'|^{-2/3}$$

$$+\mu\chi_2(\sigma')|\partial_{\sigma'}\psi(x',\sigma')-\partial_{\sigma'}\psi(y',\sigma')|^2|x'-y'|^2$$

$$+O(|\sigma_1|^{3/2}+\mu^{3/2}\chi_2(\sigma')|x'-y'|^{9/2}+x_n^{3/2})$$

$$+O(\mu^2\chi_1(y')|\theta'-\sigma'|^{5/3}+\mu^3\chi_1(y')\chi_2(\sigma')|\theta'-\sigma'|^{2/3}|x'-y'|^3$$

$$+\mu^2\chi_2(\sigma')|x'-y'|^6+\mu^2\chi_1(y')\chi_2(\sigma')|\theta'-\sigma'|^{1/3}|x'-y'|^3).$$

The four latest terms appear as second order terms in the Taylor's expansion of $\psi(y',\theta')-\psi(y'-\sigma')$ in μ. Of course we have

$$|\partial_{y'}\psi(y',\theta') - \partial_{y'}\psi(y',\sigma')| \geqslant c_o|\theta'-\sigma'|$$

and

$$|\partial_{\sigma'}\psi(x',\sigma') -\partial_{\sigma'}\psi(y',\sigma') \geqslant c_o|x'-y'|$$

for some $c_o > 0$. Using the inequality

$$\mu^3\chi_1(y')\chi_2(\sigma')|\theta'-\sigma'|^{2/3}|x'-y'|^3$$

$$\leqslant \mu^2(\mu\chi_1(y')|\theta'-\sigma'|^{4/3})^{1/2} (\mu\chi_2(\sigma')|x'-y'|^4)^{1/2}$$

$$\leqslant \frac{\mu^2}{2} (\mu\chi_1(y')|\theta'-\sigma'|^{4/3} + \mu\chi_2(\sigma')|x'-y'|^4)$$

and a similar one for the latest term, we easily see that the four latest terms can be omitted when μ and ρ are small. Finally we get

$$c(\mu\chi_1(y')|\theta'-\sigma'|^{4/3} + \mu\chi_2(\sigma')|x'-y'|^4) + O(x_n^{3/2}+|\sigma_1|^{3/2}).$$

To prove (2.49) we consider three cases. If $|y'| < \rho/4$, $|\sigma'-\theta_o'|\leqslant 2s/3$ then $|x'-y'|\rho/4$ and $\chi_2(\sigma')=1$. We choose s and δ small with respect to μ and ρ to get the exponential decay in this case. If $|y'| < \rho/4$, $|\sigma'-\theta_o'| > 2s/3$ then $\chi_1(y') = 1$ and $|\theta'-\sigma'| > s/2$ if $\delta < s/6$. We choose δ small with respect to s and use the inequality

(2.51)

$$|\sigma_1|^{3/2} \leqslant C(|\theta_1|^{3/2}+|\theta_1-\sigma_1|^{3/2}) \leqslant C|\theta_1|^{3/2}+c_s|\theta_1-\sigma_1|^{4/3}, |\theta_1|, |\sigma_1| < s,$$

where C is an universal constant and c_s converges to 0 with s since

$4/3 < 3/2$. Finally if $|y'| > \rho/2$ we use the exponential decrease of w outside the ball $\{y' : |y'| \leqslant s\}$.

Proposition 2.11. *Let* V *be an open neighbourhood of* $\partial_{x'}\psi(0,\theta'_o)$. *If* s *and* ρ *are small enough and* $\alpha \in D(\{x' \in \mathbb{R}^{n-1} : |x'| < \rho\})$ *is equal to* 1 *in the ball* $\{x' : |x'| \leqslant \rho/2\}$ *then one can find constants* $C,\varepsilon,\delta > 0$ *such that*

$$(2.52) \qquad \left|\int \alpha(x')\binom{1}{D_{x_n}} U_\pm(x,\theta',\lambda)e^{-i\lambda x'.\xi'}dx'\right| \leqslant Ce^{-\varepsilon\lambda}$$

if x_n,θ',ξ' *are real,* $|\theta'-\theta'_o| < \delta$, $\xi' \notin V$ *and* $0 \leqslant x_n < \delta$.

Proof. By definition of $U_\pm$, we have

$$\int \alpha(x')\binom{1}{D_{x_n}} U_\pm(x,\theta',\lambda)e^{-i\lambda x'.\xi'}dx'$$

$$= \left(\frac{\lambda}{2\pi}\right)^{n-1} \iint_{\substack{|y'|<\rho \\ |\sigma'-\theta'_o|<s}} \alpha(x')\binom{1}{D_{x_n}}u_\pm(x,\sigma',\lambda)e^{i\lambda(\psi(y',\theta')-\psi(y',\sigma')-x'.\xi')}$$

$$f(y',\theta',\lambda)w(y',\lambda)dx'dy'd\sigma'.$$

We switch the real domain into

$$y' \to y' + i\mu\chi_1(y')(\partial_{y'}\psi(y',\theta')-\partial_{y'}\psi(y',\sigma'))|\theta'-\sigma'|^{-2/3}$$

$$\sigma' \to \sigma' + i\mu\chi_2(\sigma')(\partial_{\sigma'}\psi(x',\sigma')-\partial_{\sigma'}\psi(y',\sigma'))|x'-y'|^2$$

$$x' \to x' + i\mu\chi_3(x')(\partial_{x'}\psi(x',\sigma')-\xi')|\partial_{x'}\psi(x',\sigma')-\xi'|.$$

As above the cutoff function χ_1 (resp. χ_2, χ_3) $\in D(\{x' \in \mathbb{R}^{n-1} : |x'| < \rho$ (resp. $|x'-\theta'_o| < s$, $|x'| < \rho/2)\})$ is equal to 1 in a neighbourhood of $\{x' \in \mathbb{R}^{n-1} : |x'| \leqslant \rho/2$ (resp. $|x'-\theta'_o| \leqslant 2s/3$, $|x'| \leqslant \rho/3)\}$. To fix the ideas we only consider U_+. Here again we use Proposition 2.8 and Lemma 2.6 to estimate $u_\pm$ on the new contours. These results show that the constribution of u_+ to the imaginary part of the exponential behavior is

$$\mu\chi_2(\sigma')\mathcal{R}(\partial_{\sigma'}\phi)(x',\sqrt{x_n+a\sigma_1},\sqrt{\sigma_1},\sigma'').(\partial_{\sigma'}\psi(x',\sigma')-\partial_{\sigma'}\psi(y',\sigma'))|x'-y'|^2$$

$$+\mu\chi_3(x')\mathcal{R}(\partial_{x'}\phi)(x',\sqrt{x_n+a\sigma_1},\sqrt{\sigma_1},\sigma'').(\partial_{x'}\psi(x',\sigma')-\xi')|\partial_{x'}\psi(x',\sigma')-\xi'|$$

$$+ O(|\sigma_1|^{3/2} + \mu^{3/2}\chi_2(\sigma')|x'-y'|^{9/2} + \mu^{3/2}\chi_3(x')|\partial_{x'}\psi(x',\sigma')-\xi'|^3)$$

where we can choose all the real parts of the square roots positive.

The Lemma 2.2 shows that

$$|\mathcal{R}(\partial_{x'},\phi)(x',\sqrt{x_n+a(x',\sqrt{\sigma_1},\sigma'')\sigma_1},\sqrt{\sigma_1},\sigma'')-\partial_{\sigma'}\psi(x',\sigma')|$$
$$\leqslant C(x_n^{1/2}+|\sigma_1|^{1/2}).$$

Hence using also (2.50), we may replace $\mathcal{R}(\partial_{\sigma'},\phi)$ and $\mathcal{R}(\partial_{x'},\phi)$ by $\partial_{\sigma'}\psi$ and $\partial_{x'}\psi$ as in Proposition 2.10. To estimate $\psi(y',\theta')-\psi(y',\sigma')-x'.\xi'$ on the new contour we use a Taylor's expansion in μ. Summing up, the imaginary part of the exponential behavior of the whole phase is

$$\mu\chi_1(y')|\partial_{y'}\psi(y',\theta')-\partial_{y'}\psi(y',\sigma')|^2|\theta'-\sigma'|^{-2/3}$$

$$+\mu\chi_2(\sigma')|\partial_{\sigma'}\psi(x',\sigma')-\partial_{\sigma'}\psi(y',\sigma')|^2|x'-y'|^2+\mu\chi_3(x')|\partial_{x'}\psi(x',\sigma')-\xi'|^3$$

$$+\ \mathcal{O}(|\sigma_1|^{3/2}+\mu^{3/2}\chi_2(\sigma')|x'-y'|^{9/2}+\mu^{3/2}\chi_3(x')|\partial_{x'}\psi(x',\sigma')-\xi'|^3+x_n^{3/2})$$

$$+\ \mathcal{O}(\mu^2\chi_1(y')|\theta'-\sigma'|^{5/3}+\mu^3\chi_1(y')\chi_2(\sigma')|\theta'-\sigma'|^{2/3}|x'-y'|^3$$

$$+\ \mu^2\chi_2(\sigma')\ |x'-y'|^6+\mu^2\chi_1(y')\ \chi_2(\sigma')|\theta'-\sigma'|^{1/3}|x'-y'|^3).$$

The four latest terms can be omitted as in Proposition 2.10. Hence we get the following estimate of the imaginary part of the phase function

$$c(\mu\chi_1(y')|\theta'-\sigma'|^{4/3}+\mu\chi_2(\sigma')|x'-y'|^4+\mu\chi_3(x')|\partial_{x'}\psi(x',\sigma')-\xi'|^3)$$
$$+\ \mathcal{O}(|\sigma_1|^{3/2}+x_n^{3/2}).$$

We choose $\varepsilon > 0$ such that $|x'| < \varepsilon$ and $|\sigma'-\theta'_o| < \varepsilon$ imply

$$|\xi'-\partial_{x'}\psi(x',\sigma')| > \varepsilon\ .$$

To prove (2.52) we consider four cases. If $|x'| < \rho/3$ and $|\sigma'-\theta'_o| < s$ then we have an exponential decrease in $\mu\varepsilon^3$ if s,δ are small with respect to μ and ε. If $|x'| > \rho/3$, $|y'| < \rho/4$ and $|\sigma'-\theta'_o| \leqslant 2s/3$ we have a decrease in $\mu\rho^4$. We choose s,δ small with respect to μ and ρ . If $|x'| > \rho/3$, $|y'| < \rho/4$ and $|\sigma'-\theta'_o| > 2s/3$ we use (2.51) and conclude as in Proposition 2.10. Finally, if $|y'| > \rho/4$ we use the exponential decrease of w.

Proposition 2.12. *There are constants* C_α *and* $c > 0$ *such that*

(2.53) $$|D_x^\alpha U_\pm(x,\theta',\lambda)| \leqslant C_\alpha\lambda^{n+|\alpha|}e^{c\lambda|\theta_1|^{3/2}}$$

if x,θ' *are real,* $|x'| < \rho$, $0 \leqslant x_n < \rho$, $|\theta'-\theta'_o| < s$ *and* $\lambda > 1$.

Proof. First, we consider the integral

$$I(\theta',\sigma',\lambda) = \int_{|y'|<\rho} e^{i\lambda(\psi(y',\theta')-\psi(y',\sigma'))} f(y',\sigma',\lambda)w(y',\lambda)dy'.$$

We switch to the contour

$$y' \to y' + i\mu|\theta'-\sigma'|^{-2/3}\chi(y')(\partial_{y'}\psi(y',\theta')-\partial_{y'}\psi(y',\sigma'))\ ,\mu > 0,$$

where $\chi \in D(\{y' \in \mathbb{R}^{n-1} : |y'| < \rho\})$ is equal to 1 in $\{y' : |y'| \leq \rho/2\}$. The imaginary part of the exponential behavior becomes greater than

$$c\mu\chi(y')|\theta'-\sigma'|^{4/3} - C\mu^2\chi(y')(|\theta'-\sigma'|^{5/3} + \mu^2|\theta'-\sigma'|^{4/3}) \geq \frac{c}{2}\mu\chi(y')|\theta'-\sigma'|^{4/3}$$

if μ is small enough. Outside the ball $\{y' : |y'| \leq \rho/2\}$, w is exponentially decreasing, hence

$$|I(\theta',\sigma',\lambda)| \leq C\ e^{-c_1\lambda|\theta'-\sigma'|^{4/3}}$$

where the constant c_1 depend on ρ but not on s. Using Proposition 2.8 we obtain

$$|D_x^\alpha U_\pm(x,\theta',\lambda)| \leq C_\alpha \lambda^{n+|\alpha|} \int_{|\sigma'-\theta'_o|<s} e^{C\lambda|\sigma_1|^{3/2} - c_1\lambda|\theta'-\sigma'|^{4/3}} d\sigma'.$$

The conclusion follows from (2.5:).

3. CONCLUSION

Proof of the theorem 1.3. First we prove that there is an open neighbourhood V of $(0,\xi'_o)$ and $\delta > 0$ such that

(2.54) $\quad (x,\xi) \notin WF_a u$ if $0 < x_n \leq \delta$ and $(x',\xi') \in V$.

If $x_n > 0$ and $(x,\xi) \in WF_a u$ it follows that $p(x,\xi) = 0$, hence $r(x,\xi') \geq 0$ and $\xi_n = \pm\sqrt{r(x,\xi')}$. Let $s_o \in]0,\varepsilon[$ and $\rho > 0$ such that the open balls $B_\pm$ with center $\exp(\pm s_o H_p)(0,\xi'_o,0)$ and radius ρ do not intersect $WF_a u$. If V and δ are small enough it follows that

$$|\exp(\pm s_o H_p)(x,\xi',\pm\sqrt{r(x,\xi')})-\exp(\pm s_o H_p)(0,\xi'_o,0)| < \rho$$

when $0 < x_n \leq \delta$, $(x',\xi') \in V$ and $r(x,\xi') \geq 0$. Since $\partial_{x_n} r < 0$ we can always join a point $(x,\xi',\pm\sqrt{r(x,\xi')})$ to a point of a ball $B_\pm$ by a bicharacteristic curve of p that entirely lies in $x_n > 0$. This proves (2.54) by Hörmander's theorem.

Let K be a compact subset of V and $U_\pm$ the asymptotic solutions constructed in 2.4 but for ${}^tP$ and $(0',-\xi'_o)$ instead of P and $(0,\xi'_o)$. We may assume that (2.52) is valid when $\xi' \notin K$. Now, if $\alpha \in D(\{x' \in \mathbb{R}^{n-1} : |x'| < \rho\}$ is equal to 1 in $\{x' : |x'| \leqslant 2\rho/3\}$ we can write

(2.55)

$$\int_0^\delta u_{(x')^{x_n}}(\alpha(x')\,{}^tP(x,D)U(x,\theta',\lambda)dx_n = \iint_{0\leqslant x_n\leqslant\delta} f(x)\alpha(x')U(x,\theta',\lambda)dx'dx_n$$

$$+ u_\delta(\alpha D_{x_n}U(.,\delta,\theta',\lambda)) - u_o(\alpha D_{x_n}U(.,0,\theta',\lambda))$$

$$-(D_{x_n}u_\delta)(\alpha U(.,\delta,\theta',\lambda)) + (D_{x_n}u_o)(\alpha U(.,0,\theta',\lambda))$$

$$+v(U(.,\theta',\lambda)).$$

In this formula $f = Pu$, $U = U_+ + U_-$ and v is a distribution collecting all the terms where some derivatives act on α. Therefore the support of v is included in $\{x \in \mathbb{R}^n : \rho/2 < |x'| < \rho,\ 0 \leqslant x_n \leqslant \delta\}$.

Proposition 2.9 shows that the terms computed at $x_n = 0$ are equal to 0 or

$$\left\{\begin{matrix} u_o \\ D_{x_n}u_o \end{matrix}\right\} (\alpha w(.,\lambda)e^{-i\lambda\psi(\,.\,,\theta')})$$

modulo an exponentially decreasing term. If we prove that all the other terms are exponentially decreasing if $|\theta'-\theta'_o|$ is small enough, we conclude that $(0,\xi'_o) \notin WF_a(u_o) \cup WF_a(D_{x_n}u_o)$.

Using the Proposition 2.9 and 2.10 we see that the only fact to prove is that

$$\tau(\alpha(\begin{smallmatrix}1\\ D_{x_n}\end{smallmatrix})\ U(.,x_n,\theta',\lambda)) \tag{2.56}$$

is uniformly exponentially decreasing if $0 \leqslant x_n \leqslant \delta$, $|\theta'-\theta'_o| < \varepsilon$ and τ is a distribution in $\{x' \in \mathbb{R}^{n-1} : |x'| < a\}$ satisfying $WF^o_a\tau \cap V = \emptyset$. Of course, (2.56) is equal to

$$\left(\frac{\lambda}{2\pi}\right)^{n-1} \int_{\mathbb{R}^{n-1}} \tau_{(x')}\,(\alpha(x')e^{-\frac{\lambda}{2}|x'-u'|^2}(\begin{smallmatrix}1\\ D_{x_n}\end{smallmatrix})\ U(x,\theta',\lambda))du'.$$

Proposition 2.10 shows that the integral on $|u'| \geqslant 2\rho/3$ is exponentially decreasing. Hence we have to estimate

$$\int \Big(\int_{|u'|<2\rho/3} \tau_{(x')} (\alpha(x') e^{-i\lambda x'.\xi' - \frac{\lambda}{2}|x'-u'|^2}) du' \Big) \Big(\int \alpha(x') (\tfrac{1}{D_{x_n}}) U(x,\theta',\lambda) e^{i\lambda x'.\xi'} dx' \Big) d\xi'.$$

The first factor is exponentially decreasing if $\xi' \in K$ and ρ is small enough because we assume that $WF_a \tau \cap V = \emptyset$. Using (2.53), we obtain that the integral on $\xi' \in K$ is exponentially decreasing if s is small. Now if $\xi' \notin K$, the first factor has a polynomial growth in λ and $|\xi|$. By Proposition 2.11 the second one is exponentially decreasing in λ. Moreover performing integration by parts and using (2.53), we obtain

$$\Big|\int \alpha(x') (\tfrac{1}{D_{x_n}}) U(x,\theta',\lambda) e^{i\lambda x'.\xi'} dx'\Big| \leqslant C_k \lambda^n |\xi'|^{-k} e^{c\lambda|\theta_1|^{3/2}}$$

if $0 \leqslant x_n \leqslant \rho$, $|\theta'-\theta'_o|<s$ and $\lambda > 1$. From Proposition 2.11 it follows that

$$\Big|\int \alpha(x') (\tfrac{1}{D_{x_n}}) U(x,\theta',\lambda) e^{i\lambda x'\xi'} dx'\Big| \leqslant C_k (1+|\xi'|)^{-k} e^{-\varepsilon\lambda}.$$

if $|\theta'-\theta'_o| < \delta$. Hence the whole integral is exponentially decreasing.

APPENDIX

Let $f(t,w)$ be an holomorphic function in a neighbourhood of $(0,0)$ in $\mathbb{C} \times \mathbb{C}^k$ satisfying

(A.1) $$\partial_t f(0,0)=0 \;,\; \partial_t^2 f(0,0)=0 \text{ and } \partial_t^3 f(0,0)<0 \;.$$

Hence $t \to f(t,0)$ has a degenerated critical point at 0. The conditions A.1 are fullfilled by the functions H and G of section 2. Our purpose is to describe the curves of steepest descent of $-If$ in the t-plane.

By a theorem of N. Levinson, [2],[7] , there exists an holomorphic function $T(z,w)$ with

$$T(0,0)=0 \;,\; \partial_z T(0,0)>0$$

and functions A,ζ such that

$$f(T(z,w),w) = A(w) + \zeta(w)z - \frac{z^3}{3} \;.$$

Hence we have to study the curves of steepest descent of $-\mathcal{I}(\zeta z-z^3/3)$, $\zeta \in \mathbb{C}$. These are paths of constant level for $\mathcal{R}(\zeta z-z^3/3)$. We first study the ones that contain a critical point.

If $\zeta = 0$ the only critical point is 0. The contour lines containing it are three straight lines, see figure 1.

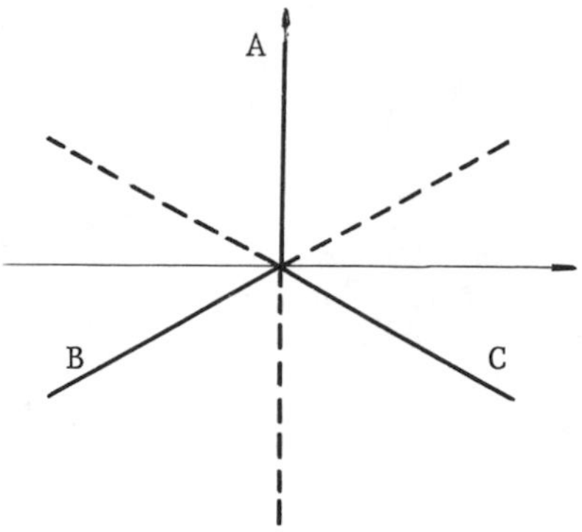

Figure 1

The continuous lines are the paths of steepest descent starting at 0, and the stippled lines are the paths of steepest ascent.

If $\zeta \neq 0$ we may assume $|\zeta| = 1$. Indeed we have only to choose $w = |\zeta|^{-1/2}z$ as new coordinate. Let $\zeta = \exp(i\theta)$ with $\theta \in]-\pi,\pi]$. The critical points are $z_\pm = \pm \exp(i\theta/2)$ and the critical values $\pm 2/3 \exp(3i\theta/2)$. Write $z = x + iy$. The paths of constant level for $\mathcal{R}(\zeta z-z^3/3)$ containing $z_\pm$ are given by

$$\text{(A.2)} \qquad x\cos\theta - y\sin\theta - \frac{x^3}{3} + xy^2 = \pm\frac{2}{3}\cos\frac{3\theta}{2}.$$

Such a curve has always the six half straight lines of figure 1 as asymptotic lines. If $\theta = 0$ we have the situation of figure 2. Here again the continuous lines are the paths of steepest descent containing $z_\pm$, and the stippled lines are the paths of steepest ascent. The stippled lines which do not contain a critical point, are contour lines of $\mathcal{R}(\zeta z-z^3/3)$ which are at the same level as z_+ or z_-. If $0 < \theta < \pi/3$ we have the situation of figure 3. If $\theta = \pi/3$ it degenerates to figure 4.

All the pictures for $\theta \in]-\pi,\pi]$ can be obtained from figures 2-4. Indeed, if C_θ is the curve defined by (A.2) we have

$$x+iy \in C_\theta \Leftrightarrow x-iy \in C_{-\theta}$$
$$x+iy \in C_\theta \Leftrightarrow e^{i\pi/3}(x+iy) \in C_{\theta+2\pi/3}.$$

Figure 5 gives the paths of steepest descent which contain the critical points for $\theta \in]-\pi,\pi]$. The critical point which corresponds to the lowest (resp. highest) critical value is denoted by L(resp. B).

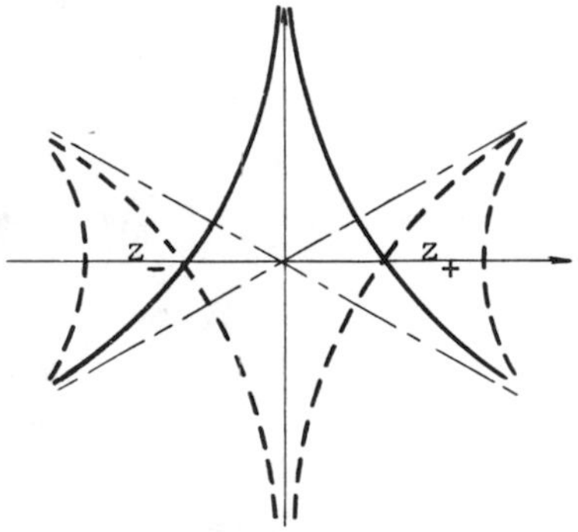

Figure 2

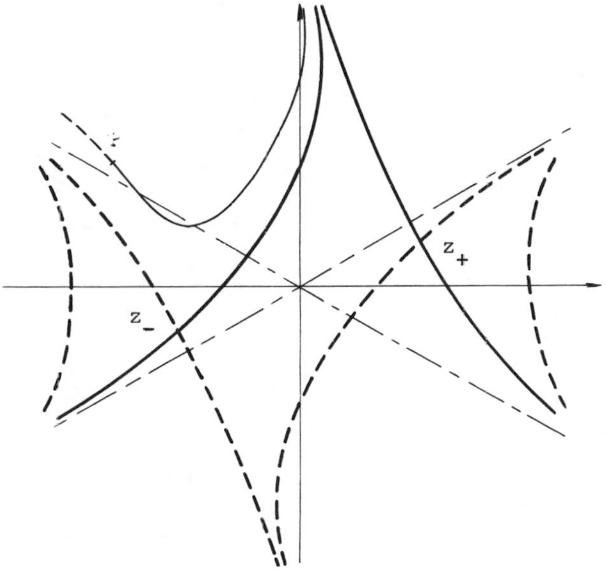

Figure 3

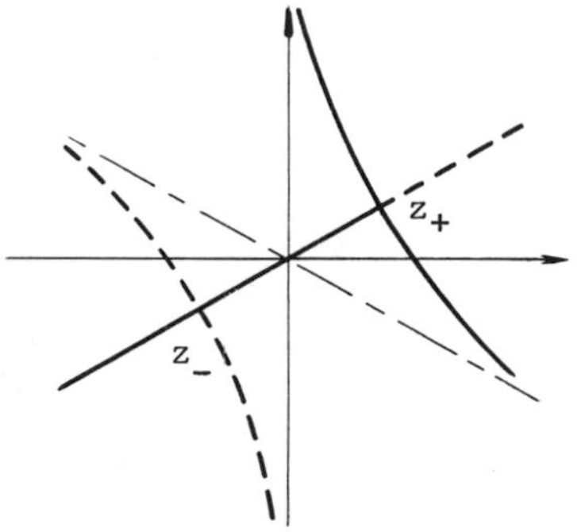

Figure 4

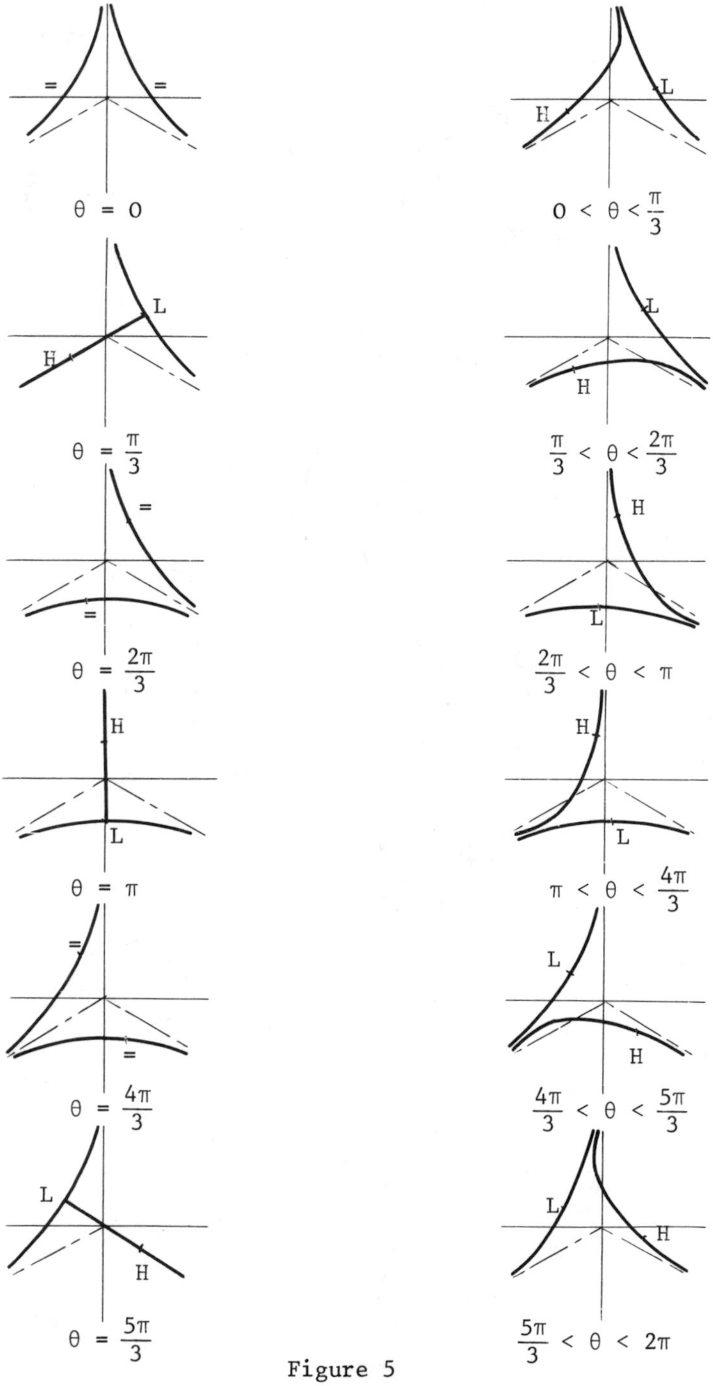
=
=
θ = 0
H
L
0 < θ < π/3
L
H
θ = π/3
L
H
π/3 < θ < 2π/3
=
=
θ = 2π/3
H
L
2π/3 < θ < π
H
L
θ = π
H
L
π < θ < 4π/3
=
=
θ = 4π/3
L
H
4π/3 < θ < 5π/3
L
H
θ = 5π/3
L
H
5π/3 < θ < 2π

Figure 5

All the other curves of steepest descent are easily obtained from the previous ones. They are curves of constant level for the real part, hence they cannot meet C_θ. Moreover they are always asymptotic to the lines of figure 1. For example we have set out the curve of steepest descent starting at a point P in figure 3.

We can estimate the rate of decrease of $-\mathcal{I}(\zeta z-z^3/3)$ on the curves of steepest descent which contain the critical points.

Lemma. *Let* ζ *be a complex number different from* 0 *and* $\zeta^{1/2}$ *one of its square roots. If* z *belongs to the curve of steepest descent containing* $\zeta^{1/2}$ *, then we have*

$$\mathcal{I}(\zeta z - \frac{z^3}{3} - \frac{2}{3}\zeta^{3/2}) \geqslant \frac{1}{12}|z-\zeta^{1/2}|^3 . \tag{A.3}$$

Proof. By homogeneity we may assume $|\zeta| = 1$. Write $z = x + iy$, $\zeta = \exp(i\theta)$ and $\zeta^{1/2} = \exp(i\theta/2)$. Since

$$\zeta z - \frac{z^3}{3} + \frac{2}{3}\zeta^{3/2} = -\frac{1}{3}(z-\zeta^{1/2})^2(z+2\zeta^{1/2})$$

we have

$$\mathcal{R}((z-\zeta^{1/2})^2(z+2\zeta^{1/2})) = 0. \tag{A.4}$$

First we shall prove that

$$|z+2\zeta^{1/2}| \geqslant 1. \tag{A.5}$$

The minimum of $|z+2\zeta^{1/2}|$ is reached on the curve when

$$\begin{pmatrix} x+2\cos\frac{\theta}{2} \\ y+2\sin\frac{\theta}{2} \end{pmatrix} = \lambda \begin{pmatrix} \cos\theta - x^2+y^2 \\ -\sin\theta + 2xy \end{pmatrix}, \quad \lambda \in \mathbb{R}.$$

It follows that

$$(x+2\cos\frac{\theta}{2})(2xy-\sin\theta)+(y+2\sin\frac{\theta}{2})(\cos\theta-x^2+y^2) = \mathcal{I}((z+2\zeta^{1/2})(z^2-\zeta))=0.$$

The point $-2\zeta^{1/2}$ does not belong to the curve of steepest descent through $\zeta^{1/2}$, hence using (A.4) we obtain $z = \zeta^{1/2}$ or

$$(z+\zeta^{1/2}) = ir(z-\zeta^{1/2}) \ , \ r \in \mathbb{R}.$$

It follows that $|z| = 1$. Now (A.5) is obvious.

Let

$$(z-\zeta^{1/2})^2(z+2\zeta^{1/2}) = -3it \ , \ t \in \mathbb{R}.$$

If $z \neq \zeta^{1/2}$, we have

$$\frac{|z+2\zeta^{1/2}|}{|z-\zeta^{1/2}|} \geqslant \frac{1}{|z-\zeta^{1/2}|} = \left|\frac{1}{3} - \frac{1}{3}\,\frac{z+2\zeta^{1/2}}{z-\zeta^{1/2}}\right| \geqslant \frac{1}{3}\left(1 - \left|\frac{z+2\zeta^{1/2}}{z-\zeta^{1/2}}\right|\right).$$

Therefore

$$\left|\frac{z+2\zeta^{1/2}}{z-\zeta^{1/2}}\right| \geqslant \frac{1}{4}$$

and

$$I\left(\zeta z - \frac{z^3}{3} - \frac{2}{3}\,\zeta^{3/2}\right) = t = \frac{1}{3}\left|\frac{z+2\zeta^{1/2}}{z-\zeta^{1/2}}\right| . |z-\zeta^{1/2}|^3 \geqslant \frac{1}{12}\,|z-\zeta^{1/2}|^3.$$

REFERENCES

1. F.G. Friedlander, R.B. Melrose, The wave front set of the solution of a simple initial boundary value problem with glancing rays, II, Math. Proc. Comb. Phil. Soc., 81, 1977, 97-120.

2. L. Hörmander, The analysis of linear partial differential operators, I-IV, Springer Verlag, 1983-1985.

3. K. Kataoka, Microlocal theory of boundary value problems, I-II, J. Fac. Sci. Univ. Tokyo 27(2), 1980, 355-399, and preprint.

4. K. Kataoka, Microlocal analysis of boundary value problems with application to diffraction, Proceeding of the Nato ASI on Singularities in boundary value problems, D. Reidel, 1980, 121-131.

5. P. Laubin, Analyse micolocale des singularités analytiques, Bull. Soc. Roy. Sc. Liège, 2, 1983, 103-212.

6. G. Lebeau, Deuxième microlocalisation sur les sous-variétés isotropes, Thèse, Orsay, 1983.

7. N. Levinson, Transformation of an analytic function of several variables to a canonical form, Duke Math. J. 28, 345-353, 1961.

8. P. Schapira, Propagation at the boundary and reflexion of analytic singularities of solutions of linear partial differential equations, I and II, Publ. RIMS, Kyoto Univ. 12, 1977, 441-453 and Sem. Goulaouic-Schwartz, IX, 1976-77.

9. J. Sjöstrand, Propagation of analytic singularities for second order Dirichlet problems, I and II, Comm in P.D.E. 5, 1980, 41-94 and 187-207.

10. J. Sjöstrand, Analytic singularities of solutions of boundary value problems, Proceeding of the Nato ASI on Singularities in boundary value problems, D. Reidel, 1980, 235-269.

11. J. Sjöstrand, Singularités analytiques microlocales, Astérisque 95, 1982.

PROPAGATION DES SINGULARITÉS GEVREY POUR LE PROBLÈME DE DIRICHLET

G. Lebeau
C.M.A. - Ecole Normale Supérieure
45 rue d'Ulm
750230 Paris Cedex 05
France

ABSTRACT. Depuis les travaux de Friedlander - Melrose, Melrose-Sjöstrand [4], Sjöstrand [7], on sait que la propagation des singularités pour les problèmes aux limites, près des points "glancing" du bord, est différente suivant qu'on s'intéresse au point de vue C^∞ ou analytique. Cet article est consacré à l'étude de la propagation des singularités Gevrey G^σ : en utilisant les techniques élaborées par J. Sjöstrand, on y montre que le comportement des singularités Gevrey G^σ est identique au comportement C^∞, [resp. analytique] pour $\sigma \geq 3$ [resp. $1 \leq \sigma < 3$] .

1. INTRODUCTION

Soit M une variété analytique réelle à bord, de dimension n , et P un opérateur différentiel linéaire à coefficients analytiques sur M , du second ordre, de symbole principal réel p . On suppose

1.1. La 1-forme canonique $\omega = \Sigma\, \xi_j\, dx_j$ et dp sont linéairement indépendants sur $p^{-1}(0) \cap (T^* M \setminus 0)$.

1.2. Le bord de M , ∂M , est non caractéristique pour P et $\frac{\partial p}{\partial \xi}|\, T^* M \setminus 0 \neq 0$ sur $p^{-1}(0) \cap \partial T^* M \setminus 0$.

On notera Π la projection de $T^*\ M$ sur M , et en suivant [4], bM l'espace topologique construit à partir de $(T^* M \setminus 0) \setminus T^*_{\partial M}$ en identifiant les points de $\pi^{-1}(\partial M)$ qui ont même projection dans $T^* \partial M \setminus 0$. Alors bM s'identifie à $(T^* \partial M \setminus 0) \cup (T^* \mathring{M} \setminus 0)$ et on a une projection canonique b de $(T^* M \setminus 0) \setminus T^*_{\partial M}$ sur bM . On notera aussi

1.3 $$\Sigma_b = b[p^{-1}(0) \cap T^* M \setminus 0] .$$

Soit alors u une distribution prolongeable sur M vérifiant

H. G. Garnir (ed.), Advances in Microlocal Analysis, 203–223.

1.4. $Pu = 0$ dans $\mathring{M}$; $u|_{\partial M} = 0$.

Désignons par $WF_b(u)$ le front d'onde C^∞ de u [4], et par $SS_b(u)$ le spectre analytique de u [5], [6] ; ce sont deux fermés de bM , contenus dans Σ_b d'après les théorèmes de régularité elliptique (voir [4], [5], [6]), et on a les deux résultats suivants :

THÉORÈME 1.1. [Melrose - Sjöstrand] [4].-
Sous les hypothèses 1.4., $WF_b(u)$ est réunion de "rayons-C^∞" maximaux.

THÉORÈME 1.2. [Sjöstrand] [7].-
Sous les hypothèses 1.4., $SS_b(u)$ est réunion de "rayons-analytiques" maximaux.

On renvoie aux articles [4] et [7] pour les définitions des "rayons-C^∞", et "rayons-analytiques".
Rappelons seulement que par chaque point de Σ_b il passe un *unique* "rayon-C^∞", mais qu'il peut exister une infinité de "rayons-analytiques" issus d'un point de Σ_b .

Le résultat principal de cet article est de montrer que le théorème 1.2. reste vrai en remplaçant singularités analytiques par singularités Gevrey d'indice s , pour tout $s \geq 1$ (le spectre Gevrey d'indice s de u , $SS_b^s(u)$ sera défini au § 2).

THÉORÈME 1.3.-
Sous les hypothèses 1.4., $SS_b^s(u)$ est réunion de "rayons-analytiques" maximaux pour tout $s \geq 1$.
En utilisant alors le résultat de [3] sur la régularité Gevrey-3 pour la diffraction, on obtient comme corollaire du théorème 1.3. la version Gevrey du théorème 1.1.

THÉORÈME 1.4.-
Sous les hypothèses 1.4., $SS_b^s(u)$ est réunion de "rayons-C^∞" maximaux, pour tout $s \geq 3$.
Rappelons que, d'après l'étude de Friedlander et Melrose [1], on sait que le théorème 1.4. est faux dans toutes les classes de Gevrey d'indice s , $s \in [1,3[$.

La démonstration du théorème 1.3. (une fois qu'est défini $SS_b^s(u)$) est en fait implicitement contenue dans la démonstration de J. Sjöstrand [7] relative au cas analytique $s = 1$. La preuve donnée ici n'est donc qu'une variante de sa méthode. Le plan de l'article est le suivant : dans le § 2, on définit $SS_b^s(u)$, dans le § 3, on rappelle les résultats relatifs aux résolutions de l'identité utiles par la suite, le § 4 est consacré à la preuve du théorème 1.3. et le § 5 à la preuve du théorème 1.4.

2. MICRO-SUPPORTS GEVREY

DÉFINITION 2.1.- Soit Ω ouvert de $\mathbb{R}^n$; une fonction $f \in C^\infty(\Omega)$ est dite de classe de Gevrey s (s réel, $s \geq 1$) si pour tout compact $K \subset \Omega$, il existe deux constantes A_K et B_K , telles que

2.1 $$\forall \alpha \in \mathbb{N}^n \quad \sup_K |\partial_x^\alpha f| \leq A_K (B_K)^{|\alpha|} (\alpha!)^s .$$

On note l'ensemble des fonctions de classe de Gevrey s sur Ω , $G^s(\Omega)$; lorsque $s = 1$, ce sont les fonctions analytiques sur Ω .

A toute distribution f sur $\mathbb{R}^n$, on associe un fermé conique de $T^*\mathbb{R}^n \setminus 0$, noté $SS^s(f)$, le micro-support Gevrey s de f défini par : (SS^1 étant le spectre analytique de Sato de f [5]).

DÉFINITION 2.2.- Soit $f \in \mathcal{D}'(\mathbb{R}^n)$ et $\alpha = (\alpha_x, \alpha_\xi) \in T^*\mathbb{R}^n \setminus 0$; alors $\alpha \notin SS^s(f)$ s'il existe Ω voisinage de α_x , et f_0 , $f_1 \in \mathcal{D}'(\Omega)$, tels que

2.2. $$f|_\Omega = f_0 + f_1 \ ; \ \alpha \notin SS^1(f_0) \ ; \ f_1 \in G^s(\Omega) .$$

Soit à présent M et P comme dans l'introduction.
Pour tout point $x_0 \in \partial M$, il existe un système de coordonnées locales $(x_1,\dots,x_n)$ centré en x_0 tel que M soit défini par $x_n \geq 0$. Si u est une distribution prolongeable sur M solution de $Pu = 0$ dans $\mathring{M}$, on notera $\bar{u}$ l'unique extension de u qui vérifie $\bar{u} = u$ dans $\mathring{M}$ et près de tout point $x_0 \in \partial M$ comme précédemment, $\bar{u} = 0$ dans $x_n < 0$ et $x_n^2 P\bar{u} \equiv 0$.

DÉFINITION 2.3.- Pour u solution de $Pu = 0$ dans $\mathring{M}$, on pose

2.3. $SS_b^s(u) \overset{\text{def}}{=} b[SS^s(\bar{u}) \cap (T^* M \setminus 0) \setminus T^*_{\partial M}]$.

Lemme 2.1.- $SS_b^s(u)$ est un fermé de bM .

Preuve.- Il suffit de le prouver près d'un point $x_o \in \partial M$.

Soit $(x_1,\dots,x_n)$ le système de coordonnées locales précédent.
On a alors :

2.4. $P\bar{u} = a_o(x_1,\dots,x_{n-1}) \times \delta'_{x_n=0} + a_1(x_1,\dots,x_{n-1}) \otimes \delta_{x_n=0}$,

où les a_i sont dans $\mathcal{D}'(\mathbb{R}^{n-1})$. Comme $x_n = 0$ est non caractéristique pour P , il existe $C_o > 0$, tel que, avec $x' = (x_1,\dots,x_{n-1})$, $\xi' = (\xi_1,\dots,\xi_{n-1})$, on ait :

2.5. si $(x' , \xi' ; x_n > 0 , \xi_n) \in SS^s(\bar{u})$, alors $|\xi_n| \leq C_o|\xi'|$ et d'après 2.4.,

2.6. si $(x_o' , \xi_o' ; x_n = 0 , \xi_{n,o}) \in SS^s(\bar{u})$ et $|\xi_{n,o}| > C_o|\xi_o'|$, alors pour tout ξ_n tel que $|\xi_n| > C_o |\xi_o'|$, on a

$$(x_o' , \xi_o' ; x_n = 0 , \xi_n) \in SS^s(\bar{u}) .$$

Alors si $\alpha_p = (x_p' , \xi_p' , x_{n,p} , \xi_{n,p})$ est une suite telle que $\alpha_p \in SS^s(\bar{u})$ $x_{n,p} \longrightarrow 0$ et $\pi(\alpha_p) \longrightarrow \beta$ $T^* \in M \setminus 0$ on peut supposer, d'après 2.5., 2.6., que $\xi_{n,p}$ est borné, d'où on obtient $\beta \in SS_b^s(u)$.

3. RÉSOLUTION DE L'IDENTITÉ

Nous rappelons dans ce paragraphe ce qui nous sera utile par la suite, concernant les résolutions de l'identité (pour une étude plus générale, on renvoie à [7]). En particulier, toutes les phases considérées ici sont quadratiques.
Pour $\alpha = (\alpha_x , \alpha_\xi) \in T^* \mathbb{R}^n$, $a \in]-2 , +2[$, $\vartheta \in \mathbb{R}_+^*$ et $(x,y) \in \mathbb{R}^n \times \mathbb{R}^n$, on pose :

3.1. $\varphi(x,y,\alpha ; a,\vartheta) = (x-y)\alpha_\xi + i\vartheta[(x-\alpha_x)^2 + a(x-\alpha_x)(y-\alpha_x) + (y-\alpha_x)^2]$

et pour $\lambda \geq 1$:

3.2. $\pi_\alpha(x,y,\lambda ; a,\vartheta) = \left(\frac{\lambda}{2\pi}\right)^n \left(\frac{\lambda\vartheta(2 + a)}{\pi}\right)^{n/2} e^{i\lambda\varphi(x,y,\alpha ; a,\vartheta)}$

de sorte que, de la formule usuelle,

3.3. $$\int_{\mathbb{R}^d} e^{-hx^2 + x.y}\, dx = \left(\frac{\pi}{h}\right)^{d/2} e^{\frac{y^2}{4h}}$$

on déduit

3.4. $$\int_{\mathbb{R}^n} \pi_\alpha\, d\alpha_x = \left(\frac{\lambda}{2\pi}\right)^n e^{i\lambda(x-y)\alpha_\xi - \frac{\lambda\vartheta}{4}(2-a)(x-y)^2}$$

de sorte que

3.5. $$\int \pi_\alpha\, d_\alpha = \delta_{x=y} .$$

Les phases précédentes ont de bonnes propriétés de composition entre elles ; si a_1 , ϑ_1 , a_2 , ϑ_2 vérifient

3.6. $$(4 - a_1^2)\, \vartheta_1^2 = (4 - a_2^2)\, \vartheta_2^2 = \gamma^2 ,$$

alors on a :

3.7. $$\int_{\mathbb{R}^n} \pi_\alpha(x,z,\lambda\, ; a_1,\vartheta_1)\, \pi_\alpha(z,y,\lambda,a_2,\vartheta_2)dz = \left(\frac{\lambda}{2\pi}\right)^n C_{a_1,a_2}^{n/2} \pi_\alpha(x,y,\lambda\, ; a,\vartheta) ,$$

où C_{a_1,a_2} est une constante ne dépendant que de a_1 , a_2 , et

3.8. $$a = \frac{-2a_1 a_2 \vartheta_1 \vartheta_2}{4\vartheta_1 \vartheta_2 + \gamma^2} \qquad \vartheta = \frac{4\vartheta_1 \vartheta_2 + \gamma^2}{4(\vartheta_1 + \vartheta_2)} .$$

Les phases $\varphi(x,y,\ldots)$ sont autoadjointes, c'est-à-dire

3.9. $$\overline{\varphi(y,x, \ldots)} = - \varphi(x,y, \ldots) ,$$

et lorsque a est non nul, $\varphi(x,y, \ldots)$ définit la transformation canonique χ_α de $T^* \mathbb{C}^n$

3.10. $$\begin{cases} \chi_\alpha(y,\eta) = (x,\xi) \\ x - \alpha_x = - \frac{2}{a}(y - \alpha_x) - \frac{1}{ia\vartheta}(\eta - \alpha_\xi) \\ \xi - \alpha_\xi = - \frac{2}{a}(\eta - \alpha_\xi) - \frac{i\vartheta}{a}(4 - a^2)(y - \alpha_x) . \end{cases}$$

On désignera par $\mu \in \,]0\, , \mu_o]$ $(\mu_o > 0)$ un paramètre réel et on posera

3.11. $$\lambda' = \mu\lambda .$$

Si X est une variété analytique complexe, un symbole analytique formel sur X sera une série formelle

3.12. $\hat{\sigma}(z,\mu,\lambda') = \sum_{k=0}^{+\infty} \hat{\sigma}_k(z,\mu)(i\lambda')^{-k}$

où les $\hat{\sigma}_k(z,\mu)$ sont des fonctions de $(z,\mu) \in X \times]\ 0\ ,\mu_o]$ holomorphes en z , et vérifiant :

3.13. $\sup_{z\in X}|\hat{\sigma}_k(z,\mu)| \leq AB^k\ k\ !$

où les constantes A , B sont *indépendantes* de μ .
Un symbole analytique sur X est alors une fonction $\sigma(z,\mu,\lambda')$, définie pour $z \in X$, $\mu \in]0\ ,\mu_o]$, $\lambda' > \lambda'_0$, holomorphe en z , et uniformément bornée en z,μ,λ' . On dit que σ est classique s'il existe un symbole analytique formel $\hat{\sigma}$, tel que

3.14. $\forall N \geq 0\ \sup_{z\in X}|\sigma(z,\mu,\lambda') - \sum_{k=0}^{N} \hat{\sigma}_k(z,\mu)(i\lambda')^{-k}| \leq AC^{(N+1)}(N+1)!\,(\lambda')^{-(N)}$

où A , C sont indépendants de μ , λ' ; le symbole formel $\hat{\sigma}$ est alors uniquement déterminé ; réciproquement, si on se donne un symbole formel $\hat{\sigma}$ et si on pose

3.15. $\sigma(z,\mu,\lambda') = \sum_{k\leq\lambda'/C_o} \hat{\sigma}_k(z,\mu)(i\lambda')^{-k}$

avec $C_o > \frac{2}{e} B$, où B est la constante de 3.13., alors σ est un symbole analytique classique, et le symbole formel associé est $\hat{\sigma}$.

Remarque 3.1.- On a introduit le paramètre μ car on s'intéresse à la propagation des singularités dans les classes de Gevry G^s , $s \geq 1$; on choisira alors μ de l'ordre $\lambda^{\frac{1-s}{s}}$, de sorte qu'on aura λ' de l'ordre de $\lambda^{\frac{1}{s}}$.
Soit à présent Ω et ω deux ouverts bornés de $\mathbb{R}^n$, Δ la diagonale $x = y = \alpha_x$ de $\bar{\Omega} \times \bar{\Omega} \times \bar{\Omega}$, X un voisinage complexe de $\Delta \times \bar{\omega}$ dans $\mathbb{C}^{4n}$, et $\Omega' \subset\subset \Omega$. Soit ρ un réel strictement positif, tel que

3.16. $\forall\alpha_x \in \bar{\Omega}'\quad \{x, |x - \alpha_x| \leq \sqrt{\rho}\} \subset \Omega$

3.17. $(x,y,\alpha_x,\alpha_\xi)$, $\alpha_x \in \bar{\Omega}$, $\alpha_\xi \in \bar{\omega}$, $(x - \alpha_x)^2 + (y - \alpha_x)^2 \leq \rho\} \subset X$

et $\chi \in C_o^\infty(]-\rho\ ,+\rho[)$ égal à 1 près de zéro.
Lorsque $(x,y,\alpha,\mu,\lambda')$ est un symbole analytique sur X et pour

$\alpha = (\alpha_x, \alpha_\xi) \in \Omega' \times \omega$, on note $\pi_\alpha^\Omega\, \sigma$ l'opérateur intégral

3.18 $$\left(\pi_\alpha^\Omega\, \sigma\right)(f)(x,\mu,\lambda') \overset{\text{déf}}{=} \int_\Omega \pi_\alpha(x,y,\lambda;a,\vartheta)\sigma(x,y,\alpha,\mu,\lambda')\chi((x-\alpha_x)^2+(y-\alpha_x)^2)f(y)dy.$$

Alors, pour tout $f \in \mathscr{D}'(\Omega)$, $\left(\pi_\alpha^\Omega\, \sigma\right)(f)(x,\mu,\lambda')$ est défini pour $\lambda' \geq \lambda_0'$, $\mu \in]0,\mu_0]$, $x \in \mathbb{R}^n$ et est C^∞ à support compact dans Ω . Si on pose

3.19 $$(a\, ,\, b)_\Omega = \int_\Omega a(x)\bar{b}(x)dx\ ,$$

on a alors, pour tout f , g dans $\mathscr{D}'(\Omega)$

3.20 $$((\pi_\alpha^\Omega\, \sigma)f\, ,\, g)_\Omega = (f\, ,\, \pi_\alpha^\Omega\, \sigma^*\, g)_\Omega$$

où $\sigma^*(x,y,\ \ldots) = \overline{\sigma(y,x,\ \ldots)}$ (prolongement holomorphe à partir du réel). Dans la suite, on choisit un couple $(\tilde{a},\tilde{\vartheta})$, $\tilde{a} \neq 0$, et on pose :

3.21. $$\begin{cases} \tilde{\pi}_\alpha = \pi_\alpha(x,y,\lambda,\tilde{a},\tilde{\vartheta}) \\ \pi_\alpha = \pi_\alpha(x,y,\lambda,a,\vartheta) \end{cases}$$

avec

3.22 $$a = \frac{-2\,\tilde{a}^2\,\tilde{\vartheta}^2}{4\,\tilde{\vartheta}^2 + \gamma^2}\quad ,\quad \vartheta = \frac{4\,\tilde{\vartheta}^2 + \gamma^2}{8\tilde{\vartheta}}$$

de sorte qu'après 3.7., on a

3.23. $$\tilde{\pi}_\alpha = \tilde{\pi}_\alpha^*\, \tilde{\pi}_\alpha = \tilde{\pi}_\alpha\, \tilde{\pi}_\alpha^* = \left(\frac{\lambda}{2\pi}\right)^n C^{n/2}\, \pi_\alpha\ .$$

On en déduit, pour $f \in \mathscr{D}'(\Omega)$, $\alpha \in \Omega' \times \omega$

3.24. $$C^{n/2}\left(\pi_\alpha^\Omega\, f\, ,\, f\right)_\Omega = \left(\frac{2\pi}{\lambda}\right)^n \|\tilde{\pi}_\alpha^\Omega\, f\|_\Omega^2 + (\mathscr{N} f\, ,\, f)_\Omega$$

où $\mathscr{N} = \mathscr{N}(x,y,\alpha,\lambda)$ est un noyau C^∞ à support compact dans $\Omega \times \Omega$, à décroissance exponentielle en λ ainsi que toutes ses dérivées.

DEFINITION 3.1.- Soit $\sigma(x,y,\alpha,\mu,\lambda')$ un symbole analytique sur X . On dira qu'il est strictement positif s'il existe $m > 0$, tel que $\operatorname{Re} \sigma \geq m$ pour $(x,y,\alpha) \in X$, $\mu \in]0\, ,\mu_0]$, $\lambda' \geq \lambda_0'$.

Lemme 3.1.- Soit α près de $\alpha^0 = (\alpha_x^0\, ,\, \alpha_\xi^0) \in \Omega' \times \omega$ et σ un symbole analytique strictement positif sur X . Alors il existe un voisinage Ω'' de α_x^0 et un symbole analytique $\tilde{\sigma}(x,y,\alpha,\mu,\lambda')$ défini près de

$(\alpha_x^o,\alpha_x^o,\alpha^o)$ tel que, pour tout $f \in \mathcal{D}(\Omega)$, on ait :

3.25. $$\lambda^n \operatorname{Re} (\pi_\alpha^\Omega \sigma f , f)_\Omega = \|\tilde{\pi}_\alpha^{\Omega''} \tilde{\sigma} f\|^2_{\Omega''} + (\mathcal{N} f , f)_\Omega$$

où $\mathcal{N}(x,y,\alpha,\mu,\lambda)$ est un noyau C^∞ à support compact dans $\Omega \times \Omega$, à décroissance exponentielle en λ ainsi que toutes ses dérivées, pour α près de α_o .

Preuve.- Tout d'abord, on considérera les μ , λ' intervenant dans le symbole σ comme des paramètres indépendants du λ intervenant dans la phase des π_α . Puisqu'on a choisi $\tilde{a} \neq 0$, pour α fixé près de α_o , π_α , $\tilde{\pi}_\alpha$ sont des isomorphismes intégraux de Fourier. Il existe donc un opérateur pseudo-différentiel r de degré 0 (avec grand paramètre λ au sens de [8]), défini près de α_o tel que

3.26 $$\lambda^n \pi_\alpha(\sigma + \sigma^*) = \tilde{\pi}_\alpha \circ r \circ \tilde{\pi}_\alpha$$

et on a $r^* = r$ (où ici $*$ désigne l'adjoint dans l'anneau des (germes) d'opérateurs pseudo-différentiels en $\alpha_o \in T^* \mathbb{R}^n$). On pose alors $\tilde{\pi}_\alpha \tilde{\tilde{\sigma}} = \tilde{\pi}_\alpha \circ q$ de sorte qu'il suffit de résoudre $r = qq^*$ qui admet la solution auto-adjointe $q = \sqrt{r}$ (par hypothèse, le symbole principal de r est > 0). Ceci permet de construire $\tilde{\tilde{\sigma}}$ pour (x,y,α) près de $(\alpha_x^o,\alpha_x^o,\alpha^o)$ et on a alors, avec un C_o uniforme en μ , λ' :

3.27. $$\lambda^n \pi_\alpha(\sigma + \sigma^*) = (\tilde{\pi}_\alpha \tilde{\sigma})^* \tilde{\pi}_\alpha \tilde{\sigma} + \mathcal{O}(e^{-C_o\lambda}) \text{ , avec } \tilde{\sigma} = \tilde{\tilde{\sigma}}^*$$

Il reste à remarquer qu'en diminuant la constante ρ qui intervient dans 3.16., 3.17., on peut remplacer Ω par Ω'' , petit voisinage de α_x^o ; on commet alors une erreur qui rentre dans le noyau $\mathcal{N}$.
On laisse au lecteur le soin de vérifier la proposition suivante qui caractérise le micro-support Gevrey, à partir des transformations π_α .

PROPOSITION 3.1.- [On conserve les notations précédentes]

1) Si $\alpha^o \notin SS^s(f)$, il existe W voisinage de α^o et $C_o > 0$, tel que, pour tout $\alpha \in W$, on ait :

3.28. $$|(\pi_\alpha^\Omega \sigma)(f)(x,\mu,\lambda')| \leq \frac{1}{C_o} e^{-C_o\lambda^{1/s}}$$

pour $\mu \in]0 , \mu_o]$, $\lambda' \geq \lambda_o'$, $x \in \mathbb{R}^n$.

2) Réciproquement, si σ est un symbole strictement positif et s'il existe W voisinage de α^o $\gamma_o > 0$, $C_o > 0$, tels que

$$3.29.\qquad \int_W d\alpha\, \|\tilde{\pi}_\alpha^\Omega\, \sigma\, f\|_\Omega^2(\mu, \lambda)\Big|_{\mu=\gamma_o\lambda^{(1-s)/s}} \leq \frac{1}{C_o} e^{-C_o\lambda^{1/s}}$$

alors $\alpha^o \notin SS^s(f)$.

On se place à présent dans la situation de l'introduction, près d'un point $m_o \in \partial M$ et on choisit des coordonnées locales (t,x), telles que M soit défini près de m_o par $t \geq 0$. On a alors $x \in \mathbb{R}^{n-1}$, et on conserve les notations précédentes, en particulier les π_α sont indépendants de t.

On désigne par $u(t,x)$ une distribution prolongeable solution de $Pu = 0$ dans $t > 0$. On a alors (*cf.* [2])

$$3.30.\qquad u(t, .) \in C^\infty([0,T_o],\mathcal{D}'_x).$$

Soit $\alpha^o = (\alpha_x^o, \alpha_\xi^o) \in T^*\partial M \setminus 0$, et $\sigma(t,x,y,\alpha,\mu,\lambda')$ un symbole analytique défini près de $t = 0$, $x = y = \alpha_x^o$, $\alpha = \alpha^o$.

PROPOSITION 3.2.- 1) On suppose $\alpha^o \notin SS_b^s(u)$. Alors pour tout $j \in \mathbb{N}$ il existe $t_o > 0$, W voisinage de α^o, et $C_o > 0$, tels que, pour tout $\alpha \in W$, on ait :

$$3.31.\qquad \left|\left(\pi_\alpha^\Omega\, \sigma\right)\left(\frac{\partial^j}{\partial t^j}\, u\right)(t,x,\mu,\lambda')\right| \leq \frac{1}{C_o}\, e^{-C_o\lambda^{1/s}}$$

pour $t \in [0,t_o]$, $x \in \mathbb{R}^n$ $\mu \in]0,\mu_o]$, $\lambda' \geq \lambda'_o$.

2) Réciproquement, si σ est strictement positif et s'il existe W voisinage de α^o, $t_o > 0$, $\gamma_o > 0$, $C_o > 0$, tels que

$$3.32.\qquad \int_0^{t_o} dt \int_W d\alpha\, \left\|\left(\tilde{\pi}_\alpha^\Omega\, \sigma\right)(u)\right\|_\Omega^2 (t,\mu,\lambda')\Big|_{\mu=\gamma_o\lambda^{\frac{1-s}{s}}} \leq \frac{1}{C_o}\, e^{-C\,\lambda^{\frac{1}{s}}},$$

alors $\alpha^o \notin SS_b^s(u)$.

Preuve.- 1) Si $\alpha^o \notin SS_b^s(u)$, alors il existe $t_o > 0$ assez petit, tel que, pour tout $t \in [0,t_o]$ et tout $k \in \mathbb{N}$, on ait

$$3.33.\qquad \alpha^o \notin SS^s\left[\frac{\partial^k u}{\partial t^k}(t, .)\right].$$

Posons :

3.34. $$G^j_\alpha(t,x,\mu,\lambda') = \left(\pi^\Omega_\alpha\,\sigma\right)\left[\frac{\partial^j u}{\partial t^j}\right](t,x,\mu,\lambda')$$

3.35. $$F^j_\alpha(t,x,\tau,\mu,\lambda') = \int_0^{t_o} e^{i\lambda(t-t')\tau-\frac{\lambda}{2}(t-t')^2}\, G^j_\alpha(t',x,\mu,\lambda')\,dt' .$$

En intégrant par partie en t' dans 3.35., pour ramener l'ordre des dérivations portant sur u à zéro, on obtient :

3.36 $$F^j_\alpha(t,x,\tau,\mu,\lambda') = H^j_\alpha(t,x,\tau,\mu,\lambda') + R^j_\alpha(t,x,\tau,\mu,\lambda') ,$$

avec

3.37. $$H^j_\alpha(t,x,\tau,\mu,\lambda') = \lambda^j\int_0^{t_o} e^{i\lambda(t-t')\tau-\frac{\lambda}{2}(t-t')^2}\left(\pi^\Omega_\alpha\tilde{\sigma}\right)(u)(t',x,\mu,\lambda')\,dt'$$

où $\tilde{\sigma}$ est un nouveau symbole, polynomial en τ de degré j et où R^j_α désigne les termes qui proviennent des contributions de 0 et t_o dans les intégrations par parties. D'après 3.33., si W est un voisinage assez petit de α , il existe $C_o > 0$, $\lambda_o > 0$, tels que pour $\lambda \geq \lambda_o$

3.38. $$t \in [0,t_o],\quad x \in \mathbb{R}^n ,\quad \tau \in \mathbb{R} ,\quad \alpha \in W \Longrightarrow |R^j_\alpha| \leq \frac{\lambda^j(1+|\tau|)^j}{C_o}\, e^{-C_o\lambda^{1/s}} .$$

Maintenant, si t_o est assez petit, on a par hypothèse

3.39. $$\forall(t,\tau) \in T^*\mathbb{R}\quad (\alpha_o,t,\tau) \notin SS^s(u(t,x)\,.\,1_{[0,t_o]}) .$$

En utilisant par exemple le point 1) de la proposition 3.1., où on remplace X par (t,x) , on a donc :
Pour tout $\tau_o > 0$, il existe W_o voisinage de α_o , et $C_o > 0$, tels que

3.40 $$t \in [0,t_o]\ \tau \in [-\tau_o,+\tau_o] ,\ \alpha \in W_o ,\ x \in \mathbb{R}^n \Longrightarrow |H^j_\alpha| \leq \frac{\lambda^j}{C_o}\, e^{-C_o\lambda^{1/s}}$$

Maintenant, puisque $t = C^{te}$ est non caractéristique pour P , si τ_o est assez grand et W assez petit, il existe un symbole $\tilde{\tilde{\sigma}}(\tau,t,t',x,y,\alpha,\mu,\lambda')$, holomorphe pour $\tau \in \mathbb{C}$, $|\tau| > \tau_o$, $|\tilde{\tilde{\sigma}}| \leq \frac{C_o}{|\tau|^2}$, et tel que, pour $|\tau| \geq \tau_o$ on ait :

3.41. $$ {}^tP(t',y,D_{t'},D_y)\ [e^{i\lambda(t-t')\tau - \frac{\lambda}{2}(t-t')^2} \pi_\alpha(x,y,\lambda)\tilde{\tilde{\sigma}}] = $$

$$ \lambda^2 e^{i\lambda(t-t')\tau - \frac{}{2}(t-t')^2} \pi_\alpha(x,y,\lambda)\tilde{\sigma} + \vartheta(e^{-C\lambda|\tau|}) \ . $$

D'où, en intégrant par partie en (t',y) dans 3.37., et en utilisant à nouveau 3.33., on trouve qu'il existe W voisinage de α^o, $C_1 > 0$, tels que :

3.42. $$ t \in [0,t_o]\ ,\ \tau \in \mathbb{R}\ ,\ |\tau| \geq \tau_o\ ,\ \alpha \in W_1\ ,\ x \in \mathbb{R} \Rightarrow |H^j_\alpha| \leq \frac{(\lambda|\tau|)^{j-1}}{C_1} e^{-C_1\lambda^{1/s}} $$

En regroupant 3.36., 3.38., 3.40., 3.42., on obtient qu'il existe W voisinage de α^o et $C_1 > 0$, tels que

3.43. $$ t \in [0,t_o]\ ,\ \tau \in \mathbb{R}^n,\ x \in \mathbb{R}^n\ ,\ \alpha \in W \Rightarrow |F^j_\alpha| \leq \frac{\lambda^j (1+|\tau|)^j}{C} e^{-C\lambda^{1/s}} \ . $$

Pour obtenir l'estimation 3.31., il suffit donc d'utiliser le lemme élémentaire suivant appliqué à $G^j_\alpha(t, \dots)$:

Lemme 3.2.- Soit $f(t,\lambda)$ une fonction de classe C^2 en $t \in [0,t_o]$, définie pour $\lambda \geq \lambda_o$, telle que

3.44. $$ \max\left[\left|\frac{\partial^2 f}{\partial t^2}(t,\lambda)\right|\ ,\ \left|\frac{\partial f}{\partial t}(t,\lambda)\right|\right] \leq 1 $$

pour $t \in [0,t_o]$, $\lambda \geq \lambda_o$. On pose :

3.45. $$ F(t,\tau,\lambda) = \left(\frac{\lambda}{2\pi}\right) \int_0^{t_o} e^{i\lambda(t-t')\tau - \frac{\lambda}{2}(t-t')^2} f(t',\lambda)dt' $$

et on suppose, avec $C_o > 0$ et $M \in \mathbb{R}_+$

3.46. $$ |f(0,\lambda)| \leq e^{-C_o\lambda^{1/s}} \qquad |f(t_o,\lambda)| \leq e^{-C_o\lambda^{1/s}} $$

3.47. $$ |F(t,\tau,\lambda)| \leq (1 + |\tau|)^M e^{-C_o\lambda^{1/s}} \ . $$

Alors, il existe une constante D qui ne dépend que de t_o telle que :

3.48. $\forall t \in [0,t_o]$, $\forall\lambda \geq \lambda_o$ $|f(t,\lambda)| \leq D.\lambda\, e^{-\frac{C_o}{M+2}\lambda^{1/s}}$.

Preuve.- On désigne par D des constantes qui ne dépendent que de t , et on pose

3.49. $$h(t,\lambda) = f(t,\lambda) - \frac{(t_o - t)f(0,\lambda) + tf(t_o,\lambda)}{t_o}$$

3.50. $$H(t,\tau,\lambda) = \frac{\lambda}{2\pi}\int_0^t e^{i\lambda(t-t')\tau - \frac{\lambda}{2}(t-t')^2}\, h(t',\lambda)dt' ,$$

alors, d'après 3.46, 3.47, on a brutalement :

3.51. $$|H(t,\tau,\lambda)| \leq D\lambda(1 + |\tau|)^M e^{-C_o\lambda^{1/s}}$$

et puisque $h(0,\lambda) = h(t_o,\lambda) = 0$ et d'après 3.44. :

3.52. $$|H(t,\tau,\lambda)| \leq D\frac{\lambda}{(1 + |\tau|)^2} .$$

Or on a

3.53. $\forall t \in [0,t_o]$, $\lambda \geq \lambda_o$, $h(t,\lambda) = \int_{-\infty}^{+\infty} H(t,\tau,\lambda)d\tau$.

On obtient 3.48. en découpant l'intégrale en deux :

$$|\tau| \leq \tau_o \;,\; |\tau| \geq \tau_o \;,\; \tau_o = \exp\left[\left(\frac{C_o}{M+2}\right)\lambda^{1/s}\right] - 1 .$$

2) Réciproquement, on utilise la transformation F.B.I. usuelle [8], et on remarque que, pour $z \in \mathbb{C}^n$ près de $z_o = \alpha_x^o - i\alpha_\xi^o$, on a, avec ω petit voisinage réel de α_x^o :

3.54. $$\int_\omega dx \int_W d\alpha\, e^{-\frac{\lambda}{2}(z-x)^2}\, \tilde{\pi}_\alpha(x,y,\lambda)\sigma = \lambda^{-\frac{3n}{2}} e^{-\frac{\lambda}{2}(z-y)^2}\, \tilde{\sigma}$$

où $\tilde{\sigma}(z,y,t,\mu,\lambda')$ est un symbole elliptique pour (z,y,t) près de $(z_o,\alpha_x^o,0)$.

Par suite, on déduit de 3.32. :

3.55. $$\left|\iint dy \int_0^{t_0} e^{-\frac{\lambda}{2}(\vartheta-t)^2-\frac{\lambda}{2}(z-y)^2}\,\tilde{\sigma}\,u(t,y)\right|_{\mu=\gamma_0\lambda^{\frac{1-s}{s}}} \leq \frac{1}{C_1}\, e^{\frac{\lambda}{2}[(\mathrm{Im}\vartheta)^2+(\mathrm{Im}z)^2]}\, e^{-C\lambda^{1/s}}$$

uniformément pour $\vartheta \in \mathbb{C}$, z près de z_0 . Comme les transformations F.B.I. caractérisent le spectre Gevrey, on déduit de 3.55. :

3.56 $$\forall(t,\tau) \in T^*\mathbb{R} \quad (\alpha_0,t,\tau) \notin SS^s[u(t,x)\,.\,1_{[0,t_0]}]$$

ce qui entraîne $\alpha^0 \notin SS_b^s(u)$.

4. PREUVE DU THÉORÈME 1.3.

Dans cette preuve, on considère comme connus les résultats de propagation Gevrey à l'intérieur, ainsi que les résultats de régularité aux points elliptiques du bord, et de réflexion transversale des singularités aux points hyperboliques du bord. On se place près d'un point $m_0 \in \partial M$ où on choisit un système de coordonnées locales (t,x) avec M défini par $t \geq 0$ et on suppose que P est sous la forme (c'est toujours possible d'après [4])

4.1. $$P = D_t^2 + R(t,x,D_x) \qquad \left(D_t = \frac{1}{i}\frac{\partial}{\partial t}\right)$$

de symbole principal réel $p = \tau^2 + r(t,x,\xi)$, et on pose

4.2. $$r_0(x,\xi) = r(0,x,\xi) .$$

Par hypothèse dr_0 et $\Sigma\, \xi_j\, dx_j$ sont indépendants sur $r_0 = 0$. D'après [7], sachant que $SS_b^s(u)$ est un fermé contenu dans Σ_b , il suffit de prouver l'estimation suivante :

PROPOSITION 4.1.- Soit $\alpha^0 \in T^*\partial M \setminus 0$ tel que $r_0(\alpha^0) = 0$ et $u(t,x)$ vérifie 1.4. Il existe $\varepsilon_0 > 0$, $\delta_0 > 0$ (indépendants de u , et de α^0 variant dans un compact) tels que si $\varepsilon \in]0,\varepsilon_0]$ et si

4.3. $$SS_b^s(u) \cap \{(\alpha,t,\tau) \quad |\alpha - \alpha^0| \leq 2\varepsilon^2 \ , \ 0 \leq t \leq 2\varepsilon^2\} = \emptyset ,$$

on a :

4.4. $\exp \ell H_{r_0}(\alpha^0) \notin SS_b^s(u)$

pour $-\delta_0\varepsilon \leq \ell \leq +\delta_0\varepsilon$.

Preuve.- Suivant [7], on introduit un système de coordonnées (ℓ,β) près de $\alpha^0 = (\alpha_x^0 , \alpha_\xi^0)$ dans $T^* \partial M \setminus 0$, centré en α_0 , tel que

4.5. $H_{r_0} = \frac{\partial}{\partial \ell}$

et pour $\alpha \in T^* \partial M \setminus 0$ près de α^0 et $t \geq 0$, on pose

4.6. $\psi(\alpha , t) = 1 - \frac{\ell}{\delta_0\varepsilon} - \frac{\beta^2}{\varepsilon^4} - \frac{t}{\varepsilon^2}$

4.7. $V_\varepsilon = \{(\ell , \beta), -\varepsilon^2 \leq \ell \leq 2\delta_0\varepsilon , |\beta| \leq 2\varepsilon^2\}$

Ici, $\varepsilon \in]0 , \varepsilon_0]$ est tel que les hypothèses 4.3. sont satisfaites pour la solution $u(t,x)$; les constantes ε_0 , δ_0 restent à déterminer ; on supposera $2\varepsilon_0 \leq \delta_0$, et on choisit un petit voisinage Ω de α_x^0 , indépendant de ε , tel que $\{\alpha_x , \exists\alpha_\xi , (\alpha_x , \alpha_\xi) \in V_\varepsilon\} \subset\subset \Omega$ pour $\varepsilon \in]0 , \varepsilon_0]$.

On introduit alors les opérateurs à noyau C^∞ à support compact dans $\Omega \times \Omega$, C^∞ en t :

4.8. $(Af)(t,x,\mu,\lambda') = \int_{\alpha\in V_\varepsilon} e^{\mu\lambda\psi(\alpha,t)} (\pi_\alpha^\Omega \sigma_A)(f)d\alpha$

4.9. $(Bf)(t,x,\mu,\lambda') = \int_{\alpha\in V_\varepsilon} e^{\mu\lambda\psi(\alpha,t)} \pi_\alpha^\Omega f d\alpha$

où $\pi_\alpha = \pi_\alpha(x,y,\lambda)$ et $\sigma_A = \sigma_A(x,y,t,\alpha ; \mu,\lambda',\varepsilon)$ est un symbole analytique, classique, défini dans un voisinage U indépendant de ε de $x = y = \alpha_x = \alpha_x^0$, $t = 0$, $\alpha_\xi = \alpha_\xi^0$, et qui vérifie dans U , pour $\mu \in]0 , \mu_0]$, $\lambda' \geq \lambda_0'$

4.10. $\exists\nu , M$ (indépendants de ε) tels que $\varepsilon^\nu|\sigma_A| \leq M$.

On pose

4.11. $Q = A + \lambda^{-1} B\frac{\partial}{\partial t}$

On a alors, P^* désignant l'adjoint de P ,

4.12. $P^* Q = (P + R^* - R)Q = [P,Q] + (R^* - R)Q + QP = H_0 + H_1\frac{\partial}{\partial t} + H_2P$

avec :

4.13. $H_0 = [R,A] - 2\lambda^{-1}\,[\frac{\partial}{\partial t}, B]R - [\frac{\partial}{\partial t}[\frac{\partial}{\partial t}, A]] + \lambda^{-1}B[R, \frac{\partial}{\partial t}] + (R^* - R)A$

4.14. $H_1 = -2\,[\frac{\partial}{\partial t}, A] + \lambda^{-1}[R,B] + \lambda^{-1}(R^* - R)B - \lambda^{-1}[\frac{\partial}{\partial t}, [\frac{\partial}{\partial t}, B]]$

4.15. $H_2 = 2\lambda^{-1}[\frac{\partial}{\partial t}, B] + Q$.

Dans la suite, on désignera par $\mathcal{N}$ des noyaux $\mathcal{N}(x,y,t,\mu,\lambda,\varepsilon)$, C^∞ à support compact en $(x,y) \in \Omega \times \Omega$, C^∞ en t près de 0 , définis pour $\mu \in]0,\mu_0]$, $\lambda' \geq \lambda'_0$ et qui vérifient

4.16. $\mathcal{N}(x,y,t\,;\mu,\lambda,\varepsilon) \leq \varepsilon^{-\nu}\, e^{-C_0\lambda}$

où ν , C^∞ sont des constantes indépendantes de ε , et des estimations identiques pour leurs dérivées.

Le calcul des commutateurs est classique. On a évidemment

4.17. $[\frac{\partial}{\partial t}, A] = \int_{\alpha\in V_\varepsilon} e^{\mu\lambda\psi}\, \pi_\alpha^\Omega \left(\mu\lambda\psi'_t\, \sigma_A + \frac{\partial\sigma_A}{\partial t}\right)$

et

Lemme 4.1.- On a

4.18. $[R,A] = \mu\lambda^2 \int_{\alpha\in V_\varepsilon} e^{\mu\lambda\psi}\, \pi_\alpha^\Omega\, s_A + \int_{\alpha\in\partial V} \lambda e^{\mu\lambda\psi}\, \pi_\alpha^\Omega\, t_A + \mathcal{N}$

où $s_A(x,y,t\,;\alpha,\mu,\lambda',\varepsilon)$, $t_A(x,y,t,\alpha,\mu,\lambda',\varepsilon)$ sont des symboles analytiques, classiques dans U , vérifiant des estimations 4.10 et en désignant par s_A^0 le terme principal, on a :

4.19. $s_A^0\Big|_{x=y=\alpha_x} = \frac{1}{i}\,\{r(t,\alpha)\,,\psi(t,\alpha)\}_\alpha\, \sigma_A^0\Big|_{x=y=\alpha_x}$

Preuve.- Par définition, on a :

4.20. $$[R,A]\,(f) = \int_{\alpha\in V_\varepsilon} e^{\mu\lambda\psi(\alpha,t)} \int_{y\in\Omega} [R(t,x,D_x) - {}^tR(t,y,D_y)]\,(\Pi_\alpha\sigma_\alpha\chi)(f)dy\ .$$

Alors on remarque que les termes de degré 1 en λ dans

$(R - {}^tR)(\pi_\alpha \sigma_A \chi f)$ rentrent dans les termes d'ordre inférieur de s_A ; en effet, si σ est un symbole, on a $\lambda\sigma = \mu\lambda^2(\frac{\sigma}{\lambda'})$. Il suffit donc pour obtenir 4.18, 4.19, d'intégrer par partie en α , une seule fois, pour le terme de degré deux qui est $\lambda^2[r(t,x,\frac{\partial\varphi}{\partial x}) - r(t,y,-\frac{\partial\varphi}{\partial y})]\ \pi_\alpha \sigma_A \chi$. Le terme d'erreur provient des dérivées de la troncature χ (remarquer que $\psi \leq \frac{3}{2}$ sur V_ε , et qu'on peut choisir μ_0 petit) =

On choisit alors le symbole σ_A pour annuler le terme H_1 dans 4.12. D'après 4.14 et le calcul des commutateurs, on a :

4.21 $$H_1 = -2[\frac{\partial}{\partial t}, A] + \mu\lambda \int_{\alpha\in V_\varepsilon} e^{\mu\lambda\psi}\, \pi_\alpha^\Omega\, s_B + \int_{\alpha\in\partial V} e^{\mu\lambda\psi}\, \pi_\alpha^\Omega\, t_B + \mathcal{N}$$

avec $s_B(x,y,t,\alpha,\mu,\lambda',\varepsilon)$, $t_B(x,y,t,\alpha,\mu,\lambda',\varepsilon)$ symboles analytiques, définis sur U classiques, et vérifiant 4.10 et

4.22. $$s_B^0 \Big|_{x=y=\alpha_x} = \frac{1}{i}\{r(t,\alpha)\ ,\psi(t,\alpha)\}_\alpha - \mu(\psi_t')^2$$

d'après 4.6. on a $\psi_t' = -\frac{1}{\varepsilon^2}$, et on choisit donc d'après 4.17, σ_A solution symbolique de l'équation

4.23. $$\sigma_A - \frac{\varepsilon^2}{\lambda'}\frac{\partial\sigma_A}{\partial t} = -\frac{\varepsilon^2}{2} s_B$$

par exemple

4.24. $$\sigma_A(x,y,t,\alpha,\mu,\lambda',\varepsilon) = -\frac{\lambda'}{2}\int_0^{t_0} s_B(t+\nu,x,y,\alpha,\mu,\lambda')\, e^{-\frac{\lambda'}{\varepsilon^2}\nu}\, d\nu$$

avec $t_0 > 0$ indépendant de ε , et assez petit pour que 4.24 soit défini. On a alors

4.25. $$H_1 = \mu\lambda \int_{\alpha\in V_\varepsilon} e^{\mu\lambda\psi}\, \pi_\alpha^\Omega\, h_1 + \int_{\alpha\in\partial V_\varepsilon} e^{\mu\lambda\psi}\, \pi_\alpha^\Omega\, t_B + \mathcal{N}$$

où le symbole $h_1(x,y,t,\alpha,\mu,\lambda',\varepsilon)$ vérifie

4.26. $$|h_1| \leq C^{te}\, \lambda'\, e^{-t_0\lambda'/\varepsilon^2}\, \varepsilon^{-\nu}$$

Maintenant, d'après 4.6, 4.7, pour $\alpha \in \partial V_\varepsilon$ on a soit $\ell = 2\delta\varepsilon$ et $\psi \leq -1$, soit $|\beta| = 2\varepsilon^2$ et aussi $\psi \leq -1$, soit $\ell = -\varepsilon^2$ et $\psi \leq \frac{3}{2}$. On pose donc à présent

4.27. $$\mu = \gamma\lambda^{\frac{1-s}{s}} \qquad \gamma > 0\ ,$$

de sorte que $\lambda' = \gamma\lambda^{1/s}$, alors d'après l'hypothèse 4.3 sur $SS_b^s(u)$, et la partie 1) de la proposition 3.2, on peut choisir $\gamma = \gamma(\varepsilon,u) > 0$, assez petit pour avoir :

4.28. $$|e^{\mu\lambda\psi(\alpha,t)}(\pi_\alpha^\Omega\sigma)(u)(t,x,\alpha,\mu,\lambda',\varepsilon)| \le \varepsilon^{-\nu} e^{-C_{(\varepsilon,u)}\lambda^{1/s}}$$

avec ν indépendant de ε, u, $C_{(\varepsilon,u)} > 0$, pour : $\alpha \in \partial V_\varepsilon$, $0 \le t \le 2\varepsilon^2$, $\lambda \ge \lambda(\varepsilon,u)$, et les symboles σ qui interviennent dans le calcul.

On a alors

4.29. $$0 = \int_0^{2\varepsilon^2} dt(Qu,Pu)_\Omega = \int_0^{2\varepsilon^2} dt(P^* Qu,u)_\Omega - \left[(Qu,\frac{\partial u}{\partial t})_\Omega\right]_{t=0}^{t=2\varepsilon^2} + \left[(\frac{\partial}{\partial t} Qu,u)_\Omega\right]_{t=0}^{t=2\varepsilon^2}$$

Pour $t = 2\varepsilon^2$, on a $\psi \le -\frac{1}{2}$ pour $\alpha \in V_\varepsilon$ et pour tout $\alpha \in V_\varepsilon$, $0 \le t \le 2\varepsilon^2$, $\psi \le \frac{3}{2}$, d'où d'après 4.12, les relations $Pu = 0$, $u|_{t=0} = 0$ et 4.26 en choisissant ε_0 assez petit et en remarquant que

4.30. $$\varepsilon^{-\nu} e^{-C_\varepsilon\lambda^{1/s}} \le e^{-\frac{C_\varepsilon}{2}\lambda^{1/s}} \quad \text{pour } \lambda \ge \lambda(\varepsilon) \ ;$$

on déduit de 4.29 (en supprimant la dépendance en u des constantes)

4.31. $\exists C_\varepsilon > 0 \ \lambda(\varepsilon)$, tels que pour $\lambda \ge \lambda(\varepsilon)$
$$\int_0^{2\varepsilon^2} dt(H_o u,u)_\Omega + \frac{1}{\lambda}(B\frac{\partial u}{\partial t},\frac{\partial u}{\partial t})_{\Omega,t=0} \le e^{-C_\varepsilon\lambda^{1/s}} .$$

D'après 4.13 et le calcul des commutateurs, on a

4.32. $$H_o = \mu\lambda^2 \int_{\alpha\in V_\varepsilon} e^{\mu\lambda\psi} \pi_\alpha^\Omega h_o + \lambda \int_{\alpha\in\partial V_\varepsilon} e^{\mu\lambda\psi} \pi_\alpha^\Omega \tilde{h}_o + \mathcal{N}$$

avec, d'après 4.19, 4.22, 4.23,

4.33. $$h_o^o|_{x=y=\alpha_x} = (\frac{1}{i}\{r,\psi\} - \mu\psi_t'^2)^2(-\frac{\varepsilon^2}{2}) + \frac{2}{\varepsilon^2} r(t,\alpha) .$$

Or on a $\{r,\psi\}_\alpha = H_{r_o}(\psi) + 0(|t||\nabla\psi|) = -\frac{1}{\varepsilon\delta_o} + 0(1)$ et $r(t,\alpha) = r_o(\alpha) + 0(|t|) \in 0(\varepsilon^2)$, d'où on déduit si $\mu \le \varepsilon^4$

4.34. $$h_o^o\Big|_{x=y=\alpha_x} = \frac{1}{2\delta_o^2} + 0(1) \quad \text{pour} \quad \alpha \in V_\varepsilon \ , \ 0 \le t \le 2\varepsilon^2$$

où le terme $0(1)$ est indépendant de ε, la phase dans

$$\int_{\alpha \in V_\varepsilon} e^{\mu\lambda\psi} \pi_\alpha^\Omega h_o$$

est $\phi = \varphi(x,y,\alpha) - i\mu\psi(\alpha,t)$, et les équations $\frac{\partial\phi}{\partial\alpha} = 0$ sont donc équivalentes à $x = \alpha_x + 0(\mu)$, $y = \alpha_x + 0(\mu)$. En intégrant par partie en α pour le terme principal de h_o, on se ramène donc à supposer d'après 3.34 et les estimations 4.10 sur h_o :

4.35. $$h_o^o = h_o^o(\alpha,t,\mu,\varepsilon) = \frac{1}{2\delta^2} + 0\left(\frac{\mu}{\varepsilon^\nu}\right) .$$

Alors puisque $h_o = h_o^o + \frac{1}{(i\lambda')} \tilde{h}$ avec $\tilde{h}$ vérifiant 4.10, en choisissant δ_o assez petit (indépendant de ε), on peut supposer qu'on a :

4.36. $$\mathrm{Re}\, h_o(x,y,\alpha,t,\mu,\lambda',\varepsilon) \ge 1$$

pour x,y,α,t dans U, $0 < \mu \le \mu_o(\varepsilon)$, $\lambda' \ge \lambda'(\varepsilon)$, $\alpha \in V_\varepsilon$, $0 \le t \le 2\varepsilon^2$. Le symbole h_o est donc strictement positif, en prenant la partie réelle de 4.31, et en utilisant le lemme 3.1 et 3.24, on a :

4.37. $$\mu\lambda^2 \int_0^{2\varepsilon^2} dt \int_{\alpha \in V_\varepsilon} e^{\mu\lambda\psi} \| \tilde{\pi}_\alpha^{\Omega''} \tilde{h}_o u \|_{\Omega''} + \frac{1}{\lambda} \int_{\alpha \in V_\varepsilon} e^{\mu\lambda\psi} \left\| \tilde{\pi}_\alpha^\Omega \frac{\partial u}{\partial t}\Big|_{t=0} \right\|_\Omega^2 \le e^{-C_\varepsilon \lambda^{1/s}} ,$$

avec $\mu = \gamma\lambda^{\frac{1-s}{s}}$ et $\lambda \ge \lambda(\varepsilon)$, $C_\varepsilon > 0$ où $\tilde{h}_o$ est strictement positif. Puisque $\psi(\alpha,t) \ge 0$ pour $|\beta| \le \varepsilon^2/2$, $t \le \varepsilon^2/4$, $\ell \le \frac{\varepsilon\delta_o}{2}$, on déduit de 4.37, tous les termes étant positifs

4.38. $$\int_0^{\varepsilon^2/4} dt \int_{|\beta| \le \frac{\varepsilon^2}{2}, \ell \le \frac{\varepsilon\delta_o}{2}} \| \tilde{\pi}_\alpha^{\Omega''} \tilde{h}_o u \|_{\Omega''}^2 (t,\mu,\lambda') \Big|_{\mu=\gamma\lambda^{\frac{1-s}{s}}} \le e^{-C_\varepsilon \lambda^{1/s}}$$

et on conclut d'après le point 2) de la proposition 3.2 qu'on a $\exp \ell H_{r_o}(\alpha^o) \notin SS_b^s(u)$ pour $\ell \le \frac{\delta_o\varepsilon}{2}$

C.Q.F.D.

5. PREUVE DU THÉORÈME 1.4.

Soit donc $s \geq 3$, u solution des équations 1.4 et ρ_0 un point de Σ_b tel que $\rho_0 \in SS_b^s(u)$. Soit $s \to \rho(s)$ *le* "rayon - C^∞" maximal dans Σ_b tel que $\rho(0) = \rho_0$. Comme $SS_b^s(u)$ est fermé, il suffit de prouver que si $\rho(s) \in SS_b^s(u)$ pour $s \in [0,s_0]$, alors $\rho(s) \in SS_b^s(u)$ pour $s - s_0$ assez petit. On peut supposer que $\rho(s_0)$ est un point glancing dans $T^* \partial M \setminus 0$, et travailler localement avec P sous la forme 4.1. On pose :

5.1. $$r_0(x,\xi) = r(0,x,\xi) \quad ; \quad r_1(x,\xi) = \frac{\partial r}{\partial t}(0,x,\xi) .$$

On suppose donc $r_0[\rho(s_0)] = 0$, et on pose $\rho(s_0) = (x_0,\xi_0)$. Soit, pour $\ell \geq 0$ petit

5.2. $$(x(\ell) \; , \; \xi(\ell) \; ; \; t(\ell) \; , \; \tau(\ell))$$

un rayon analytique issu de $\rho(s_0)$, i.e. vérifiant

5.3. $$(x(0) \; , \; \xi(0) \; , \; t(0) \; , \; \tau(0)) = (x_0 \, , \, \xi_0 \, , \, 0 \, , \, 0) .$$

Alors, de deux choses l'une :

5.4. : Soit $t(\ell) > 0$ pour $\ell > 0$ petit, et alors 5.2 est une bicaractéristique de P et vérifie donc les équations

5.5. $$\frac{dt}{d\ell} = 2\tau(\ell) \qquad \frac{d\tau}{d\ell} = -\frac{\partial r}{\partial t}(t(\ell) \, , \, x(\ell) \, , \, \xi(\ell))$$
$$\frac{dx}{d\ell} = \frac{\partial r}{\partial \xi}(t(\ell) \, , \, x(\ell) \, , \, \xi(\ell)) \qquad \frac{d\xi}{d\ell} = -\frac{\partial r}{\partial x}(t(\ell) \, , \, x(\ell) \, , \, \xi(\ell))$$

5.6. : Soit $t(\ell) \equiv 0$ pour $\ell > 0$ petit, et alors par définition d'un rayon (voir [7]), en notant $(y(\ell) \, , \eta(\ell))$ les coordonnées du rayon dans $T^* \; M \setminus 0$, on a :

5.7. $$\frac{dy}{d\ell} = \frac{\partial r_0}{\partial \xi}(y(\ell) \, , \eta(\ell)) \quad , \quad \frac{d\eta}{d\ell} = -\frac{\partial r_0}{\partial x}(y(\ell) \, , \eta(\ell)) .$$

Le premier cas, 5.4, se produit si et seulement si la demi-bicaractéristique de P issue (à droite) de $\rho(s_0)$ rentre dans $\mathring{M}$ (nécessairement à un ordre fini, par analyticité). On peut supposer qu'on est dans ce cas, sinon il n'existe qu'un seul demi-rayon issu à droite de $\rho(s_0)$ (celui donné par 5.6), et c'est donc *le* "rayon - C^∞" et on conclut par le théorème 1.3. Notons Γ le demi-rayon 5.4, et γ le demi-rayon 5.6. Alors Γ est le prolongement du "rayon - C^∞ " $\rho(s)$ à droite de s_0 , et on suppose donc (par l'absurde)

5.8. $\Gamma\setminus\rho(s_0) \cap SS_b^s(u) = \phi$.

Alors le théorème 1.3. implique, puisque $\rho(s_0) \in SS_b^s(u)$

5.9. $\gamma \subset SS_b^s(u)$.

Dans 5.4, on a $t(\ell) = c_0\, \ell^k + 0(\ell^{k+1})$ avec $k \geq 2$ et $c_0 > 0$. Alors d'après 5.5, 5.7, on en déduit

5.10 $(y(\ell)\, , \eta(\ell)) = (x(\ell)\, , \xi(\ell)) + 0(\ell^{k+1})$

donc en réutilisant 5.5.

5.11. $r_1(y(\ell)\, , \eta(\ell)) = -\frac{c_0}{2}\, k(k-1)\, \ell^{(k-2)} + 0(\ell^{k-1})$.

On en déduit que pour $\ell > 0$ petit, $(y(\ell)\, , \eta(\ell))$ est un point strictement diffractif. Fixons $\ell > 0$ petit, $SS_b^s(u)$ étant fermé, 5.8 entraîne pour $\sigma > 0$ petit

5.12 $\exp \sigma H_p(y(\ell)\, , \eta(\ell)\ ;\ 0\, , 0) \cap SS_b^s(u) = \phi$.

Puisque $s \geq 3$, le théorème 0 de [3], 5.12 et 1.4 impliquent

5.13. $(y(\ell)\, , \eta(\ell)) \notin SS_b^s(u)$

ce qui contredit 5.9
[dans [3], on a montré en fait que l'on a

5.14. $(y(\ell)\, , \eta(\ell)) \notin SS^s \left[\frac{\partial u}{\partial t}\Big|_{t=0}\right]$.

Mais on a puisque $u\big|_{t=0} = 0$

5.15. $P\bar{u} = (\frac{\partial u}{\partial t})\Big|_{t=0} \otimes \delta_{t=0}$.

Alors le théorème de propagation à l'intérieur, 5.12, 5.14 et 5.15 impliquent 5.13].

BIBLIOGRAPHIE

[1] F.G. FRIEDLANDER, R.B. MELROSE - *The wave front set of the solution of a simple initial-boundary value problem with glancing rays* II, Math. Proc. Camb. Phil. Soc. (81), 1977.

[2] L. HÖRMANDER - *Linear Partial differential Operators*, Springer, 1963.

[3] G. LEBEAU - *Régularité Gevrey 3 pour la diffraction*, C.P.D.E., (915), 1984.

[4] R.B. MELROSE, J. SJÖSTRAND - *Singularities of boundary value problem* I, C.P.A.M. (31), 1978.

[5] M. SATO, T. KAWAI, M. KASHIWARA - *Hyperfonctions and Pseudo-differential equations*, Lect. Notes, 287, Springer, 1971.

[6] P. SCHAPIRA - *Propagation at the boundary and reflection of analytic singularities of solution of linear partial differential equations*, Publ. R.I.M.S., Kyoto Univ. 12, Suppl., 1977.

[7] J. SJÖSTRAND - *Propagation of analytic singularities for second order Dirichlet problems*, C.P.D.E. (5), 1980.

[8] J. SJÖSTRAND - *Singularités analytiques microlocales*, Astérisque n° 95, 1982.

ABSTRACT. Since the works of Friedlander-Melrose, Melrose-Sjöstrand [4], Sjöstrand [7], it is well know that the propagation of singularities in boundary value problems, near the boundary glancing points, differs according to the considered C^∞ or analytic cases.

This article is devoted to the study of Gevrey G^σ singularity propagation : using Sjöstrand's technics, we show that Gevrey G^σ singularity behavior is the same as the C^∞, [resp. analytic], behavior when $\sigma \geq 3$, [resp. $1 \leq \sigma < \varepsilon$].

Conormal rings and semilinear wave equations

Richard B. Melrose
Department of Mathematics
Massachusetts Institute of Technology, Room 2-180
Cambridge Mass. 02139, U.S.A.

Abstract Bounded solutions to a semilinear waver equation, with initial data conormal at a finite number of points are discussed. A new analysis of the singularities of the solutions up to triple interactions is outlined and a general conjecture covering the conormality of the solution is made.

§1. Introduction

In these lectures the conormal, or iterative, regularity of solutions to semilinear wave equations will be examined, most ofthe original results described here were obtained in collaboration with Niles Ritter. The discussion will be concentrated on problems in two space dimensions, although essentially all the results, and the methods used to prove them, have direct extensions to higher dimensions. Whereas the situation in one space dimension is well understood, following the work of Rauch and Reed, much remains to be done even in two space dimensions.

Specifically we shall consider the flat wave operator in $\mathbb{R}^3$ or sometimes $\mathbb{R}^{n+1}$ for $n\geq 2$:

$$(1.1)\qquad P = D_t^2-D_x^2-D_y^2, \text{ or } P = D_{x_0}^2-D_{x_1}^2-\ldots-D_{x_n}^2,$$

and a semilinear equation of the simplest type:

$$(1.2)\qquad Pu = f(z,u),\ z=(t,x,y)\in\Omega,\ f\in C^\infty(\Omega\times\mathbb{R}),\ \Omega\subset\mathbb{R}^3 \text{ (or } R^{n+1}\text{) open.}$$

The obvious condition under which this equation makes distributional sense is if the solution is a measurable and locally bounded function on the open set Ω:

H. G. Garnir (ed.), Advances in Microlocal Analysis, 225–251.

(1.3) $u \in L^{\infty}(\Omega)$.

Here, and subsequently we do not include "loc" subscripts, since all spaces unless otherwise mentioned will be local. The general question of interest is the location of the singularities of u. For the linear case, when f is a linear function of u, this is well known. Let us suppose that Ω is P-convex with respect to the initial surface $\{t=0\}$; analytically this is just the requirement that if $\Omega_{\delta} = \Omega\cap\{t>\delta\}$ then for some $\delta_0>0$,

(1.4) $f\in C^{\infty}(\Omega)$, $\text{supp}(f)\subset\Omega_{\delta} \Rightarrow \exists\,!\ u\in C^{\infty}(\Omega)$ with $\text{supp}(u)\subset\Omega_{\delta}$ & $Pu=f$ in Ω.

As is shown below, if (1.1)-(1.3) hold, the Cauchy data of u is well defined on the initial surface

$$S = \Omega\cap\{t=0\},\ u_i=\gamma_i(u)=\gamma(D_t^i u)\in C^{-\infty}(S),\ i=0,1,$$

where γ is the restriction operator to S. Now suppose that this initial data is conormal with respect to some finite subset $L_0\subset S$ (see §3), i.e.

(1.5) $u_i \in I(S,N^*L_0)$ $i=0,1$

in the notation for the spaces of Lagrangian distributions introduced by Hörmander [Hö1]. If f=0, or is linear in u, then any solutions of (1.1) - (1.5) has wavefront set contained in the union of all the bicharacteristic curves of P passing over N^*L. More crudely this implies:

(1.6) $\text{singsupp}(u)\cap\Omega_o \subset E_1(L) .= \bigcup \{E(\bar{z});\ \bar{z}\in L_0\}$ (f=0)

where

(1.7) $E(\bar{z}) = \{(t,x,y)\in\Omega;\ [(x-\bar{x})^2+(y-\bar{y})^2]^{1/2} = t-\bar{t},\ \bar{x}=(\bar{t},\bar{x},\bar{y})\}$,

is the forward characteristic cone of P with pole at $\bar{z}$. To extend (1.6) to general C^{∞} f is (1.1) the right side must be considerably enlarged. Observe that the intersection of any three cones in the union on the right in (1.6) is either empty (if the poles are colinear) or else consists of one point. More generally suppose that we define for any finite set $L\subset\Omega_0$ the interaction set by:

$$L' = \bigcup \{M=E(z_1)\cap E(z_2)\cap E(z_3);\ M \text{ is finite } \&\ z_1,z_2,z_3\in L\}.$$

and then successively the k-fold interaction sets:

(1.8) $$L_{k+1} = L_k \cup (L_k)'.$$

In general:

(1.9) Conjecture If (1.1) - (1.5) holc then for any $s\in\mathbb{R}$ there exists $k\in\mathbb{N}$ such that

(1.10) $$\text{sing supp}(u) \subset \bigcup \{E(z); z\in L_k\},$$

where $\text{singsupp}_s(u)$ is computed with respect to $H^s(\Omega)$.

From the work of Bony [Bo1] this conjecture, under rather stronger hypotheses, is known in case #(L)=1 or #(L)=2, i.e. L consists of one or two points. In [MR1], and also by Bony in [Bo2], the conjecture is shown, again under somewhat stronger hypotheses, in case #(L)=3. In these cases one can strengthen the consclusion (1.10) as follows:

(1.11) Theorem If u satisfies (1.1) - (1.5) and $\#(L)\leq 3$, then

(1.12) $$\text{singsupp}(u) \subset \bigcup \{E(z);\ z\in L_1\}.$$

The proof of this result is described below, the method is somewhat different to that in [MR1] and it is hoped that it will extend to prove the Conjecture 1.9 in general. Some cases in which (1.12) remains valid even though #(L)=4 are also described. The simple regularity (1.12), even with 1 replaced by some finite k cannot be expected in general, since as an example in [MR2] illustrates the L_k do not stabilize locally and as $k\to\infty$ may have points of accumulation. To prove (1.10) in general some refinement of current understanding is needed; some possibilities are suggested in [MR2] and explored towards the end of these lectures.

§2. Initial conditions

First it needs to be checked that the Cauchy data is defined.

(2.1) Lemma If (1.1) - (1.4) hold, in any dimension, then for any $\varphi\in C_c^\infty(\Omega)$,

(2.2) $$\varphi u \in C(\mathbb{R};L^2(\mathbb{R}^n)),\ D_t(\varphi u) \in C(\mathbb{R};H^{-1}(\mathbb{R}^n)).$$

Proof Set $g(z)=f(z,u(z))\in L^2(\Omega)$. Standard energy estimates for the wave equation show that there is a unique solution to:

(2.3) $$Pv = g \text{ in } \Omega,\ \varphi v\in C^1(\mathbb{R};L^2(\mathbb{R}^n))\ \forall\ \varphi\in C_c^\infty(\Omega),\ \gamma_i v=0,\ i=0,1.$$

Let $w=u-v$ be the difference, it satisfies $Pv=0$ in Ω and certainly $v\in L^2(\Omega)$. From this the regularity (2.2) follows for v, and hence using (2.3) for $u=v+w$. To see the regularity for v, consider:

(2.4) $$P(\varphi v) = h = Vv+\psi v,\ V \text{ a } C^\infty \text{ vector field, } \psi\in C_c^\infty(\Omega).$$

Now, $h\in H_c^{-1}(\Omega)$ and both h and v have wavefront set in the characteristic variety of P. Thus, h is in the space $L^2(\mathbb{R};H^{-1}(\mathbb{R}^n))$; standard energy estimates, and the fact that v has compact support, show that (2.2) holds for v. This proves the Lemma.

Next we turn to the condition (1.6). This can be restated in a variety of ways. For present purposes the following iterative definition is useful, since it corresponds to the definition of the more general spaces treated below. Let $\mathcal{V}(L)$ be the space of C^∞ vector fields on S which vanish at L_0, i.e. with all coefficients vanishing at each of the finite number of points in L_0.

(2.5) Lemma For $u\in C^{-\infty}(S)$ $u\in I(S,N^*L_0)$ if and only if there exists $s\in\mathbb{R}$ such that:

(2.6) $$\mathcal{V}(L_0)^k \subset H^s(\Omega)\quad \forall\ k\in\mathbb{N}.$$

Proof See [Hö2].

Notice that we are not making any *a priori* assumption on the regularity of the initial data $\gamma_i(u)$, in the sense of the Sobolev regularity which is preserved, in the sense of (2.4), under the repeated action of vector fields vanishing at L_0. As will be shown below, this is not necessary because a certain amount of regularity is automatic.

§3. C^∞ initial data

The simplest case of Theorem 1.11 is when $L_0=\phi$, since this just correspond to smooth initial data.

(3.1) <u>Proposition</u> If u satisfies (1.1) - (1.5) and has $\gamma_i(u)\in C^\infty(S)$ for $i=0,1$, then $u\in C^\infty(\Omega)$.

The main reason that this elementary result is included here is that the proof, outlined below, serves as a primitive model for some of the later, more intricate, arguments. The proof, by induction, relies on two lemmas, the first of which is a standard form of the well-posedness of the Cauchy problem for P.

(3.2) <u>Lemma</u> If $s\geq 0$ and $Pu = f\in H^s(\Omega)$ has $\gamma_i(u)\in H^{s+i-1}(S)$, $i=0,1$, then $u\in H^{s+1}(\Omega)$.

To handle the nonlinear terms some multiplicative properties are needed. A linear subspace:

$$\mathcal{R} \subset L^\infty(\Omega),$$

will be called a C^∞ algebra if for any M and any $F\in C^\infty(\Omega\times\mathbb{R}^M)$,

$$(3.3) \qquad u_j\in\mathcal{R},\ j=1,..,M \Rightarrow F(z,u_1(z),\dots,u_M(z))\in\mathcal{R}.$$

As a corollary of the Gagliardo-Nirenberg inequalities:

(3.4) <u>Lemma</u> $L^\infty(\Omega)\cap H^s(\Omega)$ is a C^∞ algebra for any $s\geq 0$ and any open set $\Omega\subset\mathbb{R}^n$.

Using these two basic results we proceed to the proof of Proposition 3.1 by proving the relative version:

$$(3.5) \qquad (1.1) - (1.4),\ \gamma_i u \in H^{s-i}(S),\ i=0,1\ s\geq 0 \Rightarrow u\in H^s(\Omega).$$

Indeed this is trivially valid for $s=0$, since $u\in L^\infty(\Omega)$ by assumption hence $u \in L^2(\Omega)$. From Lemma 3.2 the solution to $Pv=0$ with the initial data $\gamma_i u$ lies in $H^s(\Omega)$. It therefore suffices to show that $w=v-u$, which satisfies (2.3), also lies in $H^s(\Omega)$. Now observe that

$$(3.6) \qquad t \leq S,\ w \in H^t(\Omega) \Rightarrow g=f(z,u) \in H^t(\Omega) \Rightarrow w \in H^{t+1}(\Omega)$$

where Lemma 3.5 is used in the first step and Lemma 3.2 for the second. Applying (3.6) repeatedly we conclude that $w \in H^{s+1}(\Omega)$. This completes the proof of the Proposition.

It is natural to enquire as to whether Proposition 3.1 remains valid without the strong hypothesis (1.1). The most immediate difficulty with this question is that the meaning in (1.2) of $f(z,u)$ is not then clear. Certain cases in which $f(z,u)$ nevertheless has a clear meaning ar discussed by Iwasaki & Nakamura [IN]. Here is a simple example, with $\Omega=\mathbb{R}^{n+1}$. Set

$$(3.7) \qquad u(t,x) = (t-1-i0)^{-1}.$$

Since $WF(u) \subset \{(1,x;\tau,0);\ \tau>0\}$ all finite powers of u are well-defined by continuous operations ([Höl]). Moreover,

$$(3.8) \qquad Pu = D_t^2 u = -2(t-1-i0)^{-3} = -2u^3.$$

Clearly the Cauchy data $\gamma_i u$, $i=0,1$, is C^∞, being constant, but u is not C^∞ throughout Ω. Thus the hypothesis (1.1) cannot be completely dropped in Proposition 3.1 and hence not in Theorem 1.11.

An interesting related question to which some of the analysis below, or more likely generalizations of it, might be applicable, concerns to global regularity of weak solutions to semilinear wave equations with positive conserved energy. This global existence was first observed by I. E. Segal. Suppose that

$$(3.9) \qquad f \in S^m(\mathbb{R}) \text{ is real and elliptic}$$

with

$$(3.10) \qquad F(T) = \int f(\tau)\ dt > 0 \quad \forall\ T \neq 0.$$

It is straightforward to show that:

(3.11) <u>Proposition</u> If u_0, $u_1 \in C_c^\infty(\mathbb{R}^n)$ $\exists$ $u \in L^{m+1}(\mathbb{R}^{n+1}) \cap H^1(\mathbb{R}^{n+1})$ (global spaces) such that (1.2) holds and $\gamma_i u = u_i$, $i=0,1$.

If m is sufficiently small with respect to the dimension, n, it follows that $u \in C^\infty(\mathbb{R}^{n+1})$. Is this always the case or can such a solution become singular in a finite time ?

§4. Conic expanding waves

In case $\#(L_0)=1$ in Theorem 1.11 the solution u is an expanding wave, with singular support contained in the characteristic cone with pole $\bar{z} \in L_0$, $E(\bar{z})$. To prove this it will be shown that u is actually conormal with respect to $E(\bar{z})$, or more precisely with respect to the corresponding C^∞ variety:

(4.1) $$\mathcal{E}(\bar{z}) = \{E(\bar{z})\backslash\{\bar{z}\},\{\bar{z}\}\},$$

consisting of the C^∞ submanifolds into which $E(\bar{z})$ naturally decomposes.

First recall ([DH]) that the solution to the linear equation Pv=0 with conormal initial data is a Lagrangian distribution:

(4.2) $$Pv=0, \text{ in } \Omega, \ \gamma_i v \in I(S,T_{\bar{z}}^*S) \Rightarrow v \in I(\Omega,\Lambda(\bar{z}))$$

where $\Lambda = \Lambda(\bar{z})$ is the closure of the conormal bundle to $E(\bar{z})\backslash\{\bar{z}\}$:

(4.3) $$\Lambda = \{(z;\tau,\xi,\eta); z \in E(\bar{z})\backslash\{\bar{z}\} \ (\tau,\xi,\eta) \in N_{\bar{z}}^* E(\bar{z}) \text{ or } z=\bar{z} \text{ and } \tau^2=\xi^2+\eta^2\}.$$

Consider the space $\mathcal{V}(\mathcal{E}(\bar{z}))$ of those C^∞ vector field in Ω which are tangent to $\mathcal{E}(\bar{z})$:

(4.4) $$V \in \mathcal{V}(\mathcal{E}(\bar{z})) \Leftrightarrow V \text{ is } C^\infty \text{ and tangent to } E(\bar{z})\backslash\{\bar{z}\},$$

since then V automatically vanishes at $\bar{z}$. Each of these vector fields is characteristic on Λ, so from (4.2) it follows that there exists some $s \in \mathbb{R}$ such that:

(4.5) $\mathcal{V}(\varepsilon(\bar{z}))^k v \subset H^s(\Omega) \quad \forall\ k.$

This type of iterated regularity is strictly weaker than the Lagrangian condition (4.2). It is frequently encountered in the discussion below so we introduce notation to describe such spaces (se [MR1]).

Let $\mathcal{V}$ be a linear space of C^∞ vector fields on Ω such that

(4.6) $[V,W] \in \mathcal{V} \quad \forall\ V,W \in \mathcal{V},$

i.e. $\mathcal{V}$ is a Lie algebra. In addition consider the conditions:

(4.7) $F\in C^\infty(\Omega),\ V\in\mathcal{V} \Rightarrow \varphi V\in\mathcal{V}$

that $\mathcal{V}$ be a C^∞ module and

(4.8) $\forall\ K\subset\subset\Omega\ \exists\ V_i\in\mathcal{V},\ i=1,..,k \text{ s.t. } \{V\in\mathcal{V};\operatorname{supp}(V)\subset K\} \subset C_c^\infty(\Omega).\mathcal{V},$

that $\mathcal{V}$ be locally finitely generated as a C^∞ module. These three conditions on $\mathcal{V}$ will be denoted <u>LA</u>, i.e.

(4.9) $\mathcal{V}$ is <u>LA</u> if (4.5) - (4.8) hold.

For the main spaces of iterated regularity with respect to $\mathcal{V}$:

(4.10) $I_kL^2(\Omega,\mathcal{V}) = \{u\in L^2(\Omega);\ \mathcal{V}^j u \subset L^2(\Omega)\ \forall\ j\leq k\},$

it is natural to suppose that $\mathcal{V}$ satisfies (4.5) and (4.6) at least. As a replacement for Lemma 3.4 the following result was proved in [MR1].

(4.11) <u>Lemma</u> If $\mathcal{V}$ is <u>LA</u> in Ω then

(4.12) $L^\infty I_kL^2(\Omega,\mathcal{V}) = L^\infty(\Omega)\cap I_kL^2(\Omega,\mathcal{V})$

is a C^∞ algebra for any $k\in\mathbb{N}$, as is:

(4.13) $L^\infty IL^2(\Omega,\mathcal{V}) = \bigcap L^\infty I_k L^2(\Omega,\mathcal{V})$.

It is also straightforward to check that

(4.14) $\mathcal{V}(\mathcal{E}(\bar{z}))$ is LA.

Using these definitions we can now state a strenthened form of Theorem 1.11 in case $\#(L_0)=1$, again we drop any restriction on the dimension of the space.

(4.15) Proposition If (1.1) - (1.5) hold and $\#(L_0)=1$, $L_0=\{\bar{z}\}$, then

(4.16) $u \in H^{\frac{1}{2}n-\epsilon}(\Omega) \cap L^\infty IL^2(\Omega,\mathcal{E}(\bar{z})) \ \forall\ \epsilon>0$.

The proof of this result is outlined in §6 below, after some preparatory embedding and interpolation results for conormal functions have been gived.

§5. Interpolation and embedding for conormal spaces

Conormal spaces can be defined by iterative regularity with respect to some Sobolev space, on the other hand for the main part we shall work only with the L^2 based spaces (4.13). The reason that the simplification is reasonable is that little is lost in the way of regularity. Consider the weak Sobolev space of order s:

(5.1) $H^{s-}(\Omega) = \bigcap_{r<s} H^r(\Omega)$.

(5.2) Proposition For any s>0, and any LA $\mathcal{V}$,

(5.3) $u \in H^s(\Omega) \cap I_k L^2(\Omega,\mathcal{V}) \Rightarrow \mathcal{V}^j u \subset H^t(\Omega)$, $t+j/k=s$,

hence in particular,

(5.4) $H^{s-}(\Omega) \cap IL^2(\Omega,\mathcal{V}) = IH^{s-}(\Omega,\mathcal{V})$.

Similar results can be found for L^p spaces, however for present purposes the following improved form, and partial inverse, of the Sobolev embedding theorem is more important.

(5.5) Proposition If $G \subset \Omega \subset \mathbb{R}^n$ is a C^∞ submanifold of codimension d and $IL^2(\Omega,G)$ is the space defined by (4.10) with $\mathcal{V} = \mathcal{V}(G)$ the space of all C^∞ vector fields tangent to G, then:

(5.6) $$IL^2(\Omega,G) \cap L^q(\Omega) \subset H^s(\Omega),\ IL^2(\Omega,G) \cap H^s(\Omega) \subset L^q(\Omega),$$

provided that $2<q<\infty$, $0<s<\frac{1}{2}d$ and

(5.7) $$1/q + s/d < 1/2.$$

§6 Proof of Proposition 4.15

There is a notion, introduced in [MR1], which is very useful in proving the second part of (4.16), the iterated regularity. Namely if $\mathcal{V}$ satisfies the condition LA then it is said to be P-complete if it has the following commutation property with respect to P:

(6.1) $$\mathcal{V} \text{ is P-complete} \iff [P,\mathcal{V}] \subset \Psi^0(\Omega)\cdot P + \Psi^1(\Omega)\cdot\mathcal{V} + \Psi^1(\Omega).$$

Here $\Psi^m(\Omega)$ is the space of pseudodifferential operators of order m on Ω and $\Psi^1(\Omega)\cdot\mathcal{V}$ is the space of operators which are locally finite sums of products. Again it is readily shown (see [MR1]) that:

(6.2) $\mathcal{V}(\mathcal{E}(\bar{z}))$ is P-complete, with P given by (1.1).

To start the proof of Propostion 4.15 we first deduce some Sobolev regularity for the solution by examining the splitting $u=v+w$, where v is fixed as the solution of (2.3), $g=f(\cdot,u(\cdot))$, thus $v \in H^1(\Omega)$. From the standard Sobolev embedding theorem this implies:

(6.3) $$v \in L^q(\Omega),\ 1/q > 1/2 - 1/(n+1).$$

Since $u \in L^\infty(\Omega)$, w has the same regularity. Now w satisfies the equation:

(6.4) $$Pw = 0 \text{ in } \Omega,\ \gamma_i(w) = \gamma_i(u),\ i=0,1.$$

By assumption the initial data is conormal at the point $\bar{z}$, so (4.2) applies. In particular, w is a conormal distribution with respect to the C^∞ hypersurface $E(\bar{z})$ in Ω_0, i.e. $\{t>0\}$. From Proposition 5.5 it follows that:

$$w \in H^s(\Omega_0), \text{ for } s<s_1=1/2(n+1). \tag{6.5}$$

Since H^s regularity is preserved (i.e. propagated) for solutions of the wave equation it follows from (6.5) and (6.4) that $w \in H^s(\Omega)$ for $s<s_1$. Since $s_1<1$, this implies the same regularity for u, hence $g \in H^s(\Omega)$, using Lemma 3.4. Starting again from (2.3) this gives improved regularity for v, etc. The argument can be iterated until the upper limit of regularity in (5.7), for d=1, provides an obstruction. Thus,

$$w \in H^{½-}(\Omega),\ v \in H^{3/2-}(\Omega), \tag{6.6}$$

independent of dimension.

To show that v has iterated regularity with respect to $\mathcal{V}(\mathcal{E}(\bar{z}))$ the commutation property P-complete can be used. Indeed, $\mathcal{V}(\mathcal{E}(\bar{z}))$ is finitely generated. Let $V_1,..,V_M$ be a set of generators and for a multiindex $\alpha=(\alpha(1),..,\alpha(M))$ set

$$V^\alpha = V_1^{\alpha(1)}\dots V_M^{\alpha(M)},\quad v_k = \{V^\alpha v;\ |\alpha|\le k\}.$$

Then (see [MR1]) v_k satifies a system of equations:

$$\mathcal{P}_k v_k = g_k,. \tag{6.7}$$

Here $\mathcal{P}_k$ is a square array of pseudodifferential operators of second order, with diagonal principal part PxId. By an inductive argument it will be shown that

$$v_k \in H^1(\Omega) \text{ and so } g_{k+1} \in L^2(\Omega). \tag{6.8}$$

That is, we need to show that (6.8), with k=p, implies the same regularity with k increased to p+1 throughout. Since Lemma 3.2 applies equally well to the system (6.7), with p=k+1, it is only necessary to find suitable regularity for the initial data, namely to show that:

$$(6.9)\qquad (6.8) \Rightarrow \gamma_i(v_{k+1}) \in H^{1-i}(S),\ i=0,1.$$

There are two components to demonstrating (6.9). First, the differential equation (2.3) gives a formula for the initial data of v_{k+1}. Namely, for any j,

$$(6.10)\qquad \gamma\left[D_t^{2j+1}v\right]_S = \gamma\left[D_t^{2j-1}+D_t^{2j-3}\Delta+\ldots+D_t\Delta^{j-1})g\right],$$

and

$$(6.11)\qquad \gamma\left[D_t^{2j}v\right] = \gamma\left[(D_t^{2j-2}+D_t^{2j-4}\Delta+\ldots+\Delta^{j-1})g\right].$$

Here the terms $D_t\Delta^j v$ and $\Delta^j v$ are missing because they both vanish on S. Now, g is a C^∞ function of z and u, and (6.8) implies that

$$(6.12)\qquad v \in I_{k+1}L^2(\Omega,\mathcal{V}(\mathcal{E}(\bar{z}))),$$

so the same is obviously true of u, and hence of g. All the vector fields in $\mathcal{V}(\mathcal{E}(\bar{z}))$ vanish at $\bar{z}$, so from (6.10) and (6.11) it follows that the Cauchy data of v_{k+1} can be written in the form:

$$(6.13)\qquad \gamma_i(v_{k+1}) = \gamma\left[\sum_{|\beta|\le k-2+i} c_{i,\beta}(z)D_z^\beta g\right],\ i=0,1,$$

where the coefficients $c_{i,\beta}(z)$ vanish at $\bar{z}$ to order $|\beta|+2-i$ at least. The expression in parentheses in (6.13) is a finite sum of terms of the form $c_r e_r$ where each c_r is C^∞ and vanishes at $\bar{z}$ to order 2-i, and $e_r \in I_{2-i}L^2(\Omega,\mathcal{V}(\mathcal{E}(\bar{z})))$. Thus consider the following Lemma. which can proved by an application of blow up techniques as in §9 below.

(6.14) <u>Lemma</u> If $e\in I_kL^2(\Omega,\mathcal{E}(\bar{z}))$ and S is a space-like hypersurface through $\bar{z}$ then there exists $e'\in I_kL^2(\Omega,\{\bar{z}\})$ such that e=e' in a conic neighbourhood of $S\backslash\{\bar{z}\}$.

Applying this to (6.13) we conclude that:

$$\gamma_i(v_{k+1}) = \sum \gamma[c_{r,i}(z)e'_{r,i}], \text{ in } S\backslash\{\bar{z}\}, \tag{6.15}$$

with $e'_{r,i} \in I_{2-i}L^2(\Omega,\{\bar{z}\})$, where the sum is finite and each $c_{r,i}$ is C^∞ and vanishes to order 2-i at $\bar{z}$. Notice however that such a product $c_{r,i}e'_{r,i} \in H^{2-i}(\Omega)$, so the standard restriction properties of Sobolev spaces give the desired conclusion, that (6.9) holds, in fact even a little more, except for the unlikely possibility of some additional term supported at $\bar{z}$ in the initial data. However since the inductive hypeothesis implies that $v_{k+1} \in L^2(\Omega)$, it is easy to rule this out on regularity grounds. Thus the inductive step (6.9) is completed and so the second part of (4.16), the iterative regularity, has been established.

The proof of the Sobolev regularity also involves Lemma 6.14. Indeed, we have now shown that $u \in IL^2(\Omega,\mathcal{E}(\bar{z}))$, so applying Lemma 6.14 there exists u' such that u=u' in a conic neighbourhood of $S\backslash\{\bar{z}\}$ and $u' \in IL^2(\Omega,\{\bar{z}\})$. Taking a smaller conic cutoff, which does not affect the conormal regularity, u' can be assumed bounded as well. Then applying Proposition 5.5, $u' \in H^{\frac{1}{2}(n+1)-}(\Omega)$. Thus, the restrictions $\gamma_i(u') = \gamma_i(u) \in H^{\frac{1}{2}n-i-}(S)$, since again there can be no terms supported at $\bar{z}$. Applying the results of §3, the first part of (4.16) follows. This completes the proof of Proposition 4.15.

§7 Cutoff near t=0

Notice that the Sobolev regularity for the solution deduced in (4.16) implies that

$$\rho u \in C(\mathbb{R}_t; H^{\frac{1}{2}n-}(\mathbb{R}^n)) \ \forall \ \rho \in C_c^\infty(\Omega). \tag{7.1}$$

This is almost, but not quite, enough to imply the original assumption of boundedness. This regularity for u in the case $\#(L_0)=1$ applies in some strip $|t|<\delta$, $\delta>0$, in the general case, by the finite speed of propagation. Thus the regularity (7.1) holds near each $\bar{z}_i \in L_0$. By cutting off near $t=\delta$ this allows the initial data to be conveniently removed.

(7.2) <u>Proposition</u> If (1.1) - (1.5) hold then, given $\epsilon>0$, $\exists\ \bar{w}\in H^{\frac{1}{2}n+1-}(\Omega)$ and

$$(7.3)\qquad \bar{v}\in L^{\infty}(\Omega)\cap H^{\frac{1}{2}n-}(\Omega)\cap\Big[\sum_{j=1}^{N} IL^{2}(\Omega,\{\bar{z}_j\})\Big],\quad L_0=\{\bar{z}_1,\dots,\bar{z}_N\},$$

both supported in $\Omega_0 = \Omega\cap\{t>0\}$ such that

$$(7.4)\qquad P\bar{w} = f(z,\bar{v}+\bar{w}) \text{ in } \Omega$$

and

$$(7.5)\qquad u = \bar{v}+\bar{w} \text{ in } \Omega_{\epsilon}=\Omega\cap\{t>\epsilon\}.$$

<u>Proof</u> Choose $\varphi\in C^{\infty}(\mathbb{R})$ with $\varphi(t)=0$ in $t<\delta$, $\varphi(t)=1$ in $t>2\delta$, where $\delta>0$ is chosen so that no two cones $E(\bar{z}_j)$ meet in Ω in $|t|<4\delta$ and $2\delta<\epsilon$. Then set $u_1=\varphi(t)u$ and note that

$$(7.6)\qquad Pu_1 = \varphi(t)f(z,u)+g_1$$

where g_1 has support in $\delta\le t\le 2\delta$ and

$$(7.7)\qquad g_1 \in H^{\frac{1}{2}n-1-}(\Omega)\cap\Big[\sum_{j=1}^{N} IL^{2}(\Omega,\{\bar{z}_j\})\Big],$$

by Proposition 4.15. Set $g = g_1+\varphi(t)f(z,u)-f(z,u_1)$, this has the same regularity (7.7). Then solve:

$$(7.8)\qquad P\bar{v} = g,\ \bar{v} = 0 \text{ in } t<0.$$

Now, the Sobolev reguarity for $\bar{v}$ is immediate from (7.8) and (7.7), for g. Moreover, since both u and u_1 are bounded, by hypothesis, the iterative argument used in §4 applies, with the considerable simplification in that there is no initial data. This proves (7.3) except for the local boundedness. Notice however that (7.4) certainly holds, so $\bar{w}\in H^{\frac{1}{2}n+1-}(\Omega)$, which implies in particular that it is locally bounded. Since $u_1 = \bar{v}+\bar{w}$ is bounded so is $\bar{v}$. This completes the proof of the Proposition.

§8 P-propagative algebras

The simple commutation methods used above are basic to the proof of iterative regularity, but need to be extended somewhat. Let us extract a slightly abstract form of the argument (see [MR1], [MR2]). Suppose that $J_k \subset L^2(\Omega)$ is a decreasing sequence of spaces each of which is a $C^\infty(\Omega)$-module:

(8.1) $$J_k \supset J_{k+1},\ C^\infty(\Omega)J_k = J_k \ \forall\ k\in\mathbb{N}.$$

Then the J_k are said to be P-propagative if

(8.2) $$Pu = f \in J_k,\ \operatorname{supp}(u) \subset \Omega_0 = \Omega\cap\{t>0\} \Rightarrow u \in J_{k+1} \ \forall\ k\in\mathbb{N}.$$

This condition is connected with the notion of P-completness for a Lie algebra in a simple way, reflected in the proofs above:

(8.3) Proposition If $\mathcal{V}$ is P-complete then the spaces $I_k L^2(\Omega,\mathcal{V})$ form a P-propagative sequence.

Thus, the commutation argument used in the preceeding section, in the proof of Proposition 7.2, is a consequence of this more general result applied locally. This uses the finite speed of propagation of the wave operator. There is another, more direct argument by superposition which is important subsequently.

(8.4) Lemma If $J_k^{(i)}$ are for $i=1,\ldots,m$ P-propagative then the sum:

(8.5) $$J_k = \sum_{i=1}^{m} J_k^{(i)}$$

is also P-propagative.

Proof If $Pu = f\in J_k$, where J_k is defined by (8.5) then:

$$(8.6)\qquad f = \sum_{i=1}^{m} f_i,\ f_i \in J_k^{(i)}.$$

Now by assumption u has support in Ω_0, hence so does f. Since the $J_k^{(i)}$ are C^∞ modules it can be ensured, by multiplying through (8.6) by a suitable C^∞ function $\varphi(t)$, that all the f_i have supports in Ω_0. Then the well-posedness of the Cauchy problem for P shows that there is a unique solution, u_i, with support in Ω_0 of the each equation $Pu_i = f_i$ in Ω. The P-propagative condition on the component spaces implies that $u_i \in J_{k+1}^{(i)}$, hence $f \in J_{k+1}$ as required to prove the Lemma.

Of course to deduce that the solution of a semilinear equation, (1.2), lies in such a space J_k some multiplicative propery is needed. We shall say that the J_k form a <u>P-propagative algebra</u> if in addition to being P-propagative each of the spaces defines a C^∞ ring $L^\infty J_k = L^\infty(\Omega) \cap J_k$. An easy argument of the type above now gives:

(8.8) <u>Lemma</u> If J_k is a P-propagative algebra, $\bar{v} \in L^\infty J_k$ and $\bar{w} \in L^\infty(\Omega)$ have support in Ω_0 and (7.4) holds then $\bar{w} \in L^\infty J_{k+1}$.

Combining these two ideas of a P-propagative algebra and the additivity of P-propagative spaces leads to the following simple way of constructing new P-propagative algebras:

(8.9) <u>Proposition</u> If $J_k^{(i)}$ are for $i=1,\ldots,m$ P-propagative algebras and each sum J_k defined by (8.5) is such that $L^\infty J_k = L^\infty(\Omega) \cap J_k$ is a C^∞ ring then J_k is a P-propagative algebra.

Although this is trivial, since the definition of a P-propagative algebra is essentially being assumed here there is a small point to watch. Namely in most practical circumstances we shall need the decomposition property

$$(8.10)\qquad L^\infty J_k = L^\infty(\Omega) \cap \left[\sum_{i=1}^{m} J_k^{(i)} \right] = \sum_{i=1}^{m} L^\infty J_k^{(i)},$$

in order to show that $L^\infty J_k$ is a C^∞ algebra.

§9. Transversal interaction

The case of Theorem 1.11 with $\#(L_0)=2$ is reduced to little more than a routine calculuation by using the notion of a P-propagative algebra. If $\mathcal{E}(\bar{z}_1)$ and $\mathcal{E}(\bar{z}_2)$ are the two cones, then $\mathcal{V}(\mathcal{E}(\bar{z}_1)\sqcup\mathcal{E}(\bar{z}_2))$ consists of all the vector fields tangent to both $\mathcal{E}(\bar{z}_1)$ and $\mathcal{E}(\bar{z}_2)$; the space defined by (4.10) for this Lie algebra is denoted $IL^2(\Omega,\mathcal{E}(\bar{z}_1)\sqcup\mathcal{E}(\bar{z}_2))$.

(9.1) Proposition If (1.1) – (1.5) hold with $\#(L_0)=2$, in any dimension, then u = v+w where:

(9.2) $$v \in L^\infty(\Omega)\cap H^{\frac{1}{2}n-}(\Omega)\cap\left[\sum_{j=1}^{2} IL^2(\Omega,\{\bar{z}_j\})\right]$$

and

(9.3) $$w \in H^{\frac{1}{2}n+1-}(\Omega)\cap IL^2(\Omega,\mathcal{E}(\bar{z}_1)\sqcup\mathcal{E}(\bar{z}_2)).$$

Proof The routine calculation referred to above is to show that $\mathcal{V}(\mathcal{E}(\bar{z}_1)\sqcup\mathcal{E}(\bar{z}_2))$ is P-complete, see [MR1]. Then it is only necessary to apply Lemma 4.11 and Proposition 8.3 to conclude that the spaces $IL^2(\Omega,\mathcal{E}(\bar{z}_1)\sqcup\mathcal{E}(\bar{z}_2))$ is a P-propagative algebra, then the desired result follows by applying Lemma 8.8 to the form of the solution obtained in Proposition 7.2

Of course Theorem 1.11 is an immediate consequence of this result in case $\#(L_0)=2$. There are other cases in which it can be applied. For example for any number of initial points all lying on one straight line there are never any intersections of cones more than in pairs, i.e.

(9.4) If L_0 is contained in a line then $L_1=L_0$.

The arguments above can then be applied locally, or else Proposition 8.9 can be applied directly to show that the sum:

$$(9.5)\qquad J_k = \sum_{r\neq j=1}^{N} I_k L^2(\Omega, \varepsilon(\bar z_j)\sqcup\varepsilon(\bar z_r)),$$

is a P-propagative algebra.

Before examining the interesting case of $\#(L_0)=3$, just note briefly a decomposition of the space $I_k L^2(\Omega,\varepsilon(\bar z_1)\sqcup\varepsilon(\bar z_2))$. The intersection of these two cones, $L_{12} = E(\bar z_1)\cap E(\bar z_2)$, is a C^∞ manifold of codimension two, since the intersection is transversal. Consider then the two C^∞ varieties:

$$(9.6)\qquad \mathscr{C}_i = \{\varepsilon(\bar z_i)\backslash L_{12}, L_{12}\},\ i=1,2,$$

and define the space $I_k L^2(\Omega,\mathscr{C}_i)$ by iterated regularity with respect to the vector fields tangent to $\mathscr{C}_i$. Then,

$$(9.7)\qquad I_k L^2(\Omega,\varepsilon(\bar z_1)\sqcup\varepsilon(\bar z_2)) = I_k L^2(\Omega,\mathscr{C}_1) + I_k L^2(\Omega,\mathscr{C}_2).$$

Of course a similar decomposition follows for the space (9.5). We shall briefly describe two proofs of (9.7). We do this to illustrate the relationship between a purely microlocal approach and a more geometric approach using ideas of blow up. In the discussion of the case $\#(L_0)=3$ below these two approaches become intermingled.

The microlocal proof of (9.7) starts from the observation that all these spaces are microlocal (so define fine sheaves over $T^*\Omega$). In particular if Q is a properly supported pseudodifferential operator of order zero then Q acts on each of the three spaces in (9.7). This just involves the L^2-boundedness of Q and a simple commutation argument. Of course (9.7) needs only to be verified near each point, and is trivial away from L_{12}, so we can think of the two cones as smooth hypersurfaces.

Next observe that, in the language of [MR1], each of the three C^∞ varieties involved is microlocally complete. That is, the space of C^∞ functions on $T^*\Omega'$, where Ω' is some small open set meeting L_{12}, which vanish on the conormal bundle of each element of the variety is generated, as a C^∞ module, by the vector fields tangent to the variety. Choose a properly supported pseudodifferential operator Q_1 of order zero in Ω' with

$$(9.8)\qquad WF'(Q_1)\cap N^*(E(\bar{z}_2)) = \phi,\ WF'(Q_2)\cap N^*(E(\bar{z}_1)) = \phi,\ Q_2=\mathrm{Id}-Q_1,$$

which is possible since these sets are disjoint. Then it follows from these observations that:

$$(9.9)\qquad Q_i : I_k L^2(\Omega', \mathcal{E}(\bar{z}_1)\sqcup\mathcal{E}(\bar{z}_2)) \to I_k L^2(\Omega', \mathcal{C}_i),$$

from which (9.7) follows directly.

The proof using blow up techniques involves the introduction of polar coordinates. To simplify the discussion we can introduce local coordinates $x,y\in\mathbb{R}$, $z\in\mathbb{R}^q$ near any given point of L_{12} so that:

$$(9.10)\qquad L_{12} = \{x=y=0\},\ C_1 = \{x=0\},\ C_2 = \{y=0\},$$

a model variety with normal crossings. Then consider the polar coordinate map in the first two variables:

$$(9.11)\qquad B: X=[0,\infty)\times S^1\times\mathbb{R}^q \to \mathbb{R}^{q+2},\ B(r,\theta,z) = (r\cos\theta, r\sin\theta, z),\ \theta\in\mathbb{R}/\mathbb{Z}.$$

The lifts of the three surfaces in (9.10), under B, into the manifold with boundary X can be described in terms of the submanifolds:

$$(9.12)\qquad R=\{r=0\}=B^{-1}(L_{12}),\ G_i=\mathrm{cl}[B^{-1}(C_i)\backslash R)],\ G_1=\{\cos\theta=0\},\ G_2=\{\sin\theta=0\}.$$

Namely C_i lifts to $R\cup G_i$. Thus we can interpret the C^∞ varieties:

$$(9.13)\qquad \mathcal{G}_i = \{G_i\backslash R, R\backslash G_i, G_i\cap R\},\ i=1,2,\ \mathcal{G} = \mathcal{G}_1\sqcup\mathcal{G}_2,$$

as the lifts to X, under B, of the three C^∞ varieties $\mathcal{C}_i$, i=1,2 and $\mathcal{C}_1\sqcup\mathcal{C}_2$. Under B, $L^2(\Omega')$ lifts to $L^2_B(X)$, the L^2 space with respect to the polar measure $r dr d\theta$. Moreover it is easily verified (cf. computations in [MR2]) that the Lie algebras $\mathcal{V}(\mathcal{C}_i)$ and $\mathcal{V}(\mathcal{C}_1\sqcup\mathcal{C}_2)$ lift under B to C^∞ vector fields which span, as $C^\infty(\Omega')$ modules, the Lie algebras $\mathcal{V}(\mathcal{G}_i)$ and $\mathcal{V}(\mathcal{G})$, respectively. From this it follows that:

(9.14) $$B^*: I_kL^2(\Omega', \mathcal{C}_i) = I_kL^2_B(X, \mathcal{G}_i), i=1,2, \; B^*: I_kL^2(\Omega', \mathcal{C}_1 \sqcup \mathcal{C}_2) = I_kL^2_B(X, \mathcal{G}).$$

Thus the proof of (9.7) is reduced to the corresponding identity on X:

(9.15) $$I_kL^2_B(X, \mathcal{G}) = I_kL^2_B(X, \mathcal{G}_1) + I_kL^2_B(X, \mathcal{G}_2).$$

Now (9.15) is trivial, since near each point of X one of the spaces on the right contains the other and is equal, locally to the space on the left. This completes the second proof of (9.7).

One advantage of the second proof of (9.7) is that it shows directly that the sum on the right in (9.7) is a C^∞ algebra, which might not be apparent without the identity itself. This is the type of proof used below to show that sums of rings are rings.

§10. The case $\#(L_0)=3$

First we consider the geometry of the triple intersection in order to define the space which will be shown to be a P-propagative algebra. If n=2 and L_0 consists of three points which are not colinear then:

(10.1) $$L_0 = \{\bar{z}_1, \bar{z}_2, \bar{z}_3\} \Rightarrow L_1 = L_0 \cup \{\bar{z}_4\}.$$

Near $\bar{z}_4$ the three cones $E(\bar{z}_i)$, i=1,2,3 meet transversally, thus:

(10.2) $$E(\bar{z}_i) \cap E(\bar{z}_j) = L_{ij}, \; i \neq j, \; E(\bar{z}_1) \cap E\bar{z}_2) \cap E(\bar{z}_3) = \{\bar{z}_4\},$$

where the $L_{ij} = L_{ji}$ give three distinct lines meeting at $\bar{z}_4$, with tangents spanning $\mathbb{R}^3$ there. The discussion of (9.7) above can easily be extended to handle this case of normal intersection. In particular we have the obvious C^∞ varieties locally near $\bar{z}_4$:

(10.3) $$\mathcal{H}_{ij} = \{E(\bar{z}_i) \backslash L_{ij}, L_{ij} \backslash M, M\}, \; 1 \leq i < j \leq 3, \; M = E(\bar{z}_1) \cap E(\bar{z}_2) \cap E(\bar{z}_3),$$

where we use the notation $\mathcal{H}$ to express the smoothness of $E(\bar{z}_i)$ locally. Of course if n=2 then $M=\{\bar{z}_4\}$, but if n>2 then M is a submanifold of codimension three, but in any case (10.3) gives six distinct C^∞ varieties having normal crossings.

For n=2 we must add the extra cone $E(\bar{z}_4)$ which we denote CM; this leads to the extra three lines of intersection:

$$(10.4)\qquad L_{i4} = CM\cap E(\bar{z}_i),\ i=1,2,3$$

and hence another three C^∞ varieties:

$$(10.5)\qquad \mathcal{D}_i = \{CM\backslash L_{i4}, E(\bar{z}_i)\backslash L_{i4}, L_{i4}\backslash M, M\},\ i=1,2,3.$$

The support of these varieties is

$$(10.6)\qquad E(\bar{z}_1)\cup E(\bar{z}_2)\cup E(\bar{z}_3)\cup CM = \bigcup_{1\le i<j\le 3} \mathcal{H}_{ij} \cup \bigcup_{i=1}^{3} \mathcal{D}_i .$$

The space we study is just the sum of the spaces of iterative regularity associated to these C^∞ varieties:

$$(10.7)\qquad J_k = \sum_{1\le i<j\le 3} I_k L^2(\Omega,\mathcal{H}_{ij}) + \sum_{i=1}^{3} I_k L^2(\Omega,\mathcal{D}_i).$$

In case n>2, let CM be the Huygens cone over M, obtained as the projection to the base of the flow out under the Hamilton vector field of the symbol, p, of P in $\Sigma = \{p=0\}$, of N^*M. Diffeomorphically it is a cone in $\mathbb{R}^3$ cross $\mathbb{R}^{n-2}$ near M. It is the carrier, in the linear theory, of singularities of the forward solution of Pu=f produced by conormal singularities of f on M. In any case the definitions above make sense and the resulting geometry is just that for n=2 cross Euclidian space of the appropriate dimension, for this reason we shall emphasize the case n=2.

The astute reader will observe that there is a much more attractive decomposition of the four cones than (10.6). Namely, instead of the three varieties $\mathcal{D}_i$ consider the six varieties:

$$(10.8)\qquad \mathcal{H}_{i4} = \{E(\bar{z}_i)\backslash L_{i4}, L_{i4}\backslash M, M\},\ \mathcal{M}_i = \{CM\backslash L_{i4}, L_{i4}\backslash M, M\},\ i=1,2,3.$$

Then again:

$$(10.9) \qquad E(\bar{z}_1)\cup E(\bar{z}_2)\cup E(\bar{z}_3)\cup CM = \bigcup_{1\le i<j\le 4} \mathcal{H}_{ij} \cup \bigcup_{i=1}^{3} \mathcal{M}_i .$$

Corresponding to this decomposition one has the smaller space:

$$(10.10) \qquad J_k' = \sum_{1\le i<j\le 4} I_k L^2(\Omega,\mathcal{H}_{ij}) + \sum_{i=1}^{3} I_k L^2(\Omega,\mathcal{M}_i) \subset J_k,$$

which will be discussed further below.

(10.11) <u>Proposition</u> The space J_k associated by (10.7) to L_0 with $\#(L_0)=3$, is a P-propagative algebra.

As an immediate corollary to this the following result, implying in particular Theorem 1.11, can be obtained:

(10.12) <u>Proposition</u> If (1.1) – (1.5) hold, for any n, then $u=v+w$ where, in $t>0$, v satisfies (9.2) and with J_k defined by (10.7)

$$(10.13) \qquad w \in H^{\frac{1}{2}n+1-}(\Omega)\cap J_k .$$

The proof is left to the reader. We proceed to discuss the proof of Proposition 10.11, but before doing so, consider also the following:

(10.14) <u>Question</u> Does Proposition 10.12 remain valid when J_k in (10.13) is replaced by J_k', defined by (10.10). In particular, is J_k' P-propagative (it can be seen to be an algebra).

§11. <u>For $\#(L_0)=3$, J_k and J_k' are algebras</u>

We start with the easier part of the proof of Proposition 10.11, showing that J_k in (10.7) is a C^∞ ring. This is very similar to the discussion of (9.7), so will be kept brief; we shall also concentrate on the case n=2, since the extra variables are essentially parameters in the higher dimensional cases. Let x,y,z be local coordinates around the triple point $\bar{z}_4$. Denote by B the polar coordinate map, now in codimension three, in these variables:

$$(11.1) \qquad B: Y=[0,\infty)\times S^2 \to (x,y,z) = r\omega .$$

It is easy to see how the various manifolds involved in the full C^∞ variety lift under B. Let

$$(11.2)\qquad R=B^{-1}(M),\ F_i=\mathrm{cl}[B^{-1}(E(\bar{z}_i)\backslash M)],\ i=1,2,3,\ F_4=\mathrm{cl}[B^{-1}(CM\backslash M)],$$

be the distinguished hypersurfaces in Y. Then the lifted varieties are obviously:

$$(11.3)\qquad \mathcal{F}_{ij} = \{F_i\backslash F_j, (F_i\cap F_j)\backslash R, R\backslash F_i, (R\cap F)\backslash F_j, R\cap F_i\cap F_j\}$$

lifting $\mathcal{H}_{ij}$ for $1\le i<j\le 3$ and

$$(11.4)\qquad \mathcal{K}_i=\{F_4\backslash(F_i\cup R), F_i\backslash(F_4\cup R), R\backslash(F_4\cup F_i),$$

$$(F_4\cap F_i)\backslash R, (F_4\cap R)\backslash F_i, (f_i\cap R)\backslash F_4, F_4\cap F_i\cap R\}$$

lifting $\mathcal{D}_i$ for i=1,2,3. The main point to observe is that where any two of these C^∞ varieities meet, one of them contains the other, in the sense that each element of the smaller is locally an element of the larger. Moreover direct computation shows that the various Lie algebras of vector fields, $\mathcal{V}(\mathcal{H}_{ij})$, $\mathcal{V}(\mathcal{D}_i)$ for i=1,2,3, i<j≤3, lift under B to span the corresponding $\mathcal{V}(\mathcal{F}_{ij})$ and $\mathcal{V}(\mathcal{K}_i)$. Denoting again by $L^2_B(Y)$ the lift of L^2 under B it follows first that:

$$(11.5)\qquad B^*J_k = \sum_{1\le i<j\le 3} I_k L^2_B(Y,\mathcal{F}_{ij}) + \sum_{i=1}^{3} I_k L^2_B(Y,\mathcal{K}_i)$$

and hence that $L^\infty J_k$ is a C^∞ algebra.

Without entering in to details we observe that the same type of proof can be applied to show that J'_k, defined by (10.10) is also a C^∞ algebra. The main part of the, relatively simple, computation involved can be found in [MR2].

§12. J_k is P-propagative

We only need to show this locally, i.e. if Ω is some small P-convex set, in the sense of (1.4) with respect to some time variable, then:

$$(12.1) \qquad \mathrm{supp}(u)\subset\Omega_\delta,\ Pu\in J_k \Rightarrow u \in J_{k+1}.$$

Using the fact that each of the components in (10.7) is a C^∞ module it suffices to show that:

$$(12.3) \qquad \mathrm{supp}(u)\subset\Omega_\delta,\ Pu\in I_kL^2(\Omega,\mathcal{A}) \Rightarrow u \in J_{k+1},\ \mathcal{A} = \mathcal{H}_{ij} \text{ or } \mathcal{D}_i,\ 1\leq i<j\leq 3.$$

There is of course considerable symmetry here, so only two distinct cases arise, say for $\mathcal{A} = \mathcal{H}_{12}$ and $\mathcal{A} = \mathcal{D}_i$. The geometry of the second variety is much more complicated than the first but (12.3) is easier to prove in the second case:

(12.4) Lemma The Lie algebras $\mathcal{V}(\mathcal{D}_i)$ of C^∞ vector fields tangent to $\mathcal{D}_i$ are P-complete, for i=1,2,3.

This is proved in [MR1], by showing that the C^∞ varieties $\mathcal{D}_i$ are characteristically complete, and hence that $\mathcal{V}(\mathcal{D}_i)$ is P-complete. This is the basic reason that the varieties $\mathcal{D}_i$ are introduced. Of course as noted in Proposition 8.3 the P-completness of $\mathcal{V}(\mathcal{D}_i)$ implies that (12.3) is valid for $\mathcal{A}=\mathcal{D}_i$ with J_{k+1} replaced by the much smaller space $I_kL^2(\Omega,\mathcal{D}_i)$.

The other case in (12.3) is not quite so simple. To derive it we shall discuss a model case in detail, the general case can be reduced to it by microlocal methods. Consider therefore the characteristic surface, line and point

$$(12.5) \qquad H = \{x=t\},\ L = \{t=x=y\},\ O = \{0\},\ \mathcal{H} = \{H\backslash L,\ L\backslash O, O\}.$$

We first derive estimates on the Fourier transform,

$$(12.6) \qquad \hat{u}(\xi,\eta,\tau) = \int e^{-i(x\xi+y\eta+t\tau)}u(x,y,t)\ dxdydt,$$

of an element, with compact support, of $I_kL^2(\mathbb{R}^3,\mathcal{H})$. Indeed, it is easy to characterize the Fourier image of this space but in any case:

$$(12.7) \qquad u \in I_k L^2(\mathbb{R}^3,\mathcal{H}) \cap C_c^{-\infty}(\mathbb{R}^3) \Rightarrow V^{\beta} D^{\alpha} \hat{u} \in L^2(\mathbb{R}^3), \quad |\beta| \leq k$$

where $L^2(\mathbb{R}^3)$ stands for the global space and $V^{\beta} = V_1^{\beta_1} \cdot V_2^{\beta_2} \cdot \ldots \cdot V_6^{\beta_6}$, where the V_i are the following six vector fields:

$$(12.8) \qquad \xi(D_{\xi}-D_{\tau}), \tau(D_{\xi}-D_{\tau}), \eta(D_{\xi}-D_{\tau}), \eta D_{\eta}, (\xi+\tau)D_{\eta}, (\xi+\tau)D_{\xi}.$$

This follows just by Fourier tranforming a suitable basis of $\mathcal{V}(\mathcal{H})$.

To interpret (12.7) we shall compactify $\mathbb{R}^3$, by adding a sphere at infinity. An elementary way to do this is by the diffeomorphsim to the interior of the unit ball, B^3:

$$(12.9) \qquad SP: \mathbb{R}^3 \to B^3, \quad SP(\xi,\tau,\eta) = (\xi,\tau,\eta)/[1+\xi^2+\tau^2+\eta^2]^{½} = \varsigma.$$

Then the compact Fourier tranform is defined by

$$(12.10) \qquad SP^*(F_c u) = \hat{u}, \quad F_c : L^2(\mathbb{R}^3) \leftrightarrow L^2_{\nu}(B^3),$$

where L^2_{ν} is the space of functions on B^3 square integrable with respect to the smooth measure ν, which degenerates only at the boundary and there as $\rho^4\mu$, where ρ is a derfining function for the boundary and μ is Lebesgue measure on $\mathbb{R}^3$. Using the coordinate tranformation (12.9) it is straightforward to transfer (12.7), (12.8) to B^3:

$$(12.11) \qquad u \in I_k L^2(\mathbb{R}^3,\mathcal{H}) \cap C_c^{-\infty} \Rightarrow \mathcal{V}(\mathcal{L})^j (\rho \mathcal{V}_b)^r F_c u \in L^2_{\nu}(B^3) \ \forall\, r, \ \forall\, j \leq k.$$

Here $\mathcal{V}_b$ is the space of vector fields tangent to the boundary, for which ρ is a defining function, and $\mathcal{L}$ is the C^{∞} variety described in terms of the coordinates (12.9):

$$(12.12) \qquad \mathcal{L} = \partial B^3 \sqcup T \sqcup K, \quad T = \{\varsigma_1+\varsigma_2=0\}, \quad K = \{\varsigma_1+\varsigma_2=\varsigma_3=0\}.$$

Using these estimates on the Fourier transform and a splitting result from [MR2], it can now be shown that

$$(12.13)\qquad Pu = f\in I_k L^2(\mathbb{R}^3,\mathscr{H}), u=0 \text{ in } t<-1 \Rightarrow$$

$$u \in \sum_{i=1}^{2} I_{k+1}L^2(\mathbb{R}^3,\mathscr{H}_i)+I_{k+1}L^2(\mathbb{R}^3,\mathscr{D}_i'),$$

where $\mathscr{H}_1 = \mathscr{H}$ is given by (12.5), $\mathscr{H}_2$ is defined similarly by replacing the characteristic surface by $\{y+t=0\}$ and the $\mathscr{D}_i$ are similarly defined by (10.5) for these two hypersurfaces.

Thus, (12.13) proves the desired result, (12.3) in the model case where the incoming hypersurfaces are actually flat. The general case follows by reduction to this case using a Fourier integral operator tranformation. This completes the discussion of Proposition 10.11

§13. Extension to $\#(L_0)>3$

For certain special geometries, with n=2, Theorem 1.11 can be extended by these methods to four initial points. For example if the initial points form a square then there is only one point of triple interaction, and no subsequent interactions, $L_1 = L_0\cup\{\bar{z}_5\}$, $L_k = L_1$, $k\geq 1$. If J_k is defined as in (10.7) but allowing the initial indices to run from 1 to 4 the method of proof extends directly.

The difficulty then is not with higher order interactions, but with interactions subsequent to the first triple point. Indeed, after such an interaction there are tangent sufaces, as in the C^∞ varieties $\mathscr{D}_i$. In [MR2] it is reasoned that J_k defined by (10.7) is not the space of conormal functions asssociated to the full C^∞ variety, but rather J_k' in (10.10) deserves this title. To a certain extent this is obvious, since J_k' is smaller, is defined by iterative regularity and allows all wavefront set that one would expect. A more convincing argument is provided by a normal resolution, i.e. blow up of the geometry which reduces it to normal crossings, under which J_k' lifts to the space of conormal functions. Away from the triple intersection this is shown in [MR2] and a similar discussion applies at the triple point itself.

References

[DH] J.J.Duistermmat & L.Hörmander Fourier integral operators II Acta Math. 128 (1972) 183-269.
[Bo1] J.-M.Bony Interaction des singularités pour équations aux dérivées partielles non linéares Sem. Goulaouic-Schwartz 1979-80, #22 &

Sem. Goulaouic-Meyer-Schwartz 1981-82, #2
[Bo2] J.-M.Bony Second microlocalization and propagation of singularities for semi-linear hyperbolic equations (preprint).
[Hö1] L.Hörmander Fourier integral operators I Acta Mat.127 (1971), 79-183.
[Hö2] L.Hörmander *The analysis of linear partial differential operators III* Springer-Verlag 1984.
[IN] T.Iwasaki & G.Nakamura Singular solutions for semilinear hyperbolic equations I. Amer. J. Math. (to appear)
[MR1] R.B.Melrose & N.Ritter Interaction of progressing waves for semilinear wave equations Ann. of Math. 121 (1985), 187-213.
[MR2] R.B.Melrose & N.Ritter Interaction of progressing waves for semilinear wave equations II (preprint).

GENERAL BOUNDARY VALUE PROBLEMS IN THE FRAMEWORK OF HYPERFUNCTIONS

Toshinori Ôaku

Department of Mathematics
University of Tokyo
Hongo, Tokyo, 113, Japan

ABSTRACT. First, we present a new method to formulate boundary value problems in the framework of hyperfunctions both for general systems of linear partial differential equations with non-characteristic boundary and for (single) Fuchsian equations degenerating on the boundary. Secondly, we give a theorem on propagation of micro-analyticity up to the boundary for such equations under a condition of micro-hyperbolicity. Finally, we study continuation of real analytic solutions of such equations as an application of the above theorem. These results extend previous ones for single partial differential equations with non-characteristic hypersurface.

CONTENTS

INTRODUCTION

Boundary value problem was formulated by Schapira [17] and Komatsu-Kawai [11] independently in the framework of hyperfunctions for single linear partial differential equations with non-characteristic boundary. They showed that hyperfunction solutions of such an equation have boundary values as hyperfunctions on the boundary, and that the solutions are determined locally by these boundary values (Holmgren's type

H. G. Garnir (ed.), Advances in Microlocal Analysis, 253–270.

uniqueness theorem).

After these works, Kashiwara-Kawai [5] studied boundary value problems for elliptic systems of partial differential equations, and Kashiwara-Oshima [7] for equations with regular singularities.

In Section 1, we review the new method to formulate boundary value problems both for systems with non-characteristic boundary and for Fuchsian equations, which has been developed in Ôaku [13,14].

Section 2 is the main part of this paper, where we study the propagation of micro-analyticity of solutions of boundary value problems for such equations as above under a condition of relative micro-hyperbolicity. For this purpose we introduce a sheaf $\mathcal{C}_{M+}$ on $M \times \sqrt{-1}S^{n-2}$ supported by $\bar{M}_+ \times \sqrt{-1}S^{n-2}$, where $M = \mathbb{R}^n \ni x = (x_1, x')$, $M_+ = \{x \in M; x_1 > 0\}$, and S^{n-2} is the (n-2)-sphere. It describes a cotangential decomposition of singularities of hyperfunctions on M_+. First, we present a theorem on unique continuation of $\mathcal{C}_{M+}$-solutions (Theorem 2.1). This is an analogue in the boundary value problem of a theorem of Kashiwara-Schapira for the interior problem ([8]). Theorem 2.1 implies propagation of micro-analyticity up to the boundary for hyperfunction solutions (Theorems 2.2 and 2.4). Results of this type have been proved by Kaneko [2], Schapira [18,19], Kataoka [9,10], Sjöstrand [20] for single equations (or determined systems) with non-characteristic boundary by completely different methods.

The purpose of Section 3 is to generalize results of Kaneko [2,3] on continuation of real analytic solutions of single equations. He gives sufficient conditions for continuation of real analytic solutions whose singularities (i.e. the points where the solution is not defined) are contained in a real analytic hypersurface non-characteristic for a single equation. Kaneko's idea is to reduce such continuation problem to (two-sided) boundary value problem. Hence the results in Section 2 enable us to extend his results in two directions: on the one hand, to general systems with non-characteristic hypersurface, and on the other hand, to Fuchsian partial differential equations.

Further details of Sections 1 and 2 will appear in [14] except Theorem 2.3, and those of Section 3 will appear in [15].

1. FORMULATION OF BOUNDARY VALUE PROBLEMS

1.1. New sheaves attached to the boundary

Let M be an open subset of $\mathbb{R}^n \ni x$ and N be a real analytic hypersurface of M. Since we are interested in local properties of boundary value problems, we may assume, without loss of generality, that $M = \mathbb{R}^n$, $N = \{x \in M; x_1 = 0\}$. We use the notation $x = (x_1, x')$ with $x' \in \mathbb{R}^{n-1}$. We put $M_+ = \{x \in M; x_1 > 0\}$ and let $\iota_+ : M_+ \longrightarrow M$ be the embedding. We denote by $\mathcal{B}_M$ the sheaf on M of hyperfunctions.

Definition 1.1. $\mathcal{B}_{M+} = (\iota_+)_*(\iota_+)^{-1}\mathcal{B}_M$, $\mathcal{B}_{N|M+} = \mathcal{B}_{M+}|_N$.

Put $\tilde{M} = \mathbb{R}\times \mathbb{C}^{n-1}$, $\tilde{M}_+ = \mathbb{R}_+ \times \mathbb{C}^{n-1}$, $Y = \{0\} \times \mathbb{C}^{n-1}$ with $\mathbb{R}_+ = \{x_1 \in \mathbb{R}\,;\, x_1 > 0\}$ and let $\tilde{\iota}_+ : \tilde{M}_+ \longrightarrow \tilde{M}$ be the embedding. We denote by $\mathcal{BO}$ the sheaf on $\tilde{M}$ of hyperfunctions in (x_1, z') with holomorphic parameters $z' = (z_2,\dots,z_n) \in \mathbb{C}^{n-1}$.

<u>Definition 1.2</u>. $\mathcal{BO}_{\tilde{M}+} = (\tilde{\iota}_+)_*(\tilde{\iota}_+)^{-1}\mathcal{BO}$, $\quad \mathcal{BO}_{Y|\tilde{M}+} = \mathcal{BO}_{\tilde{M}+}\big|_Y$.

By the abstract edge of the wedge theorem of Kashiwara-Laurent [6], we can verify that N is purely (n-1)-codimensional with respect to $\mathcal{BO}_{Y|\tilde{M}+}$.

<u>Definition 1.3</u>. $\tilde{\mathcal{B}}_{N|M+} = \mathcal{H}_N^{n-1}(\mathcal{BO}_{Y|\tilde{M}+})$.

On the other hand, we have a natural isomorphism

$$\mathcal{B}_{M+} \cong \mathcal{H}_M^{n-1}(\mathcal{BO}_{\tilde{M}+}).$$

Hence there exists a natural sheaf homomorphism

$$\alpha_+ : \mathcal{B}_{N|M+} \longrightarrow \tilde{\mathcal{B}}_{N|M+}.$$

Let us illustrate this homomorphism in the simplest case, i.e. when $n = 2$. Let $\mathring{x} = (0, \mathring{x}_2)$ be a point of N. Then the germs of $\mathcal{B}_{N|M+}$ and of $\tilde{\mathcal{B}}_{N|M+}$ at $\mathring{x}$ are expressed as follows:

$$\mathcal{B}_{N|M+,\mathring{x}} = \varinjlim_{\varepsilon} (\mathcal{BO}(U_\varepsilon^+)\oplus\mathcal{BO}(U_\varepsilon^-))/\mathcal{BO}(U_\varepsilon),$$

$$\tilde{\mathcal{B}}_{N|M+,\mathring{x}} = \varinjlim_{\varepsilon, V^+, V^-} (\mathcal{BO}(U_\varepsilon^+ \cap V^+)\oplus\mathcal{BO}(U_\varepsilon^- \cap V^-))/\mathcal{BO}(U_\varepsilon),$$

where $U_\varepsilon = \{(x_1,z_2) \in \mathbb{R}\times \mathbb{C};\ 0 < x_1 < \varepsilon,\ |x_2 - \mathring{x}_2| < \varepsilon,\ |y_2| < \varepsilon\}$, $U_\varepsilon^\pm = \{(x_1,z_2) \in U_\varepsilon;\ \pm y_2 > 0\}$ with the notation $z_2 = x_2 + \sqrt{-1}y_2$, and $V^\pm$ runs on the system of neighborhoods of $\{(0,z_2);\ \pm y_2 > 0\}$ in $\mathbb{R}\times \mathbb{C}$.

These expressions immediately imply that the natural homomorphism $\alpha_+ : \mathcal{B}_{N|M+} \longrightarrow \tilde{\mathcal{B}}_{N|M+}$ is injective when $n = 2$. We can prove that this is also true for $n \geqq 3$ by using the edge of the wedge theorem for $\mathcal{B}_{N|M+}$ and $\tilde{\mathcal{B}}_{N|M+}$ and the Radon transformation (curvilinear wave expansion) for hyperfunctions with holomorphic parameters. Namely we have the following:

<u>Proposition 1.1</u>. $\alpha_+ : \mathcal{B}_{N|M+} \longrightarrow \tilde{\mathcal{B}}_{N|M+}$ is injective.

Now let us microlocalize the sheaves $\mathcal{B}_{N|M+}$ and $\tilde{\mathcal{B}}_{N|M+}$ just in the same way as the sheaf of microfunctions is constructed as a microlocalization of the sheaf of hyperfunctions (cf. Sato-Kawai-Kashiwara [16]). Let

$$\pi_{M/\tilde{M}} : (\tilde{M} - M) \cup S^*_M\tilde{M} \longrightarrow \tilde{M}, \qquad \pi_{N/Y} : (Y - N) \cup S^*_N Y \longrightarrow Y$$

be the comonoidal transforms of $\tilde{M}$ and Y with centers M and N respectively, where $S^*_M\tilde{M} = (T^*_M\tilde{M} - 0)/\mathbb{R}_+$ is the conormal sphere boundle of M in $\tilde{M}$. Note that we can identify $S^*_N Y$ with the purely imaginary cosphere boundle $\sqrt{-1}S^*N$. A point of $S^*_M\tilde{M}$ is written as $(x, \sqrt{-1}\xi'\infty)$ with $x \in \mathbb{R}^n$ and $\xi' = (\xi_2, \dots, \xi_n) \in \mathbb{R}^{n-1} - \{0\}$. Let us denote by a the antipodal map.

<u>Definition 1.4</u>. $\mathcal{C}_{M+} = \mathcal{H}^{n-1}_{S^*_M\tilde{M}}((\pi_{M/\tilde{M}})^{-1}\mathcal{BO}_{\tilde{M}+})^a$, $\quad \mathcal{C}_{N|M+} = \mathcal{C}_{M+}|_{S^*_N Y}$,

$$\tilde{\mathcal{C}}_{N|M+} = \mathcal{H}^{n-1}_{S^*_N Y}((\pi_{N/Y})^{-1}\mathcal{BO}_{Y|\tilde{M}+})^a.$$

By virtue of the abstract edge of the wedge theorem of Kashiwara-Laurent [6], the argument of Sato-Kawai-Kashiwara [16, Chapter I] applies to our case. In particular we have the following:

<u>Proposition 1.2</u>. There exist exact sequences

$$0 \longrightarrow \mathcal{BO}_{\tilde{M}+}|_M \longrightarrow \mathcal{B}_{M+} \xrightarrow{sp_+} (\pi_{M/\tilde{M}})_*\mathcal{C}_{M+} \longrightarrow 0,$$

$$0 \longrightarrow \mathcal{BO}_{Y|\tilde{M}+}|_N \longrightarrow \tilde{\mathcal{B}}_{N|M+} \xrightarrow{\tilde{sp}} (\pi_{N/Y})_*\tilde{\mathcal{C}}_{N|M+} \longrightarrow 0.$$

By the definitions, it is easy to see that there exists a natural sheaf homomorphism $\alpha_+ : \mathcal{C}_{N|M+} \longrightarrow \tilde{\mathcal{C}}_{N|M+}$ compatible with $\alpha_+ : \mathcal{B}_{N|M+} \longrightarrow \tilde{\mathcal{B}}_{N|M+}$. We can prove the following:

<u>Proposition 1.3</u>. $\alpha_+ : \mathcal{C}_{N|M+} \longrightarrow \tilde{\mathcal{C}}_{N|M+}$ is injective.

1.2. Non-characteristic boundary value problem for systems

We denote by $\mathcal{D}_X$ the sheaf on $X = \mathbb{C}^n$ of rings of linear partial differential operators (of finite order) with holomorphic coefficients. It is well-known (cf. Kashiwara [4]) that systems of linear partial differential equations with holomorphic (resp. real analytic) coefficients are nothing but coherent $\mathcal{D}_X$-modules (resp. their restriction to $M = \mathbb{R}^n$). We call coherent $\mathcal{D}_X$-modules simply systems if there is no fear of confusion. For a system $\mathcal{M}$, its characteristic variety $SS(\mathcal{M})$ is defined as the support of $\mathcal{E}_X \otimes_{\pi^{-1}\mathcal{D}} \pi^{-1}\mathcal{M}$, where $\pi : T^*X \longrightarrow X$ is the projection and $\mathcal{E}_X$ denotes the sheaf on the cotangent bundle T^*X of microdifferential operators of finite order. In this subsection we assume that $Y = \{z \in X;\ z_1 = 0\}$ is non-characteristic with respect to a system $\mathcal{M}$; i.e. $(T^*_YX - 0) \cap SS(\mathcal{M}) = \phi$. Then the tangential system $\mathcal{M}_Y$ of $\mathcal{M}$ to Y is a coherent $\mathcal{D}_Y$-module defined by $\mathcal{M}_Y = \mathcal{M}/z_1\mathcal{M}$. Let $\mathcal{F}$ be a sheaf of $\mathcal{D}_X$-modules (e.g. $\mathcal{B}_M$, $\mathcal{B}_{N|M+}$, $\tilde{\mathcal{B}}_{N|M+}$). Then the sheaf $\mathcal{H}om_{\mathcal{D}_X}(\mathcal{M},\mathcal{F})$ is nothing but the sheaf of $\mathcal{F}$-solutions of $\mathcal{M}$.

<u>Proposition 1.4</u>. Let $\mathcal{M}$ be a coherent $\mathcal{D}_X$-module with respect to which Y is non-characteristic. Then there exists a sheaf isomorphism

$$\mathcal{H}om_{\mathcal{D}_X}(\mathcal{M}, \tilde{\mathcal{B}}_{N|M+}) \overset{\sim}{=} \mathcal{H}om_{\mathcal{D}_Y}(\mathcal{M}_Y, \mathcal{B}_N).$$

<u>Sketch of the proof</u>. Put $\tilde{M}_- = \tilde{M} - \tilde{M}_+$. Then from the exact sequence

$$0 \longrightarrow \Gamma_{\tilde{M}_-}(\mathcal{BO})\big|_Y \longrightarrow \mathcal{BO}\big|_Y \longrightarrow \mathcal{BO}_{Y|\tilde{M}_+} \longrightarrow 0$$

we get a triangle

$$\begin{array}{ccc} & \mathbb{R}\mathcal{H}om_{\mathcal{D}_X}(\mathcal{M}, \Gamma_{\tilde{M}_-}(\mathcal{BO})\big|_Y) & \\ \swarrow & & \nwarrow {\scriptstyle +1} \\ \mathbb{R}\mathcal{H}om_{\mathcal{D}_X}(\mathcal{M}, \mathcal{BO}|_Y) & \longrightarrow & \mathbb{R}\mathcal{H}om_{\mathcal{D}_X}(\mathcal{M}, \mathcal{BO}_{Y|\tilde{M}+}). \end{array}$$

We can show by using Theorem 2.2.1 of Kashiwara-Schapira [8] that

$$\mathbb{R}\mathcal{H}om_{\mathcal{D}_X}(\mathcal{M}, \Gamma_{\tilde{M}_-}(\mathcal{BO})\big|_Y) = 0.$$

Hence there is an isomorphism (in the derived category)

$$\mathbb{R}\mathcal{H}om_{\mathcal{D}_X}(\mathcal{M}, \mathcal{BO}\big|_Y) \xrightarrow{\sim} \mathbb{R}\mathcal{H}om_{\mathcal{D}_X}(\mathcal{M}, \mathcal{BO}_{Y|\tilde{M}+}). \tag{1.1}$$

On the other hand, since Y is non-characteristic with respect to $\mathcal{M}$, we have an isomorphism

$$\mathbb{R}\mathcal{H}om_{\mathcal{D}_X}(\mathcal{M}, \mathcal{O}_X\big|_Y) \xrightarrow{\sim} \mathbb{R}\mathcal{H}om_{\mathcal{D}_X}(\mathcal{M}, \mathcal{BO}\big|_Y), \tag{1.2}$$

where $\mathcal{O}_X$ denotes the sheaf on X of holomorphic functions. By the Cauchy-Kowalevsky theorem due to Kashiwara (cf. [4]) we also have

$$\mathbb{R}\mathcal{H}om_{\mathcal{D}_X}(\mathcal{M}, \mathcal{O}_X\big|_Y) \xrightarrow{\sim} \mathbb{R}\mathcal{H}om_{\mathcal{D}_Y}(\mathcal{M}_Y, \mathcal{O}_Y). \tag{1.3}$$

Combining the isomorphisms (1.1)-(1.3), we get an isomorphism

$$\mathbb{R}\mathcal{H}om_{\mathcal{D}_X}(\mathcal{M}, \mathcal{BO}_{Y|\tilde{M}+}) \xrightarrow{\sim} \mathbb{R}\mathcal{H}om_{\mathcal{D}_Y}(\mathcal{M}_Y, \mathcal{O}_Y).$$

Applying the right derived functor $\mathbb{R}\Gamma_N$ to this isomorphism we get an isomorphism

$$\mathbb{R}\mathcal{H}om_{\mathcal{D}_X}(\mathcal{M}, \tilde{\mathcal{B}}_{N|M+}) \xrightarrow{\sim} \mathbb{R}\mathcal{H}om_{\mathcal{D}_Y}(\mathcal{M}_Y, \mathcal{B}_N).$$

This completes the proof.

By this proposition and the injectivity of α_+ we get

Theorem 1.1. Let $\mathcal{M}$ be a coherent $\mathcal{D}_X$-module with respect to which Y is non-characteristic. Then there exists an injective sheaf homomorphism (boundary value map)

$$\gamma_+ : \mathcal{H}om_{\mathcal{D}_X}(\mathcal{M}, \mathcal{B}_{N|M+}) \longrightarrow \mathcal{H}om_{\mathcal{D}_Y}(\mathcal{M}_Y, \mathcal{B}_N).$$

This theorem means that any hyperfunction solution $u(x)$ of $\mathcal{M}$ on M_+ has a boundary value which is a hyperfunction solution on N of $\mathcal{M}_Y$ and that $u(x)$ is determined locally by its boundary value. In the same way we can microlocalize this boundary value map:

Theorem 1.2. There exists an injective sheaf homomorphism

$$\gamma_+ : \mathcal{H}om_{\mathcal{D}_X}(\mathcal{M}, \mathcal{C}_{N|M+}) \longrightarrow \mathcal{H}om_{\mathcal{D}_Y}(\mathcal{M}_Y, \mathcal{C}_N)$$

compatible with γ_+ in Theorem 1.1 (i.e commuting with the spectral maps), where $\mathcal{C}_N$ denotes the sheaf on $S^*_N Y$ of microfunctions.

There exists a natural map

$$p : S^*_M X - S^*_{\tilde{M}} X \longrightarrow S^*_M \tilde{M}.$$

Put $L_+ = (\pi_{M/\tilde{M}})^{-1}(M_+)$. There exists a natural sheaf homomorphism

$$\psi : p^{-1}\mathcal{C}_{M+} \longrightarrow \mathcal{C}_M$$

defined on $p^{-1}(L_+)$ such that $\psi(sp_+(f)) = sp(f)$ for a section f of $\mathcal{B}_{M+}\big|_{M+} = \mathcal{B}_M\big|_{M+}$, where $sp : \pi^{-1}\mathcal{B}_M \longrightarrow \mathcal{C}_M$ is the spectral map. Theorem 1.2 implies a microlocal version of Holmgren's uniqueness theorem:

Corollary 1.1. Under the same assumption as in Theorem 1.1, let u be a $\mathcal{B}_{N|M+}$-solution of $\mathcal{M}$. Assume that $\gamma_+(u)$ is micro-analytic at $(\mathring{x},\sqrt{-1}\mathring{\xi}'\infty) \in S^*_N Y$. Then, as a hyperfunction, u is micro-analytic on

$$\{(x,\sqrt{-1}\xi\infty) \in S^*_M X;\ 0 < x_1 < \varepsilon,\ |x' - \mathring{x}'| < \varepsilon,\ \xi_1 \in \mathbb{R},\ |\xi' - \mathring{\xi}'| < \varepsilon\}$$

for some $\varepsilon > 0$.

Let us give more concrete meaning of the boundary value map. For this purpose, let us recall the notion of F-mild hyperfunctions defined by Ôaku [12], which is a generalization of mild hyperfunctions introduced by Kataoka [9,10].

Definition 1.5. Let f be a germ of $\mathcal{B}_{N|M+}$ at $\mathring{x} \in N$. Then f is called F-mild at $\mathring{x}$ if and only if f has a boundary value expression

$$(1.4) \qquad f(x) = \sum_{j=1}^{J} F_j(x_1, x'+\sqrt{-1}\Gamma_j 0)$$

as a hyperfunction on $\{x \in M_+;\ |x - \mathring{x}| < \varepsilon\}$, where J is a positive integer, ε is a positive number, Γ_j is an open convex cone in $\mathbb{R}^{n-1}$, F_j is a holomorphic function defined on a neighborhood in $\mathbb{C}^n$ of $\{z = (z_1, z') \in \mathbb{C}^n;\ |z-\mathring{x}| < \varepsilon,\ \mathrm{Re}\, z_1 \geqq 0,\ \mathrm{Im}\, z_1 = 0,\ \mathrm{Im}\, z' \in \Gamma_j\}$. We denote by $\mathcal{B}^F_{N|M+}$ the subsheaf of $\mathcal{B}_{N|M+}$ consisting of its sections F-mild at each point of N.

For a F-mild hyperfunction $f(x)$, its boundary value $f(+0,x')$ is naturally defined as a section of $\mathcal{B}_N$. There is a natural homomorphism

$$\gamma_+^F : \mathcal{H}om_{\mathcal{D}_X}(\mathcal{M}, \mathcal{B}^F_{N|M+}) \longrightarrow \mathcal{H}om_{\mathcal{D}_Y}(\mathcal{M}_Y, \mathcal{B}_N)$$

induced by the boundary value map for F-mild hyperfunctions,

$$\mathcal{B}^F_{N|M+} \ni f(x) \longmapsto f(+0,x') = \sum_{j=1}^{J} F_j(0, x'+\sqrt{-1}\Gamma_j 0) \in \mathcal{B}_N,$$

where $f(x)$ is defined by (1.4).

Theorem 1.3. Let $\mathcal{M}$ be a coherent $\mathcal{D}_X$-module with respect to which Y is non-characteristic. Then there is a sheaf isomorphism

$$\mathcal{H}om_{\mathcal{D}_X}(\mathcal{M}, \mathcal{B}^F_{N|M+}) \xrightarrow{\ \sim\ } \mathcal{H}om_{\mathcal{D}_X}(\mathcal{M}, \mathcal{B}_{N|M+}).$$

Moreover, under this isomorphism, γ_+ of Theorem 1.1 coincides with γ_+^F.

Corollary 1.2. Under the condition of Theorem 1.1, γ_+ is independent of local coordinate systems.

1.3. Boundary value problem for Fuchsian equations

We use the notation $D' = (D_2, \dots, D_n)$ with $D_j = \partial/\partial z_j$ (or $D_j = \partial/\partial x_j$ in the real domain). In this subsection we assume that a linear partial differential operator P with analytic coefficients is a Fuchsian operator of weight $m-k$ with respect to N in the sense of Baouendi-Goulaouic [1]; i.e. we assume that P is written in the form

$$P = x_1^k D_1^m + A_1(x,D')x_1^{k-1}D_1^{m-1} + \cdots + A_k(x,D')D_1^{m-k} + \cdots + A_m(x,D')$$

with the following conditions:

(i) $k, m \in \mathbb{Z}$, $0 \leqq k \leqq m$,

(ii) $A_j(x,D')$ is of order $\leqq j$ for $1 \leqq j \leqq m$,

(iii) $A_j(0,x',D')$ is of order 0; i.e. equals a function $a_j(x')$ for $1 \leqq j \leqq k$.

The roots $\lambda = 0,1,\ldots,m-k-1,\ \lambda_1(x'),\ldots,\lambda_k(x')$ of the equation

$$\lambda(\lambda-1)\cdots(\lambda-m+1) + a_1(x')\lambda(\lambda-1)\cdots(\lambda-m+2) + \cdots + \\ + a_k(x')\lambda(\lambda-1)\cdots(\lambda-m+k+1) = 0$$

are called the <u>characteristic exponents</u> of P. We introduce a condition $C(\mathring{x})$ for a point $\mathring{x} = (0,\mathring{x}')$ of N as follows:

$$C(\mathring{x}) : \lambda_i(\mathring{x}') \notin \mathbb{Z},\ \lambda_i(\mathring{x}') - \lambda_j(\mathring{x}') \notin \mathbb{Z} - \{0\} \quad \text{for} \quad 1 \leqq i,j \leqq k.$$

In this subsection we put $\mathcal{M} = \mathcal{D}_X/\mathcal{D}_X P$ with a Fuchsian operator P; i.e. $\mathcal{M}$ is the equation $Pu = 0$.

<u>Theorem 1.4</u>. Let $\mathcal{M}$ be as above and assume the condition $C(\mathring{x})$ for a point $\mathring{x}$ of N. Then there exist injective sheaf homomorphisms

$$\gamma_+ : \mathcal{H}om_{\mathcal{D}_X}(\mathcal{M}, \mathcal{B}_{N|M+}) \longrightarrow (\mathcal{B}_N)^m,$$
$$\gamma_+ : \mathcal{H}om_{\mathcal{D}_X}(\mathcal{M}, \mathcal{C}_{N|M+}) \longrightarrow (\mathcal{C}_N)^m$$

defined on a neighborhood of $\mathring{x}$ and of $\pi_{N/Y}^{-1}(\mathring{x})$ respectively compatible with each other. Moreover, for a $\mathcal{B}_{N|M+}$-solution $u(x)$ of $\mathcal{M}$, $\gamma_+(u)$ is decomposed in the form

$$\gamma_+(u) = (\gamma_{+reg}(u),\ \gamma_{+sing}(u)) \in (\mathcal{B}_N)^{m-k} \oplus (\mathcal{B}_N)^k$$

so that $u(x)$ is F-mild if and only if $\gamma_{+sing}(u) = 0$. If $u(x)$ is F-mild, we have

$$\gamma_{+reg}(u) = (u(+0,x'),\ldots,D_1^{m-k-1}u(+0,x')).$$

In order to prove this theorem, let us introduce a new class of (formal) linear partial differential operators of infinite order:

<u>Definition 1.6</u>. The sections of the sheaf $\mathcal{O}\hat{\mathcal{D}}_{Y|\tilde{M}}$ over an open subset Ω of Y are the operators

$$B(z,D') = \sum_{j=0}^{\infty} B_j(z,D')$$

satisfying the following conditions:

(i) $B_j(z,\zeta')$ is holomorphic on a neighborhood of $\Omega \times \mathbb{C}^{n-1}$ in $X \times \mathbb{C}^{n-1}$, and homogeneous of degree j with respect to ζ';

(ii) for any compact set K of Ω, and for any $\varepsilon > 0$, there exist $C > 0$ and $\delta > 0$ such that

$$|B_j(z,\zeta')| \leqq \frac{C}{j!}(\varepsilon|\zeta'|)^j$$

for any j and ζ' if $|z_1| < \delta$, $z' \in K$.

Note that $\widetilde{\mathcal{OD}}_{Y|\tilde{M}}$ has a natural ring structure and that it acts on $\mathcal{BO}|_Y$ and $\mathcal{BO}_{Y|\tilde{M}+}$, and hence on $\tilde{\mathcal{B}}_{N|M+}$ and on $\tilde{\mathcal{C}}_{N|M+}$.

Sketch of the proof of Theorem 1.4. Let $u(x)$ be a $\mathcal{B}_{N|M+}$-solution of $\mathcal{M}$ and put $v(x) = {}^t(v_1(x),\cdots,v_m(x))$ with

$$v_1(x) = x_1^{-1}u(x),\cdots,\ v_{m-k}(x) = x_1^{-1}D_1^{m-k-1}u(x),$$

$$v_{m-k+1}(x) = D_1^{m-k}u(x),\cdots,\ v_m(x) = x_1^{k-1}D_1^{m-1}u(x).$$

Then $v(x)$ satisfies a system $(x_1D_1 - A)v(x) = 0$ with

$$A = \begin{pmatrix} -1 & x_1 & & & & & & \\ & \ddots & \ddots & & & & & \\ & & \ddots & x_1 & & & & \\ & & & -1 & 1 & & & \\ & & & & 0 & 1 & & \\ & & & & & 1 & \ddots & \\ & & & & & & \ddots\ k-2 & 1 \\ -x_1A_m & \cdots & -x_1A_{k+1}, & -A_k & \cdots & \cdots & & -A_1+k-1 \end{pmatrix}.$$

Since this is a Fuchsian system in the sense of Tahara [21], there exists an invertible $m \times m$ matrix $Q(x,D')$ of $\widetilde{\mathcal{OD}}_{Y|\tilde{M}}$ such that

$$Q^{-1}(x_1D_1 - A)Q = x_1D_1 - \begin{pmatrix} \begin{matrix}-1 & & \\ & \ddots & \\ & & -1\end{matrix} & 0 \\ 0 & E(x') \end{pmatrix},$$

where $E(x')$ is a $k \times k$ matrix of analytic functions such that none of the eigenvalues of $E(\mathring{x})$ is an integer (cf. Theorem 1.3.6 of [21]). Since Q acts on $(\tilde{\mathcal{B}}_{N|M+})^m$, $\alpha_+(v)$ is uniquely written in the form

$$\alpha_+(v) = Q(x,D')\begin{pmatrix} x_1^{-1}a(x') \\ x_1^{E(x')}b(x') \end{pmatrix} = x_1^{-1}Q(x,D')\begin{pmatrix} a(x') \\ x_1^{E(x')+1}b(x') \end{pmatrix}$$

with $a(x') \in (\mathcal{B}_N)^{m-k}$ and $b(x') \in (\mathcal{B}_N)^k$. Hence $\alpha_+(u)$ is the first component of the vector

$$Q(x,D')\begin{pmatrix} a(x') \\ x_1^{E(x')+1}b(x') \end{pmatrix}.$$

We set $a(x') = \gamma_{+reg}(u)$ and $b(x') = \gamma_{+sing}(u)$. By the injectivity of α_+, the homomorphism $\gamma_+ = (\gamma_{+reg},\gamma_{+sing})$ is also injective. Other assertions can be also proved by the above expression of $u(x)$.

2. MICRO-HYPERBOLIC BOUNDARY VALUE PROBLEM

2.1. Relative micro-hyperbolicity

Let $H : T^*(T^*X) \xrightarrow{\sim} T(T^*X)$ be the Hamilton map defined by

$$H(dz_j) = -\frac{\partial}{\partial \zeta_j}, \qquad H(d\zeta_j) = \frac{\partial}{\partial z_j} \qquad (j = 1,\dots,n),$$

where $\zeta = (\zeta_1,\dots,\zeta_n) \in \mathbb{C}^n$ are the variables dual to z. For a point x^* and subsets S, V of T^*X, we denote by $C_{x^*}(S;V)$ the normal cone of S along V at x^* in the sense of Kashiwara-Schapira [8]. More concretely, $C_{x^*}(S;V)$ is a closed subset of the tangent space $T_{x^*}(T^*X)$ defined as follows: $v \in T_{x^*}(T^*X)$ is contained in $C_{x^*}(S;V)$ if and only if there are sequences $\{x_n\}$ in S, $\{y_n\}$ in V, and $\{a_n\}$ in $\mathbb{R}_+$ such that $x_n \longrightarrow x^*$, $y_n \longrightarrow x^*$, $a_n(x_n - y_n) \longrightarrow v$ as $n \longrightarrow \infty$.

<u>Definition 2.1</u>. A coherent $\mathcal{E}_X$-module $\mathcal{M}$ defined on a neighborhood of $x^* \in T^*_M X|_N$ in T^*X is called <u>micro-hyperbolic relative to $\tilde{M}_+$ in the direction</u> $\theta \in T^*_{x^*}(T^*X)$ <u>at</u> x^* if and only if

$$H(\theta) \notin C_{x^*}(SS(\mathcal{M}) \cap T^*X|_{\tilde{M}+};\ T^*_M X),$$

where $SS(\mathcal{M})$ denotes the characteristic variety (i.e. the support) of $\mathcal{M}$. We remark that this condition is weaker than that of micro-hyperbolicity defined in [8].

Let $\rho : T^*X \longrightarrow \mathbb{C} \times T^*Y$ be the map defined by $\rho(z,\zeta) = (z_1, z', \zeta')$ with $\zeta = (\zeta_1, \zeta')$. We denote by $\mathcal{R}$ the subsheaf of $\rho_*\mathcal{E}_X$ consisting of its sections which are polynomials with respect to D_1. It is easy to see that $\mathcal{R}$ is a sheaf of rings and that it acts on $\mathcal{C}_{M+}$.

<u>Theorem 2.1</u>. Let x^* be a point of $T^*_N Y - 0$ and $\mathcal{M}$ be a coherent $\mathcal{R}$-module defined on a neighborhood of x^*. Assume the following:

(C.1) $(T^*_Y X - 0) \cap \mathrm{cl}(SS(\mathcal{M}) \cap T^*X|_{\tilde{M}+}) = \phi$, where cl denotes the closure in T^*X, and $SS(\mathcal{M})$ denotes the support of $\mathcal{E}_X \otimes_{\rho^{-1}\mathcal{R}} \rho^{-1}\mathcal{M}$;

(C.2) $\mathcal{E}_X \otimes_{\rho^{-1}\mathcal{R}} \rho^{-1}\mathcal{M}$ is micro-hyperbolic relative to $\tilde{M}_+$ in the direction dz_1 at each point of $\rho^{-1}(x^*) \cap T^*_M X$.

(C.3) $\rho^{-1}(x^*) \cap \mathrm{cl}(SS(\mathcal{M}) \cap T^*X|_{\tilde{M}+}) \subset \{(\zeta_1, x^*) \in \rho^{-1}(x^*);\ \mathrm{Re}\,\zeta_1 \geqq 0\}$.

Under these conditions, putting $L_o = S^*_N Y \subset S^*_M \tilde{M}$, we have

$$\mathbb{R}\mathcal{H}om_{\mathcal{R}}(\mathcal{M}, \Gamma_{L_o}(\mathcal{C}_{M+}))_{x^*} = 0.$$

<u>Sketch of the proof</u>. We use the method developed by Kashiwara-Schapira [8]. First let us recall the definition of $\mathcal{E}(G;D)$ following [8]. Let G be a proper convex closed cone in $X = \mathbb{C}^n$ and let D be a

G-round open set of X, i.e. satisfying $(z+G)\cap(w-G)\subset D$ for any $z,w\in D$. Then we put

$$\mathcal{E}(G;D) = H^n_Z(D\times D;\mathcal{O}^{(0,n)}_{X\times X}),$$

where $Z=\{(z,w)\in X\times X;\ w-z\in Z\}$, and $\mathcal{O}^{(0,n)}_{X\times X}$ denotes the sheaf of n-forms with respect to w with coefficients holomorphic in (z,w). Note that $\mathcal{E}(G;D)$ has a natural ring structure (cf. [8]).

Now let us proceed to the proof of the theorem. We may assume $x^* = (0, -\sqrt{-1}dx_n)$. There are a proper convex closed cone $G_0\subset\{0\}\times\mathbb{C}^{n-1}$, a G-round open set $D\subset X$, and a complex

$$\mathcal{M}^\bullet : 0 \longleftarrow \mathcal{E}(G_0;D)^{n_0} \xleftarrow{P_1} \mathcal{E}(G_0;D)^{n_1} \longleftarrow \cdots \xleftarrow{P_r} \mathcal{E}(G_0;D)^{n_r} \longleftarrow 0$$

of free $\mathcal{E}(G_0;D)$-modules, where P_j are matrices of elements of $\mathcal{E}(G_0;D)$ such that

$$0\longleftarrow \mathcal{E}^{\mathbb{R}}_X\otimes_{\rho^{-1}\mathcal{R}}\rho^{-1}\mathcal{M} \longleftarrow (\mathcal{E}^{\mathbb{R}}_X)^{n_0} \xleftarrow{P_1} (\mathcal{E}^{\mathbb{R}}_X)^{n_1} \longleftarrow\cdots\xleftarrow{P_r} (\mathcal{E}^{\mathbb{R}}_X)^{n_r}\longleftarrow 0$$

is exact on a neighborhood of $\rho^{-1}(x^*)$ (for the definition of the extension ring $\mathcal{E}^{\mathbb{R}}_X$ of $\mathcal{E}_X$, see [16,8]). By the conditions (C.1)-(C.3), there exist C_0 and c_0 with $C_0>4$ and $0<c_0<1/4$ such that

(2.1) $\{z\in X;\ |z|\leqq 4nc_0\}\subset D,$

(2.2) $G_0-\{0\}\subset\{(0,z')\in Y;\ \operatorname{Im} z_n < -c_0(|\operatorname{Re} z'|+|\operatorname{Im} z''|)\},$

(2.3) $$\{(x_1,z',\zeta)\in T^*X|_{\tilde{M}_+};\ 0<x_1\leqq c_0,\ |z'|\leqq c_0,\ \zeta\in\mathbb{C}^n-\{0\},$$
$$|\zeta_1|\geqq C_0|\zeta'|\}\cap SS(\mathcal{M}) = \phi,$$

(2.4) $$\{(x_1,z',\zeta);\ 0<x_1\leqq c_0,\ |z'|\leqq c_0,\ |\operatorname{Re}\zeta'|\leqq c_0|\operatorname{Im}\zeta_n|,$$
$$|\operatorname{Im}\zeta''|\leqq c_0|\operatorname{Im}\zeta_n|,\ \operatorname{Im}\zeta_n<0,$$
$$\operatorname{Re}\zeta_1 < -C_0(|\operatorname{Re}\zeta'|+|\operatorname{Im} z'||\operatorname{Im}\zeta_n|)\}\cap SS(\mathcal{M}) = \phi,$$

where we use the notation $z=(z_1,z')$, $\zeta=(\zeta_1,\zeta')$, $z'=(z'',z_n)$, etc. Let a, b, C be parameters such that

(2.5) $0<b<\dfrac{c_0}{8},\quad C\geqq 2C_0,\quad 0<a\leqq a_0=a_0(b)=\dfrac{bc_0}{8C_0},$

and put

$$\Omega=\Omega(a,C)=\{(x_1,z');\ x_1>0,\ x_1+a>\frac{1}{4C_0}(|x'|+\frac{1}{c_0}y_n),$$
$$x_1+a > C(|y''|+\frac{1}{c_0}y_n)\}$$

with the notation $z=x+\sqrt{-1}y$, $y''=(y_2,\ldots,y_{n-1})$. We denote by

$H(a,C)$ the set of C^1-functions h on $\mathbb{R}$ such that

$$\begin{cases} h(0) = 0, \quad 0 < h(x_1) \leqq C_o \quad \text{for} \quad x_1 > 0, \\ 0 \leqq h'(x_1) \leqq C_o \quad \text{for} \quad x_1 \geqq 0, \quad 2\frac{c_o a}{C} < h(a). \end{cases}$$

For $h \in H(a,C)$ put

$$\omega = \omega(a,b,C,h) = \{(x_1,z') \in \Omega(a,b,C);\ x_1 > a \quad \text{or}$$
$$y_n < h(x_1)\exp(\frac{8C_o}{b}x_1) \quad \text{or} \quad y_n < \frac{b}{a_o}(x_1 + a_o)|y''|\}.$$

Let I be the set of (a,b,C,h) with a, b, C satisfying (2.5) and $h \in H(a,C)$. We define an order $\geqq$ in I by

$$(a_1,b_1,C_1,h_1) \geqq (a_2,b_2,C_2,h_2)$$

if and only if $a_1 \leqq a_2$, $b_1 \leqq b_2$, $C_1 \geqq C_2$, $h_j \in H(a_j,C_j)$, and

$$\frac{1}{b_1}h_1(x_1)\exp(\frac{8C_o}{b_1}x_1) \leqq \frac{1}{b_2}h_2(x_1)\exp(\frac{8C_o}{b_2}x_1) \quad \text{for} \quad 0 \leqq x_1 \leqq a_1.$$

Note that in this case we have $\Omega(a_1,C_1) \supset \Omega(a_2,C_2)$ and $\omega(a_1,b_1,C_1,h_1) \supset \omega(a_2,b_2,C_2,h_2)$.

By modifying the argument in [8] so as to apply to cohomology groups with $\mathcal{BO}$-coefficients instead of $\mathcal{O}_X$-coefficients, we can show that $\mathbb{R}\Gamma_{\Omega-\omega}(\Omega;\mathcal{BO})$ is well-defined as an $\mathcal{E}(G_o;D)$-module. We have

$$\mathbb{R}\mathcal{H}om_{\mathcal{R}}(\mathcal{M}, \Gamma_{L_o}(\mathcal{C}_{M+}))_{x^*} \cong \mathbb{R}\mathrm{Hom}_{\mathcal{E}(G_o;D)}(\mathcal{M}^\bullet, \Gamma_{L_o}(\mathcal{C}_{M+})_{x^*})$$
$$\cong \varinjlim_{I} \mathbb{R}\mathrm{Hom}_{\mathcal{E}(G_o;D)}(\mathcal{M}^\bullet, \mathbb{R}\Gamma_{\Omega-\omega}(\Omega;\mathcal{BO}))[n-1].$$

By the argument similar to [8], we can prove

$$\mathbb{R}\mathrm{Hom}_{\mathcal{E}(G_0;D)}(\mathcal{M}^\bullet, \mathbb{R}\Gamma_{\Omega-\omega}(\Omega;\mathcal{BO})) = 0$$

for any $(a,b,C,h) \in I$ (see [14] for the detailed arguments).

2.2. Propagation of micro-analyticity up to the boundary

Theorem 2.1 implies the propagation of micro-analyticity of solutions of boundary value problems both for systems with non-characteristic boundary and for Fuchsian equations.

<u>Theorem 2.2</u>. Let $\mathcal{M}$ be a coherent $\mathcal{D}_X$-module defined on a neighbor-

hood of $\pi_{M/\tilde{M}}(x^*)$ with a point $x^* = (\mathring{x},\sqrt{-1}\mathring{\xi}'\infty)$ of $S^*_N Y$ such that Y (i.e. N) is non-characteristic with respect to $\mathcal{M}$. Suppose moreover that $\mathcal{M}$ satisfies (C.2) and (C.3) of Theorem 2.1. Let $u(x)$ be a $\mathcal{B}_{N|M+}$-solution of $\mathcal{M}$ micro-analytic on

$$\{(x,\sqrt{-1}\xi\infty) \in S^*_M X;\ |x - \mathring{x}| < \varepsilon,\ x_1 > 0,\ |\xi' - \mathring{\xi}'| < \varepsilon,\ \xi_1 \in \mathbb{R}\}$$

with an $\varepsilon > 0$. Then its boundary value $\gamma_+(u)$ is micro-analytic at x^*.

This theorem is an immediate consequence of Theorem 2.1 and the following lemma.

<u>Lemma 2.1</u>. Let $\mathcal{M}$ be a coherent $\mathcal{D}_X$-module satisfying the condition (C.1) of Theorem 2.1. Then the sheaf homomorphism

$$\psi : p^{-1}(\mathcal{H}om_{\mathcal{D}_X}(\mathcal{M}, \mathcal{C}_{M+})|_{L_+}) \longrightarrow \mathcal{H}om_{\mathcal{D}_X}(\mathcal{M}, \mathcal{C}_M)|_{p^{-1}(L_+)}$$

is injective on $p^{-1}(L_+)$.

<u>Example 2.1</u>. The system

$$\mathcal{M} : ((D_1 + \sqrt{-1}x_1^k D_2)^m + D_3^m)u = (D_3 + \sqrt{-1}D_4)u = 0$$

in $M = \mathbb{R}^4$ with positive integers k and m satisfies the conditions of Theorem 2.2 with $x^* = (0,\sqrt{-1}dx_2) \in S^*_N Y$.

We can also treat the systems satisfying the conditions of Theorem 2.2 except (C.3) by reformulating the arguments in Section 1 for $\mathcal{R}$-modules.

<u>Theorem 2.3</u>. Let $\mathcal{M}$ be a coherent $\mathcal{D}_X$-module defined on a neighborhood of $\pi_{M/\tilde{M}}(x^*)$ with $x^* = (\mathring{x},\sqrt{-1}\mathring{\xi}'\infty) \in S^*_N Y$. Assume that Y is non-characteristic with respect to $\mathcal{M}$ and that $\mathcal{M}$ satisfies (C.2) of Theorem 2.1. Then there are coherent $\mathcal{E}_X$-modules $\mathcal{M}_1$ and $\mathcal{M}_2$ such that

$$\mathcal{E}_X \otimes_{\pi^{-1}\mathcal{D}_X} \pi^{-1}\mathcal{M} = \mathcal{M}_1 \oplus \mathcal{M}_2,$$

$$SS(\mathcal{M}_1) \cap \rho^{-1}(x^*) \subset \{(\zeta_1,x^*) \in \rho^{-1}(x^*);\ \mathrm{Re}\,\zeta_1 \geqq 0\},$$

$$SS(\mathcal{M}_2) \cap \rho^{-1}(x^*) \subset \{(\zeta_1,x^*) \in \rho^{-1}(x^*);\ \mathrm{Re}\,\zeta_1 < 0\}.$$

Put $\mathcal{N}_j = \rho_*(\mathcal{M}_j/z_1\mathcal{M}_j)$. Then as an $\mathcal{E}_Y$-module, we have

$$\mathcal{E}_Y \otimes_{\pi'^{-1}\mathcal{D}_Y} \pi'^{-1}\mathcal{M}_Y \cong \mathcal{N}_1 \oplus \mathcal{N}_2,$$

where $\pi' : T^*Y \longrightarrow Y$ is the projection. Let $u(x)$ be a $\mathcal{B}_{N|M+}$-solution of $\mathcal{M}$ micro-analytic on

$$\{(x,\sqrt{-1}\xi\infty) \in S^*_M X;\ |x - \mathring{x}| < \varepsilon,\ x_1 > 0,\ |\xi' - \mathring{\xi}'| < \varepsilon,\ \xi_1 \in \mathbb{R}\}$$

with an $\varepsilon > 0$. Then the image of its boundary value $\gamma_+(u)$ (as a microfunction solution of $\mathcal{M}_Y$) under the homomorphism

$$\mathcal{H}om_{\mathcal{E}_Y}(\mathcal{M}_Y, \mathcal{C}_N) \longrightarrow \mathcal{H}om_{\mathcal{E}_Y}(\mathcal{N}_1, \mathcal{C}_N)$$

(this is induced naturally by the injective homomorphism $\mathcal{N}_1 \longrightarrow \mathcal{M}_Y$) vanishes at x^*. (i.e. $\gamma_+(u)$ satisfies a system of microdifferential equations stronger than $\mathcal{M}_Y$.)

Next let us study Fuchsian partial differential equations.

Theorem 2.4. Let P be a Fuchsian partial differential operator of weight m-k with respect to N defined on a neighborhood of $\pi_{M/\tilde{M}}(x^*)$ with $x^* = (\mathring{x}, \sqrt{-1}\mathring{\xi}'\infty) \in S^*_N Y$ satisfying the condition $C(\mathring{x})$. Assume that there exists $\varepsilon > 0$ such that the principal symbol $\sigma(P)(x;\zeta_1,\sqrt{-1}\xi')$ never vanishes if $x \in \mathbb{R}^n$, $|x - \mathring{x}| < \varepsilon$, $x_1 > 0$, $\mathrm{Re}\,\zeta_1 < 0$, $\xi' \in \mathbb{R}^{n-1}$, $|\xi' - \mathring{\xi}'| < \varepsilon$. Under these conditions, if $u(x)$ is a $\mathcal{B}_{N|M+}$-solution of $Pu = 0$ micro-analytic on

$$\{(x,\sqrt{-1}\xi\infty) \in S^*_M X;\ |x - \mathring{x}| < \varepsilon,\ x_1 > 0,\ |\xi' - \mathring{\xi}'| < \varepsilon,\ \xi_1 \in \mathbb{R}\},$$

then its boundary value $\gamma_+(u) \in (\mathcal{B}_N)^m$ is micro-analytic at x^*.

This is also an immediate consequence of Theorem 2.1 and Lemma 2.1.

Example 2.2. Put $M = \mathbb{R}^3$ and

$$P = x_1(D_1^2 - x_1^k(D_2^2 - D_3^2)) + \sum_{j=1}^{3} a_j(x)D_j + b(x),$$

where a_j and b are analytic on a neighborhood of 0 with $a_1(0) \notin \mathbb{Z}$ and k is a positive integer. Then P satisfies the conditions of Theorem 2.4 with $x^* = (0,\sqrt{-1}dx_2)$.

3. CONTINUATION OF REAL ANALYTIC SOLUTIONS

3.1. Systems with non-characteristic boundary

In this subsection, we assume that N is a real analytic hypersurface of M with respect to which a system $\mathcal{M}$ is non-characteristic. Our aim is to extend Kaneko's theorem ([2]) to such systems. Since the problems treated here are again of local character, we may assume that $N = \{x \in M;\ x_1 = 0\}$. The following lemma is a generalization of a theorem of Komatsu-Kawai [11] to systems. We use the notation $M_\pm = \{x \in M;\ \pm x_1 > 0\}$. For example, $\mathcal{B}_{N|M-}$ and the boundary value map $\gamma_-^{\pm}$ are defined by reversing the sign of x_1. Let $\mathring{x}$ be a point of N.

Lemma 3.1. Let $\mathcal{M}$ and N be as above and let u_+ and u_- be $\mathcal{B}_{N|M+}$- and $\mathcal{B}_{N|M-}$-solution of $\mathcal{M}$ respectively. Then there exists a hyperfunction solution u of $\mathcal{M}$ on a neighborhood of $\mathring{x}$ such that $u = u_\pm$ on $M_\pm$ if and only if $\gamma_+(u_+) = \gamma_-(u_-)$. Moreover, such u is unique.

We denote by $V^+_{N,A}(\mathcal{M})$ the complement in $S^*_N Y$ of the set of points x^* such that conditions (C.2),(C.3) of Theorem 2.1 are satisfied with x^*. We define $V^-_{N,A}(\mathcal{M})$ by reversing the sign of x_1 (hence also of dz_1) and put

$$V_{N,A}(\mathcal{M}) = V^+_{N,A}(\mathcal{M}) \cup V^-_{N,A}(\mathcal{M}).$$

Then by the same argument as in [2], we get the following theorem from Theorem 2.2.

Theorem 3.1. Let $\mathcal{M}$ be a coherent $\mathcal{D}_X$-module with respect to which N is non-characteristic. Let $\mathring{x}$ be a point of N and ϕ be a real valued C^1-function on N such that $\phi(\mathring{x}) = 0$ and $d\phi(\mathring{x}) \neq 0$. Assume that K is a closed subset of N such that $\phi \leqq 0$ on K and that $V_{N,A}(\mathcal{M})$ does not contain both of the points $(\mathring{x}, \pm\sqrt{-1}d\phi(\mathring{x})\infty) \in \sqrt{-1}S^*N$. Then any real analytic solution u of $\mathcal{M}$ on $U - K$ with an open neighborhood U of $\mathring{x}$ is uniquely continued to a neighborhood of $\mathring{x}$ as a hyperfunction solution of $\mathcal{M}$. Moreover, if for any point y^* of $\pi^{-1}(\mathring{x}) \cap (T^*_M X - 0)$ there exist $a, b \in \mathbb{R}$ such that $\mathcal{M}$ is micro-hyperbolic in the direction $adx_1 + bd\phi$ at y^* in the sense of [8], then the continued solution is real analytic at $\mathring{x}$.

Example 3.1. Put $M = \mathbb{R}^4$ and define a system $\mathcal{M}$ by

$$\mathcal{M} : (D_1^3 + D_3^3)u = D_2(D_3^2 + D_4^2)u = 0.$$

Then any real analytic solution of $\mathcal{M}$ defined on $U - \{x \in U;\ x_1 = 0,\ x_2 \leqq 0\}$ with an open neighborhood U of 0 is continued to a neighborhood of 0 as a real analytic solution of $\mathcal{M}$. In fact, it is easy to see that $(0,\sqrt{-1}dx_2)$ is not contained in $V_{N,A}(\mathcal{M})$ and that $\mathcal{M}$ is micro-hyperbolic in the direction dx_1 if $\xi_2 \neq 0$, and in the direction dx_2 if $\xi_2 = 0$, at $y^* = (0,\sqrt{-1}\xi)$.

3.2. Fuchsian equations

In this subsection we generalize theorems of Kaneko in [2,3] to Fuchsian equations with respect to a real analytic hypersurface N of $M = \mathbb{R}^n$. Again, we may assume that $N = \{x \in M;\ x_1 = 0\}$.

Lemma 3.2. Let P be a Fuchsian partial differential operator of weight $m-k$ with respect to N defined on a neighborhood of $\mathring{x} \in N$ satisfying $C(\mathring{x})$. Let u_+ and u_- be $\mathcal{B}_{N|M+}$- and $\mathcal{B}_{N|M-}$-solution of $Pu = 0$ respectively. Then there exists a hyperfunction solution u of $Pu = 0$ on a neighborhood of $\mathring{x}$ such that $u = u_\pm$ on $M_\pm$ if and only if $\gamma_{+reg}(u_+) = \gamma_{-reg}(u_-)$. Moreover such u is unique. In addition, if $\gamma_{+sing}(u_+) = \gamma_{-sing}(u_-) = 0$, then u has x_1 as a real analytic parameter.

We define a closed subset $V_{N,A}(P)$ of S_N^*Y as follows: a point $x^* = (\mathring{x}, \sqrt{-1}\mathring{\xi}'\infty)$ is not contained in $V_{N,A}^+(P)$ if and only if there exists $\varepsilon > 0$ such that $\sigma(P)(x; \zeta_1, \sqrt{-1}\xi') \neq 0$ for $|x - \mathring{x}| < \varepsilon$, $x_1 > 0$, $\mathrm{Re}\,\zeta_1 < 0$, $|\xi' - \mathring{\xi}'| < \varepsilon$. We define $V_{N,A}^-(P)$ by reversing the sign of x_1 and of $\mathrm{Re}\,\zeta_1$. We put

$$V_{N,A}(P) = V_{N,A}^+(P) \cup V_{N,A}^-(P).$$

Then by Theorem 2.4 we get the following:

Theorem 3.2. Let P be a Fuchsian partial differential operator of weight $m-k$ with respect to N defined on a neighborhood of $\mathring{x} \in N$ with the condition $C(\mathring{x})$. Let ϕ be a real valued C^1-function on N such that $\phi(\mathring{x}) = 0$, $d\phi(\mathring{x}) \neq 0$ and K be a closed subset of N such that $\phi \leqq 0$ on K. Assume that $V_{N,A}(P)$ does not contain both of the points $(\mathring{x}, \pm\sqrt{-1}d\phi(\mathring{x}))$. Then any real analytic solution $u(x)$ of $Pu = 0$ on $U - K$ with a neighborhood U of $\mathring{x}$ is uniquely continued to a neighborhood of $\mathring{x}$ as a hyperfunction solution of $Pu = 0$. Moreover, the continued solution has x_1 as a real analytic parameter.

By studying the propagation of micro-analyticity in the interior for the continued solution, we get the following theorem on removable singularities of real analytic solutions.

Theorem 3.3. Let P be a Fuchsian partial differential operator of weight $m-k$ with respect to N satisfying $C(\mathring{x})$ with a point $\mathring{x}$ of N. Assume that the principal symbol of P is written in the form

$$\sigma(P)(x,\xi) = x_1^k p(x,\xi)$$

with a real valued real analytic function $p(x,\xi)$ such that

(i) $\mathrm{grad}_\xi p(\mathring{x},\xi) \neq 0$ for any $\xi \in \mathbb{R}^n - \{0\}$ such that $p(\mathring{x},\xi) = 0$,

(ii) there exists $\xi' \in \mathbb{R}^{n-1} - \{0\}$ such that the equation $p(\mathring{x},\zeta_1,\xi') = 0$ in ζ_1 has m real distinct roots.

Under these assumptions, any real analytic solution of $Pu = 0$ defined on $U - \{\mathring{x}\}$ with a neighborhood U of $\mathring{x}$ is continued to U as a real analytic function.

Example 3.2. Let U be an open subset of $\mathbb{R}^n$ containing 0 and put

$$P = x_1(D_1^2 + \cdots + D_r^2 - D_{r+1}^2 - \cdots - D_n^2) + \sum_{j=1}^{n} a_j(x)D_j + b(x),$$

where a_j and b are real analytic functions on U with $a_1(0) \notin \mathbb{Z}$, and $1 \leqq r \leqq n-1$. Then any real analytic solution of $Pu = 0$ defined on $U - \{0\}$ is continued to U as a real analytic function.

Example 3.3. Let U be an open subset of $\mathbb{R}^3$ containing 0 and put

$$P = x_1(D_1^3 - 3D_1D_2^2 + D_2^3 + D_3^3) + \sum_{1\leqq i,j\leqq 3} a_{ij}(x)D_iD_j + \sum_{j=1}^{3} a_j(x)D_j + b(x),$$

where a_{ij}, a_j, b are real analytic on U with $a_{11}(0) \notin \mathbb{Z}$. Then any real analytic function $u(x)$ on $U - \{0\}$ satisfying $Pu(x) = 0$ is continued to U as a real analytic function.

ACKNOWLEDGEMENT

This work has been supported in part by the Japan Association for Mathematical Sciences.

REFERENCES

[1] Baouendi, M. S., Goulaouic, C., Cauchy problems with characteristic initial hypersurface, Comm. Pure Appl. Math. 26 (1973), 455-475.

[2] Kaneko, A., Singular spectrum of boundary values of solutions of partial differential equations with real analytic coefficients, Sci. Pap. Coll. Gen. Educ. Univ. Tokyo, 25 (1975), 59-68.

[3] Kaneko, A., On continuation of regular solutions of linear partial differential equations, Publ. Res. Inst. Math. Sci. 12 Suppl. (1977), 113-121.

[4] Kashiwara, M., Systems of microdifferential equations, Birkhäuser, Boston-Basel-Stuttgart, 1983.

[5] Kashiwara, M., Kawai, T., On the boundary value problem for elliptic system of linear differential equations, I, II, Proc. Japan Acad. 48 (1972), 712-715; 49, 164-168 (1973).

[6] Kashiwara, M., Laurent, Y., Théorèmes d'annulation et deuxième microlocalisation, Prépublications, Univ. Paris-Sud, 1983.

[7] Kashiwara, M., Oshima, T., Systems of differential equations with regular singularities and their boundary value problems, Ann. of Math. 106 (1977), 145-200.

[8] Kashiwara, M., Schapira, P., Micro-hyperbolic systems, Acta Math. 142 (1979), 1-55.

[9] Kataoka, K., Micro-local theory of boundary value problems, I, II, J. Fac. Sci. Univ. Tokyo 27 (1980), 355-399; 28 (1981), 31-56.

[10] Kataoka, K, Microlocal analysis of boundary value problems with applications to diffraction, Singularities in Boundary Value Problems, ed. H. G. Garnir, pp.121-131, D. Reidel, Dordrecht-Boston-London, 1980.
[11] Komatsu, H., Kawai, T., Boundary values of hyperfunction solutions of linear partial differential equations, Publ. Res. Inst. Math. Sci. 7 (1971), 95-104.
[12] Ôaku, T., F-mild hyperfunctions and Fuchsian partial differential equations. Advanced Studies in Pure Math. 4 (1984), 223-242.
[13] Ôaku, T., A new formulation of local boundary value problem in the framework of hyperfunctions, I, II, III, Proc. Japan Acad. 60 (1984), 283-286; 61(1985), 129-132; 61(1985) (in press).
[14] Ôaku, T., Boundary value problems for systems of linear partial differential equations and propagation of micro-analyticity, Preprint (1985).
[15] Ôaku, T., Removable singularities of solutions of linear partial differential equations — Systems and Fuchsian equations —, Preprint (1985).
[16] Sato, M., Kawai, T., Kashiwara, M., Microfunctions and pseudo-differential equations, Lecture Notes in Math. No. 287, pp.265-529, Springer, Berlin-Heidelberg-New York, 1973.
[17] Schapira, P., Problème de Dirichlet et solutions hyperfonctions des équations elliptiques, Boll. Un. Mat. Ital. (Serie 4) 2 (1969), 367-372.
[18] Schapira, P., Propagation at the boundary and reflection of analytic singularities of solutions of linear partial differential equations I, Publ. Res. Inst. Math. Sci. 12 Suppl. (1977), 441-453.
[19] Schapira, P., Propagation at the boundary of analytic singularities, Singularities in Boundary Value Problems, ed. H. G. Garnir, pp.185-212, D. Reidel, Dordrecht-Boston-London, 1980.
[20] Sjöstrand, J., Analytic singularities and microhyperbolic boundary value problems. Math. Ann. 254 (1980), 211-256.
[21] Tahara, H., Fuchsian type equations and Fuchsian hyperbolic equations, Japan. J. Math. 5 (1979), 245-347.

STUDY OF SHEAVES OF SOLUTIONS OF MICRODIFFERENTIAL SYSTEMS

Pierre SCHAPIRA
Université Paris-Nord - CSP
Avenue J.B. Clément,
93430 VILLETANEUSE FRANCE

ABSTRACT. We recall some constructions and results of Kashiwara-Schapira ([8],[9]) with emphasis on the applications to the study of microdifferential systems.

1. MICRO-SUPPORT

Let X be a real manifold of class C^{α} ($2 \leq \alpha \leq \infty$ or $\alpha = \omega$, i.e. X is real analytic). We denote by $\pi : T^*X \longrightarrow X$ the cotangent bundle to X, by ω_X the canonical 1-form on T^*X. If M is a submanifold we denote by T_MX (resp. T_M^*X) the normal (resp. conormal) bundle to M. In particular T_X^*X denotes the zero section of T^*X, that one identifies to X.

We denote by $\underline{\omega}_X$ the orientation sheaf on X. If M is a submanifold of X, we write $\underline{\omega}_{M/X}$ for $\underline{\omega}_M \otimes (\underline{\omega}_X|_M)$.

We denote by $D^+(X)$ the derived category of the category of complexes, bounded from below, of sheaves of abelian groups on X. We denote by $D^b(X)$ the subcategory consisting of complexes with bounded cohomology, (cf. [3]).

Thus an object $\underline{F}$ of $D^+(X)$ is represented by a complex of sheaves :

$$\underline{F} \cong \; \ldots \longrightarrow \underline{F}^i \xrightarrow[d]{} \underline{F}^{i+1} \longrightarrow \ldots$$

H. G. Garnir (ed.), Advances in Microlocal Analysis, 271–289.

with $\underline{F}^i = 0$ for $i \ll 0$. Moreover two complexes which are quasi-isomorphic are identified in $D^+(X)$, and any object $\underline{F}$ of $D^+(X)$ may be represented by a complex of flabby sheaves.

We denote by $\underline{F}[k]$ the shifted complex : $(\underline{F}[k])^i = \underline{F}^{i+k}$, the differential d being replaced by $(-1)^k d$.

We identify a sheaf $\underline{F}$ with the complex $\ldots \longrightarrow 0 \longrightarrow \underline{F} \longrightarrow 0 \longrightarrow \ldots$ concentrared in degree 0 . Remark that one often incorrectly call an object of $D^+(X)$ "a sheaf on X" .

Definition 1.1. : Let $\underline{F} \in Ob(D^+(X))$. The micro-support of $\underline{F}$, denoted $SS(\underline{F})$, is the subset of T^*X defined by :

$p \notin SS(\underline{F}) \Longleftrightarrow$ there exists an open neighborhood U of p in T^*X such that for any $x_1 \in X$, any real function ϕ of class C^α , defined in a neighborhood of x_1 with $\phi(x_1) = 0$, $d\phi(x_1) \in U$, we have :

$$(\mathbb{R}\Gamma_{\{x;\phi(x) \geqq 0\}}(\underline{F}))_{x_1} = 0 .$$

Recall that if Z is a locally closed subset of X (here $Z = \{x;\phi(x) \geqq 0\}$) , the complex $\mathbb{R}\Gamma_Z(\underline{F})$ is calculated by representing $\underline{F}$ by a complex of flabby sheaves and applying the functor $\Gamma_Z(\cdot)$, where $\Gamma_Z(\underline{F})$ is the subsheaf of $\underline{F}$ of sections with support in Z .

In this paper, we shall write $H^j_Z(\underline{F})$ instead of $H^j(\mathbb{R}\Gamma_Z(\underline{F}))$.

Roughly speaking, when $\underline{F}$ is a sheaf, $p \notin SS(F)$ means that $\underline{F}$ has no section, and no "cohomology" supported by "half-spaces" whose conormal lies in a neighborhood of p .

Similarly if $u : \underline{F} \longrightarrow \underline{G}$ is a morphism in $D^+(X)$, we define SS(u) as $SS(\underline{H})$, where $\underline{H}$ is the simple complex associated to the double complex $\underline{F} \longrightarrow \underline{G}$ (i.e. : the "mapping cone" of u).

It follows immediately by the definition that :

- $SS(\underline{F})$ is a closed cone in T^*X ,
- $SS(\underline{F}) \cap T^*_X X = \mathrm{supp}(\underline{F})$, where $\mathrm{supp}(F) = \overline{\bigcup_j \mathrm{supp}\, H^j(\underline{F})}$ is the support of the complex $\underline{F}$,
- If $0 \longrightarrow \underline{F}_1 \longrightarrow \underline{F}_2 \longrightarrow \underline{F}_3 \longrightarrow 0$ is an exact sequence of sheaves (or more generally if we have a distinguished triangle $\underline{F}_1 \longrightarrow \underline{F}_2 \longrightarrow \underline{F}_3 \longrightarrow \underline{F}_1[+1]$ in $D^+(X)$), then :

$$SS(\underline{F}_i) \subset SS(\underline{F}_j) \cup SS(\underline{F}_k) \quad \text{if} \quad \{i,j,k\} = \{1,2,3\} .$$

Theorem 1.2. : $SS(\underline{F})$ is an involutive subset of T^*X .

This result is proved by making contact transformations operate on sheaves.

Proposition 1.3. : Assume X is open in a vector space E . Let $p = (x_o,\xi_o) \in T^*X \cong X \times E^*$, and let $\underline{F} \in \mathrm{Ob}(D^+(X))$. Then :
$p \notin SS(\underline{F}) \iff$ there exists a neighborhood V of x_o , $\varepsilon > 0$ and a closed proper convex cone G in E , with $0 \in G$, $G \setminus \{0\} \subset \{\gamma \in E ; \langle\gamma,\xi_o\rangle < 0\}$, such that, setting $H = \{x \in E ; \langle x-x_o,\xi_o\rangle \geq -\varepsilon\}$, $L = \{x \in E ; \langle x-x_o,\xi_o\rangle = -\varepsilon\}$, the natural morphism :

$$\mathbb{R}\Gamma((x+G) \cap H ; \underline{F}) \longrightarrow \mathbb{R}\Gamma((x+G) \cap L ; \underline{F})$$

is an isomorphism for any $x \in V$.

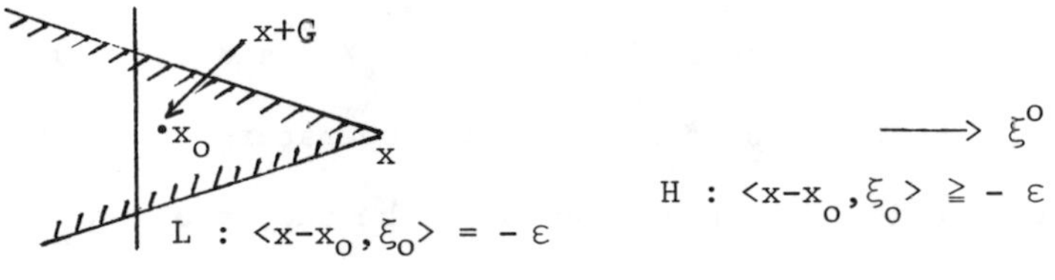

Example 1.4. i) Assume X is a vector space. Let C be a closed convex cone with vertex at 0. Let C^{o} be the dual cone in X^{*} :

$$C^{o} = \{\xi \in X^{*} \; ; \; \langle x,\xi\rangle \geqq 0 \quad \forall x \in C\} .$$

Let $\underline{\mathbb{Z}}_{C}$ be the constant sheaf on C in X (recall that $(\underline{\mathbb{Z}}_{C})_{x} = \mathbb{Z}$ if $x \in C$, $= 0$ if $x \notin C$). Then :

$$SS(\underline{\mathbb{Z}}_{C}) \cap \pi^{-1}(0) = C^{o}$$

In particular if now X is a manifold and M is a closed submanifold, we have :

$$SS(\underline{\mathbb{Z}}_{M}) = T^{*}_{M}X .$$

ii) Let ϕ be a real C^{1}-function on X,

$$M^{+} = \{x \; ; \; \phi(x) > 0\}, \quad \overline{M}^{+} = \{x \; ; \; \phi(x) \geqq 0\} ,$$

and assume $d\phi \neq 0$ on $\{x \; ; \; \phi(x) = 0\}$.

Then :

$$SS(\underline{\mathbb{Z}}_{\overline{M}^{+}}) = \{(x,\xi) \in T^{*}X \; ; \; \phi(x) \geqq 0, \; \xi = 0 \text{ or } \phi(x) = 0, \; \xi = \lambda d\,\phi(x), \; \lambda \geqq 0\}$$

$$SS(\underline{\mathbb{Z}}_{M^{+}}) = \{(x,\xi) \in T^{*}X \; ; \; \phi(x) \geqq 0 , \; \xi = 0 \text{ or } \phi(x) = 0, \; \xi = \lambda d\,\phi(x), \; \lambda \leqq 0\} .$$

2. MICRODIFFERENTIAL SYSTEMS

In this section we denote by X a complex manifold. We also make use of $X^{\mathbb{R}}$, the real underlying manifold, but we often identify X and $X^{\mathbb{R}}$. Similarly we often identify $T^{*}X$ and $(T^{*}X)^{\mathbb{R}} \cong T^{*}X^{\mathbb{R}}$.

Let $\mathcal{O}_{X}$ be the sheaf of holomorphic functions on X, and let $\mathcal{D}_{X}$ be the sheaf of rings of holomorphic differential operators of finite order on X. Let us recall a few basic facts of the theory of $\mathcal{D}_{X}$-

modules, and refer to [2] or [14] for a detailed exposition.
The Ring $\mathcal{D}_X$ is coherent and noetherian. If $\mathcal{M}$ is a coherent left $\mathcal{D}_X$-module then locally on X, $\mathcal{M}$ is quasi-isomorphic to a bounded finite-free complex $\mathcal{M}^\bullet$:

$$\mathcal{M}^\bullet : 0 \longrightarrow \mathcal{D}_X^{N_p} \xrightarrow[P_{p-1}]{} \cdots \longrightarrow \mathcal{D}_X^{N_1} \xrightarrow[P_o]{} \mathcal{D}_X^{N_o} \longrightarrow 0$$

The P_j's are matrices of differential operators acting on the right, the complex is exact except in degree 0, and $\mathcal{M} \cong \mathcal{D}_X^{N_o} / \mathcal{D}_X^{N_1} \cdot P_o$.

The Ring $\mathcal{D}_X$ is filtered by the subgroups $\mathcal{D}_X(k)$ consisting of differential operators of order at most k, and the associated graded Ring,

$$\mathrm{gr}\ \mathcal{D}_X = \bigoplus_k \mathcal{D}_X(k) / \mathcal{D}_X(k-1) ,$$

is identified to the subring of $\pi_* \mathcal{O}_{T^*X}$ of sections which are polynomials in the fibers of T^*X.
Let us endow locally $\mathcal{M}$ with a good filtration. For example in the preceding situation one can endow $\mathcal{M}$ with the filtration

$$\mathcal{M}_k = \mathcal{D}_X^{N_o}(k) / (\mathcal{D}_X^{N_1} \cdot P_o \cap \mathcal{D}_X^{N_o}(k)) .$$

Let $\mathrm{gr}(\mathcal{M})$ be the associated graded Module.
Then $\mathrm{gr}(\mathcal{M})$ is a coherent $\mathrm{gr}(\mathcal{D}_X)$-module whose support in T^*X is a closed conic analytic subset which depends only on $\mathcal{M}$, not on the choice of the good filtration. This is the characteristic variety of $\mathcal{M}$, denoted $\mathrm{char}(\mathcal{M})$.

Now consider the "sheaf of holomorphic solutions" of $\mathcal{M}$, that is, the complex $\mathbb{R}\mathcal{H}om_{\mathcal{D}_X}(\mathcal{M}, \mathcal{O}_X)$. If $\mathcal{M}$ is represented by the complex $\mathcal{M}^{\bullet}$, then $\mathbb{R}\mathcal{H}om_{\mathcal{D}_X}(\mathcal{M}, \mathcal{O}_X) \cong \mathcal{H}om_{\mathcal{D}_X}(\mathcal{M}^{\bullet}, \mathcal{O}_X)$ is represented by the complex :

$$0 \longrightarrow \mathcal{O}_X^{N_o} \xrightarrow[P_o]{} \mathcal{O}_X^{N_1} \longrightarrow \dots \longrightarrow \mathcal{O}_X^{N_p} \longrightarrow 0$$

where now the P_j's operate on the left.

Theorem 2.1. : One has the equality :

$$SS(\mathbb{R}\mathcal{H}om_{\mathcal{D}_X}(\mathcal{M}, \mathcal{O}_X)) = \mathrm{char}(\mathcal{M}) .$$

For the proof of the inclusion "$\subset$", we proceed as follows : by a standart argument, we may reduce the proof to the case where $\mathcal{M} = \mathcal{D}_X / \mathcal{D}_X P$, that is, to the case where $\mathbb{R}\mathcal{H}om_{\mathcal{D}_X}(\mathcal{M}, \mathcal{O}_X)$ is simply the complex :

$$0 \longrightarrow \mathcal{O}_X \xrightarrow[P]{} \mathcal{O}_X \longrightarrow 0$$

where P is a single differential operator. Then the result follows immediately from the Cauchy-Kowalewski theorem, in the refined formulation of Leray [11], by applying Proposition 1.3. .

Remark 2.2. : By this theorem, we obtain a new proof of the involutivity of $\mathrm{char}(\mathcal{M})$.

3. SPECIALIZATION AND MICROLOCALIZATION

If $E \longrightarrow Z$ is a vector bundle over a manifold Z, we denote by $D^+_{conic}(E)$ the subcategory of $D^+(E)$ consisting of complexes of sheaves whose cohomology groups are locally constant on the half-lines of E (the orbits of the action of $\mathbb{R}^+$).

Let X be a (real) manifold, M a submanifold, $\underline{F}$ an object of $D^+(X)$. We shall associate, with M. Sato and Sato-Kashiwara-Kawaï [12] an object $\nu_M(\underline{F})$ of $D^+(T_MX)$ and an object $\mu_M(\underline{F})$ of $D^+(T^*_MX)$, respectively called the "specialization" and the "microlocalization" of $\underline{F}$ along M. These sheaves will be characterized by Propositions 3.2. and 3.3. below.

In order to give precise statements we need to recall the notion of "normal cones".

Definition 3.1. Let S and Y be two subsets of X. The normal cone of S along Y, denoted $C(S,Y)$ is the subset of TX defined by

$$C(S,Y) = \bigcup_{x \in X} C_x(S,Y)$$

$\theta \in C_x(S,Y) \iff$ there exists (in a choice of local coordinates on X) a sequence $\{(c_n, s_n, y_n)\}$ in $\mathbb{R}^+ \times S \times Y$ such that $s_n \xrightarrow[n]{} x$, $y_n \xrightarrow[n]{} x$, $c_n(s_n - y_n) \xrightarrow[n]{} \theta$.

If Y is a submanifold of X, then $C(S,Y) + TY \subset C(S,Y)$, and one denotes by $C_Y(S)$ the image of $C(S,Y)$ in T_YX.

Proposition 3.2. : Let $\underline{F} \in \mathrm{Ob}(D^+(X))$. Then :

i) $\nu_M(\underline{F}) \in \mathrm{Ob}(D^+_{conic}(T_MX))$.

ii) $\nu_M(\underline{F})|_M \cong \underline{F}|_M$.

iii) Let V be a conic open subset of T_MX. Then :

$$H^j(V, \nu_M(\underline{F})) = \varinjlim_U H^j(U,F)$$

where U runs over the family of open subsets of X such that $C_M(X \setminus U) \cap V = \emptyset$.

iv) Let A be a closed conic subset of $T_M X$. Then :

$$H^j_A(T_M X , \nu_M(\underline{F})) = \varinjlim_{Z,U} H^j_Z(U,\underline{F})$$

where Z runs over the family of closed subsets of X such that $C_M(Z) \subset A$, and U runs overs the family of open neighborhoods of M in X .

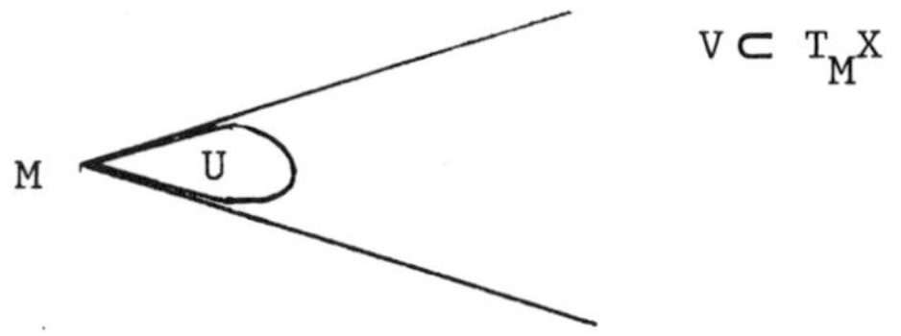

Proposition 3.3. : Let $\underline{F} \in \mathrm{Ob}\,(D^+(X))$. Then :

i) $\mu_M(F) \in \mathrm{Ob}(D^+_{conic}(T^*_M X))$.

ii) $\mu_M(\underline{F})\big|_M \cong \mathbb{R}\Gamma_M(\underline{F})\big|_M$.

iii) Let U be a convex open proper cone of $T^*_M X$. Then :

$$H^j(U,\mu_M(\underline{F})) = \varinjlim_{Z,V} H^j_Z(V,\underline{F}) ,$$

where (Z,V) runs over the family of closed subsets Z of X and open subsets V of X such that :

$$V \cap M = \pi(U) , \quad C_M(Z) \subset U^o .$$

iv) Let Z be a closed convex proper cone of $T^*_M X$. Then :

$$H^j_Z(T^*_M X,\mu_M(\underline{F})) = \varinjlim_{U} H^{j-\ell}(U,\underline{F}) \otimes \underline{\omega}_{M/X}$$

where U runs over the family of open subsets of X such that $C_M(X \setminus U) \cap \mathrm{Int}\, Z^{oa} = \emptyset$, and $\ell = \mathrm{codim}\, M$ (here $Z^{oa} = - Z^o$).

Now we shall evaluate $SS(\mu_M(\underline{F}))$. We identify TT^*X and T^*T^*X by the Hamiltonian isomorphism $-H$. If (x) is a system of local coordinates on X , $(x;\xi)$ the associated coordinates on T^*X , with $\omega_X = \sum_i \xi_i\, dx_i$, then :

$$-H(<\lambda,dx> + <\mu,d\xi>) \;=\; <\lambda\,,\frac{\partial}{\partial\xi}> - <\mu\,,\frac{\partial}{\partial x}> \;.$$

In particular if Λ is a Lagrangean conic submanifold of T^*X , we shall identify $T_\Lambda T^*X$ and $T^*\Lambda$ by $-H$.

Finally let us remark that if M is a submanifold of X , the two bundles T^*T_MX and $T^*T^*_MX$ are naturally isomorphic, and T^*M is naturally embedded into $T^*T^*_MX$. If (y,t) is a system of local coordinates on X , with $M = \{(y,t)\ ;\ t = 0\}$ $(y,t\ ;\ \eta,\tau)$ the associated coordinates on T^*X , $(y,\tau\ ;\ <\eta,dy> + <t,d\tau>)$ the coordinates on $T^*T^*_MX$, then the isomorphism $T^*T^*_MX \cong T^*T_MX$ is given by $(y,\tau\ ;\ \eta,t) \longrightarrow (y,-t\ ;\ \eta,\tau)$ and T^*M is embedded into $T^*T^*_MX$ by $(y\ ;\ \eta) \longmapsto (y,0\ ;\ \eta,0)$.

Theorem 3.4. : Let $\underline{F} \in Ob(D^+(X))$. Then :

i) $SS(\mu_M(\underline{F})) = SS(\nu_M(\underline{F}))$.

ii) $SS(\mu_M(\underline{F})) \subset C_{T^*_MX}(SS(\underline{F}))$.

iii) $SS(\underline{F}|_M) \subset T^*M \cap C_{T^*_MX}(SS(\underline{F}))$.

iv) $SS(\mathbb{R}\Gamma_M(\underline{F})|_M) \subset T^*M \cap C_{T^*_MX}(SS(\underline{F}))$.

Example 3.5. : Let us come back to the situation of §2., and let $\mathcal{M}$ be a coherent $\mathcal{D}_X$-module (where now X is a complex manifold).
Let M be a real C^2-submanifold of X . Applying Proposition 2.1. we find :

$$SS(\mathbb{R}\mathcal{H}om_{\mathcal{D}_X}(\mathcal{M}, \mu_M(\mathcal{O}_X)) \subset C_{T^*_M X}(\mathrm{char}(\mathcal{M})) .$$

In particular assume M is a real analytic manifold and X is a complexification of M. Let $n = \dim M$. Recall that the sheaves B_M and C_M of Sato's hyperfunctions and Sato's microfunctions respectively, are defined by :

$$B_M = \mathbb{R}\Gamma_M(\mathcal{O}_X)[n] \otimes \underline{\omega}_M$$
$$C_M = \mu_M(\mathcal{O}_X)[n] \otimes \underline{\omega}_M$$

and these sheaves are concentrated in degree 0.

Thus we get :

$$SS(\mathbb{R}\mathcal{H}om_{\mathcal{D}_X}(\mathcal{M}, C_M)) \subset C_{T^*_M X}(\mathrm{char}(\mathcal{M}))$$
$$SS(\mathbb{R}\mathcal{H}om_{\mathcal{D}_X}(\mathcal{M}, B_M)) \subset T^*M \cap C_{T^*_M X}(\mathrm{char}(\mathcal{M})) .$$

Since $C_{T^*_M X}(\mathrm{char}(\mathcal{M}))$ is the set of non micro-hyperbolic directions, we recover classical results (cf. [8 Theorem 10.5.1.] for generalizations to $\mathcal{E}^{\mathbb{R}}_X$-modules).

4. COMPLEX CONTACT TRANSFORMATIONS

In order to state our next result, let us recall the classical notion of "Maslov index" associated to three Lagrangean planes, (cf. [5],[10]). Let (E,σ) be a (real) symplectic vector space, $\lambda_1, \lambda_2, \lambda_3$ three Langrangean planes (plane = linear subspace). The index $\tau(\lambda_1,\lambda_2,\lambda_3)$ is the signature of the quadratic form q on $\lambda_1 \oplus \lambda_2 \oplus \lambda_3$ defined by :

$$q(x_1,x_2,x_3) = \sigma(x_1,x_2)+\sigma(x_2,x_3)+\sigma(x_3,x_1), (x_1,x_2,x_3) \in \lambda_1 \oplus \lambda_2 \oplus \lambda_3 .$$

Now let X be a complex manifold, M a real C^2-submanifold. Let

$p \in T^*_M X$. We set :

$$s(M,p) = \frac{1}{2}\,\tau(\lambda_M(p),\ i\lambda_M(p),\ \lambda_o(p))$$

where $\lambda_M(p) = T_p\ T^*_M X$, $\lambda_o(p) = T_p\ \pi^{-1}\pi(p)$, and τ is associated to the real symplectic structure of T^*X , i.e. to $\mathrm{Re}\, d\omega_X$.

Theorem 4.1. : Let X and Y be two complex manifolds, ϕ a complex contact transformation from $\Omega_X \subset T^*X$ to $\Omega_Y \subset T^*Y$. Let M and N be two real submanifolds of X and Y respectively, and assume ϕ induces an isomorphism $T^*_M X \cap \Omega_X \xrightarrow[\phi]{\sim} T^*_N Y \cap \Omega_Y$. Then :

i) the function $s(M,p) - s(N,\phi(p))$ is locally constant on $T^*_M X \cap \Omega_X$,

ii) locally on Ω_X , we may quantize ϕ as an isomorphism :

$$\phi_*\ \mu_M(\mathcal{O}_X) \cong \mu_N(\mathcal{O}_Y)\ [d]$$

where $d = \frac{1}{2}[(\dim M + s(M,p)) - (\dim N + s(N,\phi(p))]$ and $p \in \Omega_X \cap T^*_M X$.

We can compare the index $s(M,p)$ to the index of the Levi form of M.

For that purpose, set :

$$E(p) = T_p T^*X\ , \qquad \sigma(p) = d\omega_X(p)$$

$$\lambda_M(p) = T_p\ T^*_M X$$

$$\lambda_o(p) = T_p\ \pi^{-1}\pi(p)$$

$$\dim_{\mathbb{C}} X = n$$

$$\dim(\lambda_M(p) \cap \lambda_o(p)) = m (= \mathrm{codim}_{\mathbb{R}} M)$$

$$\dim_{\mathbb{C}}(\lambda_M(p) \cap i\lambda_M(p) \cap \lambda_o(p)) = \delta(p)$$

$$\dim_{\mathbb{C}}(\lambda_M(p) \cap i\lambda_M(p)) = d(p)$$

We have already defined the integer $s(M,p)$ as $\frac{1}{2}\,\tau(\lambda_M(p),\ i\lambda_M(p),\ \lambda_o(p))$ where τ is the index associated to the symplectic form

Re $\sigma(p)$ on $E(p)$. Now we define $s^+(M,p)$ and $s^-(M,p)$ by :

$$s^+(M,p) - s^-(M,p) = s(M,p)$$

$$s^+(M,p) + s^-(M,p) = n - m + 2\delta(p) - d(p)$$

Remark that :

$$\delta(p) = \operatorname{codim}_{\mathbb{C}}(T_{\pi(p)}M + i\, T_{\pi(p)}M)$$

This number is of course equal to zero if M is a real hypersurface. More generally $\delta(p) = 0$ is equivalent to saying that the submanifold M is non characteristic in $X^{\mathbb{R}}$ for the Cauchy-Riemann system $\bar{\partial}$.

Example 4.2. : Assume M is a real hypersurface, $\phi(x) = 0$ an equation of M , with $d\phi \neq 0$ on M . Let :

$$T_x^{\mathbb{C}}M = \{v \in T_xX \; ; \; \langle v \, , \, \partial\phi(x)\rangle = 0\}$$

where $\partial\phi$ is the differential of ϕ with respect to the holomorphic variables. Let L_ϕ be the Levi form of ϕ on $T_x^{\mathbb{C}}M$. Recall that if $(x_1,\ldots,x_n)$ is a system of holomorphic coordinates on X , $(\bar{x}_1,\ldots,\bar{x}_n)$ the complex conjugate coordinates, then L_ϕ is represented by the matrix $\left(\frac{\partial^2\phi}{\partial x_i \partial \bar{x}_j}\right)_{(1\leq i,j\leq n)}$ on $T_x^{\mathbb{C}}M$.

Proposition 4.3. : In the situation of Example 4.2., $s^+(M,d\phi(x))$ and $s^-(M,d\phi(x))$ are respectively the number of positive and negative eigenvalues of L_ϕ on $T_x^{\mathbb{C}}M$.

(cf. [13] for a related result).

Proposition 4.4. : Let M be a real submanifold of class C^2 of X, $p \in T_M^*X$. Let $\nu(p)$ be the complex line of T_pT^*X, the tangent space at p of $\mathbb{C}^\times p$. Assume $\dim_{\mathbb{R}}(T_p T_M^*X \cap \nu(p)) = 1$.

Then :

i) $H^j(\mu_M(\mathcal{O}_X))_p = 0$ for $j \notin [m+s^-(M,p)-\delta(p), n-s^+(M,p)+\delta(p)]$.

ii) Assume moreover that $s^-(M,p) - \delta(p)$ is constant in a neighborhood of p_o . Set $j_o = \operatorname{codim} M + s^-(M,p_o) - \delta(p_o)$.
Then $H^j(\mu_M(\mathcal{O}_X))_{p_o} = 0$ for $j \neq j_o$, and for $j = j_o$ this space is infinite dimensional.

Sketch of the proof

i) Let ϕ be a complex contact transformation which interchanges $(T^*_M X, p)$ and $(T^*_N Y, q)$ where N is a real submanifold of Y. Applying Theorem 4.1. we get, by a simple calculation :

$$(4.1)\quad \mu_M(\mathcal{O}_X)_p[\operatorname{codim} M+s^-(M,p)-\delta(p)] \cong \mu_N(\mathcal{O}_Y)_q[\operatorname{codim} N + s^-(N,q)-\delta(q)]$$

and

$$(4.2)\quad \mu_M(\mathcal{O}_X)_p[n - s^+(M,p) + \delta(p)] \cong \mu_N(\mathcal{O}_Y)_q[n-s^+(N,q)+\delta(q)]$$

By the assumption we can choose for N a real hypersurface. Then $\delta(q) = 0$. Moreover we can choose N such that $s^-(N,q) = 0$. Since $\mu_N(\mathcal{O}_Y)$ is a complex concentrated in degree $[1,n]$ (more generally, $(\mathcal{H}^j_Z(\mathcal{O}_Y))_x = 0$ for $j \notin [1,n]$, if Z is a locally closed subset of Y and $x \notin \operatorname{int}(Z)$), we get that the left-hand side of (4.1) is concentrated in degree $\geqq 0$. Similarly by choosing N such that $s^+(N,q) = 0$, we find that the left-hand side of (4.2) is concentrated in degree $\leqq 0$.

ii) One proves easily that under this assumption, $s^-(N,q) - \delta(q)$ is locally constant. We choose for N a real hypersurface such that $s^-(N,q_o) = 0$. Then N is the boundary of a (weakly) pseudo-convex open set, and the result follows, (cf. [4]) .

Remark 4.5. : One easily deduces theorems similar to the "edge of the wedge theorem" for real submanifolds of X from Proposition 4.4. . In particular when $\delta(p) = 0$ we recover many classical results (from [1] to [15] and the bibliography quoted there). Remark moreover that when $\delta(p) = 0$ and M is real analytic, the complex $\mu_M(\mathcal{O}_X) \otimes \underline{\omega}_M[\mathrm{codim}\, M]$ is isomorphic to the complex of microfunctions on M solution of the induced Cauchy-Riemann system (cf. [6]). Let us put the emphasis on the fact that, in our proofs, we do not make use of the induced Cauchy-Riemann system, which may be arbitrarly degenerated.

5. SYSTEMS WITH SIMPLE CHARACTERISTICS

Let X be a complex manifold, of dimension n, and let $\mathcal{O}_{X\times X}^{(0,n)}$ be the sheaf of holomorphic forms of type $(0,n)$ on $X \times X$. Let Δ be the diagonal of $X \times X$. The sheaf of Ring $\mathcal{E}_X^{\mathbb{R}}$ on T^*X (identified to $T^*_\Delta(X \times X)$) is defined in [12] by :

$$\mathcal{E}_X^{\mathbb{R}} = \mu_\Delta(\mathcal{O}_{X\times X}^{(0,n)})\ [n] .$$

This Ring contains the important subring $\mathcal{E}_X$ of finite order micro-differential operators, but we do not recall its construction here, and refer to [14] for a detailed exposition.

The Ring $\mathcal{E}_X$ is coherent, flat over $\pi^{-1}\mathcal{D}_X$, and any coherent $\mathcal{E}_X$-module admits locally a bounded finite-free presentation of length at most n .

Let $\mathcal{M}$ be a coherent left $\mathcal{E}_X$-module defined on an open subset U of T^*X , and set :

$$V = \mathrm{char}(\mathcal{M}) \qquad (= \mathrm{supp}(\mathcal{M})) .$$

One says that $\mathcal{M}$ has simple characteristics along V if V is a smooth conic regular involutive manifold, and locally on V there exists a generator u of $\mathcal{M}$ such that the symbol Ideal of the annihilator of u is reduced.

Now let M be a real analytic manifold such that X is a complexification of M. We write for short :

$$\Lambda = T^*_M X .$$

Following Sato-Kashiwara-Kawaï [12] we define the Levi form of V (or $\mathcal{M}$) along Λ at $p \in \Lambda \cap V$ as follows.

Let ℓ be the complex codimension of V in T^*X, and let $\phi = (\phi_1,\ldots,\phi_\ell)$ be a local defining system for V : the ϕ_j's are holomorphic and homogeneous with respect to the action of $\mathbb{C}^\times$, and :

$$\begin{cases} V = \{p \in U \ ; \ \phi_1(p) = \ldots = \phi_\ell(p) = 0\} \\ d\phi_1 \wedge \ldots \wedge d\phi_\ell \wedge \omega_X \neq 0 \quad \text{on } V \end{cases}$$

Let ϕ_j^c be the holomorphic function defined in a neighborhood of $\Lambda \cap U$ by :

$$\phi_j^c \big|_\Lambda = \overline{\phi}_j \big|_\Lambda$$

where $\overline{}$ means complex conjugation.

Consider the Hermitian form $L(\phi,p)$ defined by the matrix :

$$(\{\phi_i,\phi_j^c\}\ (p))_{1 \leq i,j \leq \ell} .$$

It is clear that the number of positive (resp. negative) eigenvalues of this form do not depend on the choice of the defining system ϕ. We denote by $s^+(V, \Lambda, p)$ (resp. $s^-(V, \Lambda, p)$) this number.

Theorem 5.1. Let $\mathcal{M}$ be a system with simple characteristic along V, and let $p \in \Lambda \cap V$. Assume :

$$\Lambda \cap V \text{ is smooth }, \quad T(\Lambda \cap V) = T\Lambda \cap TV ,$$

and $\omega_X|_{\Lambda\cap V}$ never vanishes.

Then :

i) if $s^-(V, \Lambda, p) \geqq r$, one has :

$$\mathcal{E}xt^j_{\mathcal{E}_X}(\mathcal{M}, C_M)_p = 0 \quad \text{for } j < r$$

ii) if $s^+(V, \Lambda, p) \geqq s$, one has :

$$\mathcal{E}xt^j_{\mathcal{E}_X}(\mathcal{M}, C_M)_p = 0 \quad \text{for } j > \ell - s$$

iii) if $s^-(V, \Lambda, p) = r$ for all p in a neighborhood of p_o in $\Lambda \cap V$, one has :

$$\mathcal{E}xt^j_{\mathcal{E}_X}(\mathcal{M}, C_M)_{p_o} = 0 \quad \text{for } j \neq r$$

and for $j = r$ this space is infinite dimensional.

Sketch of the proof

By performing a quantized contact transformation (cf. [7] and [16]) we may assume from the beginning that M is the boundary to a strictly pseudo-convex open set of $\mathbb{C}^n$, and $V = \{(x,\xi) \in T^*\mathbb{C}^n ;\ \xi_1 = \ldots = \xi_\ell = 0\}$. Then $C_M \cong \mu_M(\mathcal{O}_X)$ [1] , and $\mathcal{M}$ is isomorphic to the system $\mathcal{E}_X / \mathcal{J}$ where $\mathcal{J}$ is the left Ideal generated by $(D_1, \ldots, D_\ell)$.

Let $\mathcal{I}$ be the left Ideal of $\mathcal{D}_X$ generated by $(D_1, \ldots, D_\ell)$. Then :

$$\mathcal{M} = \mathcal{E}_X \underset{\pi^{-1}\mathcal{D}_X}{\otimes} \pi^{-1}(\mathcal{D}_X/\mathcal{I})$$

and

$$\mathbb{R}\mathcal{H}om_{\mathcal{E}_X}(\mathcal{M}, C_M) \cong \mathbb{R}\mathcal{H}om_{\mathcal{D}_X}(\mathcal{D}_X/\mathcal{I}, C_M)$$

Let $Y = \mathbb{C}^{n-\ell}$ and let f be the projection from X to Y , $(x_1, \ldots, x_n) \longrightarrow (x_{\ell+1}, \ldots, x_n)$. Then $V = X \underset{Y}{\times} T^*Y$.

We have :

$$\mathbb{R}\mathcal{H}om_{\mathcal{E}_X}(\mathcal{M}, C_M) \cong \mathbb{R}\mathcal{H}om_{\mathcal{D}_X}(\mathcal{D}_X/\mathcal{I}, \mu_M(\mathcal{O}_X)) [1]$$

$$\cong \mu_M(\mathbb{R}\mathcal{H}om_{\mathcal{D}_X}(\mathcal{D}_X/\mathcal{I}, \mathcal{O}_X)) [1]$$

$$\cong \mu_M(f^{-1} \mathcal{O}_Y) [1]$$

Let us denote by ρ and ϖ the natural maps :

$$T^*X \xleftarrow[\rho]{} X \underset{Y}{\times} T^*Y \xrightarrow[\varpi]{} T^*Y \quad .$$

By performing a contact transformation on T^*Y, we may assume that the Lagrangean manifold $\varpi\rho^{-1}(\Lambda)$ is the conormal $T^*_N Y$ to a real hypersurface N of Y. Then one proves that for any $\underline{F} \in D^+(Y)$, one has :

$$\mu_M(f^{-1}\underline{F})_p \cong \mu_N(\underline{F})_{\varpi(p)} [-d] \quad ,(p \in \Lambda \cap X \underset{Y}{\times} T^*Y)$$

where

$$d = \frac{1}{2}\tau(\lambda_0(p), T_p\varpi^{-1}T^*_N Y, T_p\Lambda)$$
$$+ \frac{1}{2}(\dim_{\mathbb{R}} X - \dim_{\mathbb{R}} Y - \dim_{\mathbb{R}}(T_p\Lambda \cap T_p V)^{\perp}) .$$

Then it remains to prove :

$$d = s^-(V, \Lambda, p) - s^-(N, \varpi(p))$$
$$= \ell - n - s^+(V, \Lambda, p) + s^+(N, \varpi(p))$$

and to apply Proposition 4.4. .

Remarque 5.2. Conclusion i) and ii) of Theorem 5.1. are well-known (Sato-Kashiwara-Kawaï [12 Ch. 3, Theorem 2.3.10]) but let us emphasis that our proof is completely different. When $\mathcal{M}$ reduces to a single microdifferential operator ($\mathcal{M} \cong \mathcal{E}_X/\mathcal{E}_X P$) then Theorem 5.1. is a particular case of a theorem of Trepreau [16].

Bibliographie

[1] ANDREOTTI A., GRAUERT H. : Theorem de finitude pour la cohomologie des espaces complexes. Bull. Soc. Math. France, 90, 193-259, (1962).

[2] BJÖRK J.E. : Rings of differential operators. North-Holland Math. Library, (1979).

[3] HARTSHORNE R. : Residues and duality. Lecture Notes in Math. 20, Springer-Verlag, (1966).

[4] HÖRMANDER L. : An introduction to complex analysis in several variables, Van Norstrand, Princeton-London-Toronto (1966).

[5] HÖRMANDER L. : Fourier integral operators. Acta Math. 127, 79-183, (1971).

[6] KASHIWARA M., KAWAÏ T. : On the boundary value problem for elliptic systems of linear differential equations, I, II. Proc. Japan. Acad. 48, 712-715, (1972) and 49, 164-168, (1973).

[7] KASHIWARA M., KAWAÏ T. : Some applications of boundary value problems for elliptic systems of linear differential equations. Ann. Math. Studies, 93, Princeton (1980).

[8] KASHIWARA M., SCHAPIRA P. : Microlocal study of sheaves. Astérisque Soc. Math. France, 128 (1985).

[9] KASHIWARA M., SCHAPIRA P. A vanishing theorem for a class of systems with simple characteristics. Inventiones Math. (1985) (to appear).

[10] LERAY J. : Analyse Lagrangienne et mécanique quantique. Collège de France (1976-77).

[11] LERAY J. : Problème de Cauchy I. Bull Soc. Math. France 85, 389-430 (1957).

[12] SATO M., KASHIWARA M., KAWAÏ T.: Hyperfunctions and pseudo-differential equations. Lecture Notes in Math. 287, 265-529, Springer-Verlag, (1973).

[13] SCHAPIRA P. : Condition de positivité dans une variété symplectique complexe. Application à l'étude des microfonctions. Ann. Ec. Norm. Sup. 14, 121-139, (1981).

[14] SCHAPIRA P. : Microdifferential systems in the complex domain. Grundlehren der Math. Wissenchaften. 269. Springer-Verlag (1985).

[15] TAJIMA S. : Analyse microlocale des variétés de Cauchy-Riemann et problème de prolongement des solutions holomorphes des équations aux dérivées partielles. Publ. R.I.M.S., Kyoto Univ. Vol. 18, 911-945 (1983).

[16] TREPREAU . : Séminaire Bourbaki n° 595 (81-82). Thèse Univ. Reims (1984).

Effet tunnel pour l'opérateur de Schrödinger semi-classique

II . Résonances .

B. Helffer
Dept. de Mathématiques
Université de Nantes
2, Chemin de la Houssinière
F-44072 Nantes, France

J. Sjöstrand
Dept. de Mathématiques
Université de Lund
Box 118, S-22100 Lund, Suède.
Université de Paris Sud.

Abstract . In this second part of our survey , we outline a general theory for resonances for semiclassical Schrödinger operators and we study the case of a potential well in an island with applications to singular perturbations and in particular to the Zeeman effect .

0. Introduction.

Ceci est la deuxième et dernière partie de notre survey , commencé dans [1] et on y expose le travail [2] . On s'intéresse toujours essentiellement à l'opérateur de Schrödinger $P = -h^2\Delta + V(x)$, $V \in C^\infty(\mathbb{R}^n,\mathbb{R})$ et à ses propriétés spectrales près de 0 quand $h \to 0$. Si $\underline{\lim}_{|x|\to\infty} V(x) > 0$ on peut considérer l'extension de Friedrichs de P qui a un spectre discret près de 0 , et les résultats exposés dans [1] s'appliquent . Si l'on supprime cette hypothèse , on n'a plus nécessairement une extension autoadjointe naturelle de P , et on cherche alors à définir d'autres types de valeurs propres , qui peuvent être complexes ,

H. G. Garnir (ed.), Advances in Microlocal Analysis, 291–322.

et que l'on appelle " résonances " . Nous nous sommes inspirés de la méthode de "complex scaling" initialisée par Aguilar-Combes [3] et Balslev-Combes [4] , reprise ensuite et développée par beaucoup d'auteurs . Voir p.ex. [5-8] . L'idée de cette méthode est de remplacer $L^2(\mathbb{R}^n)$ par $L^2(\Gamma)$, où $\Gamma \subset \mathbb{C}^n$ est un sous-espace ou une sous-variété qui s'obtient par déformation de $\mathbb{R}^n$. On suppose bien entendu que V est convenablement analytique à l'infini . Dans le nouvel espace P devient elliptique à l'infini (mais pas auto-adjoint) et admet un spectre discret près de 0 .

Dans notre présentation , on développe une théorie micro-locale pour les résonances , qui dans sa forme actuelle est assez technique . On gagne cependant une plus grande généralité et aussi de la souplesse dans les applications .

Dans la première partie de cet exposé, on esquisse la théorie generale; dans la deuxième partie, on traite les " shape resonances" engendrées par un puits de potentiel dans une isle . Dans la troisième partie, on obtient des applications à la théorie des perturbations singulières .

1. Une théorie générale des résonances.

L'idée de notre méthode est de remplacer l'espace de phase $\mathbb{R}^{2n}_{x,\xi}$ par une variété I-Lagrangienne (c.a.d. Lagrangienne pour la forme symplectique réelle $- \operatorname{Im} dx \wedge d\xi$) $\Lambda \subset \mathbb{C}^{2n}$ proche de $\mathbb{R}^{2n}$, telle que $p|_{\Lambda}$ devienne elliptique à l'infini , où $p = \xi^2 + V(x)$ est le symbole principal de P . Il faut ensuite associer à Λ des " espaces de Sobolev " et des opérateurs pseudodifferentiels par un mélange de techniques de Beals, Fefferman [9,10], L. Hörmander [11] et Sjöstrand [12].

On commence par choisir des échelles à l'infini. Soient $r \geq 1$, $R > 0$ des fonctions C^∞ sur $\mathbb{R}^n_x$ telles que

$$(1.1) \qquad r\,R \geq 1 \;,\quad r \geq 1 \;,\quad \partial^\alpha R = \mathcal{O}(R^{1-|\alpha|}) \;,$$
$$\partial^\alpha r = \mathcal{O}(r\,R^{-|\alpha|}) \;,$$

pour tous $\alpha \in \mathbb{N}^n$. On pose $\tilde{r}(x,\xi) = (\xi^2 + r(x)^2)^{\frac{1}{2}}$. Si $m(x,\xi) > 0$ est de classe C^∞ et $a(x,\xi)$ est C^∞, on dit que $a \in S(m)$ si $\partial_x^\alpha \partial_\xi^\beta a = \mathcal{O}(m\,\tilde{r}^{-|\beta|} R^{-|\alpha|})$ pour tous $\alpha, \beta \in \mathbb{N}^n$. Souvent on admet que a dépend aussi de $h > 0$, et dans ce cas on exige que les estimations sur les derivées de a soient uniformes par rapport à h. On définit aussi de manière évidente les espaces $S(m\,h^{-l})$, $l \in \mathbb{R}$. On demande toujours que m soit une <u>fonction d'ordre</u>, à savoir que $m \in S(m)$. Les fonctions r, $\tilde{r}$, R sont des fonctions d'ordre, et on écrira souvent : $S^{m,k} = S(\tilde{r}^m R^k)$, $S^{m,k,l} = S(\tilde{r}^m R^k h^{-l})$.

Soit $V \in C^\infty(\mathbb{R}^n; \mathbb{R})$ tel qu'il existe un compact $K \subset \mathbb{R}^n$ et une constante $C > 0$ tels que :

(1.2) V est analytique sur $\mathbb{R}^n \setminus K$ et s'étend holomorphiquement à $x \in \mathbb{C}^n$; $\mathrm{Re}\,x \in \mathbb{R}^n \setminus K$, $|\mathrm{Im}\,x| < C^{-1} R(\mathrm{Re}\,x)$ en vérifiant $|V(x)| \leq C\,r(\mathrm{Re}\,x)^2$.

Les inégalités de Cauchy entrainent que $V \in S(r^2)$ et donc que $p \in S(\tilde{r}^2)$. Introduisons les boites de base réelles ou complexes : $B((x,\xi), \varepsilon) = \{(y,\eta) \in \mathbb{R}^{2n}\ (\mathbb{C}^{2n})$;

$|\xi-\eta| \leq \varepsilon \tilde{r}(x,\xi)$, $|x-y| \leq \varepsilon R(x) \}$. (Si (x,ξ) est complexe , on pose $R(x) = R(\operatorname{Re} x)$, $\tilde{r}(x,\xi) = \tilde{r}(\operatorname{Re}(x,\xi))$.)

On a aussi besoin d'une hypothèse d'ellipticité loin des caractéristiques réelles $\Sigma_p = \{(x,\xi) \in \mathbb{R}^{2n} ; p(x,\xi) = 0\}$:

(1.3) Pour tout $\varepsilon_0 > 0$, il existe $C_0 > 0$ t.q. si $(x,\xi) \in \mathbb{R}^{2n}$ et $B((x,\xi),\varepsilon_0) \cap \Sigma_p = \emptyset$, alors $|p(x,\xi)| \geq C_0^{-1}\tilde{r}(x,\xi)^2$.

(Notre théorie se généralise aux opérateurs elliptiques plus géneraux que ceux de Schrödinger .) Posons $\dot{S}^{1,1} =$

$$\{G(x,\xi) \in C^\infty(\mathbb{R}^{2n};\mathbb{R}) ; \partial_{\xi_j} G \in S^{0,1} , \partial_{x_j} G \in S^{1,0} , \forall j\}.$$

Si $G \in S^{1,1}$ et $H_p = \sum \frac{\partial p}{\partial \xi_j}\frac{\partial}{\partial x_j} - \frac{\partial p}{\partial x_j}\frac{\partial}{\partial \xi_j}$, alors $H_p G \in S^{2,0}$.

Définition 1.1. On dit que $G \in \dot{S}^{1,1}$ est une fonction fuite s'il existe un ensemble compact $K \subset \Sigma_p$ et une constante $C > 0$ tels que $H_p G \geq r^2/C$ sur $\Sigma_p \setminus K$.

On peut dire que la théorie du "complex scaling" est basée sur un choix de fonction fuite qui est linéaire en ξ . Très souvent on a $G = x\cdot\xi$.

Exemple . $V(x) = -1 + o(1)$ est holomorphe pour $|\operatorname{Im} x| < C^{-1}(1 + |\operatorname{Re} x|)$. Alors on peut prendre $r = 1$, $R = (1 + x^2)^{\frac{1}{2}}$, $G = x\cdot\xi$.

Il est possible de construire des potentiels vérifiant toutes les hypothèses ci-dessus, qui admettent une fonction fuite , mais aucune qui soit linéaire en ξ . Notre theorie s'étend aussi aux opérateurs d'ordre $>$ 2 , et dans ce cas la classe des fonctions fuite ξ-linéaires nous semble beaucoup trop petite .

Si $G \in \dot{S}^{1,1}$ est suffisamment proche d'une fonction $g(x) \in \dot{S}^{1,1}(\mathbb{R}^n)$ (dans la topologie de $\dot{S}^{1,1}(\mathbb{R}^{2n})$) et indépendant de ξ pour x dans un ensemble compact assez grand , alors la variété I-Lagrangienne

$$(1.4) \qquad \Lambda_G \ : \ \operatorname{Im} \xi = -\frac{\partial G}{\partial \operatorname{Re} x}(\operatorname{Re}(x,\xi)) \ , \ \operatorname{Im} x = \frac{\partial G}{\partial \operatorname{Re} \xi}(\operatorname{Re}(x,\xi))$$

est contenue dans le domaine de définition de p et $p|_{\Lambda_G} = p(\operatorname{Re} x + i\frac{\partial G}{\partial \operatorname{Re} \xi} , \operatorname{Re} \xi - i\frac{\partial G}{\partial \operatorname{Re} x})$ est de classe $S^{2,0}$.

Si G est une fonction fuite et $\Lambda_t = \Lambda_{tG}$, alors pour t assez petit , $(x,\xi) \in \Lambda_t$, $\operatorname{Re}(x,\xi) \notin$ ensemble compact , la valeur $p(x,\xi)$ appartient à un ensemble

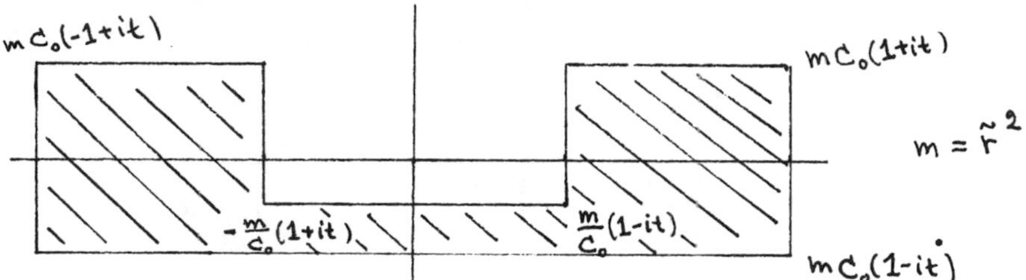

fig. 1

En particulier , on a $|p|_{\Lambda_t}| \geq t\,\tilde{r}^2/C$ en dehors d'un compact et donc $(p|_{\Lambda_t})^{-1}$ est bien défini et de classe $S^{-2,0}$ en-dehors d'un compact .(On dira que $p|_{\Lambda_t}$ est elliptique à l'infini .) On discute ensuite très brièvement les espaces de Sobolev et les opérateurs pseudodifferentiels . Toute la discussion est valable pour un $g(x) \in S^{1,1}(\mathbb{R}^n)$ et pour $G = G(x,\xi) \in \dot{S}^{1,1}(\mathbb{R}^{2n})$ avec $G(x,\xi) - g(x)$ assez petit dans $S(rR)$. On associe d'abord à $\Lambda = \Lambda_G$ une transformation FBI (cf. [12]) :

$$(1.5)\qquad Tu(\alpha,h) = \int_{\mathbb{R}^n} e^{i\varphi(\alpha,y)/h}\, t(\alpha,y,h)\,\chi_\alpha(y)\, u(y)\, dy\ ,$$

pour $\alpha \in \Lambda$, $\varphi(\alpha,y) = (\alpha_x - y)\cdot\alpha_\xi + i\lambda(\alpha)(\alpha_x - y)^2$, où $\lambda(\alpha) > 0$ est elliptique dans $S^{1,-1}$, $t \in S^{n/4,-n/4,3n/4}$ est à valeurs dans $\mathbb{C}^{n+1}$, affine linéaire en y et tel que t , $\partial t/\partial y_1$, ... , $\partial t/\partial y_n$ sont linéairement indépendants au sens elliptique naturel .(On écrit $\alpha = (\alpha_x, \alpha_\xi)$ et on étend de manière naturelle la définition de $S(m)$, $S(m\,h^{-1})$ en comptant x, α_x, y comme des variables "x" et α_ξ comme des variables " ξ ", et en se restreignant a un domaine du type $|x-\alpha_x| + |y-\alpha_x| < R(\alpha_x)$.) De plus $\chi_\alpha(y) = \chi((y - \mathrm{Re}\,\alpha_x)/R(\alpha_x))$, où $\chi \in C_0^\infty(\mathbb{R}^n)$ vaut 1 près de 0 . Grace au fait que t est vectoriel , on peut trouver une "bonne" formule d'inversion . Soit $\varphi^*(x,\alpha) = (x-\alpha_x)\cdot\alpha_\xi + i\lambda(\alpha)(x-\alpha_x)^2$. On peut alors trouver un opérateur de la forme, pour $v \in C^\infty(\Lambda, \mathbb{C}^{n+1})$:

$$(1.6)\qquad Sv(x,h) = \int_\Lambda e^{i\varphi^*(x,\alpha)/h}\, s(x,\alpha,h)\,\chi_\alpha(x)\, v(\alpha)\, d\alpha\ ,$$

où s est un symbole avec les mêmes propriétés que t , tel que

$$(1.7)\qquad S\,T = I + R\ ,$$

où R a un noyau distribution de la forme $\rho(x,y,h)\, e^{-(g(x)-g(y))/h}$. Ici ρ est C^∞ à support dans $|x-y| \leq$ const. $R((x+y)/2)$ et vérifie

$$(1.8)\qquad |\partial_x^\alpha \partial_y^\beta \rho| \leq C_{\alpha,\beta}\; e^{-\varepsilon_0 Rr/h}\; R^{-n-|\alpha|-|\beta|}\ ,$$

pour tous $\alpha, \beta \in \mathbb{N}^n$ avec un $\varepsilon_0 > 0$ indépendant de α,β . En particulier R est borné dans $L^2(\mathbb{R}^n;\ e^{-g(x)/h}dx)$ avec une norme $\mathcal{O}(e^{-\varepsilon_0/h})$.

Puisque Λ est I-Lagrangienne et difféomorphe à $\mathbb{R}^{2n}$, il existe $H \in \dot{S}^{1,1}(\Lambda)$ (espace qui se définit de manière évidente à l'aide de la paramétrisation (1.4)), tel que $-\mathrm{Im}\ \alpha_\xi\, d\alpha_x|_\Lambda = dH$. Si G est indépendant de ξ, on peut prendre $H = G$. Si $m(x,\xi)$ est une fonction d'ordre (que l'on considère aussi comme une fonction sur Λ à l'aide de (1.4)) alors, pour $h > 0$ assez petit, on définit grosso modo l'espace de Sobolev $H(\Lambda,m)$ comme l'espace de distributions u sur $\mathbb{R}^n$ (avec une certaine condition de croissance à l'infini) telles que $Tu \in L^2(\Lambda\ ;\ (m\, e^{-H/h})^2\, d\alpha)$. Ces espaces sont des espaces de Hilbert et si G est indépendant de ξ, on trouve $H(\Lambda,1) = L^2(\mathbb{R}^n\ ;\ e^{-2G/h}dx)$.

Si G_1, G_2 sont proches de la même fonction $g(x)$ dans $\dot{S}^{1,1}$ (avec $G_j - g_j$ petit dans $S(rR)$, $g_j = g_j(x) \in \dot{S}^{1,1}(\mathbb{R}^n)$), on dit que $\Lambda_1 \leq \Lambda_2$, si $G_1 \leq G_2$. Si de plus m_1 et m_2 sont des fonctions d'ordre avec $m_1 \geq m_2$ alors, pour h assez petit, on a $H(\Lambda_1,m_1) \subset H(\Lambda_2,m_2)$ avec une inclusion qui en norme est $\mathcal{O}(1)$ quand $h \to 0$. Si de plus $m_1/m_2 \to \infty$ quand α tend vers l'infini, alors l'inclusion est compacte.

Les opérateurs pseudodifférentiels associés à Λ sont de la forme

$$(1.9)\qquad Au(x,h) = \iint_{\alpha\in\Lambda} e^{i\varphi(x,y,\alpha)/h} a(x,y,\alpha,h)\ \chi_\alpha(x,y)\, u(y)\, dy\, d\alpha \quad ,$$

où $\varphi \in S^{1,1}$ est quadratique en (x,y) et tel que, pour $x = y = \alpha_x$: $\varphi'_x = -\varphi'_y = \alpha_\xi$, $\varphi'_\alpha = 0$, Im $\varphi''_{(x,y),(x,y)} - g''_{x,x} + g''_{y,y} \sim r/R$. La troncature $\chi_\alpha(x,y)$ appartient à $S^{0,0}$, vaut 1 dans un domaine du type $|x-\alpha_x| + |y-\alpha_x| \leq$ const. $R(\alpha_x)$ et a son support dans un domaine du même type . De plus , $a \in S(m\,\tilde{r}^{3n/2} R^{n/2} h^{-3n/2})$ est holomorphe par rapport à (x,y) près de supp χ . Ici m est une fonction d'ordre et nous dirons que A est un opérateur pseudodifférentiel d'ordre m adapté à Λ . Les résultats sur le calcul pseudodifférentiel sont alors grosso modo les suivants . (On suppose que l'on a fixé un "choix automatique " de phase qui ne depend pas de Λ , p.ex. $\varphi(x,y,\alpha) = (x-y)\cdot\alpha_\xi + i\,\lambda(\alpha)((x-\alpha_x)^2 + (y-\alpha_x)^2)$ et de même pour la fonction troncature , on suppose de plus que $G-g$ est assez petit dans $S(rR)$.)

1° Pour chaque fonction d'ordre $\tilde{m}$, si $h > 0$ est assez petit , alors l'opérateur pseudodifférentiel A ci-dessus est borné : $H(\Lambda,\tilde{m}) \to H(\Lambda,\tilde{m}/m)$ et de norme $\mathcal{O}(1)$.

Disons qu'un opérateur **K** est négligeable d'ordre m , si pour chaque fonction d'ordre $\tilde{m}$ et tout $N \geq 0$, on a $K = \mathcal{O}(1)$ comme opérateur borné : $H(\Lambda,\tilde{m}) \to H(\Lambda,(\tilde{m}/m)(R\tilde{r}/h)^N)$, pour h assez petit .

2° La classe d'opérateurs pseudodifférentiels d'ordre m associée à Λ , ne depend pas , modulo des opérateurs négligeables d'ordre m , du choix de φ et de χ .

3° Il y a un symbole principal naturel dans $S(m)/S(m(h/\tilde{r}R))$ avec les propriétés habituelles.

4° Si A_j est un opérateur pseudodifférentiel d'ordre m_j associé à Λ pour $j = 1,2$, alors modulo un opérateur négligeable d'ordre $m_1 m_2$ le composé $A_1 A_2$ est un opérateur pseudodifférentiel d'ordre $m_1 m_2$. Le symbole principal de $A_1 A_2$ est le produit des symboles principaux de A_1 et de A_2.

Si $G(x, \xi)$ est indépendant de ξ pour x dans un compact assez grand de $\mathbb{R}^n$, alors modulo un opérateur négligeable d'ordre $\tilde{r}^2$, on peut considérer P comme un opérateur pseudodifférentiel d'ordre $\tilde{r}^2$ associé à $\Lambda = \Lambda_G$ de symbole principal $p|_\Lambda$. Soit maintenant G une fonction fuite et soit $C_0 > 0$ comme dans la description des valeurs de $p|_{\Lambda_t}$. Soit Ω_t :

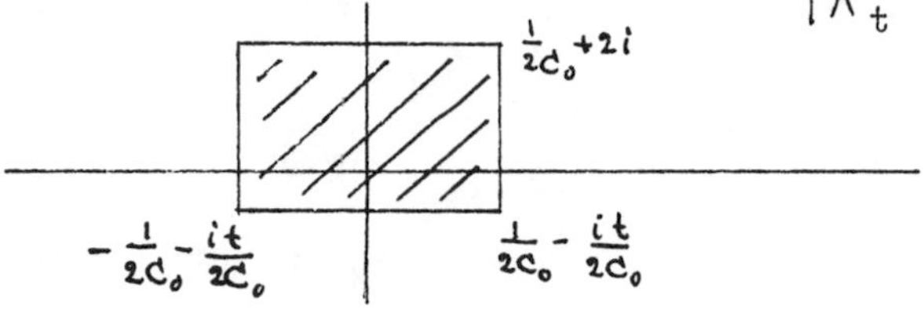

fig. 2.

Alors pour $t > 0$ assez petit, $p|_{\Lambda_t} - z$ est elliptique à l'infini pour tout $z \in \Omega_t$.

<u>Théorème 1.2.</u> Il existe $t_1 > 0$ tel que si $0 < t \leq t_1$ et si m est une fonction d'ordre, alors il existe $h_0 > 0$ tel que, pour $0 < h \leq h_0$, on a : Il existe un ensemble discret $\Gamma(h) = \Gamma(h,t,m) \subset \Omega_t$ tel que $(P - z) : H(\Lambda_t, \tilde{r}^2 m) \to H(\Lambda_t, m)$ est bijectif d'inverse borné pour $z \in \Omega_t \setminus \Gamma(h)$, tandis que pour $z \in \Gamma(h)$ l'opérateur est encore Fredholm d'indice 0, admet une décomposition en somme directe en $P : F_z \to F_z$ et

$P : G_z \cap H(\Lambda_t, \tilde{r}^2 m) \longrightarrow G_z$, où $\dim F_z < \infty$, $F_z \neq 0$,

G_z est un sous-espace fermé de $H(\Lambda_t, m)$, $H(\Lambda_t, m) = F_z \oplus G_z$.

$P-z$ est nilpotent sur F_z et un isomorphisme : $G_z \cap H(\Lambda_t, \tilde{r}^2 m) \longrightarrow G_z$.

Démonstration . Soit $z_0 = i$. Pour $t \leq t_1$ assez petit , il est clair que $p|_{\Lambda_t} - z_0$ est inversible pas seulement en-dehors d'un compact , mais partout . Donc $(p|_{\Lambda_t} - z_0)^{-1} \in S(\tilde{r}^{-2})$. Soit $Q(z_0) =$ " $\mathrm{Op}_{\Lambda_t}((p-z_0)^{-1})$ " un opérateur pseudodifférentiel correspondant (associé à Λ_t et de symbole principal $(p-z_0)^{-1}$) . Alors

$$(1.10) \qquad (P-z_0)\, Q(z_0) = I + R_{-1}(z_0) \quad ,$$

où R_{-1} est $\mathcal{O}(1) : H(\Lambda_t, m) \longrightarrow H(\Lambda_t, m\, \tilde{r}\, R/h)$ et en particulier de norme $\mathcal{O}(h)$ dans $H(\Lambda_t, m)$. On a de même :

$$(1.11) \qquad Q(z_0)\, (P-z_0) = I + L_{-1}(z_0) \quad ,$$

où L_{-1} a les mêmes propriétés . Donc, pour h assez petit , $P-z_0$: $H(\Lambda_t, m\, \tilde{r}^2) \rightarrow H(\Lambda_t, m)$ est bijectif avec un inverse borné .

Soit $K \subset \Lambda_t$ un compact assez grand et $\chi \in C_0^\infty(\Lambda_t)$ égal à 1 sur K . Pour $z \in \Omega_t$, on pose

$$(1.12) \qquad \tilde{Q}(z) = \mathrm{Op}_{\Lambda_t}\left((p-z_0)^{-1} \chi + (p-z)^{-1} (1-\chi) \right) \quad ,$$

de telle manière que Q dépend holomorphiquement de z et $\tilde{Q}(z_0) = Q(z_0)$. Alors

$$(1.13) \qquad (P-z)\, \tilde{Q}(z) = I + K(z) + \tilde{R}_{-1}(z) \quad ,$$

où $\tilde{R}_{-1}$ a les mêmes propriétés que R_{-1} ci-dessus et $K = \mathrm{Op}_{\Lambda_t}(k)$ avec $k = \frac{(p-z)}{(p-z_0)}\chi \in S^{0,0,0}$ à support compact. De plus, tous les opérateurs dépendent holomorphiquement de z, $K(z_0) = 0$. Nous avons le résultat analogue pour $\tilde{Q}(z)(P-z)$. On peut écrire $I + K(z) + \tilde{R}_{-1}(z) = (I+\tilde{R}_{-1})(I+(I+\tilde{R}_{-1})^{-1}K)$, et observer que $I+\tilde{R}_{-1}$ est un isomorphisme et que $(I+\tilde{R}_{-1})^{-1}K(z)$ est compact et s'annule pour $z = z_0$. Faisant la même chose avec $\tilde{Q}(P-z)$ on obtient, par la théorie de Fredholm, que $P-z$ est d'indice 0 partout dans Ω_t et bijectif en-dehors d'un ensemble discret. Un peu plus de travail abstrait donne aussi la décomposition de Jordan pour chaque $z \in \Gamma(h)$. #

On montre ensuite sous les hypothèses du Théorème 1.2, que si $\tilde{m}$ est une autre fonction d'ordre, alors, pour $0 < t \leq t_1$, $0 < h \leq h_0(t)$, on a $\Gamma(h,t,m) = \Gamma(h,t,\tilde{m})$, $F_z = \tilde{F}_z$, pour tout $z \in \Gamma(h,t,m)$ (où $\tilde{F}_z$ se définit comme dans le Théorème 1.2 mais par rapport à $\tilde{m}$). On vérifie aussi que Γ est un indépendant de la fonction fuite :

<u>Théorème 1.3.</u> Sous les hypothèses du Théorème 1.2, soit $\tilde{G}$ une deuxième fonction fuite. Alors $G_s = (1-s)G + s\tilde{G}$ est une fonction fuite pour tous $s \in [0,1]$ et on peut choisir C_0 assez grand pour que $p|_{\Lambda_{t G_s}}$ prenne ses valeurs dans l'ensemble décrit par la fig. 1 pour tous $s \in [0,1]$ et tous $t > 0$ assez petits, quand $\mathrm{Re}\,(x,\xi)$ est en-dehors d'un ensemble compact fixé. On définit Ω_t avec cette constante C_0. Alors, pour $0 < t \leq t_1$ et c. ..., nous avons $\Gamma(h,t,m) = \tilde{\Gamma}(h,t,m)$ et $F_z = \tilde{F}_z$ (où

$\tilde{\Gamma}$ et $\tilde{F}_z$ se rapporte à $\tilde{G}$).

Ce théorème se démontre par un argument de déformation. On écrira désormais $\Gamma(h)$ à la place de $\Gamma(h,t,m)$ et les élements de cet ensemble seront appelés " résonances ".

<u>Théorème 1.4.</u> Sous les hypothèses du Théorème 1.2, si $0 < t \le t_1$ et c. ..., alors $\Gamma(h) \subset \{z \in \Omega_t ; \text{Im } z \le 0\}$, et si $x \in \Gamma(h)$ **est réel** et $u \in F_x$, alors $(P-x)u = 0$ et $u \in H(\Lambda_0, \tilde{r})$.

Ce théorème se démontre par des inégalités du type de Gårding.

Il semble aussi possible de montrer que le nombre de résonances dans Ω_t est $\mathcal{O}(h^{-n})$ et nous espérons présenter (des versions raffinées de) ce resultat dans un travail futur.

<u>2. Le cas d'un puits dans une isle.</u>

Soit $U \subset \ddot{O} \subset \mathbb{R}^n$, où U est compact, $\ddot{O}$ ouvert et connexe. On suppose que V est analytique partout et que les hypothèses générales de la section 1 sont vérifiées. De plus, on suppose que

(2.1) $V \le 0$ sur U, $V > 0$ sur $\ddot{O}\setminus U$, $V < 0$ sur $\complement \overline{\ddot{O}}$.

(2.2) Il existe $G \in \dot{S}^{1,1}$ tel que $H_p(G) \ge C_0^{-1} r^2$ sur $\Sigma_p |_{\complement \ddot{O}}$.

(2.3) Pour tout $\varepsilon > 0$ il existe $C_0(\varepsilon) > 0$ tel que $V(x) \ge C_0(\varepsilon)^{-1} r^2$ sur $\ddot{O} \setminus \bigcup_{y \in \partial \ddot{O}} B(y, \varepsilon)$, où $B(y, \varepsilon) = \{x \in \mathbb{R}^n ; |x-y| < \varepsilon R(y)\}$.

Soit d la distance d'Agmon sur $\ddot{O}$ associée à la métrique $\max(V(x),0)\,dx^2$, et posons $B_d(U,r) = \{x \in \ddot{O} \ ; \ d(x,U) < r\}$. On suppose :

(2.4) $\qquad \mathrm{diam}_d(U) = 0$.

Soit $S_0 = d(U, \partial \ddot{O})$. On peut alors montrer que $\overline{B_d(U,S_0)}$ est compact . Soit $\eta > 0$ petit et indépendant de h , et posons : $M_0 = \overline{B_d(U,S_0-\eta)}$. Soit P_{M_0} la réalisation de Dirichlet correspondante dans $L^2(M_0)$. Soit $I(h)$ un intervalle compact qui tend vers $\{0\}$ quand $h \to 0$ et soit $a(h) > 0$ une fonction qui tend vers 0 avec h , telle que $a(h) \geq C_\varepsilon^{-1}\, e^{-\varepsilon/h}$ pour tout $\varepsilon > 0$. On suppose que :

(2.5) $\qquad \sigma(P_{M_0}) \cap ((I(h)+[-2a,2a]) \setminus I(h)) = \emptyset$.

Soit $b(h) \geq a(h)$ une deuxième fonction qui tend vers 0 avec h , et soit $\Omega(h)$ le rectangle :

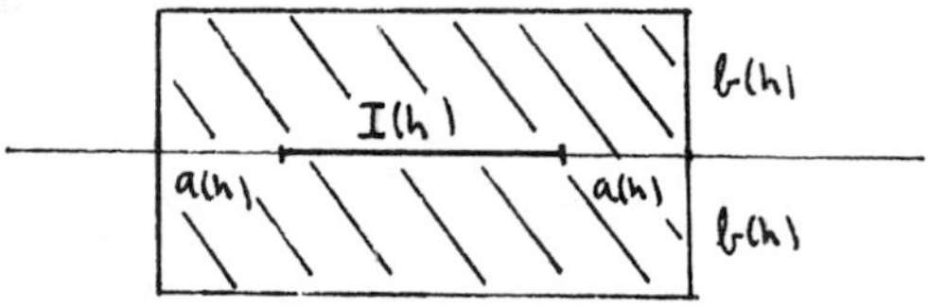

fig. 3 .

Théorème 2.1. Pour h assez petit , il y a une bijection $b : \sigma(P_{M_0}) \cap I(h) \to \Gamma(h) \cap \Omega(h)$, où les éléments des deux ensembles sont comptés avec leur multiplicité naturelle , telle que $b(\mu) - \mu = \tilde{O}(e^{-2S_0/h})$. (Ici $\tilde{O}(e^{-2S_0/h})$ désigne une quantité $O(e^{-(2S_0-\varepsilon(\eta))/h})$, où $\varepsilon(\eta) \to 0$, $\eta \to 0$.)

Démonstration (quelques idées). Soit $I(h) \cap \sigma(P_{M_0}) = \{\mu_1, \dots , \mu_m\}$ et soit $\varphi_1 , \dots , \varphi_m$ une famille O.N. correspondante . Alors on sait (voir [1]) que $\varphi_j = \tilde{\mathcal{O}}(e^{-d(U,x)/h})$. Soit $\chi \in C_0^\infty(M_0)$ égale à 1 sur $B_d(U,S_0-2\eta)$. Alors $(P-\mu_j)\,\psi_j = r_j$, où $\psi_j = \chi\,\varphi_j$ et $r_j = \tilde{\mathcal{O}}(e^{-S_0/h})$ a son support près de ∂M_0 .

On fait maintenant varier z sur $\partial\Omega(h)$, et donc toujours assez loin de $\sigma(P_{M_0})$. Alors $(P_{M_0}-z)^{-1}$ est bien défini et de norme $\mathcal{O}(1)$. Soit $0 \leq W \in C_0^\infty(\mathring{M}_0)$, > 0 sur $\bar{U}$ et à support près de U . Si l'on suppose (sans perte de généralité) que $G = 0$ au dessus de M_0 , alors notre théorie générale nous dit que $\tilde{P}-z : H(\Lambda_t,\tilde{r}^{\,2}) \rightarrow H(\Lambda_t,1)$ est un isomorphisme pour tout $z \in \overline{\Omega(h)}$. Ici $\tilde{P} = -h^2\Delta + V + W = P + W$. Par une série de perturbation avec des troncatures convenables , faisant intervenir $(P_{M_0}-z)^{-1}$ et $(\tilde{P}-z)^{-1}$, on montre ensuite le même résultat pour P-z , où z est maintenant sur $\partial\Omega(h)$. On montre aussi que le noyau de $(P-z)^{-1}$ est $\hat{\mathcal{O}}(e^{-d(x,y)/h})$ sur $M_0 \times M_0$. (Voir [1] pour la terminologie.)

Ecrivons maintenant :

$$(z-P)^{-1}\psi_j = (z-\mu_j)^{-1}\,\psi_j + (z-P)^{-1}\,(z-\mu_j)^{-1}\,r_j \quad ,$$

et intégrons sur $\partial\Omega$. A gauche,on obtient alors la projection spectrale naturelle de ψ_j dans $F \overset{\text{def.}}{=} \bigoplus_{z\in\Gamma(h)\cap\Omega(h)} F_z$;

$$v_j \overset{\text{def.}}{=} \Pi_F\,\psi_j = \psi_j + \oint_{\partial\Omega(h)} (z-P)^{-1}\,(z-\mu_j)^{-1}\,r_j\,dz \quad .$$

Ici la dernière intégrale est $\tilde{\mathcal{O}}(e^{-(d(\partial M_0,x)+S_0)/h})$ dans M_0 et on voit facilement que $v_1,\ldots,v_m$ sont linéairement indépendants. D'autres arguments semblables montrent aussi que ces vecteurs forment une base dans F .

On a :

$$(v_j \mid v_k)_{L^2(M_0)} = \delta_{j,k} + \tilde{\mathcal{O}}(e^{-2S_0/h})$$
$$(Pv_j \mid v_k) = \delta_{j,k}\, \mu_k + \tilde{\mathcal{O}}(e^{-2S_0/h}) \quad .$$

On en déduit que la matrice de $P|_F$ dans la base $v_1 ,\dots, v_m$ est diag $(\mu_j) + \tilde{\mathcal{O}}(e^{-2S_0/h})$, et il y a donc une bijection avec les propriétés voulues entre les valeurs propres de cette matrice et $\{\mu_1 ,\dots, \mu_m\}$.

Soit $s(x) = d(U,x)$ dans $\overline{B_d(U,S_0)}$, $= S_0$ en-dehors . Alors, pour tout $\varepsilon > 0$ et tout compact $K \subset \mathbb{R}^n$, on montre que :

$$(2.6) \quad \begin{cases} v_j = \mathcal{O}(e^{(\varepsilon - s(x))/h}) \ , \\ (P-\mu_j)v_j = \mathcal{O}(e^{-(\varepsilon_0 + s(x))/h}) \end{cases} ,$$

uniformément sur K . Ici $\varepsilon_0 > 0$.

On se restreint maintenant à un cas , où on peut faire des calculs BKW . On suppose ,

$$(2.7) \qquad U = \{x_0\} \ , \quad V''(x_0) > 0 \ .$$

Il résulte de [13] que les valeurs propres de P_{M_0} dans tout intervalle $[0 \, , C\, h]$ sont de la forme :

$$(2.8) \qquad h\, E(h) \sim h \left(\sum_{j=0}^{\infty} E_j\, h^j \right) \quad \text{(sommation avec } j \in \tfrac{1}{2}\mathbb{N} \text{ en général) ,}$$

où E_0 est une valeur propre de l'oscillateur harmonique $-\Delta + \frac{1}{2}\langle V''(x_0)x,x\rangle$ et $E(h)$ est une réalisation d'un symbole analytique classique . Si N_0 est la multiplicité de E_0, il y a exactement N_0 valeurs propres (comptées avec leur multiplicité) de la forme $h(E_0 + \mathcal{O}(h^{1/2}))$.

Soit μ_1 une valeur propre de P_{M_0} avec un développement asymptotique (2.8) . Soit N_1 la multiplicité asymptotique , c.à.d. le nombre de valeurs propres avec le même développement que μ_1 . Soit $E \subset L^2(M_0)$ de dimension N_1 la somme des espaces propres correspondants .

Théorème 2.2.([13] , voir aussi le Th. 3.3 de [1]). Il existe une famille orthonormale $\varphi_1 ,\dots, \varphi_{N_1}$ dans E telle que dans un voisinage de U :

$$(2.9) \qquad \varphi_j(x,h) = h^{-n/4}\, a_j(x,h)\, e^{-d(U,x)/h} \quad ,$$

où a_j est réalisation d'un symbole analytique classique ;
$a_j(x,h) \sim \sum_{-\infty}^{2m_j} a_{j,\nu}(x)\, h^{-\nu/2}$, $a_{j,\nu}(x) = \mathcal{O}(\, |x-x_0|^{\nu})$, $\nu \geq 0$.

Les fonctions φ_j ne sont pas nécessairement des fonctions propres exactes , mais on peut souvent ignorer ce fait et introduire $v_j = \pi_F(\chi\varphi_j)$ comme avant .

On s'interesse maintenant aux parties imaginaires des résonances ce qui nécessite une étude du comportement des fonctions v_j près de $\partial\ddot{O}$. On dira qu'un point de $\partial\ddot{O}$ est de type 1 s'il est dans $\overline{B_d(U,S_0)} \cap \partial\ddot{O}$, sinon on dira qu'il est de type 2 . Soit $x_2 \in \partial\ddot{O}$ un point de type 2 . Alors dans un voisinage de x_2 on a $v_j = \mathcal{O}(\, e^{(\varepsilon - S_0)/h})$ pour tous $\varepsilon > 0$. Sans utiliser le fait que x_2 soit de type 2, on montre d'autre part que $e^{S_0/h}\, v_j$

est microlocalement à décroissance exponentielle en tout point $(x_3, \xi_3) = \exp t\, H_p(x_2, 0)$ avec $t < 0$. Puisque $e^{S_0/h}(P-\mu_1)v_j$ est à décroissance exponentielle dans un voisinage de $\partial \ddot{O}$, on montre ensuite (comme pour le théorème classique sur la propagation des singularités analytiques pour des opérateurs de type principal réel), que $e^{S_0/h} v_j$ est à décroissance exponentielle microlocalement près de $(x_2, 0)$. Puisque $(x_2, 0)$ est le seul point caractéristique au-dessus de x_2, on obtient ensuite que :

(2.10) $\qquad e^{S_0/h} v_j$ est à décroissance exponentielle près de x_2.

Cet argument ne marche pas pour un point x_1 de type 1, car on n'a pas nécessairement $v_j = \mathcal{O}(e^{(\varepsilon - S_0)/h})$ pour tout $\varepsilon > 0$ dans un voisinage de x_1. Pour un tel point, il y a une géodesique minimale unique γ dans $\ddot{O} \cup \{x_1\}$ (pour la métrique d'Agmon) qui relie U à x_1. Si $q = \xi^2 - V(x)$, on peut voir γ comme projection d'une trajectoire $\tilde{\gamma}$ de H_q paramétrée sur $[-\infty, 0]$ avec $\tilde{\gamma}(-\infty) = (U, 0)$, $\tilde{\gamma}(0) = (x_1, 0)$, $\gamma(t) = \pi_x\, \tilde{\gamma}(t)$. Soit $f(x) = d(U,x) - S_0$. On montre d'abord que f est analytique dans un voisinage de $\gamma([-\infty, 0[)$ et que si $\Lambda = \Lambda_f : \xi = f'_x$, alors Λ fait partie de la variété Lagrangienne stable sortante pour H_q qui passe par $(U, 0)$. Si l'on prolonge Λ comme variété H_q-invariante jusqu'à un voisinage de $\tilde{\gamma}([-\infty, 0])$, on montre ensuite que la projection $\pi_x : \Lambda \to \mathbb{R}^n$ a une singularité du type "pli simple" en $(x_1, 0)$. Il existe donc une hypersurface analytique $C \subset \overline{\ddot{O}}$ passant par x_1 telle que f se prolonge analytiquement à un voisinage ouvert de $\gamma([-\infty, 0[)$, dont le bord près de x_1 coincide avec C . Si l'on définit C par une équation analytique $\rho = 0$, avec $d\rho \neq 0$, $\rho(\gamma(t)) > 0$, $t \in [-\varepsilon_0, 0[$, alors près de x_1 on a :

$$(2.11) \qquad f(x) = F(x, \rho(x)^{\frac{1}{2}}) \quad ,$$

où $F(x,s)$ est analytique près de $(x_1,0)$, et la singularité de $F(x, \rho(x)^{\frac{1}{2}})$ est $\mathcal{O}(\rho^{3/2})$. (On désigne maintenant par f l'extension analytique de $d(U,x)-S_0$ considérée au départ dans un voisinage de $\gamma(]-\infty,0[)$. On voit que : $f \geq d(U,x)-S_0$.) Si $\tilde{f} = f|_C$, on montre que :

$$(2.12) \qquad \tilde{f} \geq C_0^{-1} V|_C \ .$$

La formule (2.11) donne deux extensions possibles de f à l'autre coté de C . Des deux côtés de C, on a l'équation eiconale $(f')^2 = V$ et, dans la région extérieure ($\rho < 0$), on obtient donc : $(\mathrm{Re}\, f')^2 - (\mathrm{Im}\, f')^2 = V$, $\mathrm{Re}\, f' \cdot \mathrm{Im}\, f' = 0$. La dernière relation et (2.12) donnent :

$$(2.13) \qquad \mathrm{Re}\, f \geq 0 \quad \text{dans la région extérieure} ,$$

avec égalité sur la projection de la bicaractéristique de p , passant par $(x_1,0)$.

On peut maintenant étendre la forme BKW de φ_j à un voisinage de $\gamma([-\infty,0[)$ en intégrant les équations de transport qui proviennent de la relation $(P-hE)\varphi_j \sim 0$. Dans un voisinage de x_1, on étend ensuite cette expression BKW par une intégrale oscillante du type d'Airy . On obtient ainsi une fonction $\tilde{\varphi}_j$

dans un voisinage de $\gamma([-\infty,0])$, qui coincide avec φ_j près de U et qui vérifie

(2.14) $\tilde{\varphi}_j = \mathcal{O}(h^{-N_0} e^{-s(x)/h})$, $(P-\mu_1)\tilde{\varphi}_j = \mathcal{O}(e^{-(\varepsilon_0+s(x))/h})$.

De plus , par la phase stationaire , $\tilde{\varphi}_j$ reprend la forme $h^{-n/4} a_j(x,h) e^{-(S_0+f(x))/h}$ dans la région extérieure , et on peut choisir la branche de f qui correspond à une solution asymptotiquement sortante , en choisissant correctement un contour d'intégration dans l'intégrale d'Airy . Autrement dit , on peut s'arranger pour que $\Lambda_{if} = \{(x, i f'_x(x))\}$ contienne $\{\exp tH_p(x_1,0) ; t > 0\}$. Par des inégalités L^2 avec poids on montre que

(2.15) $\tilde{\varphi}_j - v_j = \mathcal{O}(e^{-(\varepsilon(t)+d(U,x))/h})$, $\varepsilon(t) > 0$, près de tout point $\gamma(t)$, $-\infty \leq t < 0$,

et que près de x_1 :

(2.16) $\tilde{\varphi}_j - v_j = \mathcal{O}(e^{-(S_0-\varepsilon)/h})$ pour tout $\varepsilon > 0$.

Le même argument qui donnait (2.10) près d'un point de type 2 montre maintenant que :

(2.17) $e^{S_0/h}(\tilde{\varphi}_j - v_j)$ est à décroissance exponentielle dans un voisinage de x_1 .

Soit W un petit voisinage de $\overline{B_d(U,S_0)}$ à bord C^∞, qui près des points de type 1 coincide avec $\{x ; V(x) = -\varepsilon\}$, $\varepsilon > 0$ est petit . Soit z(h) une résonance proche d'un des μ_j

et soit $v = \sum \omega_j v_j$, $\|\vec{\omega}\| = 1$ une fonction propre correspondante . Par la formule de Green ,

$$(2.18) \qquad (\operatorname{Im} z(h)) \|v\|^2_{L^2(W)} = -h^2 \operatorname{Im} \int_{\partial W} (\partial v/\partial n)\bar{v}\, dS ,$$

où n est la normale unitaire extérieure . Puisque $\|v\|^{2}_{L^2(W)}-1$ est à décroissance exponentielle et à cause de (2.10), on obtient

$$(2.19) \qquad \operatorname{Im} z(h) = -h^2 \operatorname{Im} \int_{\partial W \cap \Omega_I} (\partial v/\partial n)\bar{v}\, dS + \mathcal{O}(e^{-(2S_0+\varepsilon_0)/h}),$$

où Ω_I est un voisinage des points du type 1 , où (2.17) s'applique . On obtient alors les résultats suivants :

<u>Théorème 2.3.</u> On a $0 \leq -\operatorname{Im} z(h) \leq C_0 h^{1-n/2-2\max(m_j)} e^{-2S_0/h}$.

<u>Théorème 2.4.</u> Soit $N_1 = 1$, et supposons que les points du type 1 forment une sous-variété lisse Γ de codimension d dans $\partial \ddot{O}$ et que C soit tangent à $\partial \ddot{O}$ à l'ordre 2 exactement le long de Γ . Alors

$$(2.20) \qquad -\operatorname{Im} z(h) = h^{1-n/2+d/2-2m_1} f(h)\, e^{-2S_0/h} ,$$

où $f(h)$ est une réalisation d'un symbole analytique classique formel d'ordre 0 (ayant en général des demi-puissances de h dans son développement aymptotique) .

Si E_0 est la plus petite valeur propre de $-\Delta + \frac{1}{2}\langle V''(x_0)x, x\rangle$ alors a_1 est un symbole elliptique d'ordre 0 , et on trouve la minoration :

(2.21) $$- \operatorname{Im} z(h) \geq C_0^{-1} h^{\frac{1}{2}} e^{-2S_0/h} \quad , \quad C_0 > 0 \ .$$

En adaptant les méthodes de A. Martinez [14] , on montre que cette minoration est presque toujours vérifiée :

<u>Théorème 2.5.</u> On suppose , soit que $N_1 = 1$, soit que tous les polynômes d'Hermite décrivant les fonctions propres associées à E_0 de l'oscillateur harmonique localisé, sont du même degré . Alors on a la minoration (2.21) .

3. Application aux perturbations singulières et à l'effet Zeeman.

Considerons dans $\mathbb{R}^n$ l'opérateur de Schrödinger

(3.1) $$-\Delta + x^2 + \beta \, p_{2m}(x) \quad , \quad \beta > 0 \ ,$$

où $p_{2m}(x)$ est un polynome réel elliptique positif homogène de degré $2m$, $m > 1$. Il est bien connu, que pour $\beta \in \,]0, \beta_0]$ cet opérateur admet près de la premiere valeur propre : n de l'oscillateur harmonique , une valeur propre $\lambda(\beta)$ qui admet le développement asymptotique ;

(3.2) $$\lambda(\beta) \sim n + \sum_{j=1}^{\infty} a_j \beta^j \quad , \quad \beta \to 0 \ .$$

On sait de plus (voir Simon [15] et Graffi [16]) que

(3.3) Pour tout $\eta > 0$, $\lambda(\beta)$ se prolonge holomorphiquement dans un secteur de la forme $0 < |\beta| < \beta(\eta)$, $|\arg \beta| < ((m+1)/2)\pi - \eta$ (sur la surface de Riemann de $\beta^{1/(m+1)}$) et le développement (3.2) reste valable dans ce secteur .

Dans le cas $n = 1$, $m = 2$, $p_4(x) = x^4$, Bender et Wu [17] avaient conjecturé à la suite de calculs numériques , que

$$(3.4) \qquad a_j \sim (-1)^{j+1}\, 4\, \pi^{-3/2}\, (3/2)^{j+\frac{1}{2}}\, \Gamma(j+\tfrac{1}{2}) \quad , \quad j \to \infty \;,$$

et B. Simon a montré l'équivalence de cette formule avec la formule

$$(3.5) \qquad \operatorname{Im} \lambda(-\beta - i0) \sim -4\, \pi^{-\frac{1}{2}}\, |\beta|^{-\frac{1}{2}}\, e^{-2/(3|\beta|)} \quad , \ \beta \to 0+ \;.$$

Cette formule est démontré rigoureusement dans ce cas par Harrell et Simon [18] , mais la démonstration , basée sur des techniques fines d'équations différentielles , reste liée à la dimension 1 . Le cas de la dimension > 1 est resté jusqu'à présent plus mystérieux , même s'il est parfois abordé de manière partiellement heuristique dans des articles de Banks , Bender , Wu [19] , [20] , et de J. Avron [21] , [22] .

Le lien entre (3.4) , (3.5) est donné par la formule

$$(3.6) \qquad a_j = (-1)^j\, \pi^{-1} \int_0^R \beta^{-(j+1)} \operatorname{Im} \lambda(-\beta - i0)\, d\beta \; + \; \mathcal{O}_R(R^{-(j+1)})$$

pour R assez petit , et cette formule (élémentaire) reste vraie aussi pour n , m , p_{2m} arbitraires . L'objet est donc de démontrer des formules du type (3.5) aussi en dimension supérieure .

Le changement de variables $x = \rho y$, $\rho = |\beta|^{1/(2-2m)}$ transforme l'opérateur (3.1) en $h^{-1} P_\alpha(h)$, où

(3.7) $P_\alpha(h) = -h^2 \Delta_y + y^2 + \alpha\, p_{2m}(y)$,

$\alpha = \beta/|\beta|$, $h = |\beta|^{1/(m-1)}$. Considérons donc $P_\alpha(h)$ pour $\alpha \in D = \{\alpha \in \mathbb{C} ; \frac{1}{2} < |\alpha| < 2 , \arg \alpha \in [-\pi, 0]$. Pour $\alpha > 0$, on sait (voir p. ex. [1]) que la plus petite valeur propre $E(\alpha, h)$ est $= n h + \mathcal{O}(h^2)$, et admet un développement asymptotique en puissances de h quand $h \to 0$. Bien entendu , on a $E(1,h)/h = \lambda(\beta)$ pour $\beta > 0$, $h = \beta^{1/(m-1)}$. On montre que :

(a) $E(\alpha, h)$ admet une extension holomorphe à D .

(b) $E(-1,h)$ s'interprête comme une résonance pour l'opérateur $P_{-1}(h)$, et on peut appliquer les résultats de la section 2.

(c) Il est alors clair que $\lambda(-\beta - i0) = |\beta|^{-1/(m-1)} E(-1, |\beta|^{1/(m-1)})$ et dans des cas favorables on a un développement asymptotique quand $\beta \to 0$.

Montrons ici seulement comment on peut étudier $P_{-1}(h) = - h^2 \Delta + x^2 - p_{2m}(x)$. On a clairement un puits $U = \{0\}$, qui est non-dégénéré , dans l'isle $\ddot{O} = \{x \in \mathbb{R}^n ; p_{2m}(x) < x^2\} \cup \{0\}$. Le choix naturel des échelles est donné par : $r = (1+x^2)^{m/2}$, $R = (1+x^2)^{1/2}$, et on vérifie que le symbole $p = \xi^2 + x^2 - p_{2m}(x)$ est bien de classe $S^{2,0}$ .Comme fonction fuite,on choisit $G(x, \xi) = x \cdot \xi$ et on vérifie facilement l'hypothèse (2.2) . On peut donc appliquer les résultats de la section 2 avec μ_1 égale à la première valeur propre de P_{M_0} . On trouve :

(3.9) $C_0^{-1}\, h^{\frac{1}{2}}\, e^{-2S_0/h} \leq -\operatorname{Im} E(-1,h) \leq C_0\, h^{1-n/2}\, e^{-2S_0/h}$, et si on suppose en plus que les points de type 1 forment une sous-variété de codimension d dans $\partial \ddot{C}$ et que la surface caustique C a un contact d'ordre 2 exactement avec $\partial \ddot{O}$, alors

$$(3.10) \qquad - \operatorname{Im} E(-1,h) = h^{1-n/2+d/2} f(h) e^{-2S_0/h} ,$$

où f est une réalisation d'un symbole classique elliptique et analytique . A l'aide de (3.6),(3.8),(3.9) , on trouve en géneral

$$(3.11) \qquad C^{-1} \Gamma((m-1)j-\tfrac{1}{2}) \leq (-1)^{j+1} a_j (2S_0)^{(m-1)j} \leq C \Gamma((m-1)j + n/2) ,$$

et si on rajoute l'hypothèse de (3.10) :

$$(3.12) \qquad a_j = (-1)^{j+1} (2S_0)^{-(m-1)j} \Gamma((m-1)j + (n-d)/2) (\alpha + \mathcal{O}(\tfrac{1}{j})),$$

où $\alpha \neq 0$. La démonstration de (3.12) n'utilise pas le développement asymptotique complet de f , mais seulement le premier terme . On peut aussi donner un développement de a_j dont (3.12) donne le premier terme . Pour calculer la constante α il faut savoir résoudre explicitement une équation de Riccati le long des géodésiques minimales entre U et les points de type 1 , pour déterminer le Hessien de $\tilde{f}$ dans ces points . Il faut aussi savoir résoudre la première équation de transport le long des mêmes géodésiques . Cela semble difficile en général , mais dans l'exemple suivant , étudié par Banks , Bender et Wu [19] , on peut le faire , et on retrouve partiellement leur calculs heuristiques :

Exemple . $n=2$, $m=2$, $P_{-1} = -h^2\Delta + (x^2+y^2)-(x^4+y^4+2cx^2y^2)$.
Le cas $c < -1$ mène à une isle non compacte et devrait pouvoir se traiter comme l'effet Zeeman plus loin . On se restreint au cas $c > -1$. Le cas $c = 1$ est invariant par rotations et les calculs se simplifient beaucoup et on ne le discute pas ici . Le cas $c > 1$ se ramène facilement au cas $-1 < c < 1$ par une rotation de 45° , et on ne considère que le cas $-1 < c < 1$. On a alors seulement les quatre points $(\pm 1,0)$, $(0,\pm 1)$ dans $\partial \ddot{O}$ qui sont de type 1 , et les géodésiques minimales correspondantes sont les segments droits de $(0,0)$ à ces points . A l'aide de fonctions speciales assez compliquées , on arrive à résoudre l'équation de Riccati et la première équation de transport . Le résultat final (en accord avec [19]) pour la résonance proche de $2h$ est :

$$(3.13) \qquad -\operatorname{Im} z(h) = 8\, h^{\frac{1}{2}}\, e^{-2S_0/h} \left(\frac{2\,c}{\sin \pi\nu}\right)^{\frac{1}{2}} (1+\mathcal{O}(h)) \quad ,$$

où ν est donné par $\nu(\nu+1) = 2c$.

Etudions maintenant l'effet Zeeman . Dans $\mathbb{R}^3$ on considère l'opérateur

$$(3.14) \qquad Z_\beta = -\Delta - r^{-1} + \beta(x^2+y^2) \quad ,$$

où $\beta > 0$, $r = (x^2+y^2+z^2)^{\frac{1}{2}}$. Rappelons le résultat suivant :

Théorème 3.1. (i) Pour $\beta \in \mathbb{C}$, $|\arg \beta| < \pi$, l'opérateur Z_β définit une famille holomorphe d'opérateurs de domaine $H^2(\mathbb{R}^3) \cap D(x^2+y^2)$, où $D(x^2+y^2) = \{u \in L^2(\mathbb{R}^3) ; (x^2+y^2)u \in L^2(\mathbb{R}^3)\}$.

(ii) Il existe près de la valeur propre $-1/4$ de $-\Delta - r^{-1}$, une unique valeur propre $E(\beta)$ définie pour $0 < |\beta| \leq \beta_0$, $|\arg \beta| < \pi$.

(iii) Pour tout $\eta > 0$, $E(\beta)$ admet un prolongement holomorphe dans un secteur $\tilde{\Omega}_\eta = \{\beta ; |\arg \beta| < 2\pi - \eta, 0 < |\beta| \leq \beta(\eta)\}$ (dans la surface de Riemann) et dans ce secteur on a le développement asymptotique : $E(\beta) \sim -1/4 + \sum_{n=1}^{\infty} \gamma_n \beta^n$.

(iv) $|\gamma_n| \leq C^{n+1}(2n)!$.

Ces resultats sont classiques et on pourra trouver une démonstration brève dans le survey de S. Graffi [16] où chez Avron, Herbst, Simon [23]. Comme avant on veut préciser le comportement asymptotique de γ_n, qui est lié à celui de $\operatorname{Im} E(-\beta - i0)$ par (3.6) (avec les changements de notation évidents).

Théorème 3.2. $-\operatorname{Im} E(-\beta - i0) = \beta^{-3/4}\, 2^{-3/2}\, (1 + \mathcal{O}(\beta^{\frac{1}{2}})) \exp(-\frac{\pi}{8\sqrt{\beta}})$, quand $\beta \to 0+$.

Ce résultat a été conjecturé et partiellement justifié par Avron [22].

Corollaire 3.3. $\gamma_n = (-1)^{n+1} \frac{1}{2} (4/\pi)^{5/2} (16/\pi^2)^n (2n+\frac{1}{2})! (1+\mathcal{O}(\frac{1}{n}))$.

La démarche pour démontrer le Théorème 3.2 est essentiellement la même qu'avant. Mais puisque notre théorie (dans sa présentation actuelle) n'admet pas de potentiels singuliers , nous allons d'abord tranformer le problème en suivant une stratégie suggérée par S. Graffi , et qui apparait déjà dans l'ouvrage de Titschmarch. Cette strategie apparait également dans l'étude de l'effet Stark (voir [16]). Soit f une solution de $(Z_\beta - E)f = 0$, invariante par rotation autour de l'axe des z , et introduisons des nouvelles coordonées $u , v \in \mathbb{R}^2$, $\varphi \in [0,2\pi[$ par $x = \|u\| \, \|v\| \cos\varphi$, $y = \|u\| \, \|v\| \sin\varphi$, $z = \frac{1}{2}(\|u\|^2 - \|v\|^2)$. Formellement f se transforme en une solution $f = f(u,v)$, indépendante de φ , invariante par rotations en u ou en v , de l'équation

$$(3.15) \quad W(E,\beta)f = (-\Delta - 2 + \beta(u^2+v^2)u^2v^2 - E(u^2+v^2))f = 0 \ .$$

Ce changement de variables peut être rigoureusement justifié si on travaille avec des extensions de Friedrichs .

Soit $\nu(s)$, $s > 0$ la plus petite valeur propre de

$$(3.16) \quad Q(s) = -\Delta + (x^2+y^2) + s(x^2+y^2)x^2y^2 \ , \ x,y \in \mathbb{R}^2 \ .$$

Par le changement de variables $(u,v) = \rho(x,y)$, avec $(-E)\rho^4 = 1$, on trouve pour $\beta > 0$ la relation

$$(3.17) \quad 2\,(-E(\beta))^{-\frac{1}{2}} = \nu(\beta(-E(\beta))^{-2}) \ ,$$

et on en déduit que $\nu(s)$ admet un prolongement holomorphe dans des secteurs de la forme $\tilde{\Omega}_\eta = \{ s \ ; \ 0 < |s| \leq s(\eta) \ , \ |\arg s| < 2\pi - \eta \}$, pour tout $\eta > 0$, et que l'on y a le développement

$$(3.18) \quad \nu(s) \sim \sum_{n=0}^{\infty} \delta_n s^n \ , \quad \delta_0 = 4 \ .$$

La relation (3.17) est ainsi prolongée dans ces secteurs . Dans la relation $s = \beta(-E(\beta))^{-2}$, les quantités s et β ne sont pas forcément réelles et négatives simultanément , mais par un développement de Taylor , on montre que :

$$(3.18) \qquad -\operatorname{Im} E(-\beta - i0) = \frac{(1+\mathcal{O}(\beta))}{8} \operatorname{Im} \nu\,(-\beta(\operatorname{Re} E(-\beta-i0))^{-2}-i0) \ .$$

Il s'agit donc de trouver l'asymptotique de $\operatorname{Im} \nu(-s-io)$ quand $s \to 0+$. Par une nouvelle homothétie convenable, on ramène l'étude de $\nu(s)$ pour $\arg s \in]-\pi,0]$ à l'étude de la valeur propre; $\lambda(\alpha,h)$, proche de $4h$ de

$$P_\alpha(h) = -h^2\Delta + (x^2+y^2) + \alpha(x^2+y^2)x^2y^2 \quad ,$$

pour $|\alpha| = 1$, $\arg \alpha \in]-\pi,0]$. A condition de pouvoir définir $\lambda(-1,h)$ convenablement comme une résonance , on établit que

$$(3.19) \qquad h\,\nu(-h^2-i0) = \lambda(-1,h) \ .$$

Indiquons ici seulement comment définir et étudier la résonance proche de $4h$ de $P_{-1}(h)$. Nous avons donc le potentiel $V = x^2(1-x_1^2 x_2^2)$ dans $\mathbb{R}^4$, qui présente une isle $\ddot{O}$: $x_1^2 x_2^2 < 1$ qui est non-compacte . Ici on écrit $x_j = (x_{j,1}, x_{j,2}) \in \mathbb{R}^2$, $j = 1,2$. On vérifie ensuite que les poids $r(x) = (1+x^2)^{\frac{1}{2}}(1+x_1^2 x_2^2)^{\frac{1}{2}}$, $R(x) = (1+x^2)^{-\frac{1}{2}}(1+x_1^2 x_2^2)^{\frac{1}{2}}$ sont bien adaptés . Cherchons ensuite une fonction fuite de la forme $G(x,\xi) = f'_x(x)\cdot\xi$ avec $f \in S^{0,2}$ invariant par rotation en x_1 et en x_2 . En utilisant l'inégalité :

$$H_p(f'\cdot\xi) \geq -f'\cdot V' + 2\,\lambda_{min}(f'')(-V) \quad \text{sur } \Sigma_p ,$$

où $\lambda_{min}(f'')$ désigne la plus petite valeur propre de f'', on trouve au bout de quelques calculs, que $\tilde{f} = x_1^2\, x_2^2 / x^2$ est de classe $S^{0,2}$ (en-dehors de $(0,0)$) et que si $\tilde{G} = \tilde{f}'\cdot\xi$, alors sur $\Sigma_p \setminus \{(0,0)\}$ on a pour tout $\varepsilon > 0$:

$$H_p(\tilde{G}) \geq (C_\varepsilon)^{-1}\, r^2 \quad , \text{ si } \; |x_1| - |x_2| \geq \varepsilon |x| .$$

Près de la "diagonale" : $|x_1| = |x_2|$, la fonction $x\cdot\xi$ est une bonne fonction fuite, et si χ est une troncature standard, on trouve finalement une fonction f, qui convient, de la forme

$$f(x) = \tilde{f}(x) + \delta\,\chi((|x_1| - |x_2|)/\varepsilon\,|x|)(|x_1|^2 + |x_2|^2) ,$$

avec d'abord $\varepsilon > 0$ assez petit, et ensuite $\delta > 0$ assez petit. La condition d'ellipticité (2.3) se vérifie facilement et on sait donc que tous les résultats de la section 2 s'appliquent. On peut même appliquer le Théorème 2.4, car on vérifie que $\partial\ddot{O} \cap C = \{(x_1,x_2) \in \mathbb{R}^4 \; ; \; |x_1| = |x_2| = 1\}$, et que le contact entre $\partial\ddot{O}$ et C est d'ordre 2 exactement. Ensuite on trouve $S_0 = \pi/4$ et on arrive aussi à résoudre l'équation de Riccati ainsi que la première équation de transport, ce qui permet de donner le premier terme dans le développement asymptotique de $\operatorname{Im}\lambda(-1,h)$, et ceci donne le Théorème 3.2 par les réductions déjà indiquées.

Bibliographie .

1. B. Helffer , J. Sjöstrand , Effet tunnel pour l'opérateur de Schrödinger semiclassique I . Actes des Journées des E.D.P. à St Jean de Monts , Juin 1985 , à paraître .
2. B. Helffer , J. Sjöstrand , Résonances en limite semiclassique , à paraître .
3. J. Aguilar , J.M. Combes , A class of analytic perturbations for one-body Schrödinger Hamiltonians , Comm. Math. Physics , 22(1971),269-279 .
4. E. Balslev , J.M. Combes , Spectral properties of many-body Schrödinger operators with dilation analytic interactions , Comm. Math. Physics , 22(1971), 280-294 .
5. B. Simon , Resonances and complex scaling , a rigorous overview , Int. J. Quantum Chemistry , 14(1978),529-542.
6. I. Herbst , Dilation analyticity in a constant electric field, Comm. Math. Phys. 64(1979),279-298 .
7. H. Cycon , Resonances defined by modified dilations , à paraître .
8. J.M. Combes , J. Duclos , R. Seiler , On the shape resonance , à paraître .
9. R. Beals , C. Fefferman , Spatially inhomogeneous pseudodifferential operators . Comm. Pure Appl. Math., 27(1974),1-24.
10. R. Beals , A general calculus of pseudodifferential operators, Duke Math. J. 42 , n°1 (1975) , 1-42 .
11. L. Hörmander , The Weyl calculus of pseudodifferential operators , Comm. Pure Appl. Math., 32(1979),359-443 .

12. J. Sjöstrand , Singularités analytiques microlocales , Astérisque n°95(1982) .

13. B. Helffer , J. Sjöstrand , Multiple wells in the semiclassical limit I . Comm. P.D.E. , 9(4)(1984),337-408.

14. A. Martinez , **Estimations de l'effet tunnel pour le double puits , Prépubl. d'Orsay (1985) .**

15 B. Simon ,Coupling constant analyticity for the anharmonic oscillator , Ann. of Physics , 58(1970),76-136 .

16. S. Graffi , exposé au Seminaire d'Analyse Mathématique de l'Université de Bologne , 1982-83 .

17. C. Bender , T. Wu , Anharmonic oscillator , Phys. Rev. 184(1969),1231-1260 .

18. E. Harrell , B. Simon , The mathematical theory of resonances whose widths are exponentially small , Duke Math. J., 47,n°4 , Dec. 1980 .

19. T. Banks , C. Bender , T. Wu ,Coupled anharmonic oscillators I. Equal mass case . Phys. Rev. D , Vol.8 , n°10 (Nov 1973), 3346-3366.

20. T. Banks, C. Bender, Coupled ... II. Unequal mass case , Phys. Rev. D. Vol 8,n°10(Nov 1973),3366-3378.

21. J.E. Avron , Bender-Wu formulas and classical trajectories : higher dimensions and degeneracies , Int. J. of Quantum Chemistry , 21(1982),119-124.

22. J.E. Avron ,Bender-Wu formulas for the Zeeman effect in hydrogen. Ann. of Physics 131(1981),73-94.

23. J.E. Avron, I. Herbst, B. Simon, Schrödinger operators with magnetic fields III, Atoms in homogeneous magnetic field, Comm. Math. Physics 79, p. 529-572 (1981).

PROPAGATION OF SINGULARITIES FOR HAMILTON-JACOBI EQUATION

Mikio Tsuji
Kyoto Sangyo University
Department of Mathematics
Kamigamo, Kita-ku
Kyoto 603, Japan.

ABSTRACT. In the recent progress in linear partial differential equations, especially in microlocal analysis, one of the important methods is the concept of Lagrangian manifold which makes us possible to develop the global treatment of Hamilton-Jacobi equation in the cotangent space. But we consider here a solution of Hamilton-Jacobi equation from a view-point of applications, i.e., by taking projections of Lagrangian manifold to the base space, we try to get a one-valued solution, as a function defined on the base space, with the singularities satisfying certain additional condition. Our aim is to make clear the situation how the singularities of solution may appear. Then our problem is to construct the solution in a neighborhood of singularities of the above projections.

§1. Introduction.

Consider the Cauchy problem for Hamilton-Jacobi equation in two dimensional space :

$$\frac{\partial u}{\partial t} + f\left(\frac{\partial u}{\partial x}\right) = 0 \qquad \text{in } \{ t>0,\ x\in R^2 \} \tag{1}$$

$$u(0,x) = \phi(x) \in \mathcal{S}(R^2) \tag{2}$$

where $f(p) \in C^\infty(R^2)$ and $f(p)$ is convex, i.e.,

$$f''(p) \underset{\text{def}}{=} \left[\frac{\partial^2 f}{\partial p_i \partial p_j} \right]_{1\leq i,j\leq 2} \geq C > 0 .$$

The equation (1) has been studied very much from various kinds of view-points, for example, classical mechanics, the theory of optimal control, the theory of waves, and so on. Some results can be beautifully formulated in the framework of symplectic geometry. But we consider here a global solution which is one-valued on $\{t>0,\ x\in R^2\}$.

H. G. Garnir (ed.), Advances in Microlocal Analysis, 323–331.

Then, even for smooth initial data, the Cauchy problem (1)-(2) can not have a smooth solution for all t. Therefore we treat a generalized solution whose definition will be given a little later. The existence of global generalized solution for (1)-(2) is already established ([9], [10], etc). For detailed bibliography, refer to [1]. This talk is concerned with the singularities of generalized solutions.

For a single conservation law in one dimensional space, a solution satisfying the entropy condition is piecewise smooth for any smooth initial data in $\mathcal{S}(R^2)$ except for the initial data in a certain subset of the first category ([5], [6], [8] and [15]). T. Debeneix [3] treated certain systems of conservation laws which is equivalent to Hamilton-Jacobi equation (1) in R^n ($n \leqq 4$), and proved the similar results to [15] by the same method as [15]. M. Bony [2] considers the propagation of singularities for general nonlinear partial differential equations of higher order. But the singularities discussed here would not be treated by his method.

We solve the Cauchy problem (1)-(2) by the characteristic method. Then we get a flow which corresponds an initial Lagrangian manifold to another one at a time $t>0$. When we project the Lagrangian manifold at a time t to the base space, we see that, for large t, the projections have generally the singularities. As an inverse mapping of such a projection takes many values in a neighborhood of singular points, the solution is also many-valued there. As the following definition says, we look for one-valued and continuous solution. Our aim is to show that we can uniquely choose one value from many values so that the solution is one-valued and continuous. Then the condition of semi-concavity is automatically satisfied. Here we give the definition of generalized solutions.

Definition. A lipschitz continuous function $u(t,x)$ defined on $R^1 \times R^2$ is called a generalized solution of (1)-(2) if and only if
i) $u(t,x)$ satisfies the equation (1) almost everywhere in $R^1 \times R^2$ and the initial condition (2) on $\{t=0, x \in R^2\}$,
ii) $u(t,x)$ is semi-concave, i.e., there exists a constant $K > 0$ such that

$$u(t,x+y) + u(t,x-y) - 2\,u(t,x) \leqq K|y|^2 \quad \text{for any } x,\ y \in R^2 \text{ and } t>0. \tag{3}$$

Remark. Put $v_i = \partial u/\partial x_i$ $(i=1,2)$, then the equation (1) is written as a following system of conservation law :

$$\frac{\partial}{\partial t} v_i + \frac{\partial}{\partial x_i} f(v) = 0 \qquad (i=1,2) . \tag{4}$$

Then the semi-concavity condition for (1) is equivalent to the entropy condition for (4). See Remark 2 in §3.

§2. Construction of solutions.

The characteristic lines for the Cauchy problem (1)-(2) are determined by the following equations :

$$\dot{x}_i = \frac{\partial f}{\partial p_i}(p) \quad , \quad \dot{p}_i = 0 \qquad (i=1,2)$$

with initial data

$$x_i(0) = y_i \quad , \quad p_i(0) = \frac{\partial \phi}{\partial y_i}(y) \qquad (i=1,2) \ .$$

On the characteristic line $x=x(t,y)$, the value $v(t,y)$ of the solution for (1)-(2) satisfies the equation

$$\dot{v} = -f(p) + \langle p, f'(p)\rangle \ , \quad v(0) = \phi(y)$$

where $f'(p) = (\partial f/\partial p_1, \partial f/\partial p_2)$ and $\langle p,q\rangle$ is scalar product of vectors p and q. Solving these equations, we have

$$x = y + t\, f'(\phi'(y)) \underset{\text{def}}{=} H_t(y) \tag{5}$$

$$v(t,y) = \phi(y) + t\{-f(\phi'(y)) + \langle \phi'(y), f'(\phi'(y))\rangle\}. \tag{6}$$

Then H_t is a smooth mapping from R^2 to R^2 and its Jacobian is given by

$$\frac{Dx}{Dy}(t,y) = \det [\ I + t\, f''(\phi'(y))\, \phi''(y)\ ] \ .$$

We write $A(y) = f''(\phi'(y))\phi''(y)$ and the eigenvalues of $A(y)$ by $\lambda_1(y) \leq \lambda_2(y)$. As $\phi(y)\varepsilon\mathcal{S}(R^2)$, we see

$$\min_y \ \lambda_1(y) = \lambda_1(y^0) = -M < 0 \ ,$$

and put $t^0 = 1/M$. Since $Dx/Dy(t,y) \neq 0$ in $\{(t,y);\ t<t^0,\ y\varepsilon R^2\}$, we can uniquely solve the equation (5) there with respect to y and denote it by $y=y(t,x)$. Then $u(t,x)=v(t,y(t,x))$ is a unique solution of (1)-(2) in $\{t<t^0\}$. Our problem is to construct the solution for $t>t^0$.

Suppose that $t-t^0$ is positive and sufficiently small, and consider the equation (5) in a neighborhood of (t^0,y^0). The Jacobian of H_t vanishes on $\Sigma_t = \{y\varepsilon R^2;\ 1 + t\lambda_1(y) = 0\}$. Assume the conditions

<u>(A.1)</u> Σ_t is a smooth simple closed curve,
<u>(A.2)</u> The singularities of H_t are fold and cusp points only, and the number of cusp points on Σ_t is two.

<u>Remark.</u> Suppose the conditions

(C.1) $\lambda_1(y) \neq \lambda_2(y)$ and the singularities of $\lambda_1(y)$ are non-degenerate,

(C.2) The singularities of H_t are fold and cusp points only,
(C.3) $\partial v/\partial y(t^0,y^0) \neq 0$,
then the above conditions (A.1) and (A.2) are satisfied for $t>t^0$ where $t-t^0$ is sufficiently small.

H. Whitney [20] proved that the canonical forms of a fold and cusp point are expressed respectively as follows :

$$x_1 = y_1^2,\ x_2 = y_2 \quad \text{in a neighborhood of a fold point} \tag{7}$$

$$x_1 = y_1 y_2 - y_1^3,\ x_2 = y_2 \quad \text{in a neighborhood of a cusp point.} \tag{8}$$

By this result, the mapping H_t can be regarded as the mapping (7) and (8) in a neighborhood of a fold and cusp point respectively. Moreover, Whitney [20] proved that any smooth mapping from R^2 to R^2 can be approximated by smooth mappings whose singularities are fold and cusp points only.

We denote by D_t the interior of the curve Σ_t and by Ω_t the interior of $H_t(\Sigma_t)$, and solve the equation (5) with respect to y for $x \in \Omega_t$. The expressions (7) and (8) mean that we get three solutions $y=g_i(t,x)$ (i=1,2,3). Here we choose $g_2(t,x)$ so that $g_2(t,x)$ is in D_t for any $x\in\Omega_t$. When we write $u_i(t,x)=v(t,g_i(t,x))$ (i=1,2,3), the solution of (1)-(2) takes three values $u_i(t,x)$ (i=1,2,3) on Ω_t.

Lemma 1.

i) $$\frac{\partial}{\partial x} u_i(t,x) = \frac{\partial\phi}{\partial y}(g_i(t,x)) \qquad \text{for} \quad x \in \Omega_t \quad (i=1,2,3)$$

ii) $$< g_i(t,x) - g_j(t,x)\ ,\ \frac{\partial u_i}{\partial x} - \frac{\partial u_j}{\partial x} > \ < \ 0 \qquad \text{for} \quad x \in \Omega_t\ ,\ i \neq j\ ,$$

iii) $u_1(t,x) < u_2(t,x)$ and $u_3(t,x) < u_2(t,x)$ for $x \in \Omega_t$.

Proof. i) We can easily get this by simple calculation.
ii) From the definition of $g_i(t,x)$, we have

$$x = g_i(t,x) + t\, f'(\frac{\partial u_i}{\partial x}(t,x))\ ,\quad x \in \Omega_t \quad .$$

As $g_i(t,x) \neq g_j(t,x)$ for $i \neq j$, it follows $\partial u_i/\partial x(t,x) \neq \partial u_j/\partial x(t,x)$ for $i\neq j$. Using the convexity of f(p), we get ii).
iii) We prove the first inequality. Divide the simple closed curve $\partial\Omega_t$ into two curves joining two cusps of $\partial\Omega_t$, and write them C_1 and C_2. Here we introduce the family of solution curves of

$$\frac{dx}{dr} = g_1(t,x) - g_2(t,x) \quad .$$

Then the solution curves start from C_1 (or from C_2) and end at C_2 (or at C_1) and the family of these curves covers the domain Ω_t. On the curves it holds

$$\frac{d}{dr}(u_1(t,x) - u_2(t,x)) = \langle \frac{\partial u_1}{\partial x} - \frac{\partial u_2}{\partial x}, g_1 - g_2 \rangle < 0 .$$

Since $u_1(t,x) = u_2(t,x)$ on C_1 (or on C_2 respectively), we get $u_1(t,x) \lessgtr u_2(t,x)$ in Ω_t. Q.E.D.

As we are looking for a continuous solution, Lemma 1 (iii) means that we can not attain our aim by advancing from the first branch to the second one and also from the second to the third. The last choice is to pass from the first branch to the third one. Put $I(t,x) = u_1(t,x) - u_3(t,x)$.

Lemma 2. $\Gamma_t = \{x \in \overline{\Omega_t};\ I(t,x)=0\}$ is a smooth curve in Ω_t joining two cusps of $\partial\Omega_t$.

Proof. In this case we introduce the family of solution curves by

$$\frac{dx}{dr} = g_1(t,x) - g_3(t,x).$$

Then these curves start from C_1 (or from C_2) and end at C_2 (or at C_1 respectively), and the family of the curves covers the domain Ω_t. On each curve it holds

$$\frac{d}{dr} I(t,x(r)) = \langle \frac{\partial u_1}{\partial x} - \frac{\partial u_3}{\partial x}, g_1 - g_3 \rangle < 0 .$$

On the other hand, we have, by Lemma 1,

$$I(t,x)\Big|_{C_1} = u_1(t,x) - u_2(t,x)\Big|_{C_1} < 0 ,$$

$$I(t,x)\Big|_{C_2} = u_2(t,x) - u_3(t,x)\Big|_{C_2} > 0 .$$

Therefore, on each curve $\{x=x(r)\}$, $I(t,x)=0$ has a unique solution. Obviously $I(t,x)=0$ at the cusps of $\partial\Omega_t$, and Lemma 1 (ii) guarantees $\mathrm{grad}_x I(t,x) \neq 0$ in Ω_t. Hence we see that Γ_t is a smooth curve joining two cusps of $\partial\Omega_t$. Q.E.D.

Since we are seeking for a continuous and one-valued solution, we define the solution $u(t,x)$ of (1)-(2) in Ω_t as follows :

Write $\Omega_{t,\pm} = \{x \in \Omega_t;\ u_3(t,x) - u_1(t,x) \gtrless 0\}$, and define

$$u(t,x) = \begin{cases} u_1(t,x) & \text{in} \quad \Omega_{t,+} \\ u_3(t,x) & \text{in} \quad \Omega_{t,-} \end{cases} .$$

§3. Semi-concavity of the solution u(t,x).

Let $\vec{n}(t,x)$ be a unit normal of Γ_t advancing from $\Omega_{t,-}$ to $\Omega_{t,+}$, and define at $x \in \Gamma_t$

$$\frac{\partial u}{\partial x}(t,x \pm 0) \underset{\text{def}}{=} \lim_{\varepsilon \to +0} \frac{\partial u}{\partial x}(t,x \pm \varepsilon\vec{n}) .$$

For the proof of the semi-concavity condition (3), it is sufficient to consider the case where $x \in \Gamma_t$ and $y = \varepsilon\vec{n}$ ($\varepsilon > 0$), because any C^2-function satisfies (3). Then we have

$$\begin{aligned} &u(t,x+y) + u(t,x-y) - 2\,u(t,x) \\ &= \int_0^1 \left\langle \frac{\partial u}{\partial x}(t,x+sy) - \frac{\partial u}{\partial x}(t,x+0),\ y \right\rangle ds \\ &\quad + \int_0^1 \left\langle \frac{\partial u}{\partial x}(t,x-0) - \frac{\partial u}{\partial x}(t,x-sy),\ y \right\rangle ds \\ &\quad + \left\langle \frac{\partial u}{\partial x}(t,x+0) - \frac{\partial u}{\partial x}(t,x-0),\ y \right\rangle . \end{aligned}$$

As the first and second terms are estimated by $K|y|^2$, the inequality (3) is equivalent to the following one :

$$\left\langle \frac{\partial u}{\partial x}(t,x+0) - \frac{\partial u}{\partial x}(t,x-0),\ \vec{n} \right\rangle \leqq 0 . \tag{9}$$

On the other hand, since $u_3(t,x) - u_1(t,x) \gtrless 0$ in $\Omega_{t,\pm}$, we have

$$\begin{aligned} 0 &\leqq \frac{d}{ds}\{u_3(t,x+s\vec{n}) - u_1(t,x+s\vec{n})\}\Big|_{s=0} \\ &= \left\langle \frac{\partial u_3}{\partial x}(t,x) - \frac{\partial u_1}{\partial x}(t,x),\ \vec{n} \right\rangle \qquad \text{on } \Gamma_t . \end{aligned}$$

By the definition of u(t,x) in Ω_t, it holds

$$\frac{\partial u}{\partial x}(t,x+0) = \frac{\partial u_1}{\partial x}(t,x) \quad \text{and} \quad \frac{\partial u}{\partial x}(t,x-0) = \frac{\partial u_3}{\partial x}(t,x) \qquad \text{on } \Gamma_t .$$

Hence we get the inequality (9).

Summing up the above results, we have the following

Theorem 1. Assume the conditions (A.1) and (A.2). Though the solution takes many values after the time t^0, we can uniquely pick up one value so that the solution becomes one-valued and continuous. Then the condition of semi-concavity is automatically satisfied.

Remark 1. As Γ_t is smooth, it can be parametrized as $\Gamma_t=\{x=x(s)\}$. Then we get

$$\frac{d}{ds} I(t,x(s)) = < \frac{\partial u_1}{\partial x} - \frac{\partial u_3}{\partial x}, \frac{dx}{ds} >$$

$$= < \frac{\partial u}{\partial x}(t,x+0) - \frac{\partial u}{\partial x}(t,x-0), \frac{dx}{ds} > = 0 .$$

This means that, though the derivative $\partial u/\partial x(t,x)$ has jump discontinuity along the curve Γ_t (see Lemma 1 (ii)), it is continuous with respect to the tangential direction of Γ_t.

Remark 2. Putting $v = \partial u/\partial x$ in (9), we get the condition on the jump discontinuity of $v(t,x)$ which is the entropy condition for the system of conservation law (4) given in Remark in §1.

§4. Collision of singularities.

Let Γ_1 and Γ_2 be the singularities constructed in §2, and assume that Γ_1 and Γ_2 collide each other. For detailed proofs on the following discussions, refer to [19]. We will advance looking at Figure 1. We can easily see that a collision of type (i) does not happen. When Γ_1 and Γ_2 meet at first as the type (ii), then, after the

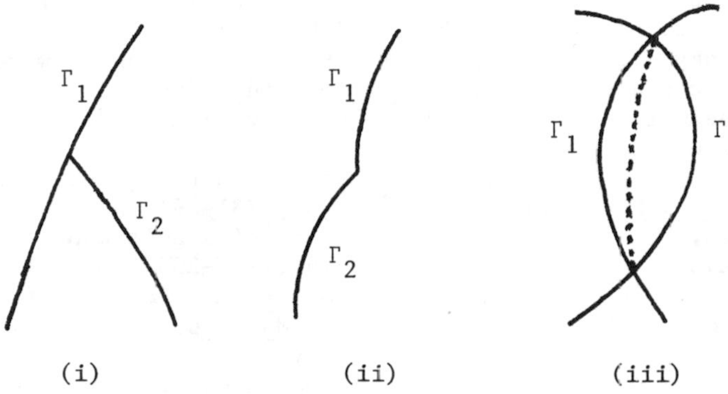

Figure 1

first contact, $\Sigma_t=\{y\varepsilon R^2;\ 1+t\lambda_1(y)=0\}$ becomes a simple closed curve. Moreover, if it satisfied the condition (A.2), we can construct the singularity of solution by the just same way as in §2.

At last we consider the case where Γ_1 and Γ_2 touch first at a point which is not an end point of Γ_1 and Γ_2 both. Then, after the collision, Γ_1 and Γ_2 intersect as (iii) of Figure 1. The solution takes two values on a domain bounded by Γ_1 and Γ_2. Doing the similar discussion as Lemma 2, we can uniquely pick up a reasonable value so that the solution is continuous and semi-concave. Then we get a new singularity which is drawn by a dotted curve in (iii) of Figure 1.

Summing up these results, we get

Theorem 2. Assume that the conditions (A.1) and (A.2) are conserved. Then, even if two singularities collide each other, we can uniquely pick up one reasonable value from many values so that the solution becomes one-valued and continuous. In this case also, the condition of semi-concavity is naturally satisfied.

Remark. Concerning the propagation of singularities for a single conservation law in several space dimensions, S. Nakane [13] is now developping the similar theory like ours.

References.

[1] S. H. Benton, Hamilton-Jacobi equation, a global approach. Academic Press, 1977.

[2] J. M. Bony, 'Calcul symbolique et propagation des singularités pour les équations aux dérivées partielles non-linéaires', Ann. Sci. Ec. Norm. Sup., 4e série, 14(1981), 209-246.

[3] T. Debeneix, 'Certains systèmes hyperboliques quasi-linéaires' (preprint), 1980.

[4] A. Douglis, 'Solutions in the large for multi-dimensional non-linear partial differential equations of first order'. Ann. Inst. Fourier Grenoble, 15(1965), 1-35.

[5] M. Golubitsky and D. G. Schaeffer, 'Stability of shock waves for single conservation law', Adv. in Math., 15(1975), 65-71.

[6] J. Guckenheimer, 'Solving a single conservation law", Lecture Notes in Math , 468(1975), 108-134.

[7] ——————, 'Shocks and rarefactions in two space dimensions Arch. Rat. Mech. Analysis, 59(1975), 281-291.

[8] G. Jennings, 'Piecewise smooth solutions of single conservation law exist', Adv. in Math., 33(1979), 192-205.

[9] S. N. Kruzkov, 'Generalized solutions of non-linear first order equations with several variables', Math. USSR Sb., 1(1967), 93-116.

[10] ——————, 'First order quasi-linear equations in several independent variables', Math. USSR Sb., 10(1970), 217-243.

[11] P. D. Lax, 'Hyperbolic systems of conservation laws II', Comm. Pure Appl. Math., 10(1957), 537-566.

[12] P. D. Lax, Hyperbolic systems of conservation laws and the mathematical theory of shock waves. SIAM Regional Conf. Ser.

[13] S. Nakane, personal communication.

[14] O. A. Oleinik, 'Discontinuous solutions of non-linear differential equation', AMS Transl. Ser., 26(1957), 95-172.

[15] D. G. Schaeffer, 'A regularity theorem for conservation law', Adv. in Math., 11(1973), 358-386.

[16] R. Thom, 'The two-fold way of catastrophe theory', Lecture Notes in Math.(Springer), 525(1976), 235-252.

[17] M. Tsuji, 'Solution globale et propagation des singularités pour l'équation de Hamilton-Jacobi', C.R.Acad. Sci. Paris, 289(1979), 397-400.

[18] ———, 'Formation of singularities for Hamilton-Jacobi equation I', Proc. Japan Acad., 59(1983), 55-58.

[19] ———, 'Formation of singularities for Hamilton-Jacobi equation II', to appear in J. Math. Kyoto Univ.

[20] H. Whitney, ' On singularities of mappings of Euclidean spaces I' Ann. Math., 62(1955), 374-410.

RAMIFICATIONS OF HOLOMORPHIC INTEGRALS

J. VAILLANT
Unité C.N.R.S. 761
Université Pierre et Marie Curie (PARIS VI)
MATHEMATIQUES, tour 45-46, 5ème étage
4, Place Jussieu 75230 PARIS CEDEX 05

ABSTRACT. We consider the integral

$$I(x) = \int_{S_q(x)} \mathcal{U}(x,\tau)\, d\tau \quad ;$$

$I(x)$ is holomorphic and ramified around $V : \varphi(x,\tau) = 0$; we integrate on the relative cycle defined by the holomorphic simplex $S_q(x)$ and its faces. We obtain the ramification of $I(x)$ using discriminants and polar manifolds. In fact, $I(x)$ is ramified around a hypersurface : $\Delta(x) = 0$; denote $V_x = \{\tau;\ \varphi(x,\tau) = 0\}$; if x is such that $\Delta(x) = 0$, either V_x has a singular point, or an edge of the simplex is tangent to V_x . Assumptions are essentially Weierstrass hypothesises on functions induced by φ on grassmannian manifolds.

1. INTRODUCTION

φ is a germ of holomorphic function at the origin of $\mathbb{C}^{n+1} \times \mathbb{C}^q$; more precisely Ω_1 (resp. Ω_2) is an open polydisk with center 0 in $\mathbb{C}^{n+1}$ (resp. $\mathbb{C}^q$) and φ is holomorphic in $\Omega_1 \times \Omega_2$: $x \in \Omega_1$, $\tau \in \Omega_2$;

$$\varphi(0, x', 0) = x^1 \ , \text{(with } x = (x^o, x^1)) \text{ and } D_\tau \varphi(0,0) = 0 \ .$$

$\mathcal{U}(x,\tau)$ is a germ of holomorphic function at the point $(y,0) \in \Omega_1 \times \Omega_2$, $(y^o = 0$, $y^1 \neq 0)$, ramified around $V : \varphi(x,\tau) = 0$; y can be chosen as near 0, as we want.

We consider the integral :

H. G. Garnir (ed.), Advances in Microlocal Analysis, 333–361.

$$I(x) = \int_{S_q(x)} \mathcal{U}(x,\tau)\, d\tau$$

$$= \int_0^{x^o} d\tau_1 \dots \int_0^{x^o - \sum_{1 \leq k \leq j-1} \tau_k} d\tau_j \dots \int_0^{x^o - \sum_{1 \leq k \leq q-1} \tau_k} \mathcal{U}(x,\tau)\, d\tau_q \quad ;$$

it defines a germ of holomorphic function at the pointy ; we intend to study its ramification.

Such an integral appears, in fact, in the representation of solutions of Cauchy's problem with singular data for holomorphic partial differential operators [3] , [7] , [8] . In [7] , [8] we had studied the case of a double integral for φ polynomial in τ and holomorphic in x . T. KOBAYASHI [5] studied the general case using the Thom's isotopy theorem ; he used a compactification necessary in order to make the projection proper. In [9] we considered the general case with assumptions of "finitude" of Weierstrass type which permit us to control zeroes ; J. LERAY [6] had treated the case where φ is polynomial in (x,τ) ; we extend his method to the case where φ is holomorphic in (x,τ).

First we define a discriminant application δ which associates, to a holomorphic germ at a point of a grassmannian manifold of j-planes $\mathbb{C}^{n+1} \times \Gamma^j$ $(\mathbb{C}^q)$, a holomorphic germ at a point of $\mathbb{C}^{n+1} \times \Gamma^{j+1}$. Next, considering φ as defined on $\mathbb{C}^{n+1} \times \Gamma^o(\mathbb{C}^q) = \mathbb{C}^{n+1} \times \mathbb{C}^q$ we define, by iterating δ, support functions Δ .

On the other hand we define germs of polar manifolds : $P_k = 0$ (these are relative polars of sections of $V : \varphi(x,\tau) = 0$, by k-planes in the form $(x, F_k))$ and, with Weierstrass assumption, we also define an operator γ which associates a germ P_{k-1} at a point of $\mathbb{C}^{n+1} \times \Gamma^{k-1}$ $(\mathbb{C}^q)$ to a germ P_k defined in a point of $\mathbb{C}^{n+1} \times \Gamma^k$ $(\mathbb{C}^q)$ and such that $\delta \circ \gamma$ is the multiplication by a invertible germ. Then we relate the notion of supporting plane to the notion of tangent plane or plane passing through a singular point, under the assumption, that, for every x , V considered as a variety V_x in τ , is not developable. For this purpose we use a lemma of [1] . These geometrical results reduce, by induction, the proof of our theorem, to the case of simple integrals.

Finally we obtain that $I(x)$ is ramified around a hypersurface $\Delta_{S_q}(x) = 0$; if x belongs to this hypersurface then either V_x has a singular point or one edge of the simplex $S_q(x)$ of integration is tangent to V_x .

Details of proofs will be published in the "Journal de Mathématiques pures et appliquées".

2 DISCRIMINANTS AND SUPPORTING PLANES

2.1. $f(u,\sigma)$ is a germ of holomorphic function at the origin of $\mathbb{C}^{\ell} \times \mathbb{C}$; $f(0,\sigma) \not\equiv 0$, $f(0,0) = 0$; $\pi(u,\sigma)$ is the Weierstrass polynomial in σ of f ; in order that f is reduced, it is necessary and sufficient that :

$$\operatorname{discr}_{\sigma} \pi(u,\hat{\sigma}) \not\equiv 0 ,$$

where $\operatorname{discr}_{\sigma}$ denote the discriminant of the polynomial π in σ .

2.2. $(x,\tau) \in \mathbb{C}^{n+1} \times \mathbb{C}^q$; the set of affine hyperplanes F of $\mathbb{C}^q$ with equations :

$$\tau_1 + a^2 \tau_2 + \ldots + a^q \tau_q + \sigma = 0 ,$$

can be identified to $\mathbb{C}^q$; F corresponds to :

$$(a,\sigma) = (a^2 ,\ldots, a^q,\sigma) \in \mathbb{C}^q .$$

$f(x, a, \sigma)$ is now a germ of a holomorphic function at the origin of $\mathbb{C}^{n+1} \times \mathbb{C}^q$ and such that : $f(0,0,\sigma) \not\equiv 0$, $f(0,0,0) = 0$; f_r is the corresponding reduced germ, with Weierstrass polynomial π_r ; $\operatorname{discr}_{\sigma} \pi_r(x, a, \hat{\sigma})$ is a germ at the origin of $\mathbb{C}^{n+1} \times \mathbb{C}^{q-1}$; it follows from 2.1 that this germ is not identically zero ; it is uniquely decomposable, except for invertible factors, by irreducible factors in the ring of germs of holomorphic functions at the origin of $\mathbb{C}^{n+1} \times \mathbb{C}^{q-1}$; the product of the distinct irreducible and independent of a factors of $\operatorname{discr}_{\sigma} \Pi_r(x, a, \hat{\sigma})$ will be denoted by :

$$\delta f(x, \hat{F}) \quad ;$$

$\delta f(x, \hat{F}) = 0$ is a germ of a reduced analytic hypersurface at the origine of $\mathbb{C}^{n+1}$ and is called the germ of the discriminant hypersurface of f. It will follow from hypothesises that furthermore, $\delta f(0, \hat{F}) = 0$, in such a way that this hypersurface is not empty.

The equality : $\delta f(\underline{x}, \hat{F}) = 0$ means that $\underline{x}$ belongs to the largest germ of a reduced analytic hypersurface, such that, at every point x of this hypersurface, the germ $f(x, F)$ generically reduced, is not reduced in x. We shall use this remark for a holomorphic in x family of polar sets : $f(x, F) = 0$.

The previous definition is easily generalized to the case of a complex manifold and a point of this manifold, instead of $\mathbb{C}^{n+1}$ and the point 0.

It can be deduced from Weierstrass theorem that there exist connected neighbourhoods of 0 in $\mathbb{C}^{n+1}$, $\mathbb{C}^{q-1}$, $\mathbb{C}$: $\Omega_1 \times \mathcal{A} \times \mathcal{S}$, such that for any $(x,a) \in \Omega_1 \times \mathcal{A}$, the number of zeroes of f, contained in $\mathcal{S}$, counted with multiplicities, is constant, equal to the order of the null zero of $f(0,0,0) = 0$ and that for any (x,a) belonging to the set $\{(x,a) \in \Omega_1 \times \mathcal{A} \ ; \ \operatorname{discr}_{\sigma} \Pi_r(x, a, \hat{\sigma}) \neq 0\}$, the number of distinct zeroes of f is constant, equal to the degree of Π_r ; such neighbourhoods will be called Weierstrass neighbourhoods. A neighbourhood Ω_1 such that δf is defined will be called an allowed neighbourhood.

Finally we denote : $\tilde{f}(x, \alpha, \sigma) = f(x, \alpha\sigma, \sigma)$ and :

$$\tilde{F} \sim (\alpha, \sigma) \in \mathbb{C}^q \ ;$$

we have the lemma :

if x is such that : $\delta \tilde{f}(x, \hat{\tilde{F}}) = 0$

then : $f(x,0,0) \ \delta f(x, \hat{F}) = 0$.

2.3. The grassmannian of affine (q-j)-planes is denoted by Γ^{q-j}, $0 \leq j \leq q$. It is a complex manifold of dimension $j(q+1-j)$. We denote in $\mathbb{C}^q$ by T_i, $1 \leq i \leq q$, the hyperplane

$$\tau_1 = 0 ,$$

and by T_o^o the hyperplane

$$\sum_{k=1}^{k=q} \tau_k = 0 \quad ;$$

for indexes : $(1_1, \ldots, i_k, \ldots, i_j)$ such that

$$1 \leq i_1 < \ldots < i_k < \ldots < i_j \leq q \quad ,$$

we set :

$$T_{i_1 \ldots i_j} = T_{i_1} \cap T_{i_2} \cap \ldots \cap T_{i_j} \in \Gamma^{q-j} \quad ,$$

$$T^o_{i_1 \ldots i_j o} = T_{i_1 \ldots i_j} \cap T^o_o \in \Gamma^{q-j-1} \quad , \; j \leq q-1 \quad .$$

$f(x, F_{i_1 \ldots i_j})$ is a germ of holomorphic function at the point : $(0, T_{i_1 \ldots i_j})$ of $\mathbb{C}^{n+1} \times \Gamma^{q-j}$; let i_ℓ one of the indexes i_k.

There exist hyperplanes : $F_{i_1}, F_{i_2}, \ldots, F_{i_{\ell-1}}, F_{i_{\ell+1}} \ldots, F_{i_j}, F_{i_\ell}$, with equations :

$$F_{i_1} : \tau_{i_1} + \sum_{k \neq i_1} a^k_{i_1} \tau_k + \sigma_{i_1} = 0$$

$$\vdots$$

$$F_{i_{\ell-1}} : \tau_{i_{\ell-1}} + \sum_{k \notin \{i_1, \ldots, i_{\ell-1}\}} a^k_{i_{\ell-1}} \tau_k + \sigma_{i_{\ell-1}} = 0$$

$$F_{i_{\ell+1}} : \tau_{i_{\ell+1}} + \sum_{k \notin \{i_1, \ldots, i_{\ell-1}, i_{\ell+1}\}} a^k_{i_{\ell+1}} \tau_k + \sigma_{i_{\ell+1}} = 0$$

$$\vdots$$

$$F_{i_j} : \tau_{i_j} + \sum_{k \notin \{i_1, \ldots, i_{\ell-1}, i_{\ell+1}, \ldots, i_j\}} a^k_{i_j} \tau_k + \sigma_{i_j} = 0$$

$$F_{i_\ell} : \tau_{i_\ell} + \sum_{k \notin \{i_1, \ldots, i_j\}} a^k_{i_\ell} \tau_k + \sigma_{i_\ell} = 0$$

and such that :

$$F_{i_1 \dots i_j} = F_{i_1} \cap F_{i_2} \cap \dots \cap F_{i_{\ell-1}} \cap F_{i_{\ell+1}} \cap \dots \quad F_{i_j} \cap F_{i_\ell} \quad .$$

In the same way, each (q-j-1) plane near $T_{i_1 \dots i_{\ell-1} \, i_{\ell+1} \dots i_j}$ can be written :

$$F_{i_1 \dots i_{\ell-1} \, i_{\ell+1} \dots i_j} = F_{i_1} \cap F_{i_2} \quad \dots \cap F_{i_{\ell-1}} \cap F_{i_{\ell+1}} \quad \dots \cap F_{i_j}$$

The germ f induces a germ f o p at the point $(0, T_{i_1}, \dots, T_{i_j})$ such that :

$$f \circ p(x, F_{i_1}, \dots, F_{i_j}) = f(x, F_{i_1} \cap \dots \cap F_{i_j}) \quad ;$$

suppose the following Weierstrass condition is satisfied

$$f(0, T_{i_1 \dots i_{\ell-1} \, i_{\ell+1} \dots i_j} \cap \dot{F}_{i_\ell}) \not\equiv 0 \quad ,$$

where : $\dot{F}_{i_\ell}$: $\tau_{i_\ell} + \sigma_{i_\ell} = 0$; Then the § 2.2 allows us to define :

$$\delta(f \circ p) \; (x, F_{i_1}, \dots, F_{i_{\ell-1}}, F_{i_{\ell+1}}, \dots, F_{i_j}, \hat{F}_{i_\ell}) \quad .$$

On the other hand, the germ f induces at the point :

$$(0, T_{i_1 \dots i_{\ell-1} \, i_{\ell+1} \dots i_j}, T_{i_\ell}) \in \mathbb{C}^{n+1} \times \Gamma^{q-j+1} \times \mathbb{C}^{q-j+1}$$

the germ f' such that :

$$f'(x, F_{i_1 \dots i_{\ell-1} \, i_{\ell+1} \dots i_j}, F_{i_\ell}) = f(x, F_{i_1} \cap \dots \cap F_{i_j})$$

and :

$$f'(0, T_{i_1 \dots i_{\ell-1} \, i_{\ell+1} \dots i_j}, \dot{F}_{i_\ell}) \neq 0 \quad ;$$

the 2.2 allows us to define :

$$\delta f'(x, F_{i_1 \dots i_{\ell-1} \, i_{\ell+1} \dots i_j}, \hat{F}_{i_\ell}) \quad .$$

Then the equality :

$$\delta f'(x, F_{i_1 \dots i_{\ell-1} \, i_{\ell+1} \dots i_j}, \hat{F}_{i_\ell}) = 0$$

is equivalent to the equality :

$$\delta(f \circ p)(x, F_{i_1}, \ldots, F_{i_{\ell-1}}, F_{i_{\ell+1}}, \ldots, F_{i_j}, \hat{F}_{i_\ell}) = 0 \quad .$$

We define :

$$\delta f(x, F_{i_1 \ldots i_{\ell-1}\, i_{\ell+1} \ldots i_j} \cap \hat{F}_{i_\ell})$$

$$\equiv \delta f'(x, F_{i_1 \ldots i_{\ell-1}\, i_{\ell+1} \ldots i_j}, \hat{F}_{i_\ell}) \quad .$$

So we have associated to the germ $f(x, F_{i_1 \ldots i_j})$ at the point $(0, T_{i_1 \ldots i_j}) \in \mathbb{C}^{n+1} \times \Gamma^{q-j}$ the germ of holomorphic fonction $\delta f(x, F_{i_1 \ldots i_{\ell-1}\, i_{\ell+1} \ldots i_j} \cap \hat{F}_{i_\ell})$ at the point :

$$(0, T_{i_1 \ldots i_{\ell-1}\, i_{\ell+1} \ldots i_j}) \in \mathbb{C}^{n+1} \times \Gamma^{q-j+1} \quad .$$

We say again that δf is defined in a allowed neighbourhood (2.2) and that : $(x, F_{i_1 \ldots i_{\ell-1}\, i_{\ell+1} \ldots i_j})$ in this neighbourhood is allowed.

In the same way we associate to the germ of holomorphic function : $f(x, F_{i_1 \ldots i_j o})$ at the point $(0, T^o_{i_1 \ldots i_j o}) \in \mathbb{C}^{n+1} \times \Gamma^{q-j-1}$, the germ of holomorphic function :

$$\delta f(x, F_{i_1 \ldots i_{\ell-1}\, i_{\ell+1} \ldots i_j o} \cap \hat{F}_{i_\ell})$$

at the point :

$$(0, T^o_{i_1 \ldots i_{\ell-1}\, i_{\ell+1} \ldots i_j o}) \in \mathbb{C}^{n+1} \times \Gamma^{q-j}$$

as long as the Weierstrass condition :

$$f(0, T^o_{i_1 \ldots i_{\ell-1}\, i_{\ell+1} \ldots i_j o} \cap \dot{F}_{i_\ell}) \not\equiv 0$$

is fulfilled

2.4 $(x,\tau) \in \Omega_1 \times \Omega_2 \mapsto \varphi(x,\tau) \in \mathbb{C}$, $(\Omega_1 = \{x \in \mathbb{C}^{n+1} ; \sup |x^i| < R_1\}$ $\Omega_2 = \{\tau \in \mathbb{C}^q ; \sup |\tau_j| < R_2\}))$, defines a germ of holomorphic function at the origin of $\mathbb{C}^{n+1} \times \mathbb{C}^q$; set $x = (x^o, x') \in \mathbb{C} \times \mathbb{C}^n$; we suppose :

$$\varphi(0,x',0) = x^1 \quad ,$$

so that :

$$\varphi(0,0) = 0 \quad , \quad D_{x_1} \varphi(0,0) = 1 \neq 0 \quad ;$$

we suppose :

$$D_{\tau} \varphi(0,0) = 0 \quad .$$

We denote by

V : the germ of hypersurface : $\varphi(x, \tau) = 0$.

Definition

(x, τ) supports V if and only if :

$$\Delta_o(x, \tau) = \varphi(x, \tau) = 0 \quad .$$

1st conditions of Weierstrass

We suppose :

$$\forall j \ , \ \Delta_o(0, T_{1\ldots(j-1)(j+1)\ldots q} \cap \dot{F}_j) \not\equiv 0 \quad ,$$

$$\forall j,k,j<k \ \ \Delta_o(0, T^o_{1\ldots(j-1)(j+1)\ldots(k-1(k+1)\ldots q\, o} \cap \dot{F}_j \not\equiv 0 \quad .$$

Definition

$$(\Delta_1)_{\hat{j}} (x, F_{1\ldots(j-1)(j+1)\ldots q})$$
$$= \delta \Delta_o(x, F_{1\ldots(j-1)(j+1)\ldots q} \cap \hat{F}_j)$$

$$(\Delta_1)_{\hat{j}} (x, F_{1\ldots(j-1)(j+1)\ldots(k-1)(k+1)\ldots q\, o})$$
$$= \delta \Delta_o(x, F_{1\ldots(j-1)(j+1)\ldots(k-1)(k+1)\ldots q\, o} \cap \hat{F}_j) \ .$$

If the first expression is zero, we say that :

$(x, F_{1\ldots(j-1)(j+1)\ldots q})$ supports V

and in the same manner, if the second is zero, we say that :

$(x, F_{1\ldots(j-1)(j+1)\ldots(k-1)(k+1)\ldots q\, o})$

supports V .

The first germ is a germ at the point :

$$(O,T_{1\ldots(j-1)(j+1)\ldots q}) \in ¢^{n+1} \times \Gamma^{1} ;$$

the second germ is a germ at the point :

$$(O,T_{1\ldots(j-1)(j+1)\ldots(k-1)(k+1)\ldots q\ o}) \in ¢^{n+1} \times \Gamma^{1} ;$$

the neighbourhoods where they are defined are called allowed, as well as the x and lines belonging to these neighbourhoods; the points of V in the Weierstrass neighbourhood corresponding to an allowed line are said allowed. These notations are not ambiguous as long as we indicate the variables in Δ_1 .

We obtain also by induction conditions of Weierstrass that we suppose satisfied and we define the corresponding supports.

Weierstrass conditions and general supports

For every sequence : $1 \leq \ell_1 \ldots < \ell_k \leq q$, we suppose :

$$(\Delta_{k-1})_{\hat{\ell}_2\ldots\hat{\ell}_k}$$

$$(O,T_{1\ldots(\ell_1-1)(\ell_1+1)\ldots(\ell_2-1)(\ell_2+1)\ldots(\ell_k-1)(\ell_k+1)\ldots q} \cap \dot{F}_{\ell_1}) \neq O ;$$

we define :

$$(\Delta_k)_{\hat{\ell}_1\ldots\hat{\ell}_k}(x,F_{1\ldots(\ell_1-1)(\ell_1+1)\ldots(\ell_k-1)(\ell_k+1)\ldots q})$$

$$=\delta(\Delta_{k-1})_{\hat{\ell}_2\ldots\hat{\ell}_k}(x,F_{1\ldots(\ell_1-1)(\ell_1+1)\ldots(\ell_k-1)(\ell_k+1)\ldots q} \cap \hat{F}_{\ell_1}) ;$$

if this expression is zero, (x, the corresponding k-plane) supports V; a neighbourhood where Δ_k is defined is called allowed as well as the x and k-planes of this neighbourhood.

We remark that we obtain at the end :

$$(\Delta_k)_{\hat{1}\ldots\hat{q}}(x) \ .$$

In the same way, for every sequence : $1 \leq \ell_1 < \ldots < \ell_{k+1} \leq q$, we suppose :

$$(\Delta_{k-1})_{\hat{\ell}_2\ldots\hat{\ell}_k}(0,\ T^{o}_{1\ldots(\ell_1-1)(\ell_1+1)\ldots(\ell_{k+1}-1)(\ell_{k+1}+1)\ldots qo}) \neq 0 \ ;$$

we define :

$$(\Delta_k)_{\hat{\ell}_1\ldots\hat{\ell}_k}(x, F_{1\ldots(\ell_1-1)(\ell_1+1)\ldots(\ell_{k+1}-1)(\ell_{k+1}+1)\ldots qo})$$

$$=\delta(\Delta_{k-1})_{\hat{\ell}_2\ldots\hat{\ell}_k}(x, F_{1\ldots(\ell_1-1)(\ell_1+1)\ldots(\ell_{k+1}-1)(\ell_{k+1}+1)\ldots qo} \cap \hat{F}_{\ell_1}) \ .$$

If this expression is zero (x, the corresponding k-plane) supports V; the neighbourhood where Δ_k is defined is called allowed, as well as the x and k-planes of this neighbourhood.

2.5 We denote T_o the hyperplane of $\mathbb{C}^q$:

$$\sum_{k=1}^{k=q} \tau_k - x^o = 0 \quad ,$$

and set, if $1 \leq i_1 < \ldots < i_j \leq q$:

$$T_{i_1 \ldots i_j o} = T_{i_1 \ldots i_j} \cap T_o \quad ;$$

at the end

$$F_{[j]} = F_{1 \ldots j} \quad , \text{ where } \quad F_k : \tau_k + \sum_{\ell > k} a_k^\ell \tau_\ell + \sigma_k = 0 \quad .$$

We define the simplex $S_{q-j}(x, F_{[j]})$, $0 \leq j \leq q$, with dimension $q-j$, if it is not degenerate, by its $q-j+1$ summits :

$$A_j = F_{[j]} \cap T_{j+1 \ldots q}$$

$$A_j^\ell = F_{[j]} \cap T_{j+1 \ldots (\ell-1)(\ell+1) \ldots q\, o} \quad , \; j < \ell \leq q \quad .$$

An edge of dimension k of S_{q-j} is an affine k-plane defined by $k+1$ summits.

If F_j takes the value T_j , A_j and A_j^ℓ become A_{j-1} and A_{j-1}^ℓ . If F_j goes through A_{j-1} , $\sigma_j = - \theta_j$, where θ_j is the jth coordinate of A_{j-1} :

$$\theta_j = x^o + \sigma_1 + \ldots + \sigma_{j-1} + h_j(a_k , \sigma_k), \; 1 \leq k \leq j-1 \quad ,$$

where h_j is a holomorphic function of (a, σ), zero for $(a) = 0$. The simplex S_q depends only on $T_1 , \ldots, T_q$ and on $T_o(x^o)$.

<u>Definition</u>

$\Delta_{S_{q-j}}(x, F_{[j]})$ is the germ at the point $(0, T_{[j]})$ obtained by taking the product of germs of supports of distinct edges of S_{q-j} . $\Delta_{S_q}(x)$ is a germ at the origin of $\mathbb{C}^{n+1}$.

3 SUPPORTING AND TANGENTS PLANES

3.1 For $x \in \Omega_1$, V_x is the analytic hypersurface of $\tau \in \Omega_2$ such that :

$$\varphi(x, \tau) = 0 .$$

In the 3.1, we suppose that $V_{\underline{x}}$ has a quadratic point $\underline{\tau}$, the tangent hyperplane of which is $\underline{F}$ and that $\underline{F} \cap \underline{G}$ is a tangent q-2 - plane passing through $\underline{\tau}$ with a quadratic contact (contact of 2nd order).

We suppose also that $\underline{F}$ has the equation :

$$\tau_1 + < \underline{a}, \tau' > + \underline{\sigma} = 0 \quad , \quad \tau' = (\tau_2 , \dots , \tau_q)$$

and that $\underline{G}$ has the equation :

$$\tau_2 + < \underline{b}, \tau'' > + \underline{s} = 0 \quad , \quad \tau'' = (\tau_3 , \dots , \tau_q) \quad .$$

It results from the implicit functions theorem, that, for (x, τ) near $(\underline{x}, \underline{\tau})$, V has equation :

$$\tau_1 + \psi(x, \tau') = 0 \quad ,$$

and $\underline{F}$ has coordinates :

$$\underline{a} = D_{\tau'} \psi(\underline{x}, \underline{\tau}') \quad , \quad \underline{\sigma} = \psi(\underline{x}, \underline{\tau}') - < \underline{a}', \underline{\tau}' > \quad .$$

By continuity, for x near $\underline{x}$, V_x has a component V'_x which is not developable (so to say having a quadratic point) ; if $\tau \in V'_x$ is quadratic, with the tangent plane $F : \tau_1 + < a, \tau' > + \sigma = 0$, it follows from the local inversion theorem, that we have an holomorphic bijection :

$$(x, \tau') \longmapsto (x,a) \quad ;$$

(since $\mathrm{Hess}_{\tau'} \psi(\underline{x}, \underline{\tau}') \neq 0$), with inverse $(x, T(x,a))$ and that V'_x has locally a polar (dual) hypersurface $\check{V}'_x$, with equation :

$$P(x, a, \sigma) \equiv \sigma + S(x,a) = 0 \quad ,$$

where :

$$S(x,a) = - \psi[x, T(x,a)] + < a, T(x,a) > \quad ;$$

We have :

$$P(\underline{x}, \underline{a}, \underline{\sigma}) = 0 \quad .$$

$F \cap G$ is a hyperplane of F near $\underline{F} \cap \underline{G}$

$$G : \tau_2 + < b, \tau'' > + s = 0 \quad .$$

We set :

$$\xi = (a, \sigma), \quad \eta = (1, b, s) \quad .$$

Given F near $\underline{F}$, we want to find the condition on G near $\underline{G}$, so that $G \cap F$ is tangent to V_x.

A simple geometric consideration shows that this condition is the following.

(x, ξ, η) is such that there exists $\rho \in \mathbb{C}$ satisfying

$$(3.1) \qquad \begin{cases} P(x, \xi + \rho\eta) = 0 \\ \dfrac{d}{d\rho} P(x, \xi + \rho\eta) = 0 \ ; \end{cases}$$

these are the equations of apparent contour of $\check{V}'_x$ seen from ξ.

Now P satisfies

$$P(\underline{x}, \underline{\xi}) = 0$$

$$\frac{d}{d\rho} P(\underline{x}, \underline{\xi} + \rho\, \underline{\eta})\ (0) = 0$$

$$\frac{d^2}{d\rho^2} P(\underline{x}, \underline{\xi} + \rho\, \underline{\eta})\ (0) \neq 0 \qquad ;$$

the second equation says that $\underline{F} \cap \underline{G}$ contains $\underline{\tau}$ and the last equation says, that the contact of $\underline{F} \cap \underline{G}$ with $V_{\underline{x}}$ is quadratic. Then we can apply the preparation theorem and obtain (with an invertible i) :

$$P(x, \xi + \rho\eta) = i\,[\pi_o(x, \xi, \eta) + \rho\,\pi_1(x, \xi, \eta) + \rho^2] \qquad ;$$

we can replace (3.1) by :

$$(3.2) \qquad \pi_1^2\ (x, \xi, \eta) - 4\,\pi_o\ (x, \xi, \eta) = 0 \ ;$$

if we set :

$$Q(x,\xi,\eta) = i^2(x,\xi,\eta,0)\,\pi_1(x,\xi,\eta) \quad ,$$

(3.2) can be written :

$$Q^2(x,\xi,\eta) - 4\,i^3(x,\xi,\eta,0)\,P(x,\xi) = 0 \quad ; \tag{3.3}$$

we use again preparation theorem :

$$Q(x,\xi,(1,b,s)) = j\,[Q_o(x,\xi,b) + s - \underline{s}\,] \ ,$$

with : $Q_o(\underline{x},\underline{\xi},\underline{b}) = 0$;

we replace (3.3) by :

$$[Q_o(x,\xi,b) + s - \underline{s}]^2 - i'(x,\xi,\eta)\,P(x,\xi) = 0 \quad , \tag{3.4}$$

i' invertible.

We denote $\gamma P(x, F\cap G)$ the first member of (3.4); then :

$$\gamma P(x, F\cap G) = 0 \quad ,$$

means that $F\cap G$ is tangent to V_x .

From this definition we deduce the following lemma :

$$P(x, F) = 0 \ \text{implies}\ \delta(\gamma P)\,(x, F\cap \hat{G}) = 0 \quad ,$$

where $\delta(\gamma P)$ is here a germ at $(\underline{x}, \underline{F})$.

3.2 Then we deduce the following propositions :

$1\leqslant k\leqslant q-1$. If the k-plane $F_{1\ldots(\ell_1-1)(\ell_1+1)\ldots(\ell_k-1)(\ell_k+1)\ldots q}$ is allowed and tangent to a component V'_x of V_x which is not developable for allowed x , at an allowed point τ, then it supports V :

$$(\Delta_k)_{\hat{\ell}_1\ldots\hat{\ell}_k}(x, F_{1\ldots(\ell_1-1)(\ell_1+1)\ldots(\ell_k-1)(\ell_k+1)\ldots q}) = 0 \ .$$

We use an induction that we shall summarize. At the beginning we obtain the result for $k = 1$. We suppose it true for $k-1$. Assume then that the allowed k-plan :

$$\underline{F}_{1\ldots(\ell_1-1)(\ell_1+1)\ldots(\ell_k-1)(\ell_k+1)\ldots q}$$

is tangent to $V'_{\underline{x}}$ at an allowed point with a contact that we first suppose to be quadratic ; the allowed k+1 plan $\underline{F}_{1\ldots(\ell_1-2)(\ell_1+1)\ldots(\ell_k-1)(\ell_k+1)\ldots q}$ transverse to $V'_{\underline{x}}$ has an intersection of dimension k with $V'_{\underline{x}}$; τ is a quadratic point of which, with for tangent plane, the preceding k-plane ; it satisfies then the local equation of the polar manifold of

$$\underline{F}_{1\ldots(\ell_1-2)(\ell_1+1)\ldots(\ell_k-1)(\ell_k+1)\ldots q} \cap V'_{\underline{x}} \quad ,$$

that is to say :

$$P_k(\underline{x}, \underline{F}_{12\ldots(\ell_1-1)(\ell_1+1)\ldots(\ell_k-1)(\ell_k+1)\ldots q}) = 0 \quad .$$

Now if $\underline{F}_{1\ldots(\ell_2-1)(\ell_2+1)\ldots(\ell_k-1)(\ell_k+1)\ldots q}$ is an allowed k-1 plane in the preceding k-plane with a quadratic contact, the neighbouring k-1-planes $F_{1\ldots(\ell_2-1)(\ell_2+1)\ldots(\ell_k-1)(\ell_k+1)\ldots q}$, tangent to neighbouring V'_x, are, from 3.1, the "points" of the germ of the irreducible hypersurface at the point :

$$(\underline{x}, \underline{F}_{1\ldots(\ell_2-1)(\ell_2+1)\ldots(\ell_k-1)(\ell_k+1)\ldots q}) \in \mathbb{C}^{n+1} \times \Gamma^{k-1} \quad ,$$

defined by :

$$\gamma P_k(x, F_{1\ldots(\ell_2-1)(\ell_2+1)\ldots(\ell_k-1)(\ell_k+1)\ldots q}) = 0$$

such that :

$$\delta(\gamma P_k)(\underline{x}, \underline{F}_{1\ldots(\ell_1-1)(\ell_1+1)\ldots(\ell_k-1)(\ell_k+1)\ldots q} \cap \hat{F}_{\ell_1}) = 0 \quad .$$

but, from the hypothesises of induction, it results that the points of this germ :

$$\gamma P_k = 0 \quad ,$$

necessarily belong to the germ defined at the same point by :

$$(\Delta_{k-1})_{\hat{\ell}_2\ldots\hat{\ell}_k}(x, F_{1\ldots(\ell_2-1)(\ell_2+1)\ldots(\ell_k-1)(\ell_k+1)\ldots q}) = 0 \quad .$$

we deduce, (cf. 2.4), that :

$$(\Delta_k)_{\hat{\ell}_1 \ldots \hat{\ell}_k}(x, \underline{F}_{1 \ldots (\ell_1 - 1)(\ell_1 + 1) \ldots (\ell_k - 1)(\ell_k + 1) \ldots q}) = 0$$

If the contact of the k-plane is not quadratic, then by density and continuity, $(x, F_{1 \ldots (\ell_1 - 1)(\ell_1 + 1) \ldots (\ell_k - 1)(\ell_k + 1) \ldots q})$ satisfies the preceding equation.

In the same way, we obtain :

$1 \leq k \leq q-1$. If the k-plane $F_{1 \ldots (\ell_1 - 1)(\ell_1 + 1) \ldots (\ell_{k+1} - 1)(\ell_{k+1} + 1) \ldots q\, o}$ is allowed and tangent to a component V'_x of V_x which is not developable at an allowed point, it supports V, which means

$$(\Delta_k)_{\hat{\ell}_1 \ldots \hat{\ell}_k}(x, F_{1 \ldots (\ell_1 - 1)(\ell_1 + 1) \ldots (\ell_{k+1} - 1)(\ell_{k+1} + 1) \ldots q\, o}) = 0 .$$

Last, we have the following proposition.

We assume that, for every x, V_x has no developable component. Let x an allowed point and $\underline{F}_{1 \ldots (\ell_1 - 1)(\ell_1 + 1) \ldots (\ell_k - 1)(\ell_k + 1) \ldots q}$ an allowed k-plane ; we assume that V_x has a singular allowed point through which, $\underline{F}_{1 \ldots (\ell_1 - 1)(\ell_1 + 1) \ldots (\ell_k - 1)(\ell_k + 1) \ldots q}$ passes, then :

$$(\Delta_k)_{\hat{\ell}_1 \ldots \hat{\ell}_k}(\underline{x}, \underline{F}_{1 \ldots (\ell_1 - 1)(\ell_1 + 1) \ldots (\ell_k - 1)(\ell_k + 1) \ldots q}) = 0 .$$

From Sard theorem, it results the density of x such that V_x is smooth and the preceding k-plane can be approached by a k-plane tangent to V'_x smooth and not developable ; by continuity, we obtain the result.

We have the analogous proposition for :

$$F_{1 \ldots (\ell_1 - 1)(\ell_1 + 1) \ldots (\ell_{k+1} - 1)(\ell_{k+1} + 1) \ldots q\, o})$$

3.3 First we give a proposition, which, in a sense, is a reciprocal of the proposition of 3.1.

$\underline{x} \in \Omega_1$, Ω_1' is an open neighbourhood of $\underline{x}$, such that, for every $x \in \Omega_1'$, every V_x is a smooth hypersurface ; $P(x, a, \sigma)$ is a Weierstrass polynomial in σ , reduced, defined in a Weierstrass neighbourhood of $(\underline{x}, \underline{a}, \underline{\sigma}) \in \mathbb{C}^{n+1} \times \mathbb{C}^q$ (more precisely the neighbourhood $\Omega_1' \times \mathcal{W}$). The part $V_x \cap \Omega_2'$ of V_x formed by the points of V_x , $x \in \Omega_1'$, such that the tangent plane F belongs to $\mathcal{W}$ is open.

We denote by P_x , the set of $F \in \mathcal{W}$ such that $P(x,F) = 0$ and we assume that P_x is the polar set of $V_x \cap \Omega_2'$; then we have :

$$\delta P(\underline{x}, \hat{F}) \neq 0 \quad .$$

We prove this proposition, with the help of the following slight adaptation of lemma [1] , [2] .

Lemma Let $\underline{\sigma}$ a zero of $P(\underline{x}, \underline{a}, \sigma) = 0$; the hyperplane : $\underline{F} \sim (\underline{a}, \underline{\sigma})$ is tangent to $V_{\underline{x}}$ and its direction is defined by $\underline{a}$. For every small enough neighbourhood of $\underline{a}$, there exists a direction a belonging to this neighbourhood, such that every hyperplane tangent with direction a is tangent to $V_{\underline{x}}$ at only one point (that is to say, the corresponding σ are distincts), with a quadratic contact.

3.4 We obtain then the following propositions/

We assume that, for every x allowed, V_x has no developable component.

$$1 \leq k \leq q-1$$

If $(\Delta_k)_{\hat{\ell}_1 \ldots \hat{\ell}_k}(x, F_{1\ldots(\ell_1-1)(\ell_1+1)\ldots(\ell_k-1)(\ell_k+1)\ldots q}) = 0$,

then, either the allowed k-plane $F_{1\ldots(\ell_1-1)(\ell_1+1)\ldots(\ell_k-1)(\ell_k+1)\ldots q}$ is tangent to V_x at an allowed point or it contains a singular allowed point of V_x .

k=q . If : $(\Delta)_{\hat{1}\ldots\hat{q}}(x) = 0$, V_x has a singular allowed point.

We prove it by induction. The result is true for $k=1$. Suppose it is true for $k-1$; we must prove that, if the allowed $F_{1\ldots(\ell_1-1)(\ell_1+1)\ldots(\ell_k-1)(\ell_k+1)\ldots q}$ is not tangent to V_x in an allowed point and contains no singular allowed point of V_x , then :

3.4 (1) $$(\Delta_k)_{\hat{\ell}_1\ldots\hat{\ell}_k}(x, F_{1\ldots(\ell_1-1)(\ell_1+1)\ldots(\ell_k-1)(\ell_k+1)\ldots q})$$
$$= \delta(\Delta_{k-1})_{\hat{\ell}_2\ldots\hat{\ell}_k}(x, F_{1\ldots(\ell_1-1)(\ell_1+1)\ldots(\ell_k-1)(\ell_k+1)\ldots q} \cap \hat{F}_{\ell_1}) \neq 0 \quad .$$

We remark first that :

$$V_x \cap F_{1\ldots(\ell_1-1)(\ell_1+1)\ldots(\ell_k-1)(\ell_k+1)\ldots q}$$

is smooth in the allowed open set and has dimension $q-1$; this set contains an open part , the tangent $k-1$ planes to this part are allowed and satisfy, according to the 3.2 of which the hypotheses are realized, the following equation :

3.4 (2) $$(\Delta_{k-1})_{\hat{\ell}_2\ldots\hat{\ell}_k}(x, F_{1\ldots(\ell_2-1)(\ell_2+1)\ldots(\ell_k-1)(\ell_k+1)\ldots q} = 0 ,$$

reciprocally, if such a $k-1$ plane satisfies this equation, it results from the induction hypothesis that it is tangent to V_x , and to the preceding intersection.

3.4 (2) is then the polar manifold of the allowed part of the intersection. The hypothesises of 3.3 are realized, replacing x by $(x, F_{1\ldots(\ell_1-1)(\ell_1+1)\ldots(\ell_k-1)(\ell_k+1)\ldots q})$ and F by F_{ℓ_1} and we have the inequality 3.4 (1).

We obtain in the same way.

<u>Assume that</u>, <u>for every allowed</u> x , V_x <u>has no developable component</u> $1 \leq k \leq q-1$. <u>If</u>

$$(\Delta_k)_{\hat{\ell}_1\ldots\hat{\ell}_k}(x, F_{1\ldots(\ell_1-1)(\ell_1+1)\ldots(\ell_{k+1}-1)(\ell_{k+1}+1)\ldots q_0}) = 0 ,$$

then, either the above allowed k-plane is tangent to V_x at an allowed point, or it contains a singular allowed point of V_x .

We can deduce the following consequences :

i) $0 \leq j \leq q-1$, $1 \leq k \leq q-j$, $\ell_1 > j+1$. If

$$\delta(\Delta_{k-1})_{\hat{\ell}_1 \ldots \hat{\ell}_{k-1}} (x, F_{[j]} \cap T_{(j+2) \ldots (\ell_1-1)(\ell_1+1) \ldots (\ell_{k-1}-1)(\ell_{k-1}+1) \ldots q} \cap \hat{F}_{j+1}) = 0 \quad ,$$

then :

$$(\Delta_k)_{\widehat{j+1}\ \hat{\ell}_1 \ldots \hat{\ell}_{k-1}} (x, F_{[j]} \cap T_{(j+2) \ldots (\ell_1-1)(\ell_1+1) \ldots (\ell_{k-1}-1)(\ell_{k-1}+1) \ldots q})$$

ii) $0 \leq j \leq q-2$, $1 \leq k \leq q-j-1$, $\ell_1 > j+1$. If :

$$\delta(\Delta_{k-1})_{\hat{\ell}_1 \ldots \hat{\ell}_{k-1}} (x, F_{[j]} \cap T_{(j+2) \ldots (\ell_1-1)(\ell_1+1) \ldots (\ell_k-1)(\ell_k+1) \ldots q\,o} \cap \hat{F}_{j+1}) = 0 \quad ,$$

then :

$$(\Delta_k)_{\widehat{j+1}\ \hat{\ell}_1 \ldots \hat{\ell}_{k-1}} (x, F_{[j]} \cap T_{(j+2) \ldots (\ell_1-1)(\ell_1+1) \ldots (\ell_k-1)(\ell_k+1) \ldots q\, o})$$

$$\equiv (\Delta_k)_{\widehat{j+1}\ \hat{\ell}_1 \ldots \hat{\ell}_{k-1}} (x, A_j^{j+1} A_j^{\ell_1} \ldots A_j^{\ell_k}) = 0 \quad .$$

To obtain i) we use again § 3.3 and remark that, if

$A_{\dot{j}}\, A_{\dot{j}}^{j+1}\, A_{\dot{j}}^{1} \ldots A_{\dot{j}}^{k-1}$ is not tangent to V_x and contains no singular point, then the polar variety of the allowed part of its intersection with V_x is :

$$(\Delta_{k-1})_{\hat{\ell}_1 \ldots \hat{\ell}_{k-1}} (x, A_{j+1} A_{j+1}^{\ell_1} \ldots A_{j+1}^{\ell_{k+1}}) = 0 \quad ,$$

according to a previous argument.

We obtain ii) in an analogous way

iii) <u>Assume that</u>, <u>for every allowed</u> x , V_x <u>has no developable component</u>. <u>If</u> : $\Delta_{S_q}(x) = 0$, <u>then either one of the edges of</u> $S_q(x)$ <u>is tangent to</u> V_x <u>at an allowed point</u>, <u>or</u> V_x <u>has an allowed singular point</u>.

3.5 We intend first to study the discriminant in F_{j+1} of $\Delta_{S_{q-j-1}}(x, F_{[j+1]})$.

<u>We assume</u>, <u>that for every allowed</u> x , V_x <u>has no developable component</u>

<u>Definition</u> – $\check{\Delta}_{S_{q-j}}(x, F_{[j]})$ is the germ at the point $(0, T_{[j]})$ obtained by taking the product of germs of supports of distinct edges of S_{q-j} issued from $A_{\dot{j}}^{j+1}$.

We get then :

<u>Lemma</u> – $0 \leq j \leq q-1$. The Condition :

$$\delta \Delta_{S_{q-j-1}}(x, F_{[j]} \cap \hat{F}_{j+1}) = 0$$

implies :

$$\check{\Delta}_{S_{q-j}}(x, F_{[j]}) = 0 \quad .$$

The end of § 2.4 and the definitions imply that under our hypothesis either

$$\overset{\vee}{\Delta}_{S_{q-j}}(x, F_{[j]}) = 0 \quad ,$$

or $(x, F_{[j]})$ is such that there exists a pair of conditions of supports of two distinct edges of S_{q-j-1} such that the two Weierstrass polynomials in σ_{j+1}, that they define have a common factor, with degree ≥ 1 in σ_{j+1}.

It is sufficient to prove that, if :

$$\overset{\vee}{\Delta}_{S_{q-j}}(x, F_{[j]}) \neq 0 \quad ,$$

then, for every pair of conditions of supports of two distinct edges of S_{q-j-1}, the two Weierstrass polynomials in σ_{j+1} that they define have no common factor with degree ≥ 1 in σ_{j+1}.

The previous inequality implies that the edges of dimension ≥ 1 of q-j hedron with summit A_j^{j+1} of the simplex S_{q-j} are not tangent to V_x at an allowed point and do not pass through an allowed singular point of V_x.

The two conditions of support of edges are of the form : (notation of 2.3)

$$(\Delta'_{k-1})_{\hat{\ell}_1 \ldots \hat{\ell}_{k-1}}$$

$$(x, F_{[j]} \cap T_{(j+2)\ldots(\ell_1-1)(\ell_1+1)\ldots(\ell_{k-1}-1)(\ell_{k-1}+1)\ldots q} \, ,$$

$$a_{j+1}^{j+2}, \ldots, a_{j+1}^{q}, \sigma_{j+1}) = 0$$

or the analogous condition containing F_o.

It results from the proofs of 3.4 that they are polar varieties of smooth intersections of V_x with edges issued from A_j^{j+1} of the simplex S_{q-j}. We deduce, with the help of lemma 3.3 that their respective hessians are identically zero on no components of them.

The consideration of the difference of their dimensions implies that the Weierstrass polynomials in σ_{j+1} that they define, have no common factor with degree ≥ 1 .

We obtain again by consideration of dimension of the polar manifold the second <u>lemma</u>.

If $\Delta_{S_{q-j-1}} (x, F_{[j]} \cap F_{j+1}) = 0$, for every hyperplane allowed F_{j+1} passing through A_j^{j+1} , then :

$$\overset{\vee}{\Delta}_{S_{q-j}} (x, F_{[j]}) = 0 \qquad .$$

4. ANALYTIC CONTINUATIONS

4.1 We shall use theorem of continuation of Hartogs, proposition 5.5 of [8] (p. 439) and the theorem on ramification in dimension 1 [6] , [8] that we shall recall

<u>Theorem</u> - $f(u,\sigma)$ is a germ of holomorphic function at the origin of $\mathbb{C}^{\ell} \times \mathbb{C}$ such that : $f(0,\sigma) \neq 0$; Ω_1 is a polydisk of $\mathbb{C}^{\ell}$ with center 0 , with "radius" R_1 and $\Omega_2 = \{\sigma \in \mathbb{C} \; ; \; |\sigma| < R_2\}$; $\Omega_1 \times \Omega_2$ is a Weierstrass neighbourhood of f and δf is holomorphic in Ω_1 ; then (cf. 2.2) :

$$\delta f(u, \hat{\sigma}) = (\operatorname*{discr}_{\sigma} \pi_r(u, \hat{\sigma}))_r \qquad .$$

$\Theta : \Omega_1 \mapsto \Omega_2$ is a holomorphic function such that : $\Theta(u_o) = 0$, $u_o \in \Omega_1$. $\mathcal{U}$ is a germ of holomorphic function at the point $(u_o, 0) \in \Omega_1 \times \Omega_2$ and has an analytic continuation to the simply connected covering of

$$\{(u, \sigma) \in \Omega_1 \times \Omega_2 \; ; \; f(u,\sigma) \neq 0\} \qquad .$$

Then, the integral (integral in the complex plane $\mathbb{C}_{\sigma}$ along the segment joining 0 and $\Theta(u)$, for u in a sufficiently small neighbourhood of u_o) :

$$I(u) = \int_o^{\theta(u)} \mathcal{U}(u, \sigma)\, d\sigma$$

defines a germ of holomorphic function at the point u_o , which has an analytic continuation to the simply connected covering of :

$$\{u \in \Omega_1 \ ; \ f(u,0).f(u, \theta(u)).\,\delta f(u, \hat{\sigma}) \neq 0\} \quad .$$

In our context, u_o can be chosen as near 0 as we want and $\theta(o) = 0$.

Next we shall obtain :

<u>Proposition</u> : $f(x, a, \sigma)$ is now a germ of holomorphic function at the origin of $\mathbb{C}^{n+1} \times \mathbb{C}^{q-1} \times \mathbb{C}$ such that :

$$f(0, 0, \sigma) \not\equiv 0 \quad , \quad f(x, 0, 0) \not\equiv 0 \quad ;$$

$\tilde{f}(x, \alpha, \sigma) = f(x, \alpha\sigma, \sigma)$ is holomorphic in the Weierstrass neighbourhood : $\Omega_1 \times \tilde{\mathcal{A}} \times \mathcal{S}$ and $\delta f(x, \hat{F})$ is holomorphic in Ω_1 (2) .

$\theta : \Omega_1 \longmapsto \mathcal{S}$ is a holomorphic function such that $\theta(y) = \theta(0) = 0$, $y \neq 0$, $y \in \Omega_1$; we assume there exists a holomorphic function $h(x)$ not identically zero such that :

$$\{x \in \Omega_1 \ , \ f(x, a, \theta(x)) = 0 \ , \ \forall a \text{ small}\} \subset \{x \in \Omega_1 \ ; \ h(x) = 0\} \quad .$$

$\mathcal{U}(x, \alpha, \sigma)$ is a germ of holomorphic function at the point :

$$(y, 0, 0) \in \Omega_1 \times \tilde{\mathcal{A}} \times \mathcal{S} \quad ,$$

and has an analytic continuation to the simply connected covering of :

$$\{(x, \alpha, \sigma) \in \Omega_1 \times \tilde{\mathcal{A}} \times \mathcal{S} \ ; \ f(x, \alpha\sigma, \sigma) \neq 0\} \quad .$$

Then the integral :

$$I(x, \alpha) = \int_o^{\theta(x)} \mathcal{U}(x, \alpha, \sigma)\, d\sigma$$

defines a germ of holomorphic function at the point $(y,0)$.

We assume now it is independent of α :

$$I(x,\alpha) \equiv I(x) \quad .$$

Then $I(x)$ has an analytic continuation to the simply connected covering of :

$$\{(x,\alpha)\in \Omega_1 \times \tilde{\mathcal{A}} \; ; \; f(x,0,0).f(x,\alpha\,\theta(x), \theta(x)).\operatorname{discr}_{\sigma} Q_r(x,\alpha,\hat{\sigma}) \neq 0\},$$

where Q_r denotes the Weierstrass polynomial of the reduced germ of $\tilde{f}(x,\alpha,\sigma)$.

The previously recalled proposition [8] (p. 439), the theorem of Hartogs and the last lemma of 2.2 imply that $I(x)$ has an analytic continuation to the simply connected covering of :

$$\{x\in \Omega_1 \; ; \; f(x,0,0).h(x).\,\delta f(x,\hat{F}) \neq 0\} \quad .$$

4.2 φ has been defined in § 1.4. $\mathcal{U}$ is a germ of holomorphic function at point $(y,0)\in \Omega_1 \times \Omega_2$, $(y^0 = 0 \;,\; y^1 \neq 0)$, having an analytic continuation to the simply connected covering of :

$$\{(x,\tau)\in \Omega_1 \times \Omega_2 \; ; \; \varphi(x,\tau) \neq 0\} \quad .$$

We consider the integral :

$$I(x) = \int_0^{x^0} d\tau_1 \ldots \int_0^{x^0 - \sum_{1\leq k\leq j-1} \tau_k} d\tau_k \ldots \int_0^{x^0 - \sum_{1\leq k\leq q-1} \tau_k} \mathcal{U}(x,\tau)\, d\tau_q \; ;$$

it defines a germ of holomorphic function, at the point y ; we want to study its ramification (that is to say the analytic continuation).

Note that this integral can be considered as an integral on a singular cycle relative to the union of faces of simplex $S_q(x)$, so that we shall denote also

$$I(x) = \int_{S_q(x)} \mathcal{U}(u, \tau)\, d\tau \quad .$$

The equations :

$$\tau_j + \sum_{j+1 \leq k \leq q} a_j^k \tau_k + \sigma_j = 0 \quad , \qquad 1 \leq j \leq q \quad ,$$

define, for small enough a , a holomorphic application s_q :

$$(x, a_1, \ldots, a_{q-1}, \sigma_1, \ldots, \sigma_q) \longmapsto (x, \tau_1, \ldots, \tau_q) \quad .$$

We denote :

$$\mathcal{U}_q = \mathcal{U} \circ s_q \qquad ;$$

it is a germ of holomorphic function at the point : $(x=y,\ a = \sigma = 0)$; we shall also consider

$$\mathcal{U}_q(x, \alpha\sigma, \sigma)$$

obtained by replacing each a_j^k by $\alpha_j^k \sigma_j$.

We obtain a formula of "change of variables" :

$$I(x) = \int_0^{-x^o} d\sigma_1 \ldots \int_0^{-\theta_j} d\sigma_j$$

$$\ldots \int_0^{-\theta_q} \mathcal{U}_q(x, \alpha\sigma, \sigma)\, i_q(\alpha, \sigma)\, d\sigma_q \qquad ,$$

for small enough α , where i_q is holomorphic and near $(-1)^q$ and where $\theta_j\ (x^o, a_1, \ldots, a_{j-1}, \sigma_1, \ldots, \sigma_{j-1})$ has been defined in 2.5.

Moreover

$$\int_0^{-\theta_j} d\sigma_j \ldots \int_0^{-\theta_q} \mathcal{U}_q\, i_q\, d\sigma_q$$

does not depend on α_k , $k \geqslant j$ and defines a holomorphic germ at the point $(x=y\ , \alpha = \sigma = 0)$.

We obtain this formula, either by using a straight forward formula of derivation of the integral of Leray [6] , or by proving it as in [8] .

It allows us to obtain :

4.3 Theorem - Assume :

H_1 $\forall x$, V_x has no developable component

H_2 Weierstrass conditions allowing to define supports are satisfied (cf. § 1.4)

H_3 $\Delta_{S_q}(x) \not\equiv 0$, (other wise the result is empty, note also that $\Delta_q(x) \not\equiv 0$).

We then obtain :

i) There exists a neighourhood $\mathcal{V}_1$ of 0 in $\mathbb{C}^{n+1}$ such that $I(x)$ is ramified around the hypersurface of $\mathcal{V}_1$ defined by :

$$\Delta_{S_q}(x) = 0 \quad ,$$

that is to say, the set of x such that a pair (x, one edge of the simplex $S_q(x)$) supports $V : \varphi(x, \tau) = 0)$;

ii) moreover, there exists a neighbourhood $\mathcal{V}_2$ of 0 in $\mathbb{C}^q$ such that :

$$\{x \in \mathcal{V}_1 \ ; \Delta_{S_q}(x) = 0\} \subset \{x \in \mathcal{V}_1 \ ; \text{ either one of the edges}$$

of $S_q(x)$ is tangent to $\mathcal{V}_x$ in $\mathcal{V}_2$ or V_x has a singular point in $\mathcal{V}_2\}$.

We proceed by recurrence. First we obtain, with the help of theorem of 4.1, taking the following condition into account

$$\varphi(0, T_{[q-1]} \cap \dot{F}_q) \not\equiv 0 \quad ,$$

that the integral :

$$\int_0^{-\theta_q} \mathcal{U}_q \, i_q \, d\sigma_q$$

defines a holomorphic germ at $(y,0)$ which has an analytic continuation to the simply connected covering of the set

$$\{ x, \alpha_i, \sigma_i \ , \ 1 \leq i \leq q-1 \ ; \ \Delta_{S_1}(x, F_{[q-1]}) \not\equiv 0 \} \quad .$$

Then, we remark that :

$\Delta_{S_{q-j}}(x, F_{[j]})$ satisfies the Weierstrass condition :

$$\Delta_{S_{q-j}}(0, T_{[j-1]} \cap \dot{F}_j) \not\equiv 0 \quad ,$$

that : $\theta_j(x, \ldots, \alpha_\ell \sigma_\ell, \sigma_\ell, \ldots, \alpha_{j-1} \sigma_{j-1}, \sigma_{j-1})$ is such that :

$$\theta_j(y,0,0) = 0 \quad ,$$

and that, moreover, because of 3.5 :

$$\{x, \ldots, \alpha_\ell, \sigma_\ell, \ldots, \alpha_{j-1}, \sigma_{j-1} \quad ;$$

$$\Delta_{S_{q-j}}(x, F_{[j]}) = 0 \ , \ \forall F_j \text{ passing through } A^j_{j-1}\}$$

$$\subset \{x, \ldots, \alpha_\ell, \sigma_\ell, \ldots, \alpha_{j-1}, \sigma_{j-1} \quad ;$$

$$\check{\Delta}_{S_{q-j}}(x, F_{[j-1]}) = 0\} \quad .$$

Finally we note that the integral :

$$\int_0^{-\theta_{j+1}} d\sigma_{j+1} \cdots \int_0^{-\theta_q} \mathcal{U}_q(x, \ldots, \alpha_\ell \sigma_\ell, \sigma_\ell, \ldots, \sigma_q) \, i_q(\ ) \, d\sigma_q$$

does not depend on α_ℓ , $\ell \geq j+1$ and defines a germ of holomorphic function at the point : $(x=y\ ,\ldots,\alpha_\ell=\sigma_\ell=0\ ,\ \ell\leq j)$.

Suppose it has an analytic continuation to the simply connected covering of :

$$\{x\ ,\ldots,\alpha_\ell\ ,\sigma_\ell\ ,\ldots,\ \alpha_j\ ,\ \sigma_j\ ;\ \Delta_{S_{q-j}}(x,\ F_{[j]})\neq 0\}\ .$$

The following integral :

$$\int_o^{-\theta_j} d\,\sigma_j \int_o^{-\theta_{j+1}} d\,\sigma_{j+1} \cdots \int_o^{-\theta_q} \mathcal{U}_q\, i_q\, d\,\sigma_q$$

does not depend on α_ℓ, $\ell \geqslant j$ and defines a germ of holomorphic function at the point :

$$(x=y\ ,\ldots,\ \alpha_{j-1}=\ \sigma_{j-1}\ =0)\ .$$

We are in the conditions of the proposition of 4.1 and this integral has an analytic continuation to the simply connected covering of :

$$\{x\ ,\ldots,\alpha_{j-1},\ \sigma_{j-1}\ ;\ \Delta_{S_{q-j}}(x,\ F_{[j-1]}\cap T_j).\check{\Delta}_{S_{q-j+1}}(x,\ F_{[j-1]})\neq 0\}\ ,$$

by using 2.3 and taking the results of the end of 3.4 and 3.5 into account ; this also means that it has an analytic continuation, to the simply connected covering of :

$$\{x\ ,\ldots,\alpha_{j-1},\ \sigma_{j-1}\ ;\ \Delta_{S_{q-j+1}}(x,\ F_{[j-1]})\neq 0\}\ .$$

Then the induction is completed and $I(x)$ is ramified around :

$$\{x\in\mathcal{V}_1\ ;\ \Delta_{S_q}(x)=0\}\ ,$$

where $\mathcal{V}_1$ is the last allowed neighbourhood that we have constructed by our induction procedure.

ii) is immediate, by definition of 2.5 and results of 3.4.

Note. We can replace assumption of "having no developable components" by an assumption of "equidimension of polar manifolds" as in [6] .

References

(1) J. BRIANCON et J.P. SPEDER - *Thèse Nice* (1976).

(2) T. FUKUDA et T. KOBAYASHI - 'A local isotopy theorem' *Tokyo Journal of Math*, Vol 5, n° 1 (1982).

(3) Y. HAMADA et G. NAKAMURA - 'On the singularities of the solution of the Cauchy problem...' *Annali Scuola Norm. Sup. di Pisa*, 4 (1977), p. 725-755.

(4) J.P. HENRY, M. MERLE, C. SABBAH - 'Sur la condition de THOM stricte pour un morphisme analytique complexe'. *Public. Centre de Math. de l'Ecole Polytechnique*, Palaiseau (1982).

(5) T. KOBAYASHI - 'On the singularities of the solution to the Cauchy problem...' *Math Annalen* (à paraître).

(6) J. LERAY - 'Un complément au théorème de N. NILSSON sur les intégrales...' *Bulletin de la Société Mathématique de France*, 95, (1967), p. 313-374.

(7) D. SCHILTZ, J. VAILLANT et C. WAGSCHAL - 'Problème de Cauchy ramifié à caractéristiques multiples en involution' *C.R. Acad. Sc. Paris*, t. 291, (1980), p. 659-662.

(8) D. SCHILTZ, J. VAILLANT et C. WAGSCHAL - 'Problème de Cauchy ramifié : racine caractéristique double ou triple en involution'. *J. Math. Pures et Appliquées*, 4, (1982), p. 423-443.

(9) J. VAILLANT - 'Intégrales singulières holomorphes (*Journées E.D.P. Saint Jean de Monts* (1983).

GENERALIZED FLOWS AND THEIR APPLICATIONS

Seiichiro Wakabayashi
Institute of Mathematics
University of Tsukuba
Ibaraki 305, Japan

ABSTRACT. We define generalized flows in order to make local results global. We study the existence domain of solutions to the Cauchy problem in the complex domain and the wave front sets of solutions to the hyperbolic Cauchy problem, using generalized flows.

1. INTRODUCTION

For microhyperbolic operators with analytic coefficients Kashiwara and Kawai [12] proved a microlocal version of Holmgren's uniqueness theorem, and we gave a global outer estimate of the singular spectra (and a result on the propagation of micro-analyticities) of solutions to the hyperbolic Cauchy problem, using their results and generalized (Hamilton) flows (see [18]). This shows that generalized flows are useful to obtain a global result from a local one. Generally, by making a local result global, the result can be improved even in the local version.

In §2 we shall define generalized flows and study some properties of generalized flows. In §3 we shall consider the Cauchy problem in the complex domain and prove theorems on the existence domain of solutions, using generalized flows. In §4 we shall investigate the wave front sets of solutions to the hyperbolic Cauchy problem in Gevrey classes or the space of real analytic functions, using generalized (Hamilton) flows.

2. GENERALIZED FLOWS

Let X be a subdomain of $\mathbb{R}^N$, and let K: $X \ni w=(w_1,\cdots,w_N) \longmapsto K(w) \in \mathcal{C}(\mathbb{R}^N)$ be a mapping, where $\mathcal{C}(\mathbb{R}^N)$ denotes the collection of all non-void closed convex cones ($\subset \mathbb{R}^N$) included in $\{\delta w=(\delta w_1,\cdots,\delta w_N)\in\mathbb{R}^N;\ \delta w_1>0\} \cup \{0\}$, where "cones" mean cones with their vertecies at the origin. We assume that $w \longmapsto K(w)$ is outer semi-continuous, i.e., for any $w^0 \in X$ and any open cone $\mathcal{C}$ in $\mathbb{R}^N$ with $K(w^0) \subset \mathcal{C} \cup \{0\}$ there is a neighborhood U of w^0 in X such

H. G. Garnir (ed.), Advances in Microlocal Analysis, 363–384.

that $K(w)\subset \mathcal{C}\cup\{0\}$ for $w\varepsilon U$. For each $w\varepsilon X$ we define

$$K_w^{\pm} = \{w(t)\varepsilon X;\ \pm t\geqq 0, \text{ and } \{w(t)\} \text{ is a Lipschitz continuous curve in } X \text{ satisfying } (d/dt)w(t)\varepsilon K(w(t)) \text{ (a.e. } t) \text{ and } w(0)=w\}.$$

We call $K_w^{\pm}$ the generalized (half) flows for K (or $\{K(w)\}$), although "flows" usually mean mappings satisfying some properties. Then it is obvious that (i) $K_{w^0}^{\pm}\subset\{w\varepsilon X;\ \pm(w_1-w_1^0)\geqq 0\}$, (ii) $K_{w^0}^{\pm}\cap\{w\varepsilon X;\ w_1=w_1^0\}=\{w^0\}$, (iii) $w^1\varepsilon K_{w^2}^{\pm}$ if $w^2\varepsilon K_{w^1}^{\mp}$, and that (iv) $K_{w^0}^{\pm}\cap\{w\varepsilon X;\ \pm(w_1-t)\geqq 0\}= \bigcup_{w\varepsilon K_{w^0}^{\pm}\cap\{w_1=t\}} K_w^{\pm}$ if $\pm(t-w_1^0)\geqq 0$. Let Ω be a compact subset of X. Set

$$K_\Omega(w) = \begin{cases} K(w) & \text{if } w\varepsilon\Omega, \\ \{0\} & \text{otherwise.}\end{cases}$$

For each h>0 and $w\varepsilon\Omega$ we choose $K_\Omega(w,h)\varepsilon\, \mathcal{C}(\mathbf{R}^N)$ such that

$$K_\Omega(w) \subset\subset K_\Omega(w,h) \subset (K_\Omega(w))_h,$$

where $\mathcal{C}_h\equiv\{\delta w\varepsilon\mathbf{R}^N;\ \delta w=0 \text{ or } ||\delta w|^{-1}\delta w-|\delta w^1|^{-1}\delta w^1|<h \text{ for some } \delta w^1\varepsilon\,\mathcal{C}\setminus\{0\}\}$ for a cone $\mathcal{C}$ and $|\delta w|=(\sum_{j=1}^N |\delta w_j|^2)^{1/2}$ for $\delta w=(\delta w_1,\cdots,\delta w_N)\varepsilon\mathbf{R}^N$. Here, for cones K_1 and K_2, $K_1\subset\subset K_2$ means that $K_1^a\subset \operatorname{Int} K_2\cup\{0\}$, where A^a and Int A denote the closure of A and the interior of A, respectively. We note that $K_\Omega(w)=(K_\Omega(w))_h$ if $K_\Omega(w)=\{0\}$. By the outer semi-continuity of $\{K(w)\}$, for each h>0 and $w^0\varepsilon\Omega$ there is $r(w^0,h)>0$ such that $r(w^0,h)<h$ and

$$K_\Omega(w^1)\subset\subset K_\Omega(w^0,h) \quad \text{for } w^1\varepsilon U(w^0,h)\equiv\{w\varepsilon\mathbf{R}^N;\ |w-w^0|<r(w^0,h)\}.$$

Since Ω is compact, there are a finite number of $w^{h,j}\varepsilon\Omega$ ($1\leqq j\leqq N(h)$) such that

$$\Omega\subset\bigcup_{j=1}^{N(h)} U'(w^{h,j},h),$$

where $U'(w^0,h)=\{w\varepsilon\mathbf{R}^N;\ |w-w^0|<r(w^0,h)/2\}$. Let us define $K_{w^0,\Omega}^{\pm}(h)$ which approximate $K_{w^0}^{\pm}$. We say $w\varepsilon K_{w^0,\Omega}^{\pm}(h)$ if $w=w^0$ or there are $j_0,\cdots,j_\nu$ and $\tilde w^1,\cdots,\tilde w^{\nu-1}\varepsilon\Omega$ such that $\tilde w^\ell\varepsilon U'(w^{h,j_\ell},h)$ ($0\leqq\ell\leqq\nu$) and $\tilde w^{\ell+1}\varepsilon\tilde w^\ell\pm\{\delta w\varepsilon K_\Omega(w^{h,j_\ell},h);\ |\delta w|\leqq\rho(h)\}$ ($0\leqq\ell\leqq\nu-1$), where $\tilde w^0=w^0$, $\tilde w^\nu=w$ and $\rho(h)=\min_{1\leqq j\leqq N(h)} r(w^{h,j},h)/2$. Then we can prove the following theorem, applying the same argument as in the proof of the existence theorem for ordinary differential equations.

<u>Theorem 2.1 ([17]).</u> Let $w^0\varepsilon X$, and let $K_{w^0,\Omega}^{\pm}$ be the generalized flows for $\{K_\Omega(w)\}$. Then

$$\bigcap_{h>0} K_{w^0,\Omega}^{\pm}(h) = \bigcap_{h>0} K_{w^0,\Omega}^{\pm}(h)^a = K_{w^0,\Omega}^{\pm}.$$

Moreover, for any neighborhood U of $K^{\pm}_{w^0,\Omega}$ there is h>0 such that

$$K^{\pm}_{w^0,\Omega}(h') \subset U \quad \text{if } 0<h'\leqq h.$$

Theorem 2.2 ([17]). Assume that $K^-_{w^0} \cap \{w\in X;\ w_1\geqq 0\} \subset\subset X$ for every $w^0\in X$, where $A\subset\subset B$ means that A^a is compact and $A^a \subset \mathrm{Int}\, B$. Then, (i) $K^-_{w^0}\cap\{w\in X;\ w_1\geqq 0\}$ is compact and K^+_w is closed in X if $w_1\geqq 0$. (ii) $K^-_{w^0,\Omega}=\Omega$ when $w^0\in X$, $w_1\geqq 0$ and $\Omega=K^-_{w^0}\cap\{w\in X;\ w_1\geqq 0\}$. (iii) For any $w^0\in X$ and any neighborhood U of $K^-_{w^0}\cap\{w\in X;\ w_1\geqq 0\}$ there is a neighborhood U_1 of w^0 such that

$$K^-_w \cap \{w\in X;\ w_1\geqq 0\} \subset U \quad \text{for } w\in U_1.$$

Applying Theorems 2.1 and 2.2, we can obtain global results (properties) from local ones. Moreover, in doing so the local result may be improved by global information.

Theorem 2.3. Assume that $K^-_{w^0}\cap\{w\in X;\ w_1\geqq 0\} \subset\subset X$ for every $w^0\in X$. Let S be a closed subset of X which satisfies the condition; for every $w^0\in S$ and $\mathcal{C}\in\mathcal{C}(\mathbb{R}^N)$ with $K(w^0)\subset\subset\mathcal{C}$ there is a sequence $\{t_j\}\subset\mathbb{R}$ such that $t_j\downarrow 0$ and

$$(\{w^0\}-\mathcal{C})\cap\{w\in X;\ w_1=w_1^0-t_j\}\cap S\neq\emptyset \quad \text{for each } j.$$

Then

$$K^-_{w^0}\cap\{w\in X;\ w_1=0\}\cap S\neq\emptyset \quad \text{for } w^0\in S \text{ with } w_1^0\geqq 0.$$

Remark. Applying Theorem 2.3, we can derive outer estimates of the wave front sets of solutions to the hyperbolic Cauchy problem from a microlocal version of Holmgren's uniqueness theorem.

Proof. Assume that $w^0\in S$, $w_1^0\geqq 0$ and $K^-_{w^0}\cap\{w\in X;\ w_1=0\}\cap S=\emptyset$. Let Ω be a compact neighborhood of $K^-_{w^0}\cap\{w\in X;\ w_1\geqq 0\}$ in X. We note that $K^-_{w^0,\Omega}\cap\{w_1\geqq 0\}=K^-_{w^0}\cap\{w_1\geqq 0\}$. Since S is closed, it follows from Theorem 2.1 that there is h>0 such that

$$K^-_{w^0,\Omega}(h)^a\cap\{w_1\geqq 0\}\subset\subset\Omega,\quad K^-_{w^0,\Omega}(h)^a\cap\{w_1=0\}\cap S=\emptyset.$$

Set

$$t_0=\inf\{t\in\mathbb{R};\ t\geqq 0 \text{ and } K^-_{w^0,\Omega}(h)^a\cap\{w_1=t\}\cap S\neq\emptyset\}.$$

Then there is $w^1\in K^-_{w^0,\Omega}(h)^a\cap S$ such that $w_1^1=t_0>0$. Therefore, there is a sequence $\{t_j\}\subset\mathbb{R}$ such that $t_j\downarrow 0$ and

$$(2.1)\qquad K^-_{w^0,\Omega}(h)^a\cap\{w_1=t_0-t_j\}\cap S\neq\emptyset \quad \text{for each } j.$$

In fact, we can choose $\{w^{(j)}\}\subset K^-_{w^0,\Omega}(h)$ such that $w^{(j)}\longrightarrow w^1$ as $j\longrightarrow\infty$. So there are $j_0 \in \mathbb{N}$ and $k \in \mathbb{N}$ with $1 \leqq k \leqq N(h)$ such that $w^1, w^{(j)} \in U'(w^{h,k},h)$ ($j \geqq j_0$). Thus we have

$$(\{w^1\}-K_\Omega(w^{h,k},h)) \cap \{w_1 \geqq t_0-\varepsilon\} \subset K^-_{w^0,\Omega}(h)^a$$

if $\varepsilon>0$ is small enough. Since $K(w^1) \subset\subset K_\Omega(w^{h,k},h)$, it follows from the properties of S that there is $\{t_j\} \subset \mathbb{R}$ such that $t_j \downarrow 0$ and

$$(\{w^1\}-K_\Omega(w^{h,k},h)) \cap \{w_1=t_0-t_j\} \cap S \neq \emptyset.$$

This proves (2.1). On the other hand, (2.1) contradicts the definition of t_0. Q.E.D.

Corollary. Assume that $K^-_{w^0} \cap \{w \in X;\ w_1 \geqq 0\} \subset\subset X$ for every $w^0 \in X$. Let S be an open subset of X (or $X_+ \equiv \{w \in X;\ w_1 \geqq 0\}$) which satisfies the condition that $w^0 \in S$ if there are $\mathcal{C} \in \mathcal{C}(\mathbb{R}^N)$ with $K(w^0) \subset\subset \mathcal{C}$ and $t_0>0$ such that

$$(\{w^0\}-\mathcal{C}) \cap \{w \in X;\ w_1^0-t_0 \leqq w_1 < w_1^0\} \subset S.$$

Then $w^0 \in S$ if $w_1^0 \geqq 0$ and $K^-_{w^0} \cap \{w \in X;\ w_1=0\} \subset S$.

Remark. The corollary is the contraposition of Theorem 2.3. This corollary gives global existence theorems of solutions to the Cauchy problem in the complex domain.

3. EXISTENCE DOMAIN OF SOLUTIONS

Let $\mathcal{D}$ be a subdomain of $\mathbb{R}\times\mathbb{C}^n$, and let $P(t,z,\tau,\zeta)=\tau^m+\sum_{j=1}^m \sum_{|\alpha|\leq j} a_{j\alpha}(t,z)\tau^{m-j}\zeta^\alpha$ be a polynomial in $(\tau,\zeta)=(\tau,\zeta_1,\cdots,\zeta_n)\in\mathbb{C}\times\mathbb{C}^n$ of degree m whose coefficients are defined for $(t,z)=(t,z_1,\cdots,z_n)\in\mathcal{D}$. We identify $\mathbb{C}^n$ with $\mathbb{R}^{2n}$, i.e., $\mathbb{R}^{2n} \cong \mathbb{C}^n$; $(x,y)\longmapsto z=x+iy$. So we also regard $\mathcal{D}$ as a subdomain of $\mathbb{R}\times\mathbb{R}^{2n}$. In this section, we assume that

(H-1) $a_{j\alpha}(t,z)\in C_A(\mathcal{D})$ for $1\leqq j\leqq m$ and $|\alpha|\leqq j$, and $a_{j\alpha}(t,z)\in C_t^{1+\sigma}(\mathcal{D})$ for $1\leqq j\leqq m$ and $|\alpha|=j$, where $0<\sigma\leqq 1$, $a(t,z)\in C_A(\mathcal{D})$ means that $a(t,z)$ is continuous in $(t,z)\in\mathcal{D}$ and analytic in z, and $a(t,z)\in C_t^{1+\sigma}(\mathcal{D})$ means that $a(t,z)\in C^1(\mathcal{D})$ and $\partial_t a(t,z)$ is locally Hölder continuous with exponent σ.

Let us consider the Cauchy problem

$$\text{(CP)} \qquad \begin{cases} P(t,z,\partial_t,\partial_z)u(t,z) = f(t,z), \\ \partial_t^{j-1}u(0,z) = u_j(z) \quad (1\leqq j\leqq m), \end{cases}$$

where $z=x+iy$, $\partial_z=(\partial_{z_1},\cdots,\partial_{z_n})$ and $\partial_{z_j}=2^{-1}(\partial_{x_j}-i\partial_{y_j})$. Here we assume that $f(t,z)\in C_A(\mathcal{D})$ and $u_j(z)\in\mathcal{O}(D)$, where D is a subdomain of $\mathbf{C}^n\cong\mathbf{R}^{2n}$ and $a(z)\in\mathcal{O}(D)$ means that $a(z)$ is analytic in D.

Definition 3.1. Let Ω be a subdomain of $\mathbf{R}\times\mathbf{C}^n\cong\mathbf{R}^{2n+1}$, and write $\Omega_+=\{(t,z)\in\Omega;\ t\geqq 0\}$ and $\Omega_0=\{z\in\mathbf{C}^n;\ (0,z)\in\Omega\}$. We say that the Cauchy problem (CP) has a solution in Ω_+ if $\Omega_+\subset\mathcal{D}$, $\Omega_0\subset D$ and $\Omega_0\neq\emptyset$, and if there is a function $u\in C_A^m(\Omega_+)$ such that $Pu=f$ in Ω_+ and $\partial_t^{j-1}u(0,z)=u_j(z)$ ($1\leqq j\leqq m$) in Ω_0, where $u\in C_A^m(\Omega_+)$ means that $u\in C^m(\Omega_+)$ and u is analytic in z.

We write

$$p(t,z,\tau,\zeta) = \Pi_{j=1}^m (\tau-\lambda_j(t,z,\zeta)),$$

where $p(t,z,\tau,\zeta)$ denotes the principal symbol of $P(t,z,\tau,\zeta)$ and the $\lambda_j(t,z,\zeta)$ are continuous. Define for $(t,x,y)\in\mathcal{D}$

$$\Gamma(t,x,y) = \{(\tau,\xi,\eta)\in\mathbf{R}^{2n+1};\ \tau>\max_{1\leqq j\leqq m} \operatorname{Re}\lambda_j(t,x+iy,\xi-i\eta)\}.$$

Let π: $\mathbf{R}^{2n}\longrightarrow\mathbf{R}^{2n}$ be a projection, and define $\tilde{\pi}$: $\mathbf{R}^{2n+1}\longrightarrow\mathbf{R}^{2n+1}$ by $\tilde{\pi}(t,x,y)=(t,\pi(x,y))$. We assume that $\mathcal{D}=\tilde{\pi}^{-1}(\tilde{\pi}(\mathcal{D}))$ and $D=\pi^{-1}(\pi(D))$, and we set

$$\Gamma(t,x',y';\pi) = \bigcap_{(x,y)\in\pi^{-1}(x',y')} \Gamma(t,x,y)$$

for $(t,x',y')\in\tilde{\pi}(\mathcal{D})$. Then

$$\operatorname{Int}\Gamma(t,x',y';\pi) = \{(\tau,\xi,\eta)\in\mathbf{R}^{2n+1};\ \tau>\mu(t,x',y',\xi,\eta;\pi)\},$$

where $\mu(t,x',y',\xi,\eta;\pi)=\sup\{\operatorname{Re}\lambda_j(t,x+iy,\xi-i\eta);\ 1\leqq j\leqq m$ and $(x,y)\in\pi^{-1}(x',y')\}$. It is obvious that $\mu(t,x',y',\xi,\eta;\pi)<\infty$ if $\vartheta\equiv(1,0,\cdots,0)\in\operatorname{Int}\Gamma(t,x',y';\pi)$. Moreover, we assume that

(H-2) $\vartheta\in\operatorname{Int}\Gamma(t,x',y';\pi)$ for $(t,x',y')\in\tilde{\pi}(\mathcal{D})$ and $\{\Gamma(t,x',y':\pi)\}$ ($(t,x',y')\in\tilde{\pi}(\mathcal{D})$) is inner semi-continuous, i.e., for any $(t_0,x^{0\prime},y^{0\prime})\in\tilde{\pi}(\mathcal{D})$ and any cone $\mathcal{C}$ in $\mathbf{R}^{2n+1}$ with $\mathcal{C}\subset\subset\Gamma(t_0,x^{0\prime},y^{0\prime};\pi)$ there is a neighborhood U of $(t_0,x^{0\prime},y^{0\prime})$ in $\tilde{\pi}(\mathcal{D})$ such that $\mathcal{C}\subset\Gamma(t,x',y')$ for $(t,x',y')\in$ U.

Proposition 3.2. (i) If $\pi=\mathrm{id}$: $(x,y)\longmapsto(x,y)$, then (H-2) is valid. (ii) If $\sup\{|a_{j\alpha}(t,x+iy)|;\ (x,y)\in\pi^{-1}(x',y'),\ 1\leqq j\leqq m$ and $|\alpha|=j\}<\infty$ for each $(t,x',y')\in\tilde{\pi}(\mathcal{D})$, and if for any compact subset K of $\tilde{\pi}(\mathcal{D})$ and $\varepsilon>0$ there is $\delta>0$ such that

$$|a_{j\alpha}(t_1,x^1+iy^1)-a_{j\alpha}(t_2,x^2+iy^2)| < \varepsilon \quad (1\leqq j\leqq m \text{ and } |\alpha|=j)$$

when $\tilde{\pi}(t_k,x^k,y^k)\in K$ ($k=1,2$), $|\tilde{\pi}(t_1,x^1,y^1)-\tilde{\pi}(t_2,x^2,y^2)|<\delta$ and $(id-\pi)(x^1, y^1)=(id-\pi)(x^2,y^2)$, then (H-2) is valid.

Let $\{\gamma(t,x',y')\}_{(t,x',y')\in\tilde{\pi}(\mathcal{D})}$ be an inner semi-continuous family of open convex cones in $\tilde{\pi}^*(\mathbb{R}^{2n+1})$ such that

$$\vartheta\in\gamma(t,x',y')\subset\Gamma(t,x',y';\pi)\cap\tilde{\pi}^*(\mathbb{R}^{2n+1}) \quad \text{for } (t,x',y')\in\tilde{\pi}(\mathcal{D}),$$

and set

$$K(t,x',y';\pi) = \tilde{\pi}(\gamma(t,x',y')^*),$$

where $\Gamma^*=\{(\delta t,\delta x,\delta y)\in\mathbb{R}^{2n+1};\ \tau\delta t+\delta x\cdot\xi+\delta y\cdot\eta\geqq 0$ for any $(\tau,\xi,\eta)\in\Gamma\}$. Then it is clear that $\{K(t,x',y';\pi)\}_{(t,x',y')\in\tilde{\pi}(\mathcal{D})}$ is outer semi-continuous. Denote by $K^{\pm}_{(t,x',y';\pi)}$ the generalized flows for $\{K(t,x',y';\pi)\}$ as defined in §2. Now we can state a main result in this section.

<u>Theorem 3.3.</u> Assume that (H-1) and (H-2) are valid. Then the Cauchy problem (CP) has a unique solution $u\in C^m_A(\Omega_+)$ in Ω_+, where $\Omega_+=\tilde{\pi}^{-1}(\{(t',x',y')\in\tilde{\pi}(\mathcal{D});\ t'\geqq 0,\ K^-_{(t',x',y';\pi)}\cap\{t\geqq 0\}\subset\subset\tilde{\pi}(\mathcal{D})$ and $K^-_{(t',x',y';\pi)}\cap\{t=0\}\subset\subset\pi(D)\}$. Here we identified $\{0\}\times A$ with A for $A\subset\pi(\mathbb{R}^{2n})$.

<u>Remark.</u> (i) If the $a_{j\alpha}(t,z)$ and $f(t,z)$ are analytic in (t,z), Theorem 3.3 easily follows from Bony and Schapira [2] or Zerner [21], using the generalized flows. At the end of this section we shall give its proof. (ii) The assumption (H-1) may be relaxed by results in Jannelli [8].

We shall prove Theorem 3.3, reducing the Cauchy problem (CP) to the Cauchy problem for strictly hyperbolic operators. We note that Garabedian [5] proved the Cauchy-Kowalewski theorem for systems of the first order, reducing that to the Cauchy problem for symmetric hyperbolic systems.

<u>Theorem 3.4.</u> Assume that (H-1) is valid, and let $(t_0,x^0,y^0)\in\mathcal{D}$. For any convex cone $\mathcal{C}\subset\subset\Gamma(t_0,x^0,y^0)$ there are a neighborhood U of (t_0,x^0, y^0) and $A>0$ such that for any vector subspace V of $\mathbb{R}^{2n+1}$ with $\dim_{\mathbb{R}} V=2$ and $\vartheta\in V$ there are homogeneous polynomials $q_j(t,x,y,\tau,\xi,\eta)$ in (τ,ξ,η) of degree $m-1$ ($1\leqq j\leqq n$), whose coefficients are infinitely differentiable in (x,y) and belong to $C^{1+\sigma}(\mathcal{D})$, i.e., they belong to $C^1(\mathcal{D})$ and their first order derivatives are locally Hölder continuous with exponent σ, such that $\tilde{p}(t,x,y,\tau,\xi,\eta)=p(t,x+iy,\tau,(\xi-i\eta)/2)+\sum_{j=1}^n(\xi_j+i\eta_j)q_j(t,x,y,\tau,$

$\xi,\eta)$ is strictly hyperbolic w.r.t. ϑ and

$$(3.1)\qquad (\mathcal{C}\cap V)\cup\{(\tau,\xi,\eta)\in\mathbf{R}^{2n+1};\ \tau>A|(\xi,\eta)|\}$$
$$\subset\Gamma(\tilde{p}(t,x,y,\cdot,\cdot,\cdot),\vartheta)\quad\text{for } (t,x,y)\in U,$$

where $\Gamma(\tilde{p}(t,x,y,\cdot,\cdot,\cdot),\vartheta)$ denotes the connected component of the set $\{(\tau,\xi,\eta)\in\mathbf{R}^{2n+1};\ \tilde{p}(t,x,y,\tau,\xi,\eta)\neq 0\}$ which contains ϑ. Conversely, if $\tilde{p}(\tau,\xi,\eta)\in\mathcal{P}(t_0,x^0,y^0)\equiv\{\tilde{p}(\tau,\xi,\eta);\ \tilde{p}(\tau,\xi,\eta)=p(t_0,x^0+iy^0,\tau,(\xi-i\eta)/2)+\sum_{j=1}^n (\xi_j+i\eta_j)q_j(\tau,\xi,\eta)$ is hyperbolic w.r.t. ϑ and the $q_j(\tau,\xi,\eta)$ are homogeneous polynomials of degree m-1$\}$, then $\Gamma(\tilde{p},\vartheta)\subset\Gamma(t_0,x^0,y^0)$. In particular,

$$\Gamma(t_0,x^0,y^0)=\bigcup_{\tilde{p}\in\mathcal{P}(t_0,x^0,y^0)}\Gamma(\tilde{p},\vartheta).$$

Proof. We may assume that $V=\{(\tau,\xi_1,0,\cdots,0)\in\mathbf{R}^{2n+1};\ (\tau,\xi_1)\in\mathbf{R}^2\}$, making a change of coordinates in $\mathbf{C}^n$. Set

$$p^1(\tau,\xi_1,\eta_1)=\Pi_{j=1}^n(\tau-c_{j1}\xi_1-c_{j2}\eta_1),$$
$$p^2(t,x,y,\tau,\xi,\eta)=2\,\mathrm{Re}\,\{p(t,x+iy,\tau,(\xi-i\eta)/2)-$$
$$-p(t_0,x^0+iy^0,\tau,(\xi_1-i\eta_1)/2,0,\cdots,0)\}+p^1(\tau,\xi_1,\eta_1),$$

where $c_{j1},c_{j2}\in\mathbf{R}$ and $c_{j1}+ic_{j2}=\lambda_j(t_0,x^0+iy^0,1,0,\cdots,0)$. Then, there are homogeneous polynomials $r_j(t,x,y,\tau,\xi,\eta)$ in (τ,ξ,η) of degree m-1 ($1\leqq j\leqq n$), whose coefficients are infinitely differentiable in (x,y) and belong to $C^{1+\sigma}(\mathcal{D})$, such that

$$p^2(t,x,y,\tau,\xi,\eta)=p(t,x+iy,\tau,(\xi-i\eta)/2)$$
$$+\sum_{j=1}^n(\xi_j+i\eta_j)r_j(t,x,y,\tau,\xi,\eta).$$

In fact,

$$p^1(\tau,\xi_1,\eta_1)-p(t_0,x^0+iy^0,\tau,(\xi_1-i\eta_1)/2,0,\cdots,0)$$
$$=\Pi_{j=1}^n(\tau-c_{j1}\xi_1-c_{j2}\eta_1)-\Pi_{j=1}^n(\tau-(c_{j1}+ic_{j2})(\xi_1-i\eta_1)/2),$$
$$c_{j1}\xi_1+c_{j2}\eta_1=2\,\mathrm{Re}\,\{(c_{j1}+ic_{j2})(\xi_1-i\eta_1)/2\}.$$

Set, for $0<\varepsilon\leqq 1$,

$$p_\varepsilon^2(t,x,y,\tau,\xi,\eta)=(1-\varepsilon^2(\xi_1^2+\eta_1^2)\partial_\tau^2)^{[m/2]}p^2(t,x,y,\tau,\xi,\eta),$$

where [a] denotes the largest integer $\leqq a$. It follows from Lemma 2.2 in [20] (or its proof) that $p_\varepsilon^2(t_0,x^0,y^0,\tau,\xi_1,0,\cdots,0,\eta_1,0,\cdots,0)\equiv$

$(1-\varepsilon^2(\xi_1^2+\eta_1^2)\partial_\tau^2)^{[m/2]}p^1(\tau,\xi_1,\eta_1)=0$ in τ has only simple real roots, and that the distance between any two roots of the above equation is more than $c(m)\varepsilon(\xi_1^2+\eta_1^2)^{1/2}$, where $c(m)>0$. In fact, the same argument as in Lemma 2.2 in [20] shows that $\tilde\alpha_{k+1}(s)-\tilde\alpha_k(s)\geqq(\alpha_{k+1}-\alpha_k)/m$, where $\alpha_1\leqq\cdots\leqq\alpha_m$, $\tilde\alpha_1(s)\leqq\cdots\leqq\tilde\alpha_m(s)$, $\Pi_{k=1}^m(\tau-\tilde\alpha_k(s))=(1+s(d/d\tau))\Pi_{k=1}^m(\tau-\alpha_k)$ and $s\in\mathbb{R}$. Therefore, from Lemma 2.1 in [20] there is $c'(\varepsilon)>0$ such that $p_\varepsilon^2(t,x,y,\tau,\xi,\eta)=0$ in τ has only simple roots if $d(t,x,y,\xi,\eta)\equiv\{|(t-t_0,x-x^0,y-y^0)|^2\times|(\xi,\eta)|^2+|\xi'|^2+|\eta'|^2\}^{1/2}<c'(\varepsilon)|(\xi,\eta)|$, where $\xi'=(\xi_2,\cdots,\xi_n)$ and $\eta'=(\eta_2,\cdots,\eta_n)$. Since p_ε^2 is a real polynomial, $p_\varepsilon^2(t,x,y,\tau,\xi,\eta)=0$ in τ has only simple real roots if $d(t,x,y,\xi,\eta)<c'(\varepsilon)|(\xi,\eta)|$. Therefore, there is $A_\varepsilon(t,x,y)\in C^\infty(\mathcal{O})$ such that $\tilde p_\varepsilon(t,x,y,\tau,\xi,\eta)\equiv(1-A_\varepsilon(t,x,y)^2d(t,x,y,\xi,\eta)^2\times\partial_\tau^2)^{[m/2]}p_\varepsilon^2(t,x,y,\tau,\xi,\eta)$ is strictly hyperbolic w.r.t. ϑ. In fact, if $A_\varepsilon(t,x,y)$ is real-valued and $d(t,x,y,\xi,\eta)<c'(\varepsilon)|(\xi,\eta)|$, then $\tilde p_\varepsilon(t,x,y,\tau,\xi,\eta)=0$ in τ has only simple real roots. If $A_\varepsilon(t,x,y)$ is real-valued and $d(t,x,y,\xi,\eta)\geqq c'(\varepsilon)|(\xi,\eta)|$, then the distance between any two roots of $(1-A_\varepsilon(t,x,y)^2d(t,x,y,\xi,\eta)^2\partial_\tau^2)^{[m/2]}\tau^m=0$ in τ is not less than $c(m)|A_\varepsilon(t,x,y)|d(t,x,y,\xi,\eta)$ and the absolute values of the roots are less than or equal to $C(m)|A_\varepsilon(t,x,y)|d(t,x,y,\xi,\eta)$, where $C(m)>0$. It is obvious that the absolute value of the coefficient of τ^{m-j} in $\tilde p_\varepsilon(t,x,y,\tau,\xi,\eta)-(1-A_\varepsilon(t,x,y)^2d(t,x,y,\xi,\eta)^2\partial_\tau^2)^{[m/2]}\tau^m$ is less than or equal to $C(t,x,y)(1+|A_\varepsilon(t,x,y)|^{j-1})|(\xi,\eta)|^j$, where $C(t,x,y)$ is continuous and independent of the choice of $A_\varepsilon(t,x,y)$. Therefore, from Lemma 2.1 in [20] it follows that $\tilde p_\varepsilon(t,x,y,\tau,\xi,\eta)$ is strictly hyperbolic w.r.t. ϑ if $A_\varepsilon(t,x,y)>>1$. We note that one can choose a fixed $A_\varepsilon(t,x,y)$ for each $\varepsilon>0$ as V changes. Since $\tilde p_\varepsilon(t_0,x^0,y^0,\tau,\xi_1,0,\cdots,0)=(1-\varepsilon^2\xi_1^2\partial_\tau^2)^{[m/2]}p^1(\tau,\xi_1,0)$, and

$$\mathcal{C}\wedge V\subset\subset\Gamma(t_0,x^0,y^0)\wedge V=\{(\tau,\xi_1,0,\cdots,0)\in\mathbb{R}^{2n+1};\quad \tau>\max_{1\leqq j\leqq m}c_{j1}\xi_1\}\quad(\text{ in } V),$$

there are $\varepsilon>0$, a neighborhood U of (t_0,x^0,y^0) and $A>0$, independent of V, such that (3.1) holds for $\tilde p=\tilde p_\varepsilon$. This proves the first part of the theorem. Let V be a vector subspace of $\mathbb{R}^{2n+1}$ with $\dim_{\mathbb{R}}V=2$ and $\vartheta\in V$, and let $\tilde p(\tau,\xi,\eta)\in\mathcal{P}(t_0,x^0,y^0)$. We can assume without loss of generality that $V=\{(\tau,\xi_1,0,\cdots,0)\in\mathbb{R}^{2n+1};\ (\tau,\xi_1)\in\mathbb{R}^2\}$. Write

$$\tilde p(\tau,\xi_1,\eta_1)=\tilde p(\tau,\xi_1,0,\cdots,0,\eta_1,0,\cdots,0),$$

$$p(\tau,\zeta_1) = p(t_0,x^0+iy^0,\tau,\zeta_1,0,\cdots,0) = \Pi_{j=1}^m (\tau-(c_{j1}+ic_{j2})\zeta_1).$$

Consider the Cauchy problems

$$(3.2)\qquad \begin{cases} p(\partial_t,\partial_{z_1})u(t,z_1) = 0, \\ \partial_t^{j-1}u(0,z_1) = u_j(z_1) \quad (1\leqq j\leqq m), \end{cases}$$

and

$$(3.3)\qquad \begin{cases} \tilde{p}(\partial_t,\partial_{x_1},\partial_{y_1})v(t,x_1,y_1) = 0, \\ \partial_t^{j-1}v(0,x_1,y_1) = u_j(x_1+iy_1) \quad (1\leqq j\leqq m), \end{cases}$$

where the $u_j(z_1)$ are analytic in a neighborhood of $\Omega_0\equiv\{z_1\in\mathbf{C};\ z_1=x_1+iy_1$ and $(0,x_1,y_1)\in\{(1,0,0)\}-\Gamma(\tilde{p}(\cdot,\cdot,\cdot),(1,0,0))^*\}$. It is easy to see that $v(t,x_1,y_1)=u(t,x_1+iy_1)$ satisfies (3.3) if $u(t,z_1)$ satisfies (3.2). On the other hand, (3.3) has a solution $v(t,x_1,y_1)$ in a neighborhood of $\Omega_+ \equiv(\{(1,0,0)\}-\Gamma(\tilde{p}(\cdot,\cdot,\cdot),(1,0,0))^*)\cap\{t\geqq 0\}$ which is analytic in (t,z_1). In fact, $w(t,x_1,y_1)=2^{-1}(\partial_{x_1}+i\partial_{y_1})v(t,x_1,y_1)$ satisfies (3.3) with $u_j\equiv 0$ there. Therefore, $u(t,x_1+iy_1)=v(t,x_1,y_1)$ satisfies (3.2). Now assume that $\Gamma(\tilde{p},(1,0,0))\subset\{(\tau,\xi_1,\eta_1);\ \tau>\max_{1\leqq j\leqq m}(c_{j1}\xi_1+c_{j2}\eta_1)\}$ is not valid. Then there is $j\in\mathbf{N}$ such that $(c_{j1},c_{j2})\notin\Omega_0$. It is obvious that $u(t,z_1)\equiv (z_1+(c_{j1}+ic_{j2})(t-1))^{-1}$ satisfies (3.2) with $u_j(z_1)=\partial_t^{j-1}u(0,z_1)$ $(1\leqq j\leqq m)$ and that it is not analytic at $(t,z_1)=(1,0)$. Since the $u_j(z_1)$ are analytic in a neighborhood of Ω_0, (3.3) with $u_j(z_1)=\partial_t^{j-1}u(0,z_1)$ has a unique solution $v(t,x_1,y_1)$ in a neighborhood of Ω_+ (in $\{t\geqq 0\}$), which is analytic in (t,z_1). So we have $u(t,x_1+iy_1)=v(t,x_1,y_1)$ in Ω_+, which is a contradiction. This completes the proof. Q.E.D.

Let $\tilde{P}(t,x,y,\tau,\xi,\eta)\equiv P(t,x+iy,\tau,(\xi-i\eta)/2)+\sum_{j=1}^n (\xi_j+i\eta_j)Q_j(t,x,y,\tau,\xi,\eta)$ be a strictly hyperbolic polynomial w.r.t. ϑ whose coefficients and their derivatives of arbitrary order with respect to (x,y) are continuous in (t,x,y). Assume that the coefficients of the principal symbol of $\tilde{P}$ belong to $C^{1+\sigma}(\mathcal{D})$. Let us consider the Cauchy problem

$$(CP)'\qquad \begin{cases} \tilde{P}(t,x,y,\partial_t,\partial_x,\partial_y)v(t,x,y) = f(t,x+iy), \\ \partial_t^{j-1}v(0,x,y) = u_j(x+iy) \quad (1\leqq j\leqq m), \end{cases}$$

where $f(t,z)\in C_A(\mathcal{D})$ and $u_j(z)\in\mathcal{O}(D)$. Let Ω be a subdomain of $\mathbf{R}\times\mathbf{C}^n$. Then $v(t,x,y)\equiv u(t,x+iy)$ satisfies (CP)' in Ω_+ if $u(t,z)\in C_A^m(\Omega_+)$ is a solution of (CP) in Ω_+. Let $(t_0,x^0,y^0)\in\mathcal{D}$, and let $\mathcal{C}$ be an open convex cone such that $\mathcal{C}\subset\subset\Gamma(\tilde{p}(t_0,x^0,y^0,\cdot,\cdot,\cdot),\vartheta)$, where $\tilde{p}(t,x,y,\tau,\xi,\eta)$ denotes the principal symbol of $\tilde{P}(t,x,y,\tau,\xi,\eta)$. Then there is a neighborhood U of (t_0,x^0,y^0) in $\mathcal{D}$ such that $\mathcal{C}\subset\Gamma(\tilde{p}(t,x,y,\cdot,\cdot,\cdot),\vartheta)$ for $(t,x,y)\in U$. Let $(t_1,x^1,y^1)\in U$, and choose $t_2\in\mathbb{R}$ so that $t_1>t_2$ and $K\equiv(\{(t_1,x^1,y^1)\}-\mathcal{C}^*)\cap\{t\geqq t_2\}\subset\subset U$. Instead of (CP)' we consider the Cauchy problem

$$\text{(CP)}'' \quad \begin{cases}\tilde{P}(t,x,y,\partial_t,\partial_x,\partial_y)v(t,x,y) = f(t,x+iy),\\ \partial_t^{j-1}v(t_2,x,y) = u_j(x+iy) \quad (1\leqq j\leqq m),\end{cases}$$

where $f(t,x+iy)$ is continuous in a neighborhood of K and analytic in $z=x+iy$, and the $u_j(x+iy)$ are analytic in a neighborhood of $K_0\equiv\{z\in\mathbb{C}^n;$ $z=x+iy$ and $(0,x,y)\in K\}$ in $\mathbb{C}^n$.

Lemma 3.5. There are a neighborhood Ω of K and $v(t,x,y)\in C^m(\Omega_+)$ such that $v(t,x,y)$ is a unique solution of (CP)" in Ω_+, where $\Omega_+=\Omega\cap\{t\geqq t_2\}$. Moreover,

$$2^{-1}(\partial_{x_j}+i\partial_{y_j})v(t,x,y) = 0 \quad (1\leqq j\leqq n) \quad \text{in } \Omega_+$$

and $u(t,x+iy)\equiv v(t,x,y)$ satisfies

$$\begin{cases}P(t,z,\partial_t,\partial_z)u(t,z) = f(t,z) \quad \text{in } \Omega_+,\\ \partial_t^{j-1}u(t_2,z) = u_j(z) \quad (1\leqq j\leqq m) \quad \text{in } \Omega_0,\end{cases}$$

where $\Omega_0=\{(x,y)\in\mathbb{R}^{2n};\ (t_2,x,y)\in\Omega\}$.

Proof. From the existence theorem of the Cauchy problem for strictly hyperbolic operators it follows that (CP)" has locally a unique solution (see, e.g., Mizohata [14]). Using the Holmgren transformation and the method of sweeping out due to John [10], we can see that (CP)" has a unique solution $v(t,x,y)\in C^m(\Omega_+)$, where $\Omega_+=(\{(t_1+\delta t,x^1+\delta x,y^1+\delta y)\}-\text{Int}\,\mathcal{C}^*)\cap\{t\geqq t_2\}$ $(\subset\subset U)$ and $(\delta t,\delta x,\delta y)\in\text{Int}\,\mathcal{C}^*$. Moreover, $v(t,x,y)$ is infinitely differentiable in (x,y). Set

$$w_k(t,x,y) = 2^{-1}(\partial_{x_k}+i\partial_{y_k})v(t,x,y) \quad (1\leqq k\leqq n),$$
$$w(t,x,y) = {}^t(w_1(t,x,y),\cdots,w_n(t,x,y)).$$

Then we have

$$\begin{cases}L(t,x,y,\partial_t,\partial_x,\partial_y)w(t,x,y) = 0 \quad \text{in } \Omega_+,\\ \partial_t^{j-1}w(t_2,x,y) = 0 \quad (1\leqq j\leqq m) \quad \text{in } \Omega_0,\end{cases}$$

where $L\equiv(L_{kj})$ is the n×n matrix such that $L_{kj}=\tilde{P}(t,x,y,\partial_t,\partial_x,\partial_y)\delta_{kj}+\sum_{\ell=1}^{m}\sum_{|\alpha|+|\beta|\leqq\ell-1} 2^{-1}(\partial_{x_k}+i\partial_{y_k})q_{j\ell\alpha\beta}(t,x,y)\partial_t^{m-\ell}\partial_x^{\alpha}\partial_y^{\beta}$ and $Q_j(t,x,y,\partial_t,\partial_x,\partial_y)=\sum_{\ell=1}^{m}\sum_{|\alpha|+|\beta|\leqq\ell-1} q_{j\ell\alpha\beta}(t,x,y)\partial_t^{m-\ell}\partial_x^{\alpha}\partial_y^{\beta}$. Let Ω be a subdomain of $\mathbb{R}^{2n+1}$ such that $\Omega_+=\Omega\cap\{t\geqq t_2\}$, and define $w(t,x,y)=0$ for $t<t_2$ and $(t,x,y)\in\Omega$. Then we have

$$\begin{cases} L(t,x,y,\partial_t,\partial_x,\partial_y)w(t,x,y) = 0 \quad \text{in } \Omega, \\ \operatorname{supp} w \subset \{(t,x,y)\in\Omega;\ t\geqq t_2\}. \end{cases}$$

Set $\hat{\Omega}_+=(\{(t_1+\delta t/2,x^1+\delta x/2,y^1+\delta y/2)\}-\operatorname{Int}\mathscr{C}^*)\cap\{t\geqq t_2\}$ $(\subset\subset\Omega)$. We can assume without loss of generality that the boundary of $\mathscr{C}^*\cap\{\delta t=1\}$ in $\{(\delta t,\delta x,\delta y)\in\mathbb{R}^{2n+1};\ \delta t=1\}$ is smooth. Then we can construct a family $\{\phi_\theta(t,x,y)\}_{0\leqq\theta<1}$ of real-valued smooth functions defined in Ω such that $S_\theta\cap\{t=t_2\}$ is coincident with the boundary of $\hat{\Omega}_+\cap\{t=t_2\}$ (in $\{t=t_2\}$), $\bigcup_{0\leqq\theta<1} S_\theta\cap\{t\geqq t_2\}=\hat{\Omega}_+$, grad $\phi_\theta(t,x,y)\in\mathscr{C}$ if $(t,x,y)\in S_\theta$ and $\theta\in[0,1)$, and $\phi_{\theta_1}(t,x,y)>\phi_{\theta_2}(t,x,y)$ if $0\leqq\theta_1<\theta_2<1$ and $(t,x,y)\in\hat{\Omega}_+$, where $S_\theta=\{(t,x,y)\in\Omega;\ \phi_\theta(t,x,y)=0\}$. Now assume that $(t_3,x^3,y^3)\in\Omega$, $\phi_{\theta_0}(t_3,x^3,y^3)=0$ and $t_3\geqq t_2$, and that $w(t,x,y)=0$ if $(t,x,y)\in U_1$ and $\phi_{\theta_0}(t,x,y)\leqq 0$, where U_1 is a small neighborhood of (t_3,x^3,y^3). We introduce the new coordinates (t',x',y') in U_1: $t'=\phi_{\theta_0}(t,x,y)+\varepsilon|x-x^3|^2+\varepsilon|y-y^3|^2$, $x'=x-x^3$, $y'=y-y^3$, where $\varepsilon>0$ is small enough. Then we have

$$\begin{cases} \tilde{L}(t',x',y',\partial_{t'},\partial_{x'},\partial_{y'})\tilde{w}(t',x',y') = 0 \quad \text{in } U_1', \\ \operatorname{supp} \tilde{w} \subset \{(t',x',y')\in U_1';\ t'\geqq\varepsilon|x'|^2+\varepsilon|y'|^2\}, \end{cases}$$

where $U_1'=\{(\phi_{\theta_0}(t,x,y)+\varepsilon|x-x^3|^2+\varepsilon|y-y^3|^2,x-x^3,y-y^3)\in\mathbb{R}^{2n+1};\ (t,x,y)\in U_1\}$, $t'\equiv\phi_{\theta_0}(\psi(t',x',y'),x'+x^3,y'+y^3)+\varepsilon|x'|^2+\varepsilon|y'|^2$ in U_1', $\tilde{w}(t',x',y')=w(\psi(t',x',y'),x'+x^3,y'+y^3)$ for $(t',x',y')\in U_1'$ and the diagonal components of the principal symbol of $\tilde{L}$ are equal to $\tilde{P}(\psi(t',x',y'),x'+x^3,y'+y^3,$ $\operatorname{grad}_{(t,x,y)}\phi_{\theta_0}(\psi(t',x',y'),x'+x^3,y'+y^3)\tau'+2\varepsilon(0,x',y')\tau'+(0,\xi',\eta'))$. We may assume that $\tilde{L}$ is defined on $\mathbb{R}^{2n+1}$ and that the diagonal components of the principal symbol of $\tilde{L}$ are strictly hyperbolic w.r.t. ϑ, choosing ε small enough. For $(t',x',y')\in(-\infty,\delta]\times\mathbb{R}^{2n}$ we define $\tilde{w}(t',x',y')$ by $\tilde{w}(t',x',y')=0$ if $t'\leqq\delta$ and $(t',x',y')\notin U_1'$, where $\delta>0$ is small so that $\tilde{L}\tilde{w}=0$ in $(-\infty,\delta]\times\mathbb{R}^{2n}$. Then, by the energy inequalities for strictly hyperbolic operators, we have

$$W(t') \equiv \sum_{j=1}^{m} \|\partial_t^{j-1}\tilde{w}(t',\cdot,\cdot)\|_{H^{m-j}(\mathbb{R}^{2n})} \leqq$$

$$\leqq C(\delta)\int_0^{t'} W(s)ds \quad \text{for } 0\leqq t'\leqq\delta.$$

Therefore, we have

$$W(t') \leqq \exp[C(\delta)t']W(0) = 0.$$

This implies that $w(t,x,y)$ vanishes near (t_3,x^3,y^3). So the set $\{\theta\varepsilon[0,1);\ w(t,x,y)=0$ on $S_{\theta'}$ for any $\theta'\leqq\theta\}$ is open and closed in $[0,1)$, which proves that $w(t,x,y)=0$ in $\hat{\Omega}_+$. Q.E.D.

Theorem 3.6. Let Ω^j ($j=1,2$) be subdomains of $\mathbb{R}\times\mathbb{C}^n$ such that $\Omega^1_+\cap\Omega^2_+\cap\{t=c\}$ is a non-void connected set if $0\leqq c<\sup\{t;\ (t,z)\varepsilon\Omega^1_+\cap\Omega^2_+\}$. If the Cauchy problem (CP) has a solution in Ω^j_+ ($j=1,2$), respectively, then the Cauchy problem (CP) has a unique solution in $\Omega^1_+\cup\Omega^2_+$.

Proof. Let $u_j\varepsilon C^m_A(\Omega^j_+)$ be a solution of (CP) in Ω^j_+ ($j=1,2$). Set $w=u_1-u_2$ in $\Omega^1_+\cap\Omega^2_+$. Then we have

$$\begin{cases} Pw = 0 & \text{in } \Omega^1_+\cap\Omega^2_+, \\ \partial_t^{j-1}w(0,z) = 0 \quad (1\leqq j\leqq m) & \text{in } \Omega^1_0\cap\Omega^2_0. \end{cases}$$

It suffices to prove that $w=0$ in $\Omega^1_+\cap\Omega^2_+$. Now assume that there is $(t_0,z^0)\varepsilon\Omega^1_+\cap\Omega^2_+$ such that $w(t_0,z^0)\neq0$. Set $t^*=\inf\{t\geqq0;\ w(t,z)\not\equiv0 \text{ in } z\}$. Then we have $t^*<t_0$. By the assumption there is a non-void open subset U of $\mathbb{C}^n$ such that $\{(t^*,z);\ z\varepsilon U\}\subset\Omega^1_+\cap\Omega^2_+$. Since $\partial_t^{j-1}w(t^*,z)\equiv0$ in z ($1\leqq j\leqq m$), from Lemma 3.5 it follows that there are $\delta>0$ and a non-void open subset U' of $\mathbb{C}^n$ such that $w(t,z)=0$ in $[t^*,t^*+\delta]\times U'$. Since $\Omega^1_+\cap\Omega^2_+\cap\{t=c\}$ is connected, by the theorem of unicity, we have $w(t,z)=0$ in $\Omega^1_+\cap\Omega^2_+\cap\{t\leqq t^*+\delta\}$, which contradicts the definition of t^*. This proves the theorem. Q.E.D.

Lemma 3.7. Let Γ_1 and Γ_2 be closed convex cones in $\mathbb{R}^{2n+1}$. Then $(\Gamma_1\cap\Gamma_2)^*=\Gamma_1^*+\Gamma_2^*$.

Proof. The bipolar theorem gives

$$\overline{ch}[\Gamma_1^*\cup\Gamma_2^*] = (\Gamma_1^*\cup\Gamma_2^*)^{**},$$

where $\overline{ch}[A]$ denotes the closed convex hull of A. We have also

$$(\Gamma_1^*\cup\Gamma_2^*)^{**} = (\Gamma_1^{**}\cap\Gamma_2^{**})^* = (\Gamma_1\cap\Gamma_2)^*.$$

On the other hand, we have

$$\overline{\mathrm{ch}}[\Gamma_1^* \cup \Gamma_2^*] = \Gamma_1^* + \Gamma_2^*,$$

which proves the lemma. Q.E.D.

Lemma 3.8. Let $(t_0, x^0, y^0) \in \mathcal{D}$, and let Γ be an open convex cone such that $\Gamma \subset\subset \Gamma(t_0, x^0, y^0)$ and $\vartheta \in \Gamma$. Then there is a neighborhood U of (t_0, x^0, y^0) in $\mathcal{D}$ such that for any $(t_1, x^1, y^1) \in U$ and $t_2 < t_1$ with $\Omega_+ \equiv (\{(t_1, x^1, y^1)\} - \operatorname{Int}\Gamma^*) \cap \{t \geqq t_2\} \subset U$ the Cauchy problem

$$\begin{cases} P(t,z,\partial_t,\partial_z)u(t,z) = f(t,z), \\ \partial_t^{j-1}u(t_2,z) = u_j(z) \quad (1 \leqq j \leqq m) \end{cases}$$

has a unique solution $u(t,z) \in C_A^m(\Omega_+)$ in Ω_+ if $f(t,z) \in C_A(\Omega_+)$ and $u_j(z) \in \mathcal{O}(\Omega_0)$ $(1 \leqq j \leqq m)$, where $\Omega_0 = \{(x,y) \in \mathbb{R}^{2n};\ (t_2,x,y) \in \Omega_+\}$.

Proof. Let $\mathcal{C}$ be an open convex cone in $\mathbb{R}^{2n+1}$ such that $\Gamma \subset\subset \mathcal{C} \subset\subset \Gamma(t_0, x^0, y^0)$, and let $(1, \nu_x, \nu_y) \in \operatorname{Int}\mathcal{C}^*$. We may assume that there is $\lambda(x,y) \in C^\infty(\mathbb{R}^{2n} \setminus \{0\})$ such that $\lambda(x,y)$ is positively homogeneous of degree 1 and $\mathcal{C}^* \cap \{t=1\} = \{(1, \delta x, \delta y) \in \mathbb{R}^{2n+1};\ \lambda(\delta x - \nu_x, \delta y - \nu_y) \leqq 1\}$. We can choose a neighborhood U of (t_0, x^0, y^0) for which the first part of Theorem 3.4 holds. Set

$$\Omega_+(t_2, x^2, y^2; a, \varepsilon, h) = \{(t,x,y) \in \mathbb{R}^{2n+1};\ \lambda(x^2 - x + (t-t_2)\nu_x, y^2 - y + (t-t_2)\nu_y) < a - (1+\varepsilon)(t-t_2) \text{ and } t_2 \leqq t \leqq t_2 + ah/(1+\varepsilon)\},$$

$$\Omega_0(t_2, x^2, y^2; a) = \{(x,y) \in \mathbb{R}^{2n};\ \lambda(x^2 - x, y^2 - y) < a\},$$

where $a, \varepsilon, h \geqq 0$. Let $(t_2, x^2, y^2) \in U$ and $\varepsilon > 0$, and choose $a > 0$ so that $\Omega_+(t_2, x^2, y^2; a, 0, 1) \subset U$. Assume that $f(t,z) \in C_A(\Omega_+(t_2, x^2, y^2; a, 0, 1))$ and $u_j(z) \in \mathcal{O}(\Omega_0(t_2, x^2, y^2; a))$. First let us prove that for any $\varepsilon > 0$ there is $h(\varepsilon) > 0$ with $h(\varepsilon) \leqq 1$ such that the Cauchy problem

$$(3.4) \qquad \begin{cases} P(t,z,\partial_t,\partial_z)u(t,z) = f(t,z) & \text{in } \Omega_+(t_2, x^2, y^2; a, \varepsilon, h(\varepsilon)), \\ \partial_t^{j-1}u(t_2,z) = u_j(z) \quad (1 \leqq j \leqq m) & \text{in } \Omega_0(t_2, x^2, y^2; a) \end{cases}$$

has a unique solution $u(t,z)$. From Theorem 3.4 and Lemma 3.5 it follows that there is $h_0 > 0$ with $h_0 \leqq 1$ such that the Cauchy problem (3.4) has a unique solution in $\Omega_+(t_2, x^2, y^2; a/2, 0, h_0)$. Let $0 < h \leqq h_0/2$ and $(t_3, x^3, y^3) \in \Omega_+(t_2, x^2, y^2; a, \varepsilon, h) \setminus \Omega_+(t_2, x^2, y^2; a/2, 0, 2h)$, and set

$$V = \{(\tau, -s \operatorname{grad} \lambda(x^2 - x^3 + (t_3 - t_2)\nu_x, y^2 - y^3 + (t_3 - t_2)\nu_y)) \in \mathbb{R}^{2n+1};\ (\tau, s) \in \mathbb{R}^2\}.$$

From Theorem 3.4 there is a strictly hyperbolic polynomial $\tilde{p}(t,x,y,\tau,\xi,\eta)$ w.r.t. ϑ which satisfies (3.1) for the above vector subspace V of dimension 2. Therefore, by Lemma 3.5, we have a unique solution of (3.4) in $\hat{\Omega}_+$, where $\hat{\Omega}$ is a neighborhood of $K\equiv\{(t_3,x^3,y^3)\}-\{(\delta t,\delta x,\delta y)\varepsilon\, \mathcal{C}^*+V^*;\ \delta t\geqq A^{-1}|(\delta x,\delta y)|\}\cap\{t\geqq t_2\}$ and $\hat{\Omega}_+=\hat{\Omega}\cap\{t\geqq t_2\}$, if $K\subset\Omega_+(t_2,x^2,y^2;a,0,1)$. In fact, by Lemma 3.7 we have

$$\Gamma(\tilde{p}(t,x,y,\cdot,\cdot,\cdot),\vartheta)^*\subset(\mathcal{C}^*+V^*)$$

$$\cap\,\{(\delta t,\delta x,\delta y)\varepsilon R^{2n+1};\ \delta t\geqq A^{-1}|(\delta x,\delta y)|\}.$$

Set $M=\sup_{\lambda(x,y)=1,\ 1\leqq|\alpha|+|\beta|\leqq 2}|\partial_x^\alpha\partial_y^\beta\lambda(x,y)|$. Since

$$|\lambda(x^2-x^3+\delta x+(t_3-t_2)\nu_x,y^2-y^3+\delta y+(t_3-t_2)\nu_y)$$
$$-\lambda(x^2-x^3+(t_3-t_2)\nu_x,y^2-y^3+(t_3-t_2)\nu_y)|\leqq(2n)^{1/2}M|(\delta x,\delta y)|,$$

we can choose $h>0$ so that

$$\lambda(x^2-x^3+\delta x,y^2-y^3+\delta y)>a/3$$

if $|(\delta x,\delta y)|\leqq(A+|(\nu_x,\nu_y)|)(t_3-t_2)$. Therefore, if $(t_3-t_2,\delta x,\delta y)\varepsilon\,\mathcal{C}^*+V^*$ and $|(\delta x,\delta y)|\leqq A(t_3-t_2)$, we have

$$\lambda(x^2-x^3+\delta x,y^2-y^3+\delta y)\leqq\lambda(x^2-x^3+(t_3-t_2)\nu_x,y^2-y^3+(t_3-t_2)\nu_y)+$$
$$+\ \mathrm{grad}\,\lambda(x^2-x^3+(t_3-t_2)\nu_x,y^2-y^3+(t_3-t_2)\nu_y)$$
$$\cdot(\delta x-(t_3-t_2)\nu_x,\delta y-(t_3-t_2)\nu_y)+6n^2a^{-1}M(A+|(\nu_x,\nu_y)|)^2(t_3-t_2)^2$$
$$<a-\varepsilon(t_3-t_2)+6n^2(1+\varepsilon)^{-1}M(A+|(\nu_x,\nu_y)|)^2h(t_3-t_2).$$

In fact, if $(t_3-t_2,\delta x^1,\delta y^1)\varepsilon\,\mathcal{C}^*$ and $(\delta x-\delta x^1,\delta y-\delta y^1)\varepsilon V^*$, then

$$\mathrm{grad}\,\lambda(x^2-x^3+(t_3-t_2)\nu_x,y^2-y^3+(t_3-t_2)\nu_y)$$
$$\cdot(\delta x-(t_3-t_2)\nu_x,\delta y-(t_3-t_2)\nu_y)$$
$$=\lambda(\delta x^1-(t_3-t_2)\nu_x,\delta y^1-(t_3-t_2)\nu_y)\leqq t_3-t_2.$$

Moreover, we have

$$\sup_{\lambda(x,y)>a/3,\ |\alpha|+|\beta|=2}|\partial_x^\alpha\partial_y^\beta\lambda(x,y)|\leqq 3a^{-1}M.$$

Therefore, taking $h(\varepsilon)\leqq n^{-2}M^{-1}(A+|(\nu_x,\nu_y)|)^{-2}(1+\varepsilon)\varepsilon/6$, we have

$$\lambda(x^2-x^3+\delta x,\ y^2-y^3+\delta y)<a$$

if $(t_3-t_2,\delta x,\delta y)\varepsilon\,\mathcal{C}^*+V^*$ and $|(\delta x,\delta y)|\leqq A(t_3-t_2)$. Since $\mathcal{C}^*\subset\mathcal{C}^*+V^*$,

Theorem 3.6 and this prove that $K\subset\Omega_+(t_2,x^2,y^2;a,0,1)$, and that the Cauchy problem (3.4) has a unique solution in $\Omega_+(t_2,x^2,y^2;a,\varepsilon,h(\varepsilon))$. Repeating the above argument, we have a unique solution of (3.4) in $\Omega_+(t_2,x^2,y^2;a,\varepsilon,1)$ for any $\varepsilon>0$, which proves the lemma. Q.E.D.

Although we shall prove Theorem 3.3 by Lemma 3.8, we can also prove it, using the following lemma which is a weaker result than Lemma 3.8.

Lemma 3.8'. Let $(t_0,x^0,y^0)\in\mathcal{D}$, and let Γ be an open convex cone such that $\Gamma\subset\subset\Gamma(t_0,x^0,y^0)$ and $\vartheta\in\Gamma$. Assume that $u\in C_A^m(\Omega_{+\varepsilon})$ satisfies $Pu=f$ in $\Omega_{+\varepsilon}$ for some $\varepsilon>0$, where $\Omega_{+\varepsilon}\equiv(\{(t_0,x^0,y^0)\}-\mathrm{Int}\,\Gamma^*)\cap\{t\geqq t_0-\varepsilon\}$ and $f\in C_A(\mathcal{D})$. Then u can be continued in a neighborhood of (t_0,x^0,y^0) in $\mathcal{D}$ as a solution of $Pu=f$.

Proof. Let $(t_1,x^1,y^1)\in\mathcal{D}$, and let $\phi(t,x,y)$ be a real-valued smooth function defined in a neighborhood U of (t_1,x^1,y^1) such that $\phi(t_1,x^1,y^1)=0$ and $\mathrm{grad}\,\phi(t_1,x^1,y^1)\in\Gamma(t_1,x^1,y^1)$. Assume that $u\in C_A^m(\hat\Omega_-)$ satisfies $Pu=f$ in $\hat\Omega_-$, where $\hat\Omega_-\equiv\{(t,x,y)\in U;\ \phi(t,x,y)<0\}$. By Theorem 3.4, we can choose a strictly hyperbolic polynomial $\tilde p(t,x,y,\tau,\xi,\eta)$ w.r.t. ϑ for V such that $\mathrm{grad}\,\phi(t_1,x^1,y^1)\in\Gamma(\tilde p(t_1,x^1,y^1,\cdot,\cdot,\cdot),\vartheta)$, where $V=\{a\vartheta+b\,\mathrm{grad}\,\phi(t_1,x^1,y^1);\ (a,b)\in\mathbb{R}^2\}$. Then, from Lemma 3.5 it follows that u can be continued in a neighborhood of (t_1,x^1,y^1) as a solution of $Pu=f$. Therefore, using the method of sweeping out, we can prove Lemma 3.8'. Q.E.D.

Proof of Theorem 3.3. Set $S=\{(t_0,x^{0\prime},y^{0\prime})\in\tilde\pi(\Omega_+);\ t_0=0$, or $t_0>0$ and there is a neighborhood $\tilde\Omega$ of $K^-_{(t_0,x^{0\prime},y^{0\prime};\pi)}$ in $\tilde\pi(\mathcal{D})$ such that (CP) has a solution $u(t,z;t_0,x^{0\prime},y^{0\prime})$ in $\tilde\pi^{-1}(\tilde\Omega_+)\equiv\tilde\pi^{-1}(\tilde\Omega)\cap\{t\geqq0\}$ and $u(t,z)=u(t,z;t_0,x^{0\prime},y^{0\prime})$ in a neighborhood of $\tilde\pi^{-1}(K^-_{(t_0,x^{0\prime},y^{0\prime};\pi)})$ (in $\{t\geqq0\}$) if $\tilde\Omega^1$ is a neighborhood of $K^-_{(t_0,x^{0\prime},y^{0\prime};\pi)}$ in $\tilde\pi(\mathcal{D})$ and $u(t,z)$ is a solution of (CP) in $\tilde\pi^{-1}(\tilde\Omega^1_+)\}$. Then, by Theorem 2.2, S is open in $\tilde\pi(\mathbb{R}^{2n+1})\cap\{t\geqq0\}$ and we can define a (unique) solution $u(t,z)$ of (CP) in $\tilde\pi^{-1}(S)$ by $u(t,z)=u(t,z;t,\ \pi(z))$. Let $(t_0,x^{0\prime},y^{0\prime})\in\tilde\pi(\Omega_+)\cap\{t>0\}$, and let $\mathcal{C}\in\mathcal{C}(\tilde\pi(\mathbb{R}^{2n+1}))$ satisfy $K(t_0,x^{0\prime},y^{0\prime};\pi)\subset\subset\mathcal{C}$. Assume that there is $\varepsilon>0$ such that $(\{(t_0,x^{0\prime},y^{0\prime})\}-\mathcal{C})\cap\{(t,x',y')\in\tilde\pi(\mathbb{R}^{2n+1});\ t_0-\varepsilon\leqq t<t_0\}\subset S$. From the assumption (H-2) there are an open convex cone Γ in $\mathbb{R}^{2n+1}$ and a neighborhood U' of $(t_0,x^{0\prime},y^{0\prime})$ in $\tilde\pi(\mathcal{D})$ such that $\tilde\pi(\Gamma^*)\subset\subset\mathcal{C}$ and

$$\vartheta \in \Gamma \subset\subset \Gamma(t,x',y';\pi) \quad \text{for } (t,x',y') \in U'.$$

It is easy to see that there are $(t_1,x^{1\prime},y^{1\prime}) \in U'$ and $t_2 \in \mathbb{R}$ such that

$$(t_0,x^{0\prime},y^{0\prime}) \in \tilde{\Omega}^1_+ \equiv (\{(t_1,x^{1\prime},y^{1\prime})\}\text{-Int } \tilde{\pi}(\Gamma^*)) \cap \{t \geqq t_2\} \subset U'$$

$$\tilde{\Omega}^1_+ \cap \{t=t_2\} \subset S.$$

Let $(x^1,y^1) \in \pi^{-1}(x^{1\prime},y^{1\prime})$. Applying Lemma 3.8 repeatedly, we can uniquely construct $u_1(t,z) \in C^m_A(\Omega^1_+)$ such that $Pu_1=f$ in Ω^1_+ and $\partial_t^{j-1}u_1(t,z)=\partial_t^{j-1}u(t,z)$ ($1 \leqq j \leqq m$) in $\Omega^1_+ \cap \{t=t_2\}$, where $\Omega^1_+=(\{(t_1,x^1,y^1)\}\text{-Int } \Gamma^*) \cap \{t \geqq t_2\}$. This proves $(t_0,x^{0\prime},y^{0\prime}) \in S$. Therefore, from Corollary of Theorem 2.3, it follows that $S=\tilde{\pi}(\Omega_+)$, since $S \cap \{t=0\}=\tilde{\pi}(\Omega_+) \cap \{t=0\}=\{0\}\times\pi(D) \cap \tilde{\pi}(\mathcal{D})$.

Q.E.D.

Theorem 3.9. Let $q_j(t,x,y,\tau,\xi,\eta)$ be homogeneous polynomials of degree m-1 such that the coefficients belong to $C(\mathcal{D})$ and $\tilde{p}(t,x,y,\tau,\xi,\eta) \equiv p(t,x+iy,\tau,(\xi-i\eta)/2)+\sum_{j=1}^n (\xi_j+i\eta_j)q_j(t,x,y,\tau,\xi,\eta)$ is hyperbolic w.r.t. ϑ. Then the Cauchy problem (CP) has a unique solution $u \in C^m_A(\Omega_+)$, where $\Omega_+=\{(t,x,y) \in \mathcal{D};\ t \geqq 0,\ K^-_{(t,x,y)} \cap \{t \geqq 0\} \subset\subset \mathcal{D}$ and $K^-_{(t,x,y)} \cap \{t=0\} \subset\subset \{0\}\times D$ (in $\{0\}\times\mathbb{C}^n)\}$ and $K^{\pm}_{(t,x,y)}$ denote the generalized flows for $\{\Gamma(\tilde{p}(t,x,y,\cdot,\cdot,\cdot),\vartheta)^*\}$.

Remark. The theorem is an immediate consequence of Theorems 3.3 and 3.4. When the coefficients of q_j belong to $C^{1+\sigma}(\mathcal{D})$ and are infinitely differentiable in (x,y), the theorem follows directly from Lemma 3.5 for a strictly hyperbolic polynomial $\tilde{p}_\varepsilon(t,x,y,\tau,\xi,\eta) \equiv (1-\varepsilon^2|(\xi,\eta)|^2\partial_\tau^2)^{[m/2]}\tilde{p}(t,x,y,\tau,\xi,\eta)$ ($\varepsilon>0$), Theorem 2.1 and Corollary of Theorem 2.3.

Corollary. Let $\mathcal{D}'$ and D' be subdomains of $\mathbb{R}^{n+1}$ and $\mathbb{R}^n$, respectively, and let $\mathcal{D}$ and D be neighborhoods of $\mathcal{D}'$ and D' in $\mathbb{R}\times\mathbb{C}^n$ and $\mathbb{C}^n$, respectively. Assume that $P(t,z,\tau,\zeta)$ satisfies (H-1) and that $p(t,x,\tau,\xi)$ is hyperbolic w.r.t. $\vartheta'=(1,0,\cdots,0) \in \mathbb{R}^{n+1}$ for $(t,x) \in \mathcal{D}'$, where $p(t,z,\tau,\zeta)$ is the principal symbol of $P(t,z,\tau,\zeta)$. Then the Cauchy problem

$$\begin{cases} P(t,x,\partial_t,\partial_x)u(t,x) = f(t,x), \\ \partial_t^{j-1}u(0,x) = u_j(x) \quad (1 \leqq j \leqq m), \end{cases}$$

where $f(t,z) \in C_A(\mathcal{D})$ and $u_j(z) \in \mathcal{O}(D)$ ($1 \leqq j \leqq m$), has a unique solution $u(t,x) \in C^m(\Omega'_+)$ in Ω'_+ which is real analytic in x, where $\Omega'_+=\{(t_0,x^0) \in \mathcal{D}'$;

$t_0 \geq 0$, $K^-_{(t_0,x^0)} \cap \{t \geq 0\} \subset\subset \mathcal{D}'$ and $K^-_{(t_0,x^0)} \cap \{t=0\} \subset\subset D'\}$, $K^{\pm}_{(t,x)}$ denote the generalized flows for $\{\Gamma(p(t,x,\cdot,\cdot),\vartheta')^*\}$ and Γ^* denotes the set $\{(\delta t, \delta x)\in\mathbb{R}^{n+1};\ \tau\delta t+\delta x\cdot\xi \geq 0$ for any $(\tau,\xi)\in\Gamma\}$ for $\Gamma\subset\mathbb{R}^{n+1}$.

Remark. (i) When the coefficients of $P(t,x,\tau,\xi)$ are real analytic in (t,x), Bony and Schapira [3] proved this corollary. (ii) The corollary follows directly from Lemma 3.5 and Theorem 2.3.

We need the following

Lemma. Let $(t_0,x^0)\in\mathcal{D}'$, and let $\mathcal{C}$ be a cone in $\mathbb{R}^{n+1}\setminus\{0\}$ such that $\mathcal{C}\subset\subset\Gamma(p(t_0,x^0,\cdot,\cdot),\vartheta')$. Then there are positive constants δ_1, δ_2 and C such that

$$\{(\tau,\xi,\eta)\in\mathbb{R}^{2n+1};\ (\tau,\xi)\in\mathcal{C} \text{ and } |(\tau,\xi)|\geq C|y||\eta|\}\subset\Gamma(t,x,y)$$

if $(t,x,y)\in\mathcal{D}$, $|(t,x)-(t_0,x^0)|\leq\delta_1$ and $|y|\leq\delta_2$.

Proof. The lemma is essentially the same as in [3], and one can prove the lemma in the same way, using the following fact instead of a local version of Bochner's tube theorem: There are $\delta_1>0$ and $s_0>0$ such that

$$p(t,x+iy,\tau+i\nu,\xi-i\eta)\neq 0$$

if $(t,x,y)\in\mathcal{D}$, $(\tau,\xi)\in\mathcal{C}$, $\nu\in\mathbb{R}$, $\eta\in\mathbb{R}^n$, $|(t,x)-(t_0,x^0)|\leq\delta_1$, $|(\nu,\eta)|\leq\delta_1$, $|(\tau,\xi)|\geq|y|$ and $|(y,\tau,\xi)|\leq s_0$. Q.E.D.

From this lemma it follows that for any $\varepsilon>0$ and any compact subset K of $\mathcal{D}'$ there are an inner semi-continuous family $\{\gamma(t,x,y)\}_{(t,x,y)\in\mathcal{D}}$ of open convex cones in $\mathbb{R}^{2n+1}$ and positive constants $C(\varepsilon,K)$ and $\delta(\varepsilon,K)$ such that

$$\gamma(t,x,y)\supset\{(\tau,\xi,\eta)\in\mathbb{R}^{2n+1};\ (\tau,\xi)\in\Gamma_\varepsilon(t,x) \text{ and and } |(\tau,\xi)|>C(\varepsilon,K)|y||\eta|\}$$

for $(t,x,y)\in K\times\{y\in\mathbb{R}^n;\ |y|\leq\delta(\varepsilon,K)\}$ and $\vartheta\in\gamma(t,x,y)\subset\Gamma(t,x,y)$ for $(t,x,y)\in\mathcal{D}$, where $\Gamma_\varepsilon(t,x)=\{(\tau,\xi)\in\mathbb{R}^{n+1};\ (\tau-\varepsilon|\xi|,\xi)\in\Gamma(p(t,x,\cdot,\cdot),\vartheta')\}$. Then it is easy to see that

$$K^-_{(t,x,0)}\cap\{t\geq 0\}\subset K^-_{(t,x),\varepsilon}\times\{0\}\cap\{t\geq 0\}$$

if $K^-_{(t,x),\varepsilon}\cap\{t\geq 0\}\subset\subset K$, where $K^{\pm}_{(t,x,y)}$ and $K^{\pm}_{(t,x),\varepsilon}$ denote the generalized flows for $\{\gamma(t,x,y)^*\}$ and $\{\Gamma_\varepsilon(t,x)^*\}$, respectively. Since

$K^-_{(t,x),\varepsilon}\cap\{t\geqq 0\}$ approximates $K^-_{(t,x)}\cap\{t\geqq 0\}$ if $K^-_{(t,x)}\cap\{t\geqq 0\}\subset\subset\mathcal{D}'$, Theorem 3.3 proves the corollary.

Example 3.10. Set, for $(t,x',y')\in\tilde{\pi}(\mathcal{D})$,

$$\gamma(t,x',y') = \{(\tau,\xi,\eta)\in\tilde{\pi}^*(\mathbb{R}^{2n+1});\ \tau>\hat{\mu}(t,x',y';\pi)|(\xi,\eta)|\},$$

where $\hat{\mu}(t,x',y';\pi)=\sup\,\{\mathrm{Re}\,\lambda_j(t,x+iy,\xi-i\eta);\ 1\leqq j\leqq m,\ (x,y)\in\pi^{-1}(x',y'),\ (\xi,\eta)\in\pi^*(\mathbb{R}^{2n})$ and $|(\xi,\eta)|=1\}$. Then we have $K(t,x',y';\pi)=\{(\delta t,\delta x',\delta y')\in\tilde{\pi}(\mathbb{R}^{2n+1});\ \delta t\geqq 0$ and $\hat{\mu}(t,x',y';\pi)\delta t\geqq|(\delta x',\delta y')|\}$ in Theorem 3.3. In particular, when $\pi(x,y)=y$ for $(x,y)\in\mathbb{R}^{2n}$, Theorem 3.3 gives a similar result to Jannelli [9] and Kajitani [11].

Proof of Theorem 3.3 in the case where the coefficients and $f(t,z)$ are analytic in (t,z). For simplicity we assume that $\pi=\mathrm{id}$. Let S be a subdomain of $\mathcal{D}$, and let $u(t,z)$ be an analytic function in S satisfying $Pu=f$. Assume that for a fixed $(t_0,x^0,y^0)\in\mathcal{D}$ there are $\mathcal{C}\in\mathcal{C}(\mathbb{R}^{2n+1})$ and $\varepsilon>0$ such that $\mathcal{C}\subset K(t_0,x^0,y^0)\equiv K(t_0,x^0,y^0;\pi)$ and $(\{(t_0,x^0,y^0)\}-\mathcal{C})\cap\{(t,x,y)\in\mathcal{D};\ t_0-\varepsilon\leqq t<t_0\}\subset S$. Choose an open convex cone Γ in $\mathbb{R}^{2n+1}$ such that $\Gamma^*\subset\subset\mathcal{C}$ and $(\vartheta\in)\ \Gamma\subset\subset\Gamma(t_0,x^0,y^0)$. Set

$$\delta_1 = \inf\,\{|(\delta x-\delta x^1,\delta y-\delta y^1)|;\ (1,\delta x,\delta y)\notin\mathcal{C} \text{ and } (1,\delta x^1,\delta y^1)\in\Gamma^*\}.$$

Then, by the Cauchy-Kowalewski theorem, there is $\delta_2>0$ such that $u(t,z)$ can be analytically continued in $\Omega\equiv\{(t+is,x+iy)\in\mathbb{C}^{n+1};\ (t,x,y)\in(\{(t_0,x^0,y^0)\}-\mathrm{Int}\,\Gamma^*),\ t_0-\varepsilon\leqq t\leqq t_0$ and $|s|<\delta_1\delta_2(t_0-t)\}$. Set $\Gamma_\delta=\{(\tau-i\sigma,\xi-i\eta)\in\mathbb{C}^{n+1};\ (\tau-\delta|\sigma|,\xi,\eta)\in\Gamma\}$, where $0<\delta\leqq\delta_1\delta_2$. It is obvious that $\Gamma_\delta\subset\subset\{(\tau-i\sigma,\xi-i\eta)\in\mathbb{C}^{n+1};\ p(t_0,x^0+iy^0,\tau-i\sigma,\xi-i\eta)\neq 0\}$ (in $\mathbb{C}^{n+1}$) and that $\Gamma^*_\delta\equiv\{(\delta t+i\delta s,\delta x+i\delta y)\in\mathbb{C}^{n+1};\ \tau\delta t+\sigma\delta s+\delta x\cdot\xi+\delta y\cdot\eta\geqq 0$ for any $(\tau-i\sigma,\xi-i\eta)\in\Gamma_\delta\}=\{(\delta t+i\delta s,\ \delta x+i\delta y)\in\mathbb{C}^{n+1};\ (\delta t,\delta x,\delta y)\in\Gamma^*$ and $|\delta s|\leqq\delta(\delta t)\}$. Therefore, there is a neighborhood U of (t_0,x^0+iy^0) in $\mathbb{C}^{n+1}$ such that $(\{(t_0,x^0+iy^0)\}-\mathrm{Int}\,\Gamma^*_\delta)\cap U\subset\Omega$. From Lemma 3.1 in [2] it follows that $u(t,z)$ can be analytically continued in a neighborhood of (t_0,x^0+iy^0) in $\mathbb{C}^{n+1}$. So we can repeat the same argument as in the proof of Theorem 3.3, without applying Lemma 3.8. In the case where $\pi\neq\mathrm{id}$, one can also prove the theorem, using results in the case where $\pi=\mathrm{id}$. We note that one can prove Lemma 3.8, using the method of sweeping out and Lemma 3.1 in [2], if the coefficients and $f(t,z)$ are analytic in (t,z) . Q.E.D.

4. SINGULARITIES OF SOLUTIONS

In this section we shall consider the Cauchy problem for general hyperbolic operators in Gevrey classes.

Definition 4.1. (i) For $1 \leqq \kappa < \infty$ we say that $f \in \mathcal{E}^{\{\kappa\}}$ if $f \in C^\infty$ and if for any compact subset K of $\mathbb{R}^n$ there are positive constants h and C such that

$$|\partial_x^\alpha f(x)| \leqq C h^{|\alpha|} (|\alpha|!)^\kappa \quad \text{for any } x \in K \text{ and } \alpha.$$

We define $\mathcal{D}^{\{\kappa\}} = \mathcal{E}^{\{\kappa\}} \cap C_0^\infty$ for $1 < \kappa < \infty$. We introduce usual locally convex topologies in these spaces. (ii) We denote by $\mathcal{D}^{\{\kappa\}'}$ the strong dual space of $\mathcal{D}^{\{\kappa\}}$ for $1 < \kappa < \infty$, and by $\mathcal{D}^{\{1\}'}$ the space of all hyperfunctions on $\mathbb{R}^n$. (iii) For $f \in \mathcal{D}^{\{\kappa\}'}$, $WF_{\{\kappa\}}(f)$ ($\subset T^*\mathbb{R}^n \setminus 0$) is defined as follows: $(x,\xi) \notin WF_{\{\kappa\}}(f)$ if $f \in \mathcal{E}^{\{\kappa\}}$ microlocally at (x,ξ).

Remark. (i) In the above definition, $WF_{\{1\}}(f) = \{(x,\xi) \in T^*\mathbb{R}^n \setminus 0;\ (x, i\xi\infty) \in S.S.\ f\}$, where S.S. f denotes the singular spectrum of f (see [15]). (ii) For $f \in \mathcal{D}^{\{\kappa\}'}$ ($1 < \kappa < \infty$) we can also define $WF_{\{\kappa_1\}}(f)$ ($1 \leqq \kappa_1 < \infty$).

Let us consider the Cauchy problems

$$(CP)_\kappa \qquad \begin{cases} P(x,D)u = f, \\ \operatorname{supp} u \subset \{x_1 \geqq 0\}, \end{cases}$$

where $1 \leqq \kappa < \infty$ and $f \in \mathcal{D}^{\{\kappa\}'}$ with $\operatorname{supp} f \subset \{x_1 \geqq 0\}$. We assume that for $1 \leqq \kappa < \infty$

$(A\text{-}1)_\kappa$ $P(x,\xi)$ is a hyperbolic polynomial w.r.t. $\vartheta = (1,0,\cdots,0) \in \mathbb{R}^n$ for every $x = (x_1,\cdots,x_n) \in \mathbb{R}^n$, i.e.,

$$p(x,\xi - it\vartheta) \neq 0 \quad \text{if } x \in \mathbb{R}^n,\ \xi \in \mathbb{R}^n \text{ and } t > 0,$$

and the coefficients of $P(x,\xi)$ belong to $\mathcal{E}^{\{\kappa\}}$, where $p(x,\xi)$ is the principal symbol of $P(x,\xi)$.

Define the localization polynomial $p_{z^0}(\delta z)$ of $p(z)$ at $z^0 = (x^0,\xi^0) \in T^*\mathbb{R}^n \simeq \mathbb{R}^n \times \mathbb{R}^n$ by

$$p(z^0 + s\delta z) = s^\mu (p_{z^0}(\delta z) + o(1)) \quad \text{as } s \longrightarrow 0,$$

$$p_{z^0}(\delta z) \not\equiv 0 \quad \text{in } \delta z \in T_{z^0}(T^*\mathbb{R}^n) \simeq \mathbb{R}^{2n}.$$

For polynomials with constant coefficients the localization polynomials were defined by Hörmander [6] and Atiyah, Bott and Gårding [1]. Then we have the following

Lemma 4.2 ([19],[20]). $p_{z^0}(\delta z)$ is hyperbolic w.r.t. $\tilde{\vartheta}=(\vartheta,0,\cdots,0)$ $\in\mathbb{R}^{2n}$, and $\{\Gamma(p_z,\tilde{\vartheta})\}_{z\in T^*\mathbb{R}^n}$ is an inner semi-continuous family of open convex cones in $\mathbb{R}^{2n}$, where $\Gamma(p_z,\tilde{\vartheta})$ is the connected component of the set $\{\delta z\in\mathbb{R}^{2n};\ p_z(\delta z)\neq 0\}$ which contains $\tilde{\vartheta}$.

Remark. Under the assumptions that $p(x,\xi)$ is hyperbolic w.r.t. ϑ and that the coefficients of $p(x,\xi)$ belong to C^m, Lemma 4.2 is valid (see [20]).

By Lemma 4.2 we can define the generalized (Hamilton) flows $K_z^{\pm}$ for $\{\Gamma(p_z,\tilde{\vartheta})^\sigma\}_{z\in T^*\mathbb{R}^n}$, where

$$\Gamma^\sigma = \{(\delta x,\delta\xi)\in\mathbb{R}^{2n};\ \delta x\cdot\delta\xi^1-\delta x^1\cdot\delta\xi\geqq 0 \text{ for any } (\delta x^1,\delta\xi^1)\in\Gamma\}.$$

We note that (i) $K_z^{\pm}=\{z\}$ if $p(z)\neq 0$, (ii) one can write $K_{(x,0)}^{\pm}=K_x^{\pm}\times\{0\}$, where $K_x^{\pm}$ are the generalized flows for $\{\Gamma(p(x,\cdot),\vartheta)^*\}_{x\in\mathbb{R}^n}$, (iii) the definition of $K_z^{\pm}$ does not depend on the choice of the canonical coordinates, and (iv) $K_z^{\pm}$ are the null bicharacteristics of p issuing from z along which $\pm x_1$ increase, if p is srtictly hyperbolic w.r.t. ϑ and $p(z)$ $=0$. For further properties of $K_z^{\pm}$ we refer to [17] and [19]. To state the results globally, we assume that

(A-2) $K_x^-\cap\{x_1\geqq 0\}$ is bounded for every $x\in\mathbb{R}^n$.

Theorem 4.3 ([17],[18]). Assume that $(A\text{-}1)_\kappa$ and (A-2) are satisfied, and that $1\leqq\kappa<r/(r-1)$, where r ($\geqq 2$) is more than or equal to the multiplicities of the roots of the equation $p(x,\xi)=0$ in ξ_1 for $x\in\mathbb{R}^n$, $\xi'=(\xi_2,\cdots,\xi_n)\in\mathbb{R}^{n-1}\setminus\{0\}$. If $u\in\mathcal{D}^{\{\kappa\}'}$ satisfies the Cauchy problem $(CP)_\kappa$, then

$$\text{supp } u\subset\{x\in\mathbb{R}^n;\ x\in K_y^+ \text{ for some } y\in\text{supp } f\},$$

$$WF_{\{\kappa\}}(u)\subset\{z\in T^*\mathbb{R}^n\setminus 0;\ z\in K_w^+ \text{ for some } w\in WF_{\{\kappa\}}(f)\}.$$

Remark. The well-posedness in $\mathcal{D}^{\{\kappa\}'}$ (or $\mathcal{E}^{\{\kappa\}}$) of $(CP)_\kappa$ was proved by Bony and Schapira [3] when $\kappa=1$, and by Bronshtein [4] when $1<\kappa<r/(r-1)$.

Theorem 4.3 can be easily proved by Theorem 2.3 and the following proposition.

Proposition 4.4. Assume that $(A\text{-}1)_\kappa$ and (A-2) are valid and that $1\leqq\kappa<r/(r-1)$. If $u\in\mathcal{D}^{\{\kappa\}'}$ and $z^0=(x^0,\xi^0)\in WF_{\{\kappa\}}(u)\setminus WF_{\{\kappa\}}(Pu)$, then for

any cone $\Gamma \subset\subset \Gamma(p_{z^0}, \mathfrak{I})$ there is $t_0>0$ such that

$$WF_{\{\kappa\}}(u) \cap \{z^0-\Gamma^\sigma\} \cap \{x_1=x_1^0-t\} \neq \emptyset \quad \text{for } 0\leqq t\leqq t_0.$$

Remark. When $\kappa=1$, Kashiwara and Kawai [12] (and, also, Sjöstrand [16]) proved Proposition 4.4, which is a microlocalization of Bony and Schapira [3]. For $1<\kappa<r/(r-1)$, Proposition 4.4 is a microlocalization of Bronshtein [4] (see [17]).

We note that for the Cauchy problem for symmetric hyperbolic systems we can obtain the same results as in Theorem 4.3 in the framework of C^∞, combining Ivrii's result [7] and Theorem 2.3 (see [13] and [19]). We may conjecture that Theorem 4.3 is still valid in the framework of C^∞ if the hyperbolic Cauchy problem is C^∞ well-posed. In fact, if the coefficients of the hyperbolic operators are real analytic (or in some Gevrey classes), then the above conjecture is valid (see [17]).

REFERENCES

1. Atiyah, M. F., Bott, R. and Gårding, L., 'Lacunas for hyperbolic differential operators with constant coefficients I' Acta Math. 124 (1970), 109-189.

2. Bony, J.-M. and Schapira, P., 'Existence et prolongement des solution holomorphes des équations aux dérivées partielles' Invent. Math. 17 (1972), 95-105.

3. ______________, 'Solution hyperfonctions du problème de Cauchy' Hyperfunctions and Pseudo-differential Equations, Lecture Notes in Math. 287, Springer, 1973, pp82-98.

4. Bronshtein, M. D., 'The Cauchy problem for hyperbolic operators with variable multiple characteristics' Trudy Moskov. Mat. Obšč. 41 (1980), 83-99.

5. Garabedian, P. R., Partial Differential Equations, John Wiley, New York, 1964.

6. Hörmander, L. , 'On the singularities of solutions of partial differential equations' International Conference of Functional Analysis and Related Topics, Tokyo, 1969.

7. Ivrii, V. Ya., 'Wave fronts of solutions of symmetric pseudo-differential systems' Sib. Mat. Zh. 20 (1979), 557-578.

8. Jannelli, E., 'Hyperbolic systems with coefficients analytic in space variables' J. Math. Kyoto Univ. 21 (1981), 715-739.

9. ______, 'Linear Kovalevskian systems with time-dependent coefficients' Comm. in Partial Differential Equations 9 (1984), 1373-1406.

10. John, F., 'On linear partial differential equations with analytic coefficients, unique continuation of data' Comm. Pure Appl. Math. 2 (1949), 209-254.

11. Kajitani, K., 'A remark on Nagumo theorem' Preprint (in Japanese).

12. Kashiwara, M. and Kawai, T., 'Micro-hyperbolic pseudo-differential operators I' J. Math. Soc. Japan 27 (1975), 359-404.

13. Melrose, R., 'The trace of the wave group' Microlocal Analysis, Comtemporary Math. 27, AMS, 1984, pp127-167.

14. Mizohata, S., The Theory of Partial Differential Equations, Cambrige, 1973.

15. Sato, M., Kawai, T. and Kashiwara, M., 'Microfunctions and pseudo-differential equations' Lecture Notes in Math. 287, Springer, 1973, pp265-529.

16. Sjöstrand, J., 'Singularités analytiques microlocales' Astérisque 95 (1982), 1-166.

17. Wakabayashi, S., 'Singularities of solutions of the Cauchy problem for hyperbolic systems in Gevrey classes' Japanese J. Math. 11 (1985), 157-201.

18. ______________, 'Analytic singularities of solutions of the hyperbolic Cauchy problem, Proc. Japan Acad. 59 (1983), 449-452.

19. ______________, 'Singularities of solutions of the Cauchy problem for symmetric hyperbolic systems' Comm. in Partial Differential Equations 9 (1984), 1147-1177.

20. ______________, 'Remarks on hyperbolic polynomials' to appear.

21. Zerner, M., 'Domaines d'holomorphie des fonctions vérifiant une équation aux dérivées partielles' C. R. Acad. Sci. Paris 272 (1971), 1646-1648.

INDEX

O

P

Q

R

S